STUDENT'S SOLUTIONS MANUAL

Judith A. Penna

CALCULUS

FIFTH EDITION

MARVIN L. BITTINGER
Indiana University — Purdue University at Indianapolis

ADDISON-WESLEY PUBLISHING COMPANY
Reading, Massachusetts • Menlo Park, California • New York
Don Mills, Ontario • Wokingham, England • Amsterdam
Bonn • Sydney • Singapore • Tokyo • Madrid
San Juan • Milan • Paris

ISBN 0-201-53259-X

3 4 5 6 7 8 9 10 AL 95949392

TABLE OF CONTENTS

Special thanks are extended to Patsy Hammond for her excellent typing and to Pam Smith for her careful proofreading. Their patience, efficiency, and good humor made the author's work much easier.

Exercise Set 1.1

1. $5^3 = \underbrace{5\cdot 5\cdot 5}_{\text{3 factors}} = 125$

3. $(-7)^2 = \underbrace{(-7)(-7)}_{\text{2 factors}} = 49$

5. $(1.01)^2 = \underbrace{(1.01)(1.01)}_{\text{2 factors}} = 1.0201$

7. $\left(\frac{1}{2}\right)^4 = \underbrace{\frac{1}{2}\cdot\frac{1}{2}\cdot\frac{1}{2}\cdot\frac{1}{2}}_{\text{4 factors}} = \frac{1}{16}$

9. $(6x)^0 = 1$ ($a^0 = 1$, for any nonzero real number a)

11. $t^1 = t$ ($a^1 = a$, for any real number a)

13. $\left(\frac{1}{3}\right)^0 = 1$ ($a^0 = 1$, for any nonzero real number a)

15. 3^{-2}

$= \frac{1}{3^2}$ ($a^{-n} = \frac{1}{a^n}$, for any nonzero real number a)

$= \frac{1}{3\cdot 3}$ ($3^2 = \underbrace{3\cdot 3}_{\text{2 factors}}$)

$= \frac{1}{9}$

17. $\left(\frac{1}{2}\right)^{-3}$

$= \frac{1}{\left(\frac{1}{2}\right)^3}$ ($a^{-n} = \frac{1}{a^n}$, for any nonzero real number a)

$= \frac{1}{\frac{1}{2}\cdot\frac{1}{2}\cdot\frac{1}{2}}$

$= \frac{1}{\frac{1}{8}}$

$= 1\cdot\frac{8}{1}$ (Multiplying by the reciprocal)

$= 8$

19. $10^{-1} = \frac{1}{10^1} = \frac{1}{10}$, or 0.1

21. $e^{-b} = \frac{1}{e^b}$

23. $b^{-1} = \frac{1}{b^1} = \frac{1}{b}$

25. To multiply when bases are the same, add the exponents.

$x^2\cdot x^3 = x^{2+3} = x^5$ ($a^n\cdot a^m = a^{n+m}$)

Think:

$x^2\cdot x^3 = \underbrace{x\cdot x}_{2}\cdot\underbrace{x\cdot x\cdot x}_{3}$ factors

$= \underbrace{x\cdot x\cdot x\cdot x\cdot x}_{\text{5 factors}}$

$= x^5$

27. $x^{-7}\cdot x$

$= x^{-7}\cdot x^1$ ($x = x^1$)

$= x^{-7+1}$ ($a^n\cdot a^m = a^{n+m}$)

$= x^{-6}$, or $\frac{1}{x^6}$

29. $5x^2\cdot 7x^3$

$= 5\cdot 7\cdot x^2\cdot x^3$

$= 35x^{2+3}$ ($a^n\cdot a^m = a^{n+m}$)

$= 35x^5$

31. $x^{-4}\cdot x^7\cdot x$

$= x^{-4}\cdot x^7\cdot x^1$ ($x = x^1$)

$= x^{-4+7+1}$ ($a^n\cdot a^m = a^{n+m}$)

$= x^4$

33. $e^{-t}\cdot e^t$

$= e^{-t+t}$ ($a^n\cdot a^m = a^{n+m}$)

$= e^0$

$= 1$ ($a^0 = 1$, for any nonzero real number a)

35. To divide when the bases are the same subtract the exponent in the denominator from the exponent in the numerator.

$\frac{x^5}{x^2} = x^{5-2} = x^3$ ($\frac{a^n}{a^m} = a^{n-m}$)

Think:

$\frac{x^5}{x^2} = \frac{x\cdot x\cdot x\cdot x\cdot x}{x\cdot x}$ (5 factors) (2 factors)

$= \frac{x\cdot x}{x\cdot x}\cdot\frac{x\cdot x\cdot x}{1}$ (Factoring the fraction)

$= 1\cdot\frac{x\cdot x\cdot x}{1}$ ($\frac{x\cdot x}{x\cdot x} = 1$)

$= \frac{x\cdot x\cdot x}{1}$

$= x^3$

37. $\frac{x^2}{x^5}$

$= x^{2-5}$ ($\frac{a^n}{a^m} = a^{n-m}$)

$= x^{-3}$, or $\frac{1}{x^3}$

39. $\frac{e^k}{e^k}$

$= e^{k-k}$ $\left(\frac{a^n}{a^m} = a^{n-m}\right)$

$= e^0$

$= 1$ ($a^0 = 1$, for any nonzero real number a)

41. $\frac{e^t}{e^4} = e^{t-4}$ $\left(\frac{a^n}{a^m} = a^{n-m}\right)$

43. $\frac{t^6}{t^{-8}}$

$= t^{6-(-8)}$ $\left(\frac{a^n}{a^m} = a^{n-m}\right)$

$= t^{14}$ $[6 - (-8) = 6 + 8 = 14]$

45. $\frac{t^{-9}}{t^{-11}}$

$= t^{-9-(-11)}$ $\left(\frac{a^n}{a^m} = a^{n-m}\right)$

$= t^2$ $[-9 - (-11) = -9 + 11 = 2]$

47. $\frac{ab(a^2b)^3}{ab^{-1}}$

$= \frac{ab(a^6b^3)}{ab^{-1}}$ $[(a^n)^m = a^{nm}]$

$= \frac{a^7b^4}{ab^{-1}}$ (Multiplying in the numerator)

$= a^{7-1}b^{4-(-1)}$ (Dividing)

$= a^6b^5$

49. $(t^{-2})^3 = t^{-2\cdot 3} = t^{-6}$, or $\frac{1}{t^6}$ $\left|(a^n)^m = a^{n\cdot m}\right|$

Think:
$(t^{-2})^3 = (t^{-2})(t^{-2})(t^{-2})$
$= t^{-2+(-2)+(-2)}$
$= t^{-6}$

51. $(e^x)^4$

$= e^{x\cdot 4}$ $[(a^n)^m = a^{n\cdot m}]$

$= e^{4x}$

53. $(2x^2y^4)^3$

$= 2^3\cdot(x^2)^3\cdot(y^4)^3$ (Each factor is raised to the third power)

$= 2^3x^6y^{12}$ $[(a^n)^m = a^{n\cdot m}]$

$= 8x^6y^{12}$ $(2^3 = 8)$

55. $(3x^{-2}y^{-5}z^4)^{-4}$

$= 3^{-4}\cdot(x^{-2})^{-4}\cdot(y^{-5})^{-4}\cdot(z^4)^{-4}$

(Each factor is raised to the -4 power)

$= 3^{-4}x^8y^{20}z^{-16}$ $[(a^n)^m = a^{n\cdot m}]$

$= \frac{1}{81}x^8y^{20}z^{-16}$, $\left(3^{-4} = \frac{1}{3^4} = \frac{1}{81}\right)$

or $\frac{x^8y^{20}}{81z^{16}}$

57. $(-3x^{-8}y^7z^2)^2$

$= (-3)^2\cdot(x^{-8})^2\cdot(y^7)^2\cdot(z^2)^2$

(Each factor is raised to the second power)

$= (-3)^2x^{-16}y^{14}z^4$ $[(a^n)^m = a^{n\cdot m}]$

$= 9x^{-16}y^{14}z^4$, $[(-3)^2 = (-3)\cdot(-3) = 9]$

or $\frac{9y^{14}z^4}{x^{16}}$

59. $5(x - 7)$

$= 5\cdot x - 5\cdot 7$ (Using a distributive law)

$= 5x - 35$

61. $x(1 - t)$

$= x\cdot 1 - x\cdot t$ (Using a distributive law)

$= x - xt$

63. $(x - 5)(x - 2)$

$= (x - 5)x - (x - 5)2$ (Using a distributive law)

$= x\cdot x - 5\cdot x - x\cdot 2 + 5\cdot 2$ (Using a distributive law)

$= x^2 - 5x - 2x + 10$

$= x^2 - 7x + 10$ (Collecting like terms)

65. $(a - b)(a^2 + ab + b^2)$

$= (a - b)a^2 + (a - b)ab + (a - b)b^2$

(Using a distributive law)

$= a\cdot a^2 - b\cdot a^2 + a\cdot ab - b\cdot ab + a\cdot b^2 - b\cdot b^2$

(Using a distributive law)

$= a^3 - a^2b + a^2b - ab^2 + ab^2 - b^3$

$= a^3 - b^3$ (Collecting like terms)

67. $(2x + 5)(x - 1)$

$= (2x + 5)x - (2x + 5)1$ (Using a distributive law)

$= 2x\cdot x + 5\cdot x - 2x\cdot 1 - 5\cdot 1$ (Using a distributive law)

$= 2x^2 + 5x - 2x - 5$

$= 2x^2 + 3x - 5$ (Collecting like terms)

69. $(a - 2)(a + 2)$

$= a^2 - 2^2$ $[(A - B)(A + B) = A^2 - B^2]$

$= a^2 - 4$

71. $(5x + 2)(5x - 2)$

$= (5x)^2 - 2^2$ $[(A + B)(A - B) = A^2 - B^2]$

$= 25x^2 - 4$

73. $(a - h)^2$
$= a^2 - 2\cdot a\cdot h + h^2$ $[(A - B)^2 = A^2 - 2AB + B^2]$
$= a^2 - 2ah + h^2$

75. $(5x + t)^2$
$= (5x)^2 + 2(5x)(t) + t^2$
$[(A + B)^2 = A^2 + 2AB + B^2]$
$= 25x^2 + 10xt + t^2$

77. $5x(x^2 + 3)^2$
$= 5x[(x^2)^2 + 2(x^2)(3) + 3^2]$
$[(A + B)^2 = A^2 + 2AB + B^2]$
$= 5x(x^4 + 6x^2 + 9)$
$= 5x\cdot x^4 + 5x\cdot 6x^2 + 5x\cdot 9$ (Using a distributive law)
$= 5x^5 + 30x^3 + 45x$

79. $(a + b)^3 = a^3 + 3a^2b + 3ab^2 + b^3$
(Using Equation 1)

81. $(x - 5)^3$
$= [x + (-5)]^3$
$= x^3 + 3x^2(-5) + 3x(-5)^2 + (-5)^3$
$[(x + h)^3 = x^3 + 3x^2h + 3xh^2 + h^3]$
$= x^3 - 15x^2 + 75x - 125$

83. $x - xt$
$= x\cdot 1 - x\cdot t$
$= x(1 - t)$ (Factoring out the common factor, x)

85. $x^2 + 6xy + 9y^2$
$= x^2 + 2\cdot x\cdot 3y + (3y)^2$
$= (x + 3y)^2$ $[A^2 + 2AB + B^2 = (A + B)^2]$

87. $x^2 - 2x - 15$

We look for two numbers whose product is -15 and whose sum is -2.

They are -5 and 3. $(-5\cdot 3 = -15, -5 + 3 = -2)$

$x^2 - 2x - 15 = (x - 5)(x + 3)$ (Factoring)

89. $x^2 - x - 20$

We look for two numbers whose product is -20 and whose sum is -1.

They are -5 and 4. $(-5\cdot 4 = -20, -5 + 4 = -1)$

$x^2 - x - 20 = (x - 5)(x + 4)$ (Factoring)

91. $49x^2 - t^2$
$= (7x)^2 - t^2$
$= (7x - t)(7x + t)$ $[A^2 - B^2 = (A - B)(A + B)]$

93. $36t^2 - 16m^2$
$= 4(9t^2 - 4m^2)$ (Factoring out the common factor, 4)
$= 4[(3t)^2 - (2m)^2]$
$= 4(3t - 2m)(3t + 2m)$ $[A^2 - B^2 = (A - B)(A + B)]$

95. $a^3b - 16ab^3$
$= ab(a^2 - 16b^2)$ (Factoring out the common factor, ab)
$= ab[a^2 - (4b)^2]$
$= ab(a - 4b)(a + 4b)$ $[A^2 - B^2 = (A - B)(A + B)]$

97. $a^8 - b^8$
$= (a^4)^2 - (b^4)^2$
$= (a^4 + b^4)(a^4 - b^4)$ $[A^2 - B^2 = (A + B)(A - B)]$
$= (a^4 + b^4)[(a^2)^2 - (b^2)^2]$
$= (a^4 + b^4)(a^2 + b^2)(a^2 - b^2)$
$= (a^4 + b^4)(a^2 + b^2)(a + b)(a - b)$

99. $10a^2x - 40b^2x$
$= 10x(a^2 - 4b^2)$ (Factoring out the common factor, 10x)
$= 10x[a^2 - (2b)^2]$
$= 10x(a - 2b)(a + 2b)$ $[A^2 - B^2 = (A - B)(A + B)]$

101. $2 - 32x^4$
$= 2(1 - 16x^4)$ (Factoring out the common factor, 2)
$= 2[1^2 - (4x^2)^2]$
$= 2(1 - 4x^2)(1 + 4x^2)$ $[A^2 - B^2 = (A - B)(A + B)]$
$= 2[1^2 - (2x)^2](1 + 4x^2)$
$= 2(1 - 2x)(1 + 2x)(1 + 4x^2)$
$[A^2 - B^2 = (A - B)(A + B)]$

103. $9x^2 + 17x - 2$

First we look for a common factor. There is none other than 1.

Next we look for pairs of numbers whose product is 9. Since it is common practice that both factors be positive, we only consider 1, 9 and 3, 3.

We have these possibilities:

(x)(9x) (3x)(3x)

Next we look for pairs of numbers whose product is -2. These are 1, -2 and -1, 2.

Using these pairs of numbers we have six possible factorizations:

$(x + 1)(9x - 2)$ $(3x + 1)(3x - 2)$
$(x - 2)(9x + 1)$ $(3x - 1)(3x + 2)$
$(x - 1)(9x + 2)$
$(x + 2)(9x - 1)$

We multiply and find that the desired factorization is $(x + 2)(9x - 1)$.

$9x^2 + 17x - 2 = (x + 2)(9x - 1)$

105. $x^3 + 8$

$= x^3 + 2^3$

$= (x + 2)(x^2 - x\cdot 2 + 2^2)$ (See Exercise 66.)

$= (x + 2)(x^2 - 2x + 4)$

107. $y^3 - 64t^3$

$= y^3 - (4t)^3$

$= (y - 4t)[y^2 + y\cdot 4t + (4t)^2]$ (See Exercise 65.)

$= (y - 4t)(y^2 + 4yt + 16t^2)$

109. a) Substituting 4 for x and 0.1 for h in

$(x + h)^2 - x^2 = h(2x + h)$

we get

$(4.1)^2 - 4^2 = 0.1(2\cdot 4 + 0.1)$

$= 0.1(8.1)$

$= 0.81$

So $(4.1)^2$ differs from 4^2 by 0.81.

b) Substituting 4 for x and 0.01 for h in

$(x + h)^2 - x^2 = h(2x + h)$

we get

$(4.01)^2 - 4^2 = 0.01(2\cdot 4 + 0.01)$

$= 0.01(8.01)$

$= 0.0801$

So $(4.01)^2$ differs from 4^2 by 0.0801.

c) Substituting 4 for x and 0.001 for h in

$(x + h) - x^2 = h(2x + h)$

we get

$(4.001)^2 - 4^2 = 0.001(2\cdot 4 + 0.001)$

$= 0.001(8.001)$

$= 0.008001$

So $(4.001)^2$ differs from 4^2 by 0.008001.

111. a) $A = P(1 + i)^t$

$A = 1000(1 + 0.16)^1$ (Substituting)

$= 1000(1.16)$

$= \$1160$

b) $A = P(1 + \frac{i}{2})^{2t}$

$= 1000(1 + \frac{0.16}{2})^{2\cdot 1}$ (Substituting)

$= 1000(1.08)^2$

$= 1000(1.1664)$

$= \$1166.40$

c) $A = P(1 + \frac{i}{4})^{4t}$

$= 1000(1 + \frac{0.16}{4})^{4\cdot 1}$ (Substituting)

$= 1000(1.04)^4$

$= 1000(1.16985856)$

$\approx \$1169.86$ (Rounding to the nearest cent)

d) $A = P(1 + \frac{i}{365})^{365t}$

$= 1000(1 + \frac{0.16}{365})^{365\cdot 1}$ (Substituting)

$= 1000(1.000438356)^{365}$

$\approx 1000(1.17346975)$

$\approx \$1173.47$

e) There are 24·365, or 8760, hours in one year.

$A = P(1 + \frac{i}{8760})^{8760t}$

$= 1000(1 + \frac{0.16}{8760})^{8760\cdot 1}$ (Substituting)

$\approx 1000(1.000018265)^{8760}$

$\approx 1000(1.17350874)$

$\approx \$1173.51$

113. $M = P\left[\dfrac{\frac{i}{12}(1 + \frac{i}{12})^n}{(1 + \frac{i}{12})^n - 1}\right]$

We substitute \$43,000 for P, $8\frac{3}{4}$ % (or 0.0875) for i, and 300 for n ($25 \times 12 = 300$).

$M = 43{,}000\left[\dfrac{\frac{0.0875}{12}(1 + \frac{0.0875}{12})^{300}}{(1 + \frac{0.0875}{12})^{300} - 1}\right]$

$\approx 43{,}000\left[\dfrac{0.00729167(1.00729167)^{300}}{(1.00729167)^{300} - 1}\right]$

$\approx 43{,}000\left[\dfrac{0.00729167(8.84245268)}{8.84245268 - 1}\right]$

$\approx 43{,}000\left[\dfrac{0.06447625}{7.84245268}\right]$

$\approx \$353.52$

Exercise Set 1.2

1. $-7x + 10 = 5x - 11$

$7x - 7x + 10 = 7x + 5x - 11$ (Adding 7x on both sides)

$10 = 12x - 11$

$10 + 11 = 12x - 11 + 11$ (Adding 11 on both sides)

$21 = 12x$

$\frac{1}{12}\cdot 21 = \frac{1}{12}\cdot 12x$ (Multiplying by $\frac{1}{12}$ on both sides)

$\frac{21}{12} = x$

$\frac{7}{4} = x$ (Simplifying)

Check:

$-7x + 10 = 5x - 11$		
$-7 \cdot \frac{7}{4} + 10$	$5 \cdot \frac{7}{4} - 11$	(Substituting)
$-\frac{49}{4} + \frac{40}{4}$	$\frac{35}{4} - \frac{44}{4}$	
$-\frac{9}{4}$	$-\frac{9}{4}$	

The number $\frac{7}{4}$ checks and is the solution.

3. $5x - 17 - 2x = 6x - 1 - x$

$3x - 17 = 5x - 1$ (Collecting like terms on each side)

$-3x + 3x - 17 = -3x + 5x - 1$ (Adding -3x on both sides)

$-17 = 2x - 1$

$-17 + 1 = 2x - 1 + 1$ (Adding 1 on both sides)

$-16 = 2x$

$\frac{1}{2} \cdot (-16) = \frac{1}{2} \cdot 2x$ (Multiplying by $\frac{1}{2}$ on both sides)

$-8 = x$

Check:

$5x - 17 - 2x = 6x - 1 - x$		
$5(-8) - 17 - 2(-8)$	$6(-8) - 1 - (-8)$	(Substituting)
$-40 - 17 + 16$	$-48 - 1 + 8$	
-41	-41	

The number -8 checks and is the solution.

5. $x + 0.8x = 216$

$1 \cdot x + 0.8x = 216$ $(x = 1 \cdot x)$

$1.8x = 216$ (Collecting like terms)

$\frac{1}{1.8} \cdot 1.8x = \frac{1}{1.8} \cdot 216$ (Multiplying by $\frac{1}{1.8}$)

$x = \frac{216}{1.8}$

$x = 120$

Check:

$x + 0.8x = 216$		
$120 + 0.8(120)$	216	(Substituting)
$120 + 96$		
216		

The number 120 checks and is the solution.

7. $x + 0.08x = 216$

$1 \cdot x + 0.08x = 216$ $(x = 1 \cdot x)$

$1.08x = 216$ (Collecting like terms)

$\frac{1}{1.08} \cdot 1.08x = \frac{1}{1.08} \cdot 216$ (Multiplying by $\frac{1}{1.08}$)

$x = \frac{216}{1.08}$

$x = 200$

Check:

$x + 0.08x = 216$		
$200 + 0.08(200)$	216	(Substituting)
$200 + 16$		
216		

The number 200 checks and is the solution.

9. $2x(x + 3)(5x - 4) = 0$

$2x = 0$ or $x + 3 = 0$ or $5x - 4 = 0$ (Principle of zero products)

$x = 0$ or $x = -3$ or $5x = 4$

$x = 0$ or $x = -3$ or $x = \frac{4}{5}$

(Solving each equation separately)

The solutions are 0, -3, and $\frac{4}{5}$.

11. $x^2 + 1 = 2x + 1$

$x^2 = 2x$ (Adding -1 on both sides)

$x^2 - 2x = 0$ (Adding -2x on both sides)

$x(x - 2) = 0$ (Factoring on the left side)

$x = 0$ or $x - 2 = 0$ (Principle of zero products)

$x = 0$ or $x = 2$

The solutions are 0 and 2.

13. $t^2 - 2t = t$

$t^2 - 3t = 0$ (Adding -t on both sides)

$t(t - 3) = 0$ (Factoring on the left side)

$t = 0$ or $t - 3 = 0$ (Principle of zero products)

$t = 0$ or $t = 3$

The solutions are 0 and 3.

15. $6x - x^2 = -x$

$0 = x^2 - 7x$ (Adding $x^2 - 6x$ on both sides)

$0 = x(x - 7)$ (Factoring on the right side)

$x = 0$ or $x - 7 = 0$ (Principle of zero products)

$x = 0$ or $x = 7$

The solutions are 0 and 7.

17. $9x^3 = x$

$9x^3 - x = 0$ (Adding $-x$ on both sides)

$x(9x^2 - 1) = 0$ (Factoring)

$x(3x - 1)(3x + 1) = 0$ (Factoring)

$x = 0$ or $3x - 1 = 0$ or $3x + 1 = 0$ (Principle of zero products)

$x = 0$ or $3x = 1$ or $3x = -1$

$x = 0$ or $x = \frac{1}{3}$ or $x = -\frac{1}{3}$

The solutions are 0, $\frac{1}{3}$, and $-\frac{1}{3}$.

19. $(x - 3)^2 = x^2 + 2x + 1$

$x^2 - 6x + 9 = x^2 + 2x + 1$ (Squaring on the left side)

$-6x + 9 = 2x + 1$ (Adding $-x^2$)

$9 = 8x + 1$ (Adding 6x)

$8 = 8x$ (Adding -1)

$1 = x$ (Multiplying by $\frac{1}{8}$)

The solution is 1.

21. $3 - x \leqslant 4x + 7$

$3 - x + x \leqslant 4x + 7 + x$ (Adding x)

$3 \leqslant 5x + 7$ (Collecting like terms)

$3 + (-7) \leqslant 5x + 7 + (-7)$ (Adding -7)

$-4 \leqslant 5x$

$\frac{1}{5} \cdot (-4) \leqslant \frac{1}{5} \cdot 5x$ (Multiplying by $\frac{1}{5}$)

$-\frac{4}{5} \leqslant x$, or $x \geqslant -\frac{4}{5}$

Any number greater than or equal to $-\frac{4}{5}$ is a solution.

This exercise can also be done as follows:

$3 - x \leqslant 4x + 7$

$-4x + 3 - x \leqslant -4x + 4x + 7$ (Adding -4x)

$3 - 5x \leqslant 7$

$-3 + 3 - 5x \leqslant -3 + 7$ (Adding -3)

$-5x \leqslant 4$

$-\frac{1}{5} \cdot -5x \geqslant -\frac{1}{5} \cdot 4$

(Multiplying by $-\frac{1}{5}$, and reversing the inequality sign)

$x \geqslant -\frac{4}{5}$

Any number greater than or equal to $-\frac{4}{5}$ is a solution.

23. $5x - 5 + x > 2 - 6x - 8$

$6x - 5 > -6x - 6$ (Collecting like terms on each side)

$6x + 6x - 5 > 6x - 6x - 6$ (Adding 6x)

$12x - 5 > -6$

$12x - 5 + 5 > -6 + 5$ (Adding 5)

$12x > -1$

$\frac{1}{12} \cdot 12x > \frac{1}{12} \cdot (-1)$ (Multiplying by $\frac{1}{12}$)

$x > -\frac{1}{12}$

Any number greater than $-\frac{1}{12}$ is a solution.

25. $-7x < 4$

$-\frac{1}{7} \cdot -7x > -\frac{1}{7} \cdot 4$

(Multiplying by $-\frac{1}{7}$ and reversing the inequality sign)

$x > -\frac{4}{7}$

Any number greater than $-\frac{4}{7}$ is a solution.

27. $5x + 2x \leqslant -21$

$7x \leqslant -21$ (Collecting like terms)

$\frac{1}{7} \cdot 7x \leqslant \frac{1}{7} \cdot (-21)$ (Multiplying by $\frac{1}{7}$)

$x \leqslant -3$

Any number less than or equal to -3 is a solution.

29. $2x - 7 < 5x - 9$

$-7 < 3x - 9$ (Adding -2x)

$2 < 3x$ (Adding 9)

$\frac{2}{3} < x$, or $x > \frac{2}{3}$ (Multiplying by $\frac{1}{3}$)

Any number greater than $\frac{2}{3}$ is a solution.

31. $8x - 9 < 3x - 11$

$5x - 9 < -11$ (Adding -3x)

$5x < -2$ (Adding 9)

$x < -\frac{2}{5}$ (Multiplying by $\frac{1}{5}$)

Any number less than $-\frac{2}{5}$ is a solution.

33. $8 < 3x + 2 < 14$

$8 + (-2) < 3x + 2 + (-2) < 14 + (-2)$ (Adding -2)

$6 < 3x < 12$

$\frac{1}{3} \cdot 6 < \frac{1}{3} \cdot 3x < \frac{1}{3} \cdot 12$ (Multiplying by $\frac{1}{3}$)

$2 < x < 4$

Any number greater than 2 and less than 4 is a solution.

35. $3 \leqslant 4x - 3 \leqslant 19$

$3 + 3 \leqslant 4x - 3 + 3 \leqslant 19 + 3$ (Adding 3)

$6 \leqslant 4x \leqslant 22$

$\frac{1}{4} \cdot 6 \leqslant \frac{1}{4} \cdot 4x \leqslant \frac{1}{4} \cdot 22$ (Multiplying by $\frac{1}{4}$)

$\frac{6}{4} \leqslant x \leqslant \frac{22}{4}$

$\frac{3}{2} \leqslant x \leqslant \frac{11}{2}$ (Simplifying)

Any number greater than or equal to $\frac{3}{2}$ and less than or equal to $\frac{11}{2}$ is a solution.

37. $-7 \leqslant 5x - 2 \leqslant 12$

$-5 \leqslant 5x \leqslant 14$ (Adding 2)

$-1 \leqslant x \leqslant \frac{14}{5}$ (Multiplying by $\frac{1}{5}$)

Any number greater than or equal to -1 and less than or equal to $\frac{14}{5}$ is a solution.

39. 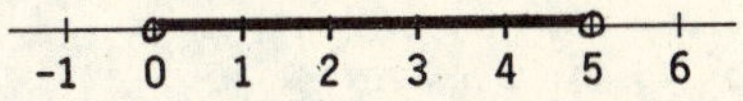

(0,5) = the set of all numbers x such that $0 < x < 5$.

The open circles and the parentheses indicate that 0 and 5 are not included.

41. 

[-9,-4) = the set of all numbers x such that $-9 \leqslant x < -4$.

The solid circle and the bracket indicate that -9 is included.

The open circle and the parenthesis indicate that -4 is not included.

43. 

[x,x + h] = the set of all numbers greater than or equal to x and less than or equal to x + h.

The solid circles and the brackets indicate that x and x + h are included.

45. 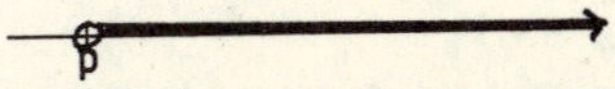

(p,∞) = the set of all numbers x such that $x > p$.

The open circle and the parenthesis on the left indicate that p is not included.

The symbol ∞ indicates that the interval is of unlimited extent in the positive direction. Since ∞ is not a number and therefore could not be an included endpoint, a parentheses is used on the right.

47. [-3,3] = the set of all numbers x such that $-3 \leqslant x \leqslant 3$.

The brackets indicate that both -3 and 3 are included.

49. [-14,-11) = the set of all numbers x such that $-14 \leqslant x < -11$.

The bracket indicates that -14 is included.

The parenthesis indicates that -11 is not included.

51. (-∞,-4] = the set of all numbers x such that $x \leqslant -4$.

The symbol -∞ indicates that the interval is of unlimited extent in the negative direction. Since -∞ is not a number and therefore could not be an included endpoint, a parenthesis is used.

The bracket indicates that -4 is included.

53. We first translate to an equation.

Investment + 11% · Investment = 721.50

↓ ↓ ↓ ↓ ↓ ↓ ↓

y + 11% · y = 721.50

Now we solve the equation.

$y + 11\%y = 721.50$

$1 \cdot y + 0.11y = 721.50$

$1.11y = 721.50$ (Collecting like terms)

$\frac{1}{1.11} \cdot 1.11y = \frac{1}{1.11} \cdot 721.50$ (Mutliplying by $\frac{1}{1.11}$)

$y = \frac{721.50}{1.11}$

$y = 650$

Check:

$650 + 11\% \cdot 650 = 650 + 71.50 = 721.50$

The number 650 checks. Thus the original investment was $650.

55. We first translate to an inequality.

Total revenue is more than $22,000.

$3x + 1000 > 22{,}000$

$3x > 21{,}000$ (Adding -1000)

$x > 7000$ (Multiplying by $\frac{1}{3}$)

More than 7000 units must be sold.

57. We first translate to an equation.

$$\underbrace{\text{Original weight}}_{w} + \underbrace{6\%}_{6\%} \cdot \underbrace{\left[\text{Original weight}\right]}_{w} = 508.8$$

$w + 6\% \cdot w = 508.8$

Now we solve the equation.

$$w + 6\%w = 508.8$$
$$1 \cdot w + 0.06w = 508.8$$
$$1.06w = 508.8 \quad \text{(Collecting like terms)}$$
$$\frac{1}{1.06} \cdot 1.06w = \frac{1}{1.06} \cdot 508.8 \quad \left(\text{Multiplying by } \frac{1}{1.06}\right)$$
$$w = \frac{508.8}{1.06}$$
$$w = 480$$

Check:

$480 + 6\% \cdot 480 = 480 + 28.8 = 508.8$

The number 480 checks. Thus the original weight was 480 pounds.

59. We first translate to an equation.

$$\underbrace{\text{Former population}}_{p} + \underbrace{2\%}_{2\%} \cdot \underbrace{\text{Former population}}_{p} = 826{,}200$$

$p + 2\% \cdot p = 826{,}200$

Now we solve the equation.

$$p + 2\%p = 826{,}200$$
$$1 \cdot p + 0.02p = 826{,}200$$
$$1.02p = 826{,}200 \quad \text{(Collecting like terms)}$$
$$p = \frac{826{,}200}{1.02} \quad \left(\text{Multiplying by } \frac{1}{1.02}\right)$$
$$p = 810{,}000$$

Check:

$810{,}000 + 2\% \cdot 810{,}000 = 810{,}000 + 16{,}200 = 826{,}200$

The number 810,000 checks. Thus the former population was 810,000.

61. Let x represent the score on the fourth test.

The average of the four scores is

$$\frac{78 + 90 + 92 + x}{4}$$

To get a B in the course, this average must be greater than or equal to 80 and less than 90.

Translate to an inequality.

$$80 \leq \frac{78 + 90 + 92 + x}{4} < 90$$
$$80 \leq \frac{260 + x}{4} < 90$$
$$320 \leq 260 + x < 360 \quad \text{(Multiplying by 4)}$$
$$60 \leq x < 100 \quad \text{(Adding -260)}$$

Thus scores greater than or equal to 60% and less than 100% will yield a B.

Exercise Set 1.3

1. a) $f(x) = 2x + 3$

$f(4.1) = 2(4.1) + 3 = 8.2 + 3 = 11.2$
$f(4.01) = 2(4.01) + 3 = 8.02 + 3 = 11.02$
$f(4.001) = 2(4.001) + 3 = 8.002 + 3 = 11.002$
$f(4) = 2(4) + 3 = 8 + 3 = 11$

Input	Output
4.1	11.2
4.01	11.02
4.001	11.002
4	11

b) $f(x) = 2x + 3$

$f(5) = 2(5) + 3 = 10 + 3 = 13$
$f(-1) = 2(-1) + 3 = -2 + 3 = 1$
$f(k) = 2(k) + 3 = 2k + 3$
$f(1 + t) = 2(1 + t) + 3 = 2 + 2t + 3 = 2t + 5$
$f(x + h) = 2(x + h) + 3 = 2x + 2h + 3$

3. $g(x) = x^2 - 3$

$g(-1) = (-1)^2 - 3 = 1 - 3 = -2$
$g(0) = 0^2 - 3 = 0 - 3 = -3$
$g(1) = 1^2 - 3 = 1 - 3 = -2$
$g(5) = 5^2 - 3 = 25 - 3 = 22$
$g(u) = u^2 - 3$
$g(a + h) = (a + h)^2 - 3 = a^2 + 2ah + h^2 - 3$
$g(1 - h) = (1 - h)^2 - 3 = 1 - 2h + h^2 - 3$
$= h^2 - 2h - 2$

5. a) $f(x) = (x - 3)^2$

$f(4) = (4 - 3)^2 = 1^2 = 1$
$f(-2) = (-2 - 3)^2 = (-5)^2 = 25$
$f(0) = (0 - 3)^2 = (-3)^2 = 9$
$f(a) = (a - 3)^2 = a^2 - 6a + 9$
$f(t + 1) = (t + 1 - 3)^2$
$= (t - 2)^2 = t^2 - 4t + 4$
$f(t + 3) = (t + 3 - 3)^2 = t^2$
$f(x + h) = (x + h - 3)^2$
$= [(x + h) - 3]^2$
$= (x + h)^2 - 6(x + h) + 9$
$= x^2 + 2xh + h^2 - 6x - 6h + 9$

b) $f(x) = x^2 - 6x + 9$

This function subtracts six times the input from the square of the input and then adds 9.

7. Graph $f(x) = 2x + 3$.

We first choose any number for x and then determine f(x), or y.

$f(-2) = 2(-2) + 3 = -4 + 3 = -1$
$f(-1) = 2(-1) + 3 = -2 + 3 = 1$
$f(0) = 2 \cdot 0 + 3 = 0 + 3 = 3$
$f(1) = 2 \cdot 1 + 3 = 2 + 3 = 5$

x	y	(x, y)
-2	-1	(-2,-1)
-1	1	(-1, 1)
0	3	(0, 3)
1	5	(1, 5)

Next we plot the input-output pairs from the table and draw the graph.

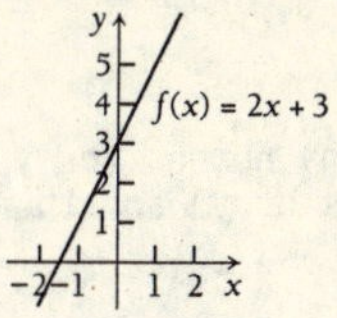

9. Graph $g(x) = -4x$.

We first choose any number for x and then determine g(x), or y.

$g(-1) = -4(-1) = 4$

$g(0) = -4 \cdot 0 = 0$

$g(1) = -4 \cdot 1 = -4$

x	y	(x, y)
-1	4	(-1, 4)
0	0	(0, 0)
1	-4	(1,-4)

Next we plot the input-output pairs from the table and draw the graph.

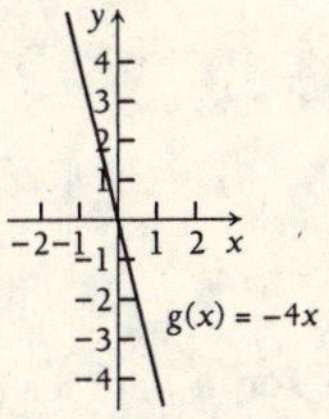

11. Graph $f(x) = x^2 - 1$.

We first choose any number for x and then determine f(x), or y.

$f(-2) = (-2)^2 - 1 = 4 - 1 = 3$

$f(-1) = (-1)^2 - 1 = 1 - 1 = 0$

$f(0) = 0^2 - 1 = 0 - 1 = -1$

$f(1) = 1^2 - 1 = 1 - 1 = 0$

$f(2) = 2^2 - 1 = 4 - 1 = 3$

x	y	(x, y)
-2	3	(-2, 3)
-1	0	(-1, 0)
0	-1	(0,-1)
1	0	(1, 0)
2	3	(2, 3)

Next we plot the input-output pairs from the table and draw the graph.

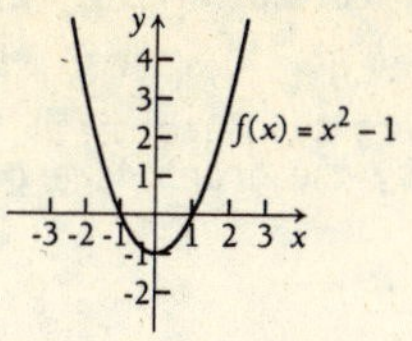

13. Graph $g(x) = x^3$.

We first choose any number for x and then determine g(x), or y.

$g(-2) = (-2)^3 = -8$

$g(-1) = (-1)^3 = -1$

$g(0) = 0^3 = 0$

$g(1) = 1^3 = 1$

$g(2) = 2^3 = 8$

x	y	(x, y)
-2	-8	(-2,-8)
-1	-1	(-1,-1)
0	0	(0, 0)
1	1	(1, 1)
2	8	(2, 8)

Next we plot the input-output pairs from the table and draw the graph.

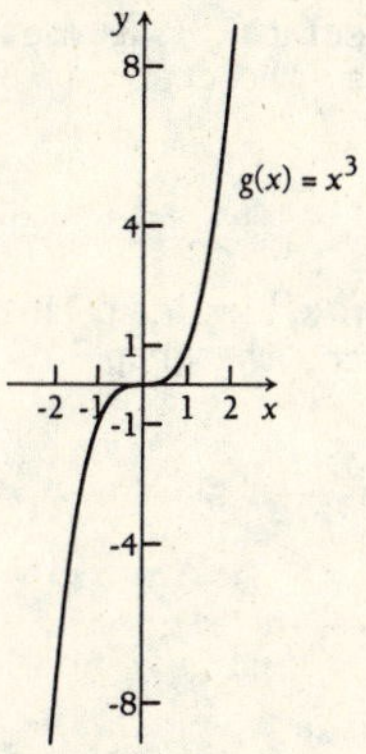

15. Vertical line test:

If it is possible for a vertical line to meet a graph more than once, the graph is not the graph of a function.

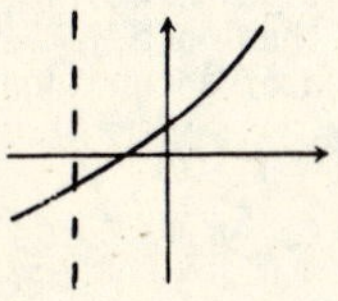

Vertical line

Visualize moving this vertical line across the graph. Ask yourself the question:

Will this line ever meet the graph more than once?

If the answer is yes, the graph <u>is not</u> a graph of a function.

If the answer is no, the graph <u>is</u> a graph of a function.

In this problem no vertical line meets the graph more than once. Thus, the graph is a graph of a function.

<u>17</u>. Vertical line test:

If it is possible for a vertical line to meet a graph more than once, the graph is not the graph of a function.

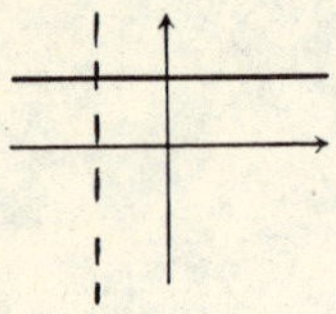

Vertical line

Visualize moving this vertical line across the graph. Ask yourself the question:

Will this line ever meet the graph more than once?

If the answer is yes, the graph <u>is</u> <u>not</u> a graph of a function.

If the answer is no, the graph <u>is</u> a graph of a function.

In this problem no vertical line meets the graph more than once. Thus, the graph is a graph of a function.

<u>19</u>. Vertical line test:

If it is possible for a vertical line to meet a graph more than once, the graph is not the graph of a function.

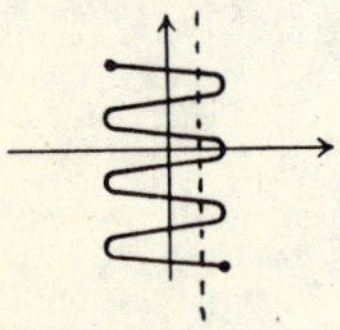

Vertical line

In this problem a vertical line (in fact many) meets the graph more than once. Therefore, the graph <u>is</u> <u>not</u> the graph of a function.

<u>21</u>.

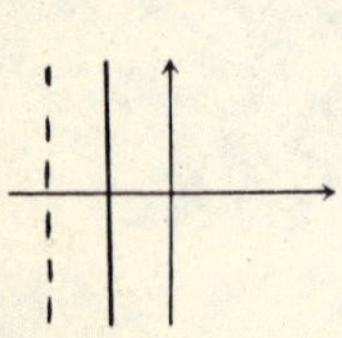

Vertical line

Visualize moving the vertical line on the left across the graph. Ask yourself the question:

Will this line ever meet the graph more than once?

The answer is yes. Thus, the graph <u>is</u> <u>not</u> the graph of a function.

<u>23</u>. We see from the graph that no vertical line meets the graph more than once. Thus, this is the graph of a function.

<u>25</u>. We see from the graph that no vertical line meets the graph more than once. Thus, this is the graph of a function.

<u>27</u>. a) Graph $x = y^2 - 1$.

We first choose any number for y (since x is expressed in terms of y) and then determine x.

For $y = -2$, $x = (-2)^2 - 1 = 4 - 1 = 3$.

For $y = -1$, $x = (-1)^2 - 1 = 1 - 1 = 0$.

For $y = 0$, $x = 0^2 - 1 = 0 - 1 = -1$.

For $y = 1$, $x = 1^2 - 1 = 1 - 1 = 0$.

For $y = 2$, $x = 2^2 - 1 = 4 - 1 = 3$.

x	y	(x, y)
3	-2	(3,-2)
0	-1	(0,-1)
-1	0	(-1, 0)
0	1	(0, 1)
3	2	(3, 2)

Next we plot the ordered pairs and draw the graph.

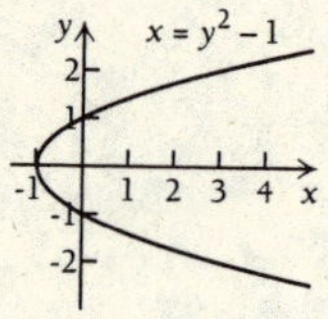

b) Vertical line test:

If it is possible for a vertical line to meet a graph more than once, the graph is not the graph of a function.

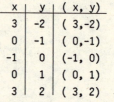

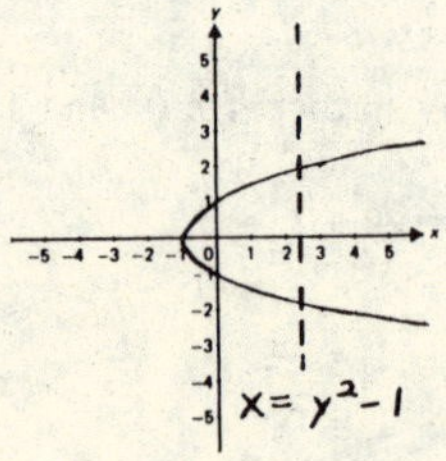

Vertical line

A vertical line (in fact, many) meets the graph more than once. Therefore the graph of $x = y^2 - 1$ <u>is</u> <u>not</u> the graph of a function.

29. $f(x) = x^2 - 3x$

$f(x + h) = (x + h)^2 - 3(x + h)$ (Substituting)

$= x^2 + 2xh + h^2 - 3x - 3h$

31. Graph: $f(x) = \begin{cases} 1 \text{ for } x < 0, \\ -1 \text{ for } x \geqslant 0 \end{cases}$

First graph $f(x) = 1$ for inputs less than 0.

x	y	(x, y)
$-\frac{1}{2}$	1	$(-\frac{1}{2}, 1)$
-1	1	(-1, 1)
-2	1	(-2, 1)
-3.21	1	(-3.21,1)
-4	1	(-4, 1)

For any input less than 0, the output is 1.

Plot the input-output pairs from the table and draw this part of the graph. Since the number 0 is not an input, the point (0,1) is not part of the graph. Therefore an open circle is used at the point (0,1).

Next graph $f(x) = -1$ for inputs greater than or equal to 0.

x	y	(x, y)
0	-1	(0,-1)
1	-1	(1,-1)
$1\frac{3}{4}$	-1	$(1\frac{3}{4},-1)$
2.7	-1	(2.7,-1)
3	-1	(3,-1)

For any input greater than or equal to 0, the output is -1.

Plot the input-output pairs from the table and draw this part of the graph. Since the number 0 is an input, the point (0,-1) is part of the graph. Therefore a solid circle is used at the point (0,-1).

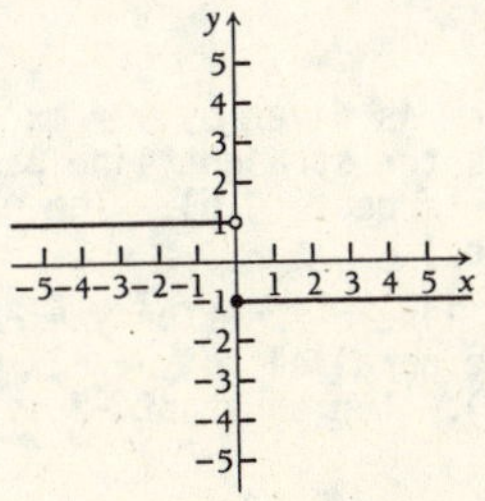

33. Graph: $f(x) = \begin{cases} -3 \text{ for } x = -2 \\ x^2 \text{ for } x \neq -2 \end{cases}$

First graph $f(x) = -3$ for $x = -2$. This graph consists of only one point, (-2,-3).

Next graph $f(x) = x^2$ for all inputs except -2.

x	y	(x, y)
-3	9	(-3, 9)
-1	1	(-1, 1)
0	0	(0, 0)
1	1	(1, 1)
2	4	(2, 4)

Plot the input-output pairs from the table and draw this part of the graph. Since the number -2 is not an input, the point (-2,4) is not part of the graph. Therefore an open circle is used at the point (-2,4).

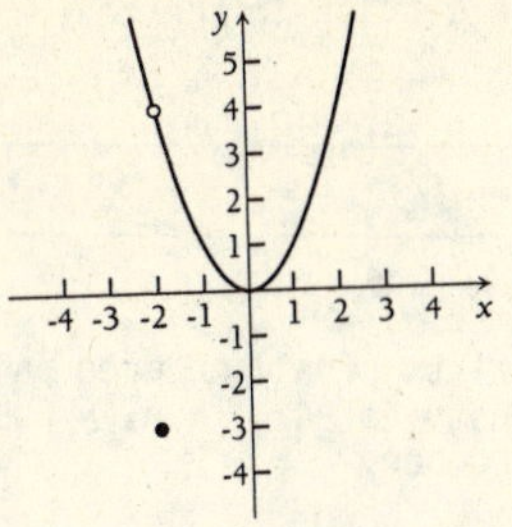

35. $R(x) = 200x + 50$

$R(10) = 200 \cdot 10 + 50 = 2000 + 50 = \$2050$

$R(100) = 200 \cdot 100 + 50 = 20{,}000 + 50 = \$20{,}050$

37. $2x + y - 16 = 4 - 3y + 2x$

$y - 16 = 4 - 3y$ (Adding -2x)

$4y - 16 = 4$ (Adding 3y)

$4y = 20$ (Adding 16)

$y = 5$ (Multiplying by $\frac{1}{5}$)

Graph $y = 5$.

x	y	(x, y)
-2	5	(-2, 5)
0	5	(0, 5)
3	5	(3, 5)

The input, x, can be any number.

The output, y, must always be 5.

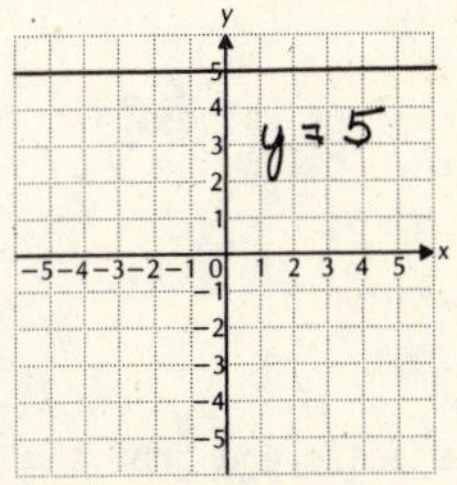

No vertical line meets the graph more than once. Thus the equation represents a function.

39. $(4y^{2/3})^3 = 64x$

$4^3 \cdot (y^{2/3})^3 = 4^3 x$

$y^2 = x$

$y = \pm \sqrt{x}$

We sketch the graph:

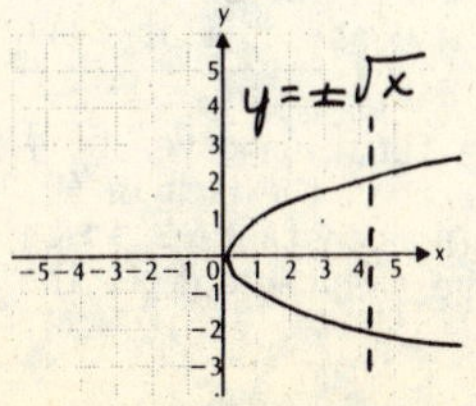

Since a vertical line (in fact, many) meets the graph more than once, this is not a function.

41. See the answer section in the text.

43. See the answer section in the text.

Exercise Set 1.4

1. Graph y = -4.

The graph consists of all ordered pairs whose second coordinate is -4. Thus, y must be -4, but x can be any number.

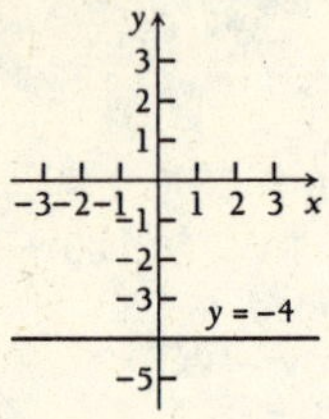

3. Graph x = 4.5.

The graph consists of all ordered pairs whose first coordinate is 4.5. Thus, x must be 4.5, but y can be any number.

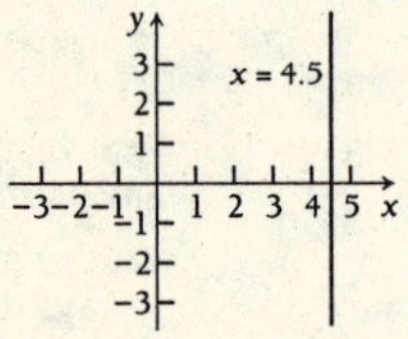

5. Graph y = -3x.

We first make a table of values. We choose any number for x and then determine y by substitution.

When x = -2, y = -3(-2) = 6.

When x = 0, y = -3·0 = 0.

When x = 1, y = -3·1 = -3.

Plot these ordered pairs and draw the graph.

x	y
-2	6
0	0
1	-3

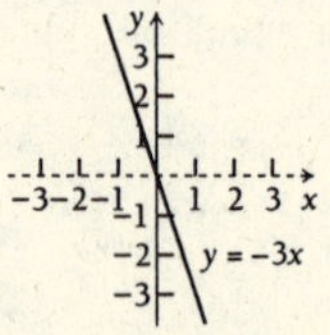

[The graph of the function given by y = mx is the straight line through the origin (0,0) and the point (1,m). The constant m is called the slope of the line.]

The graph of the function given by y = -3x is the straight line through the origin (0,0) and the point (1,-3). The constant -3 is called the slope of the line. The y-intercept is the point (0,0).

7. Graph y = 0.5x.

We first make a table of values. We choose any number for x and then determine y by substitution.

When x = -4, y = 0.5(-4) = -2.

When x = 0, y = 0.5(0) = 0.

When x = 2, y = 0.5(2) = 1.

Plot these ordered pairs and draw the graph.

x	y
-4	-2
0	0
2	1

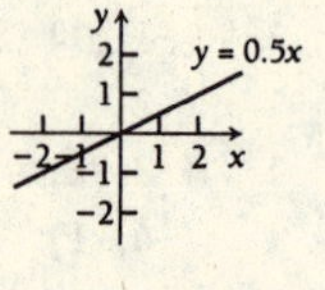

The graph of the function given by y = 0.5x is the straight line through the origin (0,0) and the point (1,0.5). The constant 0.5 is called the slope of the line. The y-intercept is the point (0,0).

9. Graph y = -2x + 3.

We first make a table of values. We choose any number for x and then determine y by substitution.

When x = -1, y = -2(-1) + 3 = 2 + 3 = 5.

When x = 0, y = -2·0 + 3 = 0 + 3 = 3.

When x = 3, y = -2·3 + 3 = -6 + 3 = -3.

Plot these ordered pairs and draw the graph.

x	y
-1	5
0	3
3	-3

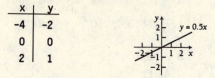

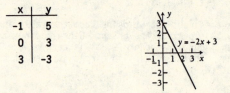

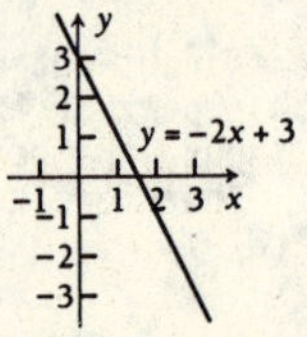

[A linear function is given by y = mx + b and has a graph which is the straight line parallel to y = mx with y-intercept (0,b). The constant m is called the slope.]

The graph of the linear function y = -2x + 3 is the straight line parallel to y = -2x with y-intercept (0,3). The constant -2 is called the slope.

11. Graph y = -x - 2.

y = -1·x - 2 (-x = -1·x)

We first make a table of values. We choose any number for x and then determine y by substitution.

When x = -5, y = -1(-5) - 2 = 5 - 2 = 3.

When x = 0, y = -1·0 - 2 = 0 - 2 = -2.

When x = 2, y = -1·2 - 2 = -2 - 2 = -4.

Plot these ordered pairs and draw the graph.

x	y
-5	3
0	-2
2	-4

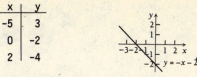

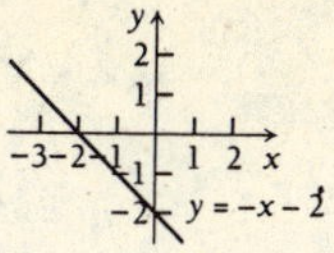

The graph of the linear function $y = -x - 2$ is the straight line parallel to $y = -x$ with y-intercept $(0,-2)$. The constant -1 is called the slope.

13. $y = mx + b$ is called the slope intercept equation of a line. The constant m is the slope of the line. The y-intercept is the point $(0,b)$.

Solve the equation for y.

$2x + y - 2 = 0$

$y = -2x + 2$ (Adding $-2x + 2$) ($m = -2$, $b = 2$)

The slope is -2.
The y-intercept is (0,2).

15. Solve the equation for y.

$2x + 2y + 5 = 0$

$2y = -2x - 5$ (Adding $-2x - 5$)

$y = -x - \frac{5}{2}$ (Multiplying by $\frac{1}{2}$)

$y = -1\cdot x + (-\frac{5}{2})$ ($m = -1$, $b = -\frac{5}{2}$)

The slope is -1.

The y-intercept is $(0,-\frac{5}{2})$.

17. $y - y_1 = m(x - x_1)$ (Point-slope equation of a line)

$y - (-5) = -5(x - 1)$ (Substituting -5 for m, 1 for x_1 and -5 for y_1)

$y + 5 = -5x + 5$ (Simplifying)

$y = -5x$ (Adding -5)

19. $y - y_1 = m(x - x_1)$ (Point-slope equation of a line)

$y - 3 = -2(x - 2)$ (Substituting -2 for m, 2 for x_1 and 3 for y_1)

$y - 3 = -2x + 4$ (Simplifying)

$y = -2x + 7$ (Adding 3)

21. $y = mx + b$ (Slope-intercept equation of a line)

$y = \frac{1}{2}x + (-6)$ (Substituting $\frac{1}{2}$ for m and -6 for b)

$y = \frac{1}{2}x - 6$

23. $y - y_1 = m(x - x_1)$ (Point-slope equation of a line)

$y - 3 = 0(x - 2)$ (Substituting 0 for m, 2 for x_1, and 3 for y_1)

$y - 3 = 0$ (Simplifying)

$y = 3$ (Adding 3)

25. Find the slope of the line containing (-4,-2) and (-2,1).

$m = \frac{y_2 - y_1}{x_2 - x_1}$ [Slope of line containing points (x_1,y_1) and (x_2,y_2)]

$m = \frac{1 - (-2)}{-2 - (-4)}$ (Substituting 1 for y_2, -2 for y_1, -2 for x_2, and -4 for x_1)

$= \frac{1 + 2}{-2 + 4}$

$= \frac{3}{2}$

It does not matter which point is taken first, as long as we subtract coordinates in the same order. We could also find m as follows.

$m = \frac{-2 - 1}{-4 - (-2)}$ (Substituting -2 for y_2, 1 for y_1, -4 for x_2, and -2 for x_1)

$= \frac{-2 - 1}{-4 + 2}$

$= \frac{-3}{-2}$

$= \frac{3}{2}$

27. Find the slope of the line containing (2,-4) and (4,-3).

$m = \frac{y_2 - y_1}{x_2 - x_1}$ [Slope of line containing points (x_1,y_1) and (x_2,y_2)]

$m = \frac{-3 - (-4)}{4 - 2}$ (Substituting -3 for y_2, -4 for y_1, 4 for x_2, and 2 for x_1)

$= \frac{-3 + 4}{4 - 2}$

$= \frac{1}{2}$

29. Find the slope of the line containing (3,-7) and (3,-9).

$m = \frac{y_2 - y_1}{x_2 - x_1}$ [Slope of line containing points (x_1,y_1) and (x_2,y_2)]

$= \frac{-9 - (-7)}{3 - 3}$ (Substituting -9 for y_2, -7 for y_1, 3 for x_2, and 3 for x_1)

$= \frac{-9 + 7}{3 - 3}$

$= \frac{-2}{0}$

Since we cannot divide by 0, the slope of the line through (3,-7) and (3,-9) is undefined. The line has no slope.

31. Find the slope of the line containing (2,3) and (-1,3).

$m = \frac{y_2 - y_1}{x_2 - x_1}$ [Slope of line containing points (x_1,y_1) and (x_2,y_2)]

$m = \frac{3 - 3}{-1 - 2}$ (Substituting 3 for y_2, 3 for y_1, -1 for x_2, and 2 for x_1)

$= \frac{0}{-3}$

$= 0$

33. Find the slope of the line containing $(x,3x)$ and $(x + h,3(x + h))$.

$m = \frac{y_2 - y_1}{x_2 - x_1}$ [Slope of line containing points (x_1,y_1) and (x_2,y_2)]

$m = \frac{3(x + h) - 3x}{x + h - x}$ (Substituting $3(x + h)$ for y_2, $3x$ for y_1, $x + h$ for x_2, and x for x_1)

$= \frac{3x + 3h - 3x}{x + h - x}$ (Removing parentheses)

$= \frac{3h}{h}$ (Collecting like terms)

$= 3$ (Simplifying)

35. Find the slope of the line containing $(x, 2x + 3)$ and $(x + h, 2(x + h) + 3)$.

$m = \frac{y_2 - y_1}{x_2 - x_1}$ [Slope of line containing points (x_1,y_1) and (x_2,y_2)]

$m = \frac{[2(x + h) + 3] - (2x + 3)}{(x + h) - x}$

(Substituting $2(x + h) + 3$ for y_2, $2x + 3$ for y_1, $x + h$ for x_2, and x for x_1)

$= \frac{2x + 2h + 3 - 2x - 3}{x + h - x}$ (Removing parentheses)

$= \frac{2h}{h}$ (Collecting like terms)

$= 2$ (Simplifying)

37. From Exercise 25, we know that the slope of the line is $\frac{3}{2}$. Using the point $(-4,-2)$, we substitute in the point-slope equation.

$y - (-2) = \frac{3}{2}[x - (-4)]$

$y + 2 = \frac{3}{2}(x + 4)$

$y + 2 = \frac{3}{2}x + 6$

$y = \frac{3}{2}x + 4$

We could also use the point $(-2,1)$:

$y - 1 = \frac{3}{2}[x - (-2)]$

$y - 1 = \frac{3}{2}(x + 2)$

Simplifying this, we also get $y = \frac{3}{2}x + 4$.

39. From Exercise 27, we know that the slope is $\frac{1}{2}$. Using the point $(2,-4)$, we substitute in the point-slope equation.

$y - (-4) = \frac{1}{2}(x - 2)$

$y + 4 = \frac{1}{2}(x - 2)$

$y + 4 = \frac{1}{2}x - 1$

$y = \frac{1}{2}x - 5$

We could also use $(4,-3)$:

$y - (-3) = \frac{1}{2}(x - 4)$

$y + 3 = \frac{1}{2}(x - 4)$

Simplifying this, we also get $y = \frac{1}{2}x - 5$.

41. From Exercise 29, we know that the line containing $(3,-7)$ and $(3,-9)$ has no slope. The graph is a line which contains all ordered pairs whose first coordinate is 3. The second coordinate can be any number. The line is vertical. The equation is $x = 3$.

43. From Exercise 31, we know that the slope is 0. The graph consists of all ordered pairs whose second coordinate is 3. The first coordinate can be any number. The line is horizontal. The equation is $y = 3$.

45. From Exercise 33, we know that the slope is 3. Using the point $(x, 3x)$, we substitute in the point-slope equation:

$y - 3x = 3(x - x)$

$y - 3x = 3\cdot 0$

$y - 3x = 0$

$y = 3x$

We could also use $(x + h, 3(x + h))$:

$y - 3(x + h) = 3[x - (x - h)]$

Simplifying this, we also get $y = 3x$.

47. From Exercise 35, we know that the slope is 2. Using the point $(x, 2x + 3)$, we substitute in the point-slope equation:

$y - (2x + 3) = 2(x - x)$

$y - 2x - 3 = 2\cdot 0$

$y - 2x - 3 = 0$

$y = 2x + 3$

We could also use $(x + h, 2(x + h) + 3)$:

$y - [2(x + h) + 3] = 2[x - (x + h)]$

Simplifying this, we also get $y = 2x + 3$.

49. a) A is directly proportional to P if there is some positive constant m such that $A = mP$.

We let P represent the investment and 14% P represent the interest earned on the investment in one year. Thus,

$A = P + 14\% P$

$= 1\cdot P + 0.14P$

$= (1 + 0.14)P$

$= 1.14P$

Since A can be expressed as a positive constant times P, we say A is directly proportional to P.

b) $A = 1.14P$

$A = 1.14(100)$ (Substituting 100 for P)

$A = \$114$

c) $A = 1.14P$

$273.60 = 1.14P$ (Substituting 273.60 for A)

$\frac{273.60}{1.14} = P$ (Multiplying by $\frac{1}{1.14}$)

$\$240 = P$

51. a) Total costs = $\text{Variable costs} + \text{Fixed costs}$

To produce x calculators it costs \$20 per calculator in addition to the fixed costs of \$100,000. That is, the variable costs are 20x dollars. The total cost is

$C(x) = 20x + 100{,}000$

b) The total revenue from the sale of x calculators is \$45 per calculator. That is, the total revenue is given by the function

$R(x) = 45x$

c) Total profit = $\text{Total revenue} - \text{Total costs}$

$P(x) = R(x) - C(x)$

$P(x) = 45x - (20x + 100{,}000)$

[Substituting 45x for R(x) and 20x + 100,000 for C(x)]

$P(x) = 45x - 20x - 100{,}000$

$P(x) = 25x - 100{,}000$

d) $P(x) = 25x - 100{,}000$ (Profit function)

$P(150{,}000) = 25(150{,}000) - 100{,}000$

(Substituting 150,000 for x)

$= 3{,}750{,}000 - 100{,}000$

$= 3{,}650{,}000$

A profit of \$3,650,000 will be realized if the expected sales of 150,000 calculators occur.

e) The profit will be \$0 if the firm breaks even. We set the profit function equal to 0 and solve for x.

$P(x) = 25x - 100{,}000$ (Profit function)

$0 = 25x - 100{,}000$

(Setting the profit function equal to 0)

$100{,}000 = 25x$ (Adding 100,000)

$4000 = x$ (Multiplying by $\frac{1}{25}$)

Thus, to break even the firm must sell 4000 calculators.

53. a) Total costs = $\text{Variable costs} + \text{Fixed costs}$

To mow x lawns, it costs \$1 per lawn in addition to the fixed cost of the lawnmower, \$250. That is, the variable cost is $1 \cdot x$, or x, dollars. Thus,

$C(x) = x + 250$

b) $P(x) = R(x) - C(x)$

$P(x) + C(x) = R(x)$ (Adding C(x) on both sides)

$(9x - 250) + x + 250 = R(x)$ (Substituting)

$10x = R(x)$ (Simplifying)

The total revenue from mowing x lawns is 10x, so the student charges \$10 per lawn.

c) We solve $P(x) = 0$ to find the break-even values of x:

$9x - 250 = 0$

$9x = 250$

$x = 27\frac{7}{9}$

$x \approx 28$ Rounding up

The student must mow 28 lawns before making a profit.

55. a) If R is directly proportional to T, then there is some positive constant m such that $R = mT$.

To find m substitute 12.51 for R and 3 for T in the equation $R = mT$ and solve for m.

$R = mT$

$12.51 = m \cdot 3$ (Substituting)

$\frac{12.51}{3} = m$ (Multiplying by $\frac{1}{3}$)

$4.17 = m$

The variation constant is 4.17.
The equation of variation is $R = 4.17T$.

b) $R = 4.17T$ (Equation of variation)

$R = 4.17(6)$ (Substituting 6 for T)

$= 25.02$

The R-factor for insulation that is 6 inches thick is 25.02.

57. a) If B is directly proportional to W, then there is some positive constant m such that

$B = mW$.

To find m substitute 200 for W and 5 for B in the equation $B = mW$ and solve for m.

$B = mW$

$5 = m \cdot 200$ (Substituting)

$\frac{5}{200} = m$ (Multiplying by $\frac{1}{200}$)

$0.025 = m$

The variation constant is 0.025.

The equation of variation is $B = 0.025W$.

b) 0.025 is equivalent to 2.5%, so the resulting resulting equation is

$B = 2.5\%\ W$.

The weight, B, of a human's brain is 2.5% of the body weight, W.

c) $B = 0.025\ W$ (Equation of variation)

$B = 0.025(120)$ (Substituting 120 for W)

$B = 3$

The weight of the brain of a person who weighs 120 lb is 3 lb.

59. a) $D(F) = 2F + 115$

$D(0°) = 2\cdot 0 + 115 = 0 + 115 = 115$ ft

$D(-20°) = 2(-20) + 115 = -40 + 115 = 75$ ft

$D(10°) = 2\cdot 10 + 115 = 20 + 115 = 135$ ft

$D(32°) = 2\cdot 32 + 115 = 64 + 115 = 179$ ft

b)

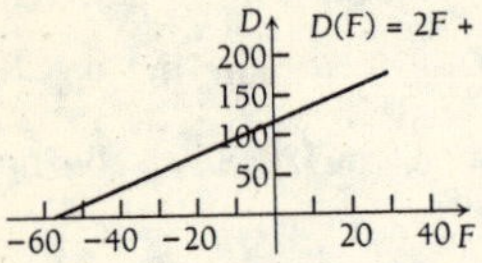

c) First solve $D(F) = 2F + 115$ for F when $D(F) = 0$.

$0 = 2F + 115$

$-115 = 2F$

$-57.5 = F$

When the temperature is -57.5° F, the stopping distance is 0 feet. Temperatures below -57.5° F would yield negative stopping distances, which have no meaning here. For temperatures above 32° F there would be no ice.

61. a) $M(x) = 2.89x + 70.64$

$M(45) = 2.89(45) + 70.64$ (Substituting)

$= 130.05 + 70.64$

$= 200.69$

The male was 200.69 cm tall.

b) $F(x) = 2.75x + 71.48$

$F(45) = 2.75(45) + 71.48$ (Substituting)

$= 123.75 + 71.48$

$= 195.23$

The female was 195.23 cm tall.

63. a) $A(t) = 0.08t + 19.7$

$A(0) = 0.08(0) + 19.7 = 0 + 19.7 = 19.7$

$A(1) = 0.08(1) + 19.7 = 0.08 + 19.7 = 19.78$

$A(10) = 0.08(10) + 19.7 = 0.8 + 19.7 = 20.5$

$A(30) = 0.08(30) + 19.7 = 2.4 + 19.7 = 22.1$

$A(40) = 0.08(40) + 19.7 = 3.2 + 19.7 = 22.9$

b) 1996 - 1950 = 46, so we find A(46):

$A(46) = 0.08(46) + 19.7 = 3.68 + 19.7 = 23.38$

The median age of women at first marriage in 1996 will be 23.38.

c)

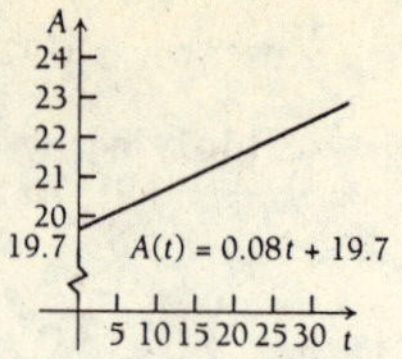

Exercise Set 1.5

1. Graph $y = \frac{1}{2}x^2$ and $y = -\frac{1}{2}x^2$ on the same set of axes.

Find some ordered pairs that are solutions of $y = \frac{1}{2}x^2$, keeping the results in a table. Since the domain of $y = \frac{1}{2}x^2$ consists of <u>all</u> real numbers, we choose <u>any</u> number for x and then find y.

When $x = 0$, $y = \frac{1}{2}\cdot 0^2 = \frac{1}{2}\cdot 0 = 0$.

When $x = 2$, $y = \frac{1}{2}\cdot 2^2 = \frac{1}{2}\cdot 4 = 2$.

When $x = -2$, $y = \frac{1}{2}\cdot (-2)^2 = \frac{1}{2}\cdot 4 = 2$.

When $x = 3$, $y = \frac{1}{2}\cdot 3^2 = \frac{1}{2}\cdot 9 = \frac{9}{2}$.

When $x = -3$, $y = \frac{1}{2}\cdot (-3)^2 = \frac{1}{2}\cdot 9 = \frac{9}{2}$.

x	y
0	0
2	2
-2	2
3	$\frac{9}{2}$
-3	$\frac{9}{2}$

(Table of values for $y = \frac{1}{2}x^2$)

Find some ordered pairs that are solutions of $y = -\frac{1}{2}x^2$, keeping the results in a table. Since the domain of $y = -\frac{1}{2}x^2$ consists of <u>all</u> real numbers, we choose any number for x and then find y.

When $x = 0$, $y = -\frac{1}{2}\cdot 0^2 = -\frac{1}{2}\cdot 0 = 0$.

When $x = 2$, $y = -\frac{1}{2}\cdot 2^2 = -\frac{1}{2}\cdot 4 = -2$.

When $x = -2$, $y = -\frac{1}{2}\cdot (-2)^2 = -\frac{1}{2}\cdot 4 = -2$.

When $x = 3$, $y = -\frac{1}{2}\cdot 3^2 = -\frac{1}{2}\cdot 9 = -\frac{9}{2}$.

When $x = -3$, $y = -\frac{1}{2}\cdot (-3)^2 = -\frac{1}{2}\cdot 9 = -\frac{9}{2}$.

x	y
0	0
2	-2
-2	-2
3	$-\frac{9}{2}$
-3	$-\frac{9}{2}$

(Table of values for $y = -\frac{1}{2}x^2$)

Plot the ordered pairs for $y = \frac{1}{2}x^2$ and draw the graph with a dashed line. Plot the ordered pairs for $y = -\frac{1}{2}x^2$ and draw the graph with a solid line. For $y = \frac{1}{2}x^2$ the coefficient of x^2 is positive, thus the graph opens upward. For $y = -\frac{1}{2}x^2$ the coefficient of x^2 is negative, thus the graph opens downward.

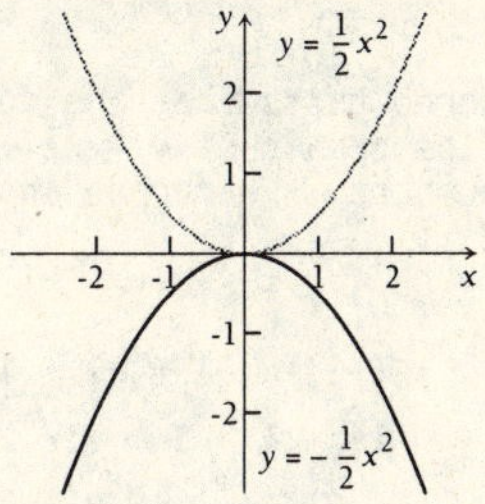

3. Graph $y = x^2$ and $y = (x - 1)^2$ on the same set of axes.

Find some ordered pairs that are solutions of $y = x^2$, keeping the results in a table. Since the domain of $y = x^2$ consists of all real numbers, we choose any number for x and then find y.

When $x = 0$, $y = 0^2 = 0$.

When $x = 1$, $y = 1^2 = 1$.

When $x = -1$, $y = (-1)^2 = 1$.

When $x = 2$, $y = 2^2 = 4$.

When $x = -2$, $y = (-2)^2 = 4$.

x	y
0	0
1	1
-1	1
2	4
-2	4

(Table of values for $y = x^2$)

Find some ordered pairs that are solutions of $y = (x - 1)^2$, keeping the results in a table. Since the domain of $y = (x - 1)^2$ consists of all real numbers, we choose any number for x and then find y.

When $x = 1$, $y = (1 - 1)^2 = 0^2 = 0$.

When $x = 0$, $y = (0 - 1)^2 = (-1)^2 = 1$.

When $x = 2$, $y = (2 - 1)^2 = 1^2 = 1$.

When $x = -1$, $y = (-1 - 1)^2 = (-2)^2 = 4$.

When $x = 3$, $y = (3 - 1)^2 = 2^2 = 4$.

x	y
1	0
0	1
2	1
-1	4
3	4

[Table of values for $y = (x - 1)^2$]

Plot the ordered pairs for $y = x^2$ and draw the graph with a dashed line. Plot the ordered pairs for $y = (x - 1)^2$, or $y = x^2 - 2x + 1$, and draw the graph with a solid line.

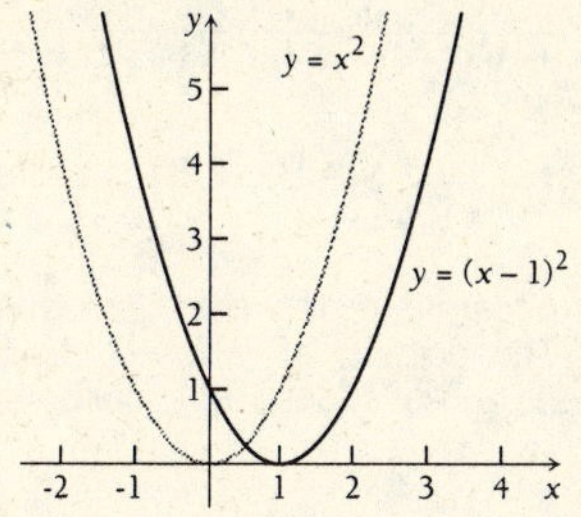

5. Graph $y = x^2$ and $y = (x + 1)^2$ on the same set of axes.

Find some ordered pairs that are solutions of $y = x^2$, keeping the results in a table. Since the domain of $y = x^2$ consists of all real numbers, we choose any number for x and then find y.

When $x = 0$, $y = 0^2 = 0$.

When $x = 1$, $y = 1^2 = 1$.

When $x = -1$, $y = (-1)^2 = 1$.

When $x = 2$, $y = 2^2 = 4$.

When $x = -2$, $y = (-2)^2 = 4$.

x	y
0	0
1	1
-1	1
2	4
-2	4

(Table of values for $y = x^2$)

Find some ordered pairs that are solutions of $y = (x + 1)^2$, keeping the results in a table. Since the domain of $y = (x + 1)^2$ consists of all real numbers, we choose any number for x and then find y.

When $x = -1$, $y = (-1 + 1)^2 = 0^2 = 0$.

When $x = -2$, $y = (-2 + 1)^2 = (-1)^2 = 1$.

When $x = 0$, $y = (0 + 1)^2 = 1^2 = 1$.

When $x = -3$, $y = (-3 + 1)^2 = (-2)^2 = 4$.

When $x = 1$, $y = (1 + 1)^2 = 2^2 = 4$.

x	y
-1	0
-2	1
0	1
-3	4
1	4

[Table of values for $y = (x + 1)^2$]

Plot the ordered pairs for $y = x^2$ and draw the graph with a dashed line. Plot the ordered pairs for $y = (x + 1)^2$, or $y = x^2 + 2x + 1$, and draw the graph with a solid line.

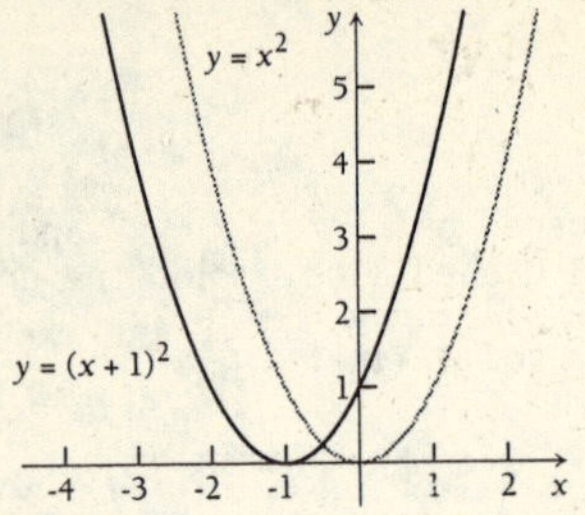

7. Graph $y = |x|$ and $y = |x + 3|$ on the same set of axes.

Find some ordered pairs that are solutions of $y = |x|$. The domain of $y = |x|$ is the set of all real numbers. We choose any number for x and then find y.

When $x = 0$, $y = |0| = 0$.

When $x = 2$, $y = |2| = 2$.

When $x = -2$, $y = |-2| = 2$.

When $x = 5$, $y = |5| = 5$.

When $x = -5$, $y = |-5| = 5$.

x	y
0	0
2	2
-2	2
5	5
-5	5

(Table of values for $y = |x|$)

Find some ordered pairs that are solutions of $y = |x + 3|$. The domain of $y = |x + 3|$ is the set of all real numbers. We choose any number for x and then find y.

When $x = -3$, $y = |-3 + 3| = |0| = 0$.

When $x = -4$, $y = |-4 + 3| = |-1| = 1$.

When $x = -2$, $y = |-2 + 3| = |1| = 1$.

When $x = -6$, $y = |-6 + 3| = |-3| = 3$.

When $x = 0$, $y = |0 + 3| = |3| = 3$.

x	y
-3	0
-4	1
-2	1
-6	3
0	3

(Table of values for $y = |x + 3|$)

Plot the ordered pairs for $y = |x|$ and draw the graph with a dashed line. Plot the ordered pairs for $y = |x + 3|$ and draw the graph with a solid line.

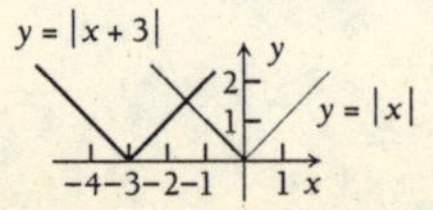

9. Graph $y = x^3$ and $y = x^3 + 1$ on the same set of axes.

Find some ordered pairs that are solutions of $y = x^3$. The domain of $y = x^3$ is the set of all real numbers. We choose any number for x and then find y.

When $x = 0$, $y = 0^3 = 0$.

When $x = -1$, $y = (-1)^3 = -1$.

When $x = 1$, $y = 1^3 = 1$.

When $x = -2$, $y = (-2)^3 = -8$.

When $x = 2$, $y = 2^3 = 8$.

x	y
0	0
-1	-1
1	1
-2	-8
2	8

(Table of values for $y = x^3$)

Find some ordered pairs that are solutions of $y = x^3 + 1$. The domain of $y = x^3 + 1$ is the set of all real numbers. We choose any number for x and then find y.

When $x = 0$, $y = 0^3 + 1 = 1$.

When $x = -1$, $y = (-1)^3 + 1 = -1 + 1 = 0$.

When $x = 1$, $y = 1^3 + 1 = 1 + 1 = 2$.

When $x = -2$, $y = (-2)^3 + 1 = -8 + 1 = -7$.

When $x = 2$, $y = 2^3 + 1 = 8 + 1 = 9$.

x	y
0	1
-1	0
1	2
-2	-7
2	9

(Table of values for $y = x^3 + 1$)

Plot the ordered pairs for $y = x^3$ and draw the graph with a dashed line. Plot the ordered pairs for $y = x^3 + 1$ and draw the graph with a solid line.

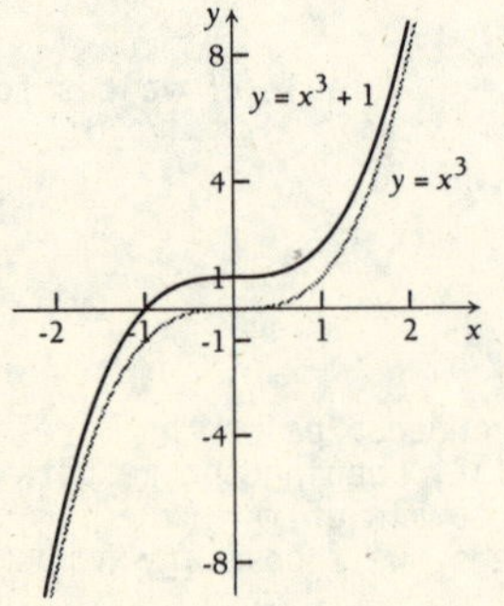

11. Graph $y = \sqrt{x}$ and $y = \sqrt{x + 1}$ on the same set of axes.

Find some ordered pairs that are solutions of $y = \sqrt{x}$. The domain consists of only the nonnegative real numbers. We choose for x any number in the interval $[0,\infty)$ and then find y.

When $x = 0$, $y = \sqrt{0} = 0$.

When $x = 1$, $y = \sqrt{1} = 1$.

When $x = 4$, $y = \sqrt{4} = 2$.

When $x = 9$, $y = \sqrt{9} = 3$.

x	y
0	0
1	1
4	2
9	3

(Table of values for $y = \sqrt{x}$)

Find some ordered pairs that are solutions of $y = \sqrt{x + 1}$. The domain of this function is restricted to those input values that result in the value of the radicand, x + 1, being greater than or equal to 0. To determine the domain, we solve the inequality $x + 1 \geqslant 0$.

$x + 1 \geqslant 0$

$x \geqslant -1$ (Adding -1)

The domain consists of all real numbers greater than or equal to -1. We choose any real number greater than or equal to -1 and then find y.

When $x = -1$, $y = \sqrt{-1 + 1} = \sqrt{0} = 0$.

When $x = 0$, $y = \sqrt{0 + 1} = \sqrt{1} = 1$.

When $x = 3$, $y = \sqrt{3 + 1} = \sqrt{4} = 2$.

When $x = 8$, $y = \sqrt{8 + 1} = \sqrt{9} = 3$.

x	y
-1	0
0	1
3	2
8	3

(Table of values for $y = \sqrt{x + 1}$)

Plot the ordered pairs for $y = \sqrt{x}$ and draw the graph with a dashed line. Plot the ordered pairs for $y = \sqrt{x + 1}$ and draw the graph with a solid line.

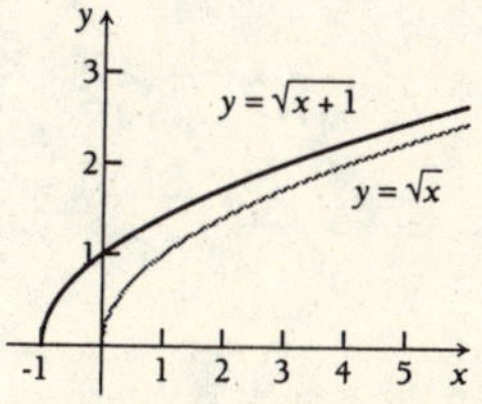

13. Graph $y = x^2 - 4x + 3$.

Before making a table of values we should recognize that $y = x^2 - 4x + 3$ is a quadratic function whose cup-shaped graph opens upward (the coefficient of x^2, 1, is positive). We first find the vertex, or turning point. The x-coordinate of the vertex is

$$x = -\frac{b}{2a} = -\frac{-4}{2\cdot 1} = 2.$$

Substituting 2 for x in the equation, we find the second coordinate of the vertex:

$$y = x^2 - 4x + 3 = 2^2 - 4\cdot 2 + 3 =$$
$$4 - 8 + 3 = -1.$$

The vertex is (2,-1).

We can find the first coordinates of points where the graph of the function intersects the x-axis, if they exist, by solving the quadratic equation

$x^2 - 4x + 3 = 0$ (Quadratic equation)

$(x - 3)(x - 1) = 0$ (Factoring)

$x - 3 = 0$ or $x - 1 = 0$ (Principle of zero products)

$x = 3$ or $x = 1$

The graph of $y = x^2 - 4x + 3$ has (3,0) and (1,0) as x-intercepts.

We choose some x-values on each side of the vertex and compute y-values.

When $x = 0$, $y = 0^2 - 4\cdot 0 + 3 = 3$.

When $x = 4$, $4^2 - 4\cdot 4 + 3 = 3$.

x	y	
3	0	(x-intercept)
1	0	(x-intercept)
2	-1	(vertex)
0	3	
4	3	

(Table of values for $y = x^2 - 4x + 3$)

Plot these ordered pairs and draw the graph.

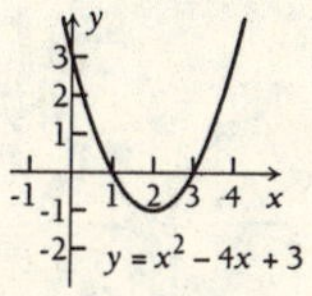

15. Graph $y = -x^2 + 2x - 1$.

We should recognize that $y = -x^2 + 2x - 1$ is a quadratic function whose graph opens downward (the coefficient of x^2, -1, is negative). We first find the vertex:

The x-coordinate is

$$x = -\frac{b}{2a} = -\frac{2}{2(-1)} = 1.$$

We substitute 1 for x in the equation to find the second coordinate of the vertex:

$$y = -(1)^2 + 2\cdot 1 - 1 = -1 + 2 - 1 = 0.$$

The vertex is (1,0). Note that this is also the x-intercept. (There cannot be another x-intercept since the vertex is the highest point on the graph.) We choose some x-values on each side of the vertex and compute y-values.

When $x = 0$, $y = -0^2 + 2\cdot 0 - 1 = -1$.

When $x = 2$, $y = -2^2 + 2\cdot 2 - 1 = -1$.

When $x = -1$, $y = -(-1)^2 + 2(-1) - 1 = -4$.

When $x = 3$, $y = -3^2 + 2\cdot 3 - 1 = -4$.

x	y	
1	0	(vertex, x-intercept)
0	-1	
2	-1	
-1	-4	
3	-4	

(Table of values for $y = -x^2 + 2x - 1$)

Plot these ordered pairs and draw the graph.

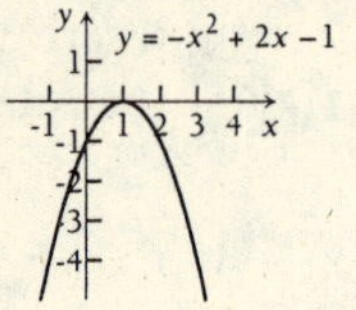

17. Graph $y = \frac{2}{x}$.

Note that 0 is not in the domain of this function since it would yield a 0 denominator.

We find some ordered pairs that are solutions of $y = \frac{2}{x}$ by choosing any real number except 0 for x and finding y.

Here we list only a few substitutions.

When $x = \frac{1}{2}$, $y = \frac{2}{\frac{1}{2}} = 2 \cdot \frac{2}{1} = 4$.

When $x = -2$, $y = \frac{2}{-2} = -1$.

When $x = 4$, $y = \frac{2}{4} = \frac{1}{2}$.

x	y
$\frac{1}{2}$	4
1	2
2	1
4	$\frac{1}{2}$
6	$\frac{1}{3}$

x	y
$-\frac{1}{2}$	-4
-1	-2
-2	-1
-4	$-\frac{1}{2}$
-6	$-\frac{1}{3}$

(Table of values for $y = \frac{2}{x}$)

Plot these ordered pairs and draw the graph.

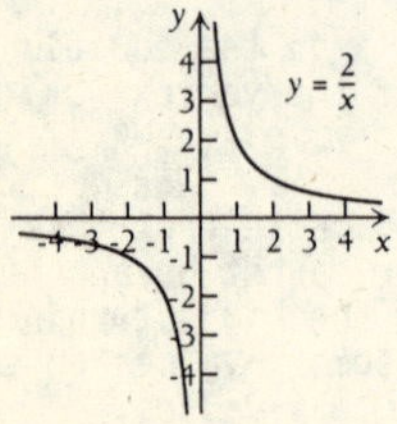

19. Graph $y = \frac{-2}{x}$.

Note that 0 is not in the domain of this function since it would yield a 0 denominator.

We find some ordered pairs that are solutions of $y = \frac{-2}{x}$ by choosing any real number except 0 for x and finding y.

Here we list only a few substitutions.

When $x = \frac{1}{2}$, $y = \frac{-2}{\frac{1}{2}} = -2 \cdot \frac{2}{1} = -4$.

When $x = 4$, $y = \frac{-2}{4} = -\frac{1}{2}$.

When $x = -\frac{1}{3}$, $y = \frac{-2}{-\frac{1}{3}} = -2 \cdot \frac{-3}{1} = 6$.

When $x = -2$, $y = \frac{-2}{-2} = 1$.

x	y
$\frac{1}{2}$	-4
1	-2
2	-1
4	$-\frac{1}{2}$
6	$-\frac{1}{3}$

x	y
$-\frac{1}{2}$	4
-1	2
-2	1
-4	$\frac{1}{2}$
-6	$\frac{1}{3}$

(Table of values for $y = \frac{-2}{x}$)

Plot these ordered pairs and draw the graph.

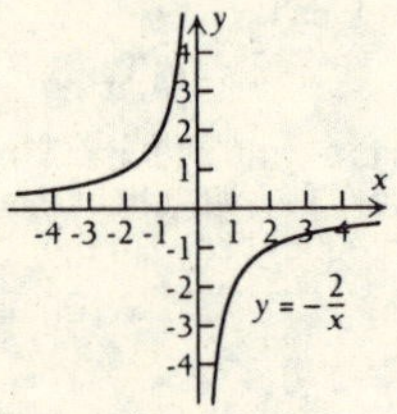

21. Graph $y = \frac{1}{x^2}$.

Note that 0 is not in the domain of this function since it would yield a 0 denominator.

We find some ordered pairs that are solutions of $y = \frac{1}{x^2}$ by choosing any real number except 0 for x and finding y.

When $x = -2$, $y = \frac{1}{(-2)^2} = \frac{1}{4}$.

When $x = -1$, $y = \frac{1}{(-1)^2} = \frac{1}{1} = 1$.

When $x = -\frac{1}{2}$, $y = \frac{1}{(-\frac{1}{2})^2} = \frac{1}{\frac{1}{4}} = 1 \cdot \frac{4}{1} = 4$.

When $x = \frac{1}{2}$, $y = \frac{1}{(\frac{1}{2})^2} = \frac{1}{\frac{1}{4}} = 4$.

When $x = 1$, $y = \frac{1}{1^2} = \frac{1}{1} = 1$.

When $x = 2$, $y = \frac{1}{2^2} = \frac{1}{4}$.

x	y
-2	$\frac{1}{4}$
-1	1
$-\frac{1}{2}$	4
$\frac{1}{2}$	4
1	1
2	$\frac{1}{4}$

(Table of values for $y = \frac{1}{x^2}$)

Plot these ordered pairs and draw the graph.

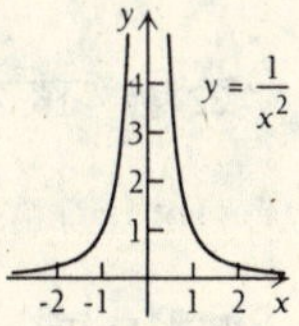

23. Graph $y = \sqrt[3]{x}$.

Find some ordered pairs that are solutions of $y = \sqrt[3]{x}$ keeping the results in a table. Since the domain of $y = \sqrt[3]{x}$ consists of all real numbers, we choose any number for x and then find y.

When $x = -6$, $y = \sqrt[3]{-6} \approx -1.817$.
When $x = -4$, $y = \sqrt[3]{-4} \approx -1.587$.
When $x = -1$, $y = \sqrt[3]{-1} \approx -1$.
When $x = -0.5$, $y = \sqrt[3]{-0.5} \approx -0.794$.
When $x = 0$, $y = \sqrt[3]{0} = 0$.
When $x = 0.5$, $y = \sqrt[3]{0.5} \approx 0.794$.
When $x = 1$, $y = \sqrt[3]{1} = 1$.
When $x = 4$, $y = \sqrt[3]{4} \approx 1.587$.
When $x = 6$, $y = \sqrt[3]{6} \approx 1.817$.

x	y
-6	-1.8
-4	-1.6
-1	-1
-0.5	-0.8
0	0
0.5	0.8
1	1
4	1.6
6	1.8

(Table of values for $y = \sqrt[3]{x}$)

Plot these ordered pairs and draw the graph.

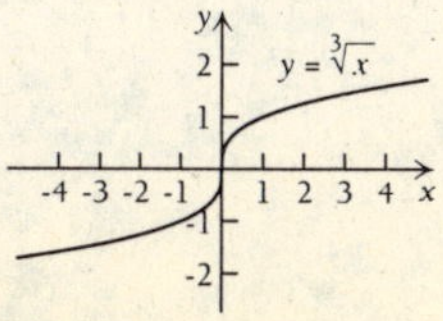

25. $f(x) = \frac{x^2 - 9}{x + 3}$

First we simplify the function.

$$f(x) = \frac{x^2 - 9}{x + 3} = \frac{(x + 3)(x - 3)}{x + 3} = \frac{x + 3}{x + 3} \cdot \frac{x - 3}{1} =$$

$$x - 3,\ x \neq -3$$

The number -3 is not in the domain of the function, because it would result in division by 0. We can express the function as

$$y = f(x) = x - 3,\ x \neq -3.$$

We substitute any value for x other than -3 and compute the corresponding y-value.

x	y
-5	-8
-4	-7
-2	-5
0	-3
1	-2
3	0
5	2

We plot these points and draw the graph. The open circle at the point (-3,-6) indicates that it is not part of the graph.

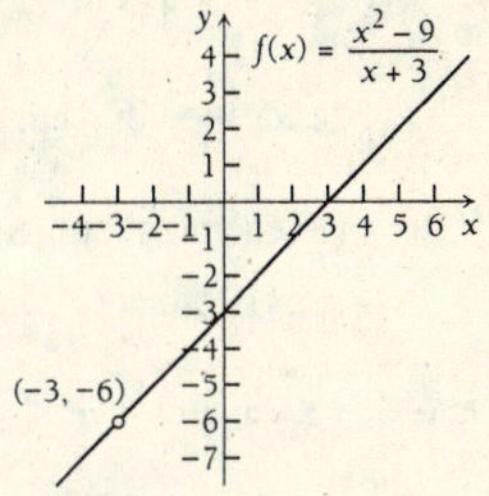

27. $f(x) = \frac{x^2 - 1}{x - 1}$

First we simplify the function.

$$f(x) = \frac{x^2 - 1}{x - 1} = \frac{(x + 1)(x - 1)}{x - 1} = \frac{x - 1}{x - 1} \cdot \frac{x + 1}{1} =$$

$$x + 1,\ x \neq 1$$

The number 1 is not in the domain of the function, because it would result in division by 0. We can express the function as

$$y = f(x) = x + 1,\ x \neq 1.$$

We substitute any value for x other than 1 and compute the corresponding y-value.

x	y
-4	-3
-2	-1
-1	0
0	1
2	3
3	4

We plot these points and draw the graph. The open circle at the point (1,2) indicates that it is not part of the graph.

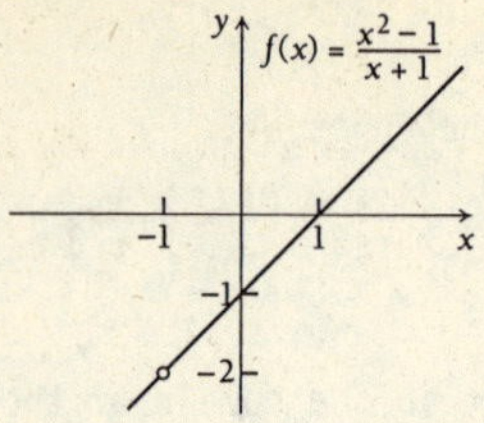

29. $x^2 - 2x = 2$

$x^2 - 2x - 2 = 0$ (Adding -2) (Standard form)

$a = 1$, $b = -2$, and $c = -2$

Then use the quadratic formula.

$x = \frac{-b \pm \sqrt{b^2 - 4ac}}{2a}$ (Quadratic formula)

$x = \frac{-(-2) \pm \sqrt{(-2)^2 - 4(1)(-2)}}{2\cdot 1}$

(Substituting 1 for a, -2 for b, and -2 for c)

$= \frac{2 \pm \sqrt{4 + 8}}{2}$

$= \frac{2 \pm \sqrt{12}}{2}$

$= \frac{2 \pm 2\sqrt{3}}{2}$ ($\sqrt{12} = \sqrt{4\cdot 3} = 2\sqrt{3}$)

$= \frac{2(1 \pm \sqrt{3})}{2}$ (Factoring in the numerator)

$= \frac{2}{2} \cdot (1 \pm \sqrt{3})$ (Factoring out a form of 1)

$= 1 \pm \sqrt{3}$ (Simplifying)

The solutions are $1 + \sqrt{3}$ and $1 - \sqrt{3}$.

31. $x^2 + 6x = 1$

$x^2 + 6x - 1 = 0$ (Adding -1) (Standard notation)

$a = 1$, $b = 6$, and $c = -1$

Then use the quadratic formula.

$x = \frac{-b \pm \sqrt{b^2 - 4ac}}{2a}$ (Quadratic formula)

$x = \frac{-6 \pm \sqrt{6^2 - 4(1)(-1)}}{2\cdot 1}$ (Substituting)

$= \frac{-6 \pm \sqrt{36 + 4}}{2}$

$= \frac{-6 \pm \sqrt{40}}{2}$

$= \frac{-6 \pm 2\sqrt{10}}{2}$ ($\sqrt{40} = \sqrt{4\cdot 10} = 2\sqrt{10}$)

$= \frac{2(-3 \pm \sqrt{10})}{2}$ (Factoring in the numerator)

$= \frac{2}{2} \cdot (-3 \pm \sqrt{10})$ (Factoring out a form of 1)

$= -3 \pm \sqrt{10}$ (Simplifying)

The solutions are $-3 + \sqrt{10}$ and $-3 - \sqrt{10}$.

33. $4x^2 = 4x + 1$

$4x^2 - 4x - 1 = 0$ (Adding -4x - 1) (Standard notation)

$a = 4$, $b = -4$, and $c = -1$

Then use the quadratic formula.

$x = \frac{-b \pm \sqrt{b^2 - 4ac}}{2a}$ (Quadratic formula)

$x = \frac{-(-4) \pm \sqrt{(-4)^2 - 4(4)(-1)}}{2\cdot 4}$ (Substituting)

$= \frac{4 \pm \sqrt{16 + 16}}{8}$

$= \frac{4 \pm \sqrt{32}}{8}$

$= \frac{4 \pm 4\sqrt{2}}{8}$ ($\sqrt{32} = \sqrt{16\cdot 2} = 4\sqrt{2}$)

$= \frac{4(1 \pm \sqrt{2})}{4\cdot 2}$ (Factoring)

$= \frac{1 \pm \sqrt{2}}{2}$ (Simplifying)

The solutions are $\frac{1 + \sqrt{2}}{2}$ and $\frac{1 - \sqrt{2}}{2}$.

35. $3y^2 + 8y + 2 = 0$ (Standard form)

$a = 3$, $b = 8$, and $c = 2$

Use the quadratic formula.

$y = \frac{-b \pm \sqrt{b^2 - 4ac}}{2a}$ (Quadratic formula)

$y = \frac{-8 \pm \sqrt{8^2 - 4\cdot 3\cdot 2}}{2\cdot 3}$ (Substituting)

$= \frac{-8 \pm \sqrt{64 - 24}}{6}$

$= \frac{-8 \pm \sqrt{40}}{6}$

$= \frac{-8 \pm 2\sqrt{10}}{6}$ ($\sqrt{40} = \sqrt{4\cdot 10} = 2\sqrt{10}$)

$= \frac{2(-4 \pm \sqrt{10})}{2\cdot 3}$ (Factoring)

$= \frac{-4 \pm \sqrt{10}}{3}$ (Simplifying)

The solutions are $\frac{-4 + \sqrt{10}}{3}$ and $\frac{-4 - \sqrt{10}}{3}$.

37. $\sqrt{x^3} = x^{3/2}$ (The index is 2; $\sqrt[n]{a^m} = a^{m/n}$.)

39. $\sqrt[5]{a^3} = a^{3/5}$ ($\sqrt[n]{a^m} = a^{m/n}$)

41. $\sqrt[7]{t} = \sqrt[7]{t^1}$ ($t = t^1$)

$= t^{1/7}$ ($\sqrt[n]{a^m} = a^{m/n}$)

43. $\frac{1}{\sqrt[3]{t^4}} = \frac{1}{t^{4/3}}$ ($\sqrt[n]{a^m} = a^{m/n}$)

$= t^{-4/3}$ ($\frac{1}{a^n} = a^{-n}$)

45. $\frac{1}{\sqrt{t}} = \frac{1}{\sqrt[2]{t^1}}$

$= \frac{1}{t^{1/2}}$ $(\sqrt[n]{a^m} = a^{m/n})$

$= t^{-1/2}$ $(\frac{1}{a^n} = a^{-n})$

47. $\frac{1}{\sqrt{x^2 + 7}} = \frac{1}{\sqrt[2]{(x^2 + 7)^1}}$

$= \frac{1}{(x^2 + 7)^{1/2}}$ $(\sqrt[n]{a^m} = a^{m/n})$

$= (x^2 + 7)^{-1/2}$ $(\frac{1}{a^n} = a^{-n})$

49. $x^{1/5} = \sqrt[5]{x^1}$ $(a^{m/n} = \sqrt[n]{a^m})$

$= \sqrt[5]{x}$ $(x^1 = x)$

51. $y^{2/3} = \sqrt[3]{y^2}$ $(a^{m/n} = \sqrt[n]{a^m})$

53. $t^{-2/5} = \frac{1}{t^{2/5}}$ $(a^{-n} = \frac{1}{a^n})$

$= \frac{1}{\sqrt[5]{t^2}}$ $(a^{m/n} = \sqrt[n]{a^m})$

55. $b^{-1/3} = \frac{1}{b^{1/3}}$ $(a^{-n} = \frac{1}{a^n})$

$= \frac{1}{\sqrt[3]{b^1}}$ $(a^{m/n} = \sqrt[n]{a^m})$

$= \frac{1}{\sqrt[3]{b}}$ $(b^1 = b)$

57. $e^{-17/6} = \frac{1}{e^{17/6}}$ $(a^{-n} = \frac{1}{a^n})$

$= \frac{1}{\sqrt[6]{e^{17}}}$ $(a^{m/n} = \sqrt[n]{a^m})$

59. $(x^2 - 3)^{-1/2} = \frac{1}{(x^2 - 3)^{1/2}}$ $(a^{-n} = \frac{1}{a^n})$

$= \frac{1}{\sqrt[2]{(x^2 - 3)^1}}$ $(a^{m/n} = \sqrt[n]{a^m})$

$= \frac{1}{\sqrt{x^2 - 3}}$

61. $9^{3/2}$

$= (9^{1/2})^3$ $(\frac{3}{2} = \frac{1}{2} \cdot 3)$

$[a^{m \cdot n} = (a^m)^n]$

$= (\sqrt{9})^3$ $(a^{1/n} = \sqrt[n]{a})$

$= 3^3$ $(\sqrt{9} = 3)$

$= 27$

63. $64^{2/3}$

$= (64^{1/3})^2$ $(\frac{2}{3} = \frac{1}{3} \cdot 2)$

$[a^{m \cdot n} = (a^m)^n]$

$= (\sqrt[3]{64})^2$ $(a^{1/n} = \sqrt[n]{a})$

$= 4^2$ $(\sqrt[3]{64} = 4)$

$= 16$

65. $16^{3/4}$

$= (16^{1/4})^3$ $(\frac{3}{4} = \frac{1}{4} \cdot 3)$

$[a^{m \cdot n} = (a^m)^n]$

$= (\sqrt[4]{16})^3$ $(a^{1/n} = \sqrt[n]{a})$

$= 2^3$ $(\sqrt[4]{16} = 2)$

$= 8$

67. The domain of a rational function is restricted to those input values that do not result in division by 0.

To determine the domain of

$f(x) = \frac{x^2 - 25}{x - 5}$

we set the denominator equal to 0 and solve.

$x - 5 = 0$

$x = 5$ (Adding 5)

Thus 5 is not in the domain. The domain consists of all real numbers except 5.

69. The domain of a rational function is restricted to those input values that do not result in division by 0.

To determine the domain of

$f(x) = \frac{x^3}{x^2 - 5x + 6}$

we set the denominator equal to 0 and solve.

$x^2 - 5x + 6 = 0$

$(x - 3)(x - 2) = 0$ (Factoring)

$x - 3 = 0$ or $x - 2 = 0$ (Principle of zero products)

$x = 3$ or $x = 2$

Thus 3 and 2 are not in the domain. The domain consists of all real numbers except 3 and 2.

71. The domain of $f(x) = \sqrt{5x + 4}$ is restricted to those input values that result in the value of the radicand, $5x + 4$, being greater than or equal to 0. To determine the domain we solve the inequality $5x + 4 \geq 0$.

$5x + 4 \geq 0$

$5x \geq -4$ (Adding -4)

$x \geq -\frac{4}{5}$ (Multiplying by $\frac{1}{5}$)

Thus the domain consists of all real numbers greater than or equal to $-\frac{4}{5}$, or $\left[-\frac{4}{5}, \infty\right]$.

73. We set $D(p) = S(p)$ and solve:

$$\frac{8 - p}{2} = p - 2$$

$$8 - p = 2(p - 2) \quad \text{(Multiplying both sides by 2)}$$

$$8 - p = 2p - 4$$

$$12 = 3p$$

$$4 = p$$

Thus, $p_E = \$4$. To find x_E we substitute p_E into either $D(p)$ or $S(p)$. We use $S(p)$:

$$x_E = S(p_E) = S(4) = 4 - 2 = 2$$

The equilibrium quantity is 2 units and the equilibrium point is (\$4,2).

75. We set $D(p) = S(p)$ and solve:

$$1000 - 10p = 250 + 5p$$

$$750 = 15p$$

$$50 = p$$

Thus, $p_E = \$50$. To find x_E we substitute p_E into either $D(p)$ or $S(p)$. We use $D(p)$:

$$x_E = D(p_E) = D(50) = 1000 - 10\cdot 50 = 1000 - 500 = 500$$

The equilibrium quantity is 500 units, and the equilibrium point is (\$50,500).

77. We set $D(p) = S(p)$ and solve:

$$\frac{5}{p} = \frac{p}{5}$$

$$25 = p^2$$

$$0 = p^2 - 25$$

$$0 = (p + 5)(p - 5)$$

$$p + 5 = 0 \quad \text{or} \quad p - 5 = 0$$

$$p = -5 \quad \text{or} \quad p = 5$$

Only 5 makes sense in this context. Thus, $p_E = \$5$. To find x_E we substitute p_E into either $D(p)$ or $S(p)$. We use $S(p)$:

$$x_E = S(p_E) = S(5) = \frac{5}{5} = 1$$

The equilibrium quantity is 1 unit, and the equilibrium point is (\$5,1).

79. We set $D(p) = S(p)$ and solve:

$$(p - 3)^2 = p^2 + 2p + 1, \; 0 \leqslant p \leqslant 3$$

$$p^2 - 6p + 9 = p^2 + 2p + 1$$

$$-6p + 9 = 2p + 1$$

$$8 = 8p$$

$$1 = p$$

Thus $p_E = \$1$. To find x_E we substitute p_E into either $D(p)$ or $S(p)$. We use $S(p)$.

$$x_E = S(p_E) = S(1) = 1^2 + 2\cdot 1 + 1 = 1 + 2 + 1 = 4$$

Thus the equilibrium quantity is 4 units, and the equilibrium point is (\$1,4).

81. We set $D(p) = S(p)$ and solve:

$$5 - p = \sqrt{p + 7}, \quad 0 \leqslant p \leqslant 5$$

$$(5 - p)^2 = (\sqrt{p + 7})^2 \quad \text{(Squaring both sides)}$$

$$25 - 10p + p^2 = p + 7$$

$$18 - 11p + p^2 = 0$$

$$(9 - p)(2 - p) = 0$$

$$9 - p = 0 \quad \text{or} \quad 2 - p = 0$$

$$9 = p \quad \text{or} \quad 2 = p$$

Since 9 is not in the domain of $D(p)$, $0 \leqslant p \leqslant 5$, the equilibrium price $p_E = \$2$. To find x_E we substitute p_E into either $D(p)$ or $S(p)$. We use $D(p)$:

$$x_E = D(p_E) = D(2) = 5 - 2 = 3$$

The equilibrium quantity is 3 units, and the equilibrium price is (\$2,3).

83. Dividends (D) are inversely proportional to the prime rate (R).

$$D = \frac{k}{R} \quad \text{(Inverse variation)}$$

$$2.09 = \frac{k}{19\%} \quad \text{(Substituting)}$$

$$2.09 = \frac{k}{0.19}$$

$$0.3971 = k \quad \text{(Multiplying by 0.19)}$$

The variation constant is 0.3971.

The equation of variation is $D = \frac{0.3971}{R}$.

$$D = \frac{0.3971}{R}$$

$$D = \frac{0.3971}{17.5\%} \quad \text{(Substituting)}$$

$$= \frac{0.3971}{0.175}$$

$$\approx \$2.27 \text{ per share}$$

85. $f(x) = \frac{1}{6}x^3 + \frac{1}{2}x^2 + \frac{1}{3}x$

$$f(7) = \frac{1}{6}\cdot 7^3 + \frac{1}{2}\cdot 7^2 + \frac{1}{3}\cdot 7$$

$$= \frac{343}{6} + \frac{49}{2} + \frac{7}{3}$$

$$= \frac{343}{6} + \frac{147}{6} + \frac{14}{6}$$

$$= \frac{504}{6}$$

$$= 84$$

There are 84 oranges when there are 7 layers.

$$f(10) = \frac{1}{6} \cdot 10^3 + \frac{1}{2} \cdot 10^2 + \frac{1}{3} \cdot 10$$
$$= \frac{1000}{6} + \frac{100}{2} + \frac{10}{3}$$
$$= \frac{1000}{6} + \frac{300}{6} + \frac{20}{6}$$
$$= \frac{1320}{6}$$
$$= 220$$

There are 220 oranges when there are 10 layers.

$$f(12) = \frac{1}{6} \cdot 12^3 + \frac{1}{2} \cdot 12^2 + \frac{1}{3} \cdot 12$$
$$= \frac{1728}{6} + \frac{144}{2} + \frac{12}{3}$$
$$= \frac{1728}{6} + \frac{432}{6} + \frac{24}{6}$$
$$= \frac{2184}{6}$$
$$= 364$$

There are 364 oranges when there are 12 layers.

87. See the answer section in the text.

89. See the answer section in the text.

Exercise Set 1.6

1. a)

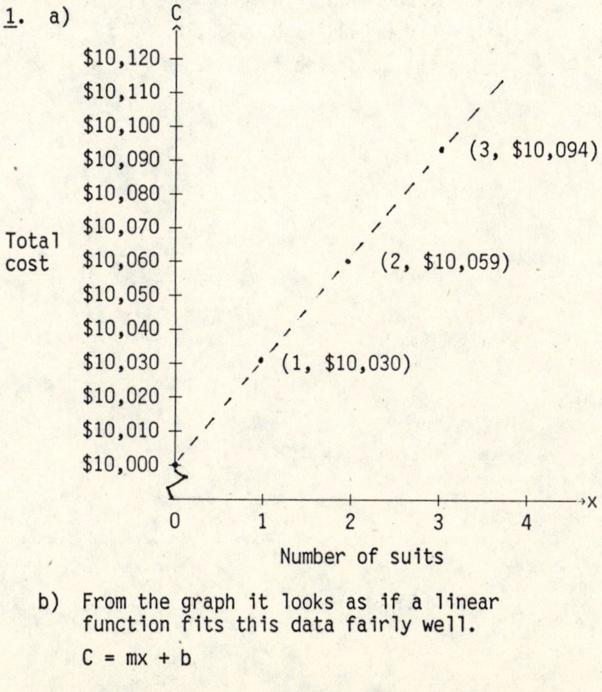

b) From the graph it looks as if a linear function fits this data fairly well.

$C = mx + b$

c) To derive the constants m and b, we use the data points (1, \$10,030) and (3, \$10,094).

It follows that

$10,030 = m \cdot 1 + b$, or $10,030 = m + b$, (1)

$10,094 = m \cdot 3 + b$, or $10,094 = 3m + b$ (2)

This is a system of equations. To eliminate b we multiply Equation (1) by -1 and then add.

$$-10,030 = -m - b$$
$$10,094 = 3m + b$$
$$64 = 2m \quad \text{(Adding)}$$
$$32 = m$$

Substituting 32 for m in Equation (1), we get

$10,030 = 32 + b$ (Substituting)

$9998 = b$ (Adding -32)

The resulting linear function is

$C = 32x + 9998$.

d) Using $C = 32x + 9998$, the total cost of producing 4 suits is given by

$C = 32 \cdot 4 + 9998$ (Substituting 4)

$= 128 + 9998$

$= \$10,126$

The total cost of producing 10 suits is given by

$C = 32 \cdot 10 + 9998$ (Substituting 10)

$= 320 + 9998$

$= \$10,318$

e) Using the data points (0, \$10,000) and (2, \$10,059) we have the system of equations

$10,000 = m \cdot 0 + b$, or $10,000 = b$, (1)

$10,059 = m \cdot 2 + b$, or $10,059 = 2m + b$ (2)

We substitute 10,000 for b in Equation (2) and solve for m.

$10,059 = 2m + b$

$10,059 = 2m + 10,000$ (Substituting)

$59 = 2m$

$29.5 = m$

The resulting linear function is

$C = 29.5x + 10,000$.

f) Using $C = 29.5x + 10,000$, the total cost of producing 4 suits is given by

$C = 29.5 \cdot 4 + 10,000$ (Substituting 4)

$= 118 + 10,000$

$= \$10,118$

The total cost of producing 10 suits is given by

$C = 29.5 \cdot 10 + 10,000$ (Substituting 10)

$= 295 + 10,000$

$= \$10,295$

3. a)

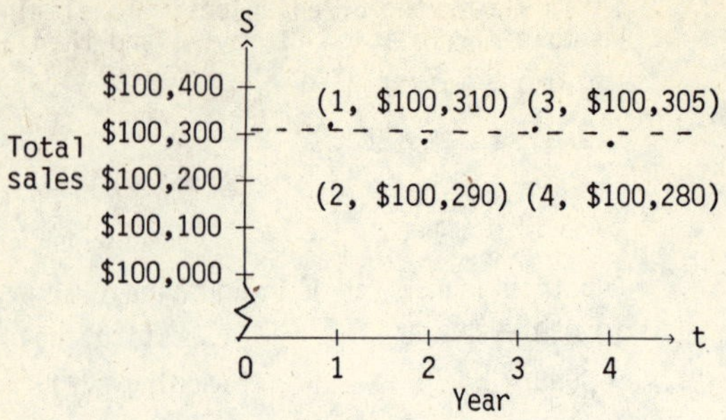

b) Both constant and linear models seem to fit.

c) We first find a linear function using the data points (1, \$100,310) and (2, \$100,290).

$S = mt + b$ (Linear function)

$100{,}310 = m\cdot 1 + b$, or $100{,}310 = m + b$, (1)

$100{,}290 = m\cdot 2 + b$, or $100{,}290 = 2m + b$ (2)

This is a system of equations. We multiply Equation (1) by -1 and then add.

$-100{,}310 = -m - b$

$100{,}290 = 2m + b$

$-20 = m$ (Adding)

Next we substitute -20 for m in either Equation (1) or Equation (2) and solve for b.

$100{,}310 = 1m + b$ (1)

$100{,}310 = -20 + b$ (Substituting)

$100{,}330 = b$ (Adding)

The resulting linear function is

$S = -20t + 100{,}330$.

d) Let S = b, where b is the average of sales totals.

$$b = \frac{100{,}310 + 100{,}290 + 100{,}305 + 100{,}280}{4}$$

$$= \frac{401{,}185}{4}$$

$$= 100{,}296.25$$

The resulting constant function is S = \$100,296.25.

5. a)

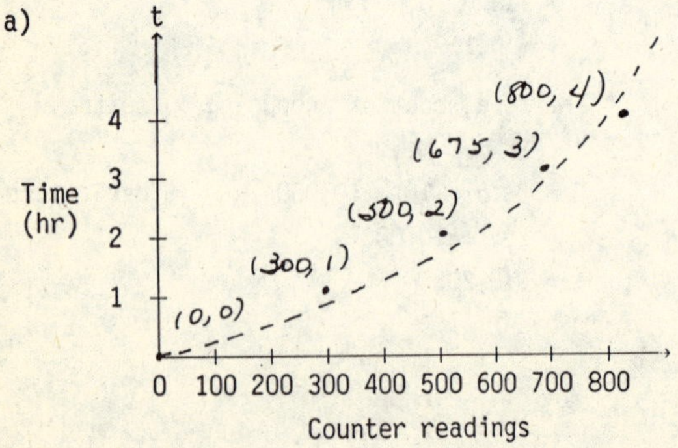

b) A quadratic function appears to fit the data.

$t = aN^2 + bN + c$

c) We substitute the data points (0,0), (300,1) and (500,2) into the equation.

$0 = a\cdot 0^2 + b\cdot 0 + c$,

$1 = a\cdot 300^2 + b\cdot 300 + c$,

$2 = a\cdot 500^2 + b\cdot 500 + c$, or

$0 = c$,

$1 = 90{,}000a + 300b + c$,

$2 = 250{,}000a + 500b + c$

Solving this system of equations, we get

$a \approx 0.0000033$, $b \approx 0.0023333$, $c = 0$.

The function is $t = 0.0000033N^2 + 0.0023333N$.

d) To find the counter reading after the tape has run 1/2 hour, we substitute 1/2 or 0.5, for t and solve for N:

$0.5 = 0.0000033N^2 + 0.0023333N$

$0 = 0.0000033N^2 + 0.0023333N - 0.5$

Using the quadratic formula, we find $N \approx 172$.

To find the counter reading after a tape has run for 1 1/2 hours, we substitute 1 1/2, or 1.5, for t and solve for N:

$1.5 = 0.0000033N^2 + 0.0023333N$

$0 = 0.0000033N^2 + 0.0023333N - 1.5$

Using the quadratic formula, we find $N \approx 408$.

To find the counter reading after a tape has run for 2 1/2 hours, we substitute 2 1/2, or 2.5, for t and solve for N:

$2.5 = 0.0000033N^2 + 0.0023333N$

$0 = 0.0000033N^2 + 0.0023333N - 2.5$

Using the quadratic formula, we find $N \approx 586$.

Exercise Set 2.1

1. Since $\lim_{x \to 1^-} f(x) = 2$ and $\lim_{x \to 1^+} f(x) = -1$, the $\lim_{x \to 1} f(x)$ does not exist. The function is discontinuous at $x = 1$ and is therefore not continuous over the whole real line $(-\infty, \infty)$. Note that the graph cannot be traced without lifting the pencil from the paper.

3. The function is continuous over the whole real line since it is continuous at each point in $(-\infty, \infty)$. Note that the graph can be traced without lifting the pencil from the paper.

5. a) As inputs x approach 1 from the right, outputs f(x) approach -1. Thus, the limit from the right is -1.

   $\lim_{x \to 1^+} f(x) = -1$

   As inputs x approach 1 from the left, outputs f(x) approach 2. Thus, the limit from the left is 2.

   $\lim_{x \to 1^-} f(x) = 2$

   Since the limit from the left, 2, is not the same as the limit from the right, -1, we say $\lim_{x \to 1} f(x)$ does not exist.

   b) When the input is 1, the output, f(1), is -1.

   $f(1) = -1$

   c) Since the limit at $x = 1$ does not exist, the function is not continuous at $x = 1$.

   d) As inputs x approach -2 from the left, outputs f(x) approach 3. Thus, the limit from the left is 3.

   $\lim_{x \to -2^-} f(x) = 3$

   As inputs x approach -2 from the right, outputs f(x) approach 3. Thus, the limit from the right is 3.

   $\lim_{x \to -2^+} f(x) = 3$

   Since the limit from the left, 3, is the same as the limit from the right, 3, we have

   $\lim_{x \to -2} f(x) = 3.$

   e) When the input is -2, the output, f(-2), is 3.

   $f(-2) = 3$

   f) The function f(x) is continuous at $x = -2$, because

   1) $f(-2)$ exists, $f(-2) = 3$,

   2) $\lim_{x \to -2} f(x)$ exists, $\lim_{x \to -2} f(x) = 3$,

   and

   3) $\lim_{x \to -2} f(x) = f(-2) = 3.$

7. a) As inputs x approach 1 from the left, outputs h(x) approach 2. Thus, the limit from the left is 2.

   $\lim_{x \to 1^-} h(x) = 2$

   As inputs x approach 1 from the right, outputs h(x) approach 2. Thus, the limit from the right is 2.

   $\lim_{x \to 1^+} h(x) = 2$

   Since the limit from the left, 2, is the same as the limit from the right, 2, we have

   $\lim_{x \to 1} h(x) = 2.$

   b) When the input is 1, the output, h(1), is 2.

   $h(1) = 2$

   c) The function h(x) is continuous at $x = 1$, because

   1) $h(1)$ exists, $h(1) = 2$,

   2) $\lim_{x \to 1} h(x)$ exists, $\lim_{x \to 1} h(x) = 2$,

   and

   3) $\lim_{x \to 1} h(x) = h(1) = 2.$

   d) As inputs x approach -2 from the left, outputs h(x) approach 0. Thus, the limit from the left is 0.

   $\lim_{x \to -2^-} h(x) = 0.$

   As inputs x approach -2 from the right, outputs h(x) approach 0. Thus, the limit from the right is 0.

   $\lim_{x \to -2^+} h(x) = 0.$

   Since the limit from the left, 0, is the same as the limit from the right, 0, we have

   $\lim_{x \to -2} h(x) = 0.$

   e) When the input is -2, the output, h(-2), is 0.

   $h(-2) = 0$

   f) The function h(x) is continuous at $x = -2$, because

   1) $h(-2)$ exists, $h(-2) = 0$,

   2) $\lim_{x \to -2} h(x)$ exists, $\lim_{x \to -2} h(x) = 0$,

   and

   3) $\lim_{x \to -2} h(x) = h(-2) = 0.$

9. The function p is not continuous at 1 since $\lim_{x \to 1} p(x)$ does not exist.

   The function p is continuous at $1\frac{1}{2}$, because

   1) $p(1\frac{1}{2})$ exists, $p(1\frac{1}{2}) = 52¢$,

   2) $\lim_{x \to 1\frac{1}{2}} p(x)$ exists, $\lim_{x \to 1\frac{1}{2}} p(x) = 52¢$,

   and

3) $\lim_{x \to 1\frac{1}{2}} p(x) = p(1\frac{1}{2}) = 52¢$.

The function p is not continuous at x = 2 since $\lim_{x \to 2} p(x)$ does not exist.

The function p is continuous at 2.53, because

1) p(2.53) exists, p(2.53) = 75¢,

2) $\lim_{x \to 2.53} p(x)$ exists, $\lim_{x \to 2.53} p(x) = 75¢$,

and

3) $\lim_{x \to 2.53} p(x) = p(2.53) = 75¢$.

11. As inputs x approach 1 from the left, outputs, p(x), approach 29¢. Thus the limit from the left is 29¢.

$\lim_{x \to 1^-} p(x) = 29¢$

As inputs x approach 1 from the right, outputs, p(x), approach 52¢. Thus the limit from the right is 52¢.

$\lim_{x \to 1^+} p(x) = 52¢$

Since the limit from the left, 29¢, is not the same as the limit from the right, 52¢, we have

$\lim_{x \to 1} p(x)$ does not exist.

13. As inputs x approach 2.3 from the left, outputs, p(x), approach 75¢. Thus the limit from the left is 75¢.

$\lim_{x \to 2.3^-} p(x) = 75¢$

As inputs x approach 2.3 from the right, outputs, p(x), approach 75¢. Thus the limit from the right is 75¢.

$\lim_{x \to 2.3^+} p(x) = 75¢$

Since the limit from the left, 75¢, is the same as the limit from the right, 75¢, we have

$\lim_{x \to 2.3} p(x) = 75¢$.

15. As inputs x approach 12 from the left, outputs C(x) approach \$20. Thus, $\lim_{x \to 12^-} C(x) = \$20$.

As inputs x approach 12 from the right, outputs C(x) approach \$28. Thus, $\lim_{x \to 12^+} C(x) = \$28$.

Since the limit from the left is not the same as the limit from the right, $\lim_{x \to 12} C(x)$ does not exist.

17. C(x) is not continuous at x = 12, because $\lim_{x \to 12} C(x)$ does not exist.

C(x) is continuous at x = 20, because

1) C(20) exists,

2) $\lim_{x \to 20} C(x)$ exists, and

3) $\lim_{x \to 20} C(x) = C(20) = \$32$.

19. As inputs t approach 20 from the right, outputs N(t) approach 100. Thus, $\lim_{t \to 20^+} N(t) = 100$.

As inputs t approach 20 from the left, outputs N(t) approach 30. Thus, $\lim_{t \to 20^-} N(t) = 30$.

Since the limit from the right, 100, is not the same as the limit from the left, 30, $\lim_{t \to 20} N(t)$ does not exist.

21. Since $\lim_{t \to 20} N(t)$ does not exist, N is not continuous at t = 20.

N is continuous at t = 30, because

1) N(30) exists,

2) $\lim_{t \to 30} N(t)$ exists, and

3) $\lim_{t \to 30} N(t) = N(30) = 100$

23. a) The "hole" may or may not appear depending on the software package or the graphing calculator.

b)

x	$\frac{\sqrt{x-2} - 3}{x - 11}$
10	0.1716
10.5	0.1690
10.7	0.1681
10.9	0.1671
10.99	0.1667
10.999	0.1667

These inputs approach 11 from the left.

x	$\frac{\sqrt{x-2} - 3}{x - 11}$
12	0.1623
11.5	0.1644
11.3	0.1653
11.1	0.1662
11.01	0.1666
11.001	0.1667

These inputs approach 11 from the right.

From the tables we see that $\lim_{x \to 11} f(x) = 0.1667$.

Exercise Set 2.2

1. Find $\lim_{x \to 1} (x^2 - 3)$.

From the continuity principles it follows that $x^2 - 3$ is continuous. Thus the limit can be found by substitution.

$\lim_{x \to 1} (x^2 - 3) = 1^2 - 3$ (Substituting)

$= 1 - 3$

$= -2$

3. Find $\lim_{x \to 0} \frac{3}{x}$.

The function $\frac{3}{x}$ is not continuous at $x = 0$, and there is no further algebraic simplification for $\frac{3}{x}$. Thus, direct substitution is not possible. We can use input-output tables or a graph. Here we use a graph.

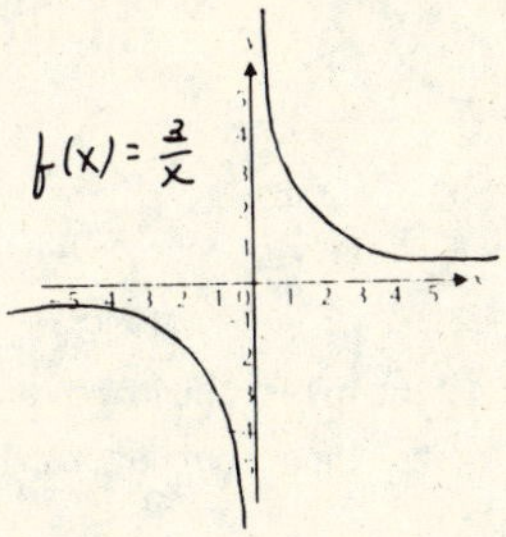

As x approaches 0 from the left, the outputs become more and more negative without bound. These numbers do not approach any real number, though it might be said that "the limit from the left is $-\infty$ (negative infinity)." As x approaches 0 from the right, the outputs get larger and larger without bound. These numbers do not approach any real number, though it might be said that "the limit from the right is ∞ (infinity)." For a limit to exist, the limits from the left and right must both exist and be the same.

Thus, $\lim_{x \to 0} \frac{3}{x}$ does not exist.

5. Find $\lim_{x \to 3} (2x + 5)$.

From the continuity principles it follows that $2x + 5$ is continuous. Thus the limit can be found by substitution.

$\lim_{x \to 3} (2x + 5) = 2 \cdot 3 + 5$ (Substituting)

$= 6 + 5$

$= 11$

7. Find $\lim_{x \to -5} \frac{x^2 - 25}{x + 5}$.

The function $\frac{x^2 - 25}{x + 5}$ is not continuous at $x = -5$. We use some algebraic simplification and then some limit principles.

$\lim_{x \to -5} \frac{x^2 - 25}{x + 5}$

$= \lim_{x \to -5} \frac{(x + 5)(x - 5)}{x + 5}$ (Factoring the numerator)

$= \lim_{x \to -5} (x - 5)$ (Simplifying, assuming $x \neq -5$)

$= \lim_{x \to -5} x - \lim_{x \to -5} 5$ (By L3)

$= -5 - 5$ (By L2 and L1)

$= -10$

9. Find $\lim_{x \to -2} \frac{5}{x}$.

The function $\frac{5}{x}$ is continuous at all real numbers except 0. Since $\frac{5}{x}$ is continuous at $x = -2$, we can substitute to find the limit.

$\lim_{x \to -2} \frac{5}{x} = \frac{5}{-2}$ (Substituting)

$= -\frac{5}{2}$

11. Find $\lim_{x \to 2} \frac{x^2 + x - 6}{x - 2}$.

The function $\frac{x^2 + x - 6}{x - 2}$ is not continuous at $x = 2$. We use some algebraic simplification and then some limit principles.

$\lim_{x \to 2} \frac{x^2 + x - 6}{x - 2}$

$= \lim_{x \to 2} \frac{(x - 2)(x + 3)}{x - 2}$ (Factoring the numerator)

$= \lim_{x \to 2} (x + 3)$ (Simplifying, assuming $x \neq 2$)

$= \lim_{x \to 2} x + \lim_{x \to 2} 3$ (By L3)

$= 2 + 3$ (By L2 and L1)

$= 5$

13. Find $\lim_{x \to 5} \sqrt[3]{x^2 - 17}$.

From the continuity principles it follows that $\sqrt[3]{x^2 - 17}$ is continuous. Thus the limit can be found by substitution.

$\lim_{x \to 5} \sqrt[3]{x^2 - 17} = \sqrt[3]{5^2 - 17}$ (Substituting)

$= \sqrt[3]{25 - 17}$

$= \sqrt[3]{8}$

$= 2$

15. Find $\lim_{x\to 1} (x^4 - x^3 + x^2 + x + 1)$.

From the continuity principles it follows that $x^4 - x^3 + x^2 + x + 1$ is continuous. Thus the limit can be found by substitution.

$\lim_{x\to 1} (x^4 - x^3 + x^2 + x + 1)$

$= 1^4 - 1^3 + 1^2 + 1 + 1$ (Substituting)

$= 1 - 1 + 1 + 1 + 1$

$= 3$

17. Find $\lim_{x\to 2} \frac{1}{x - 2}$.

The function $\frac{1}{x - 2}$ is not continuous at $x = 2$, and there is no further algebraic simplification for $\frac{1}{x - 2}$. Thus direct substitution is not possible. We can use input-output tables or a graph. Here we use a graph.

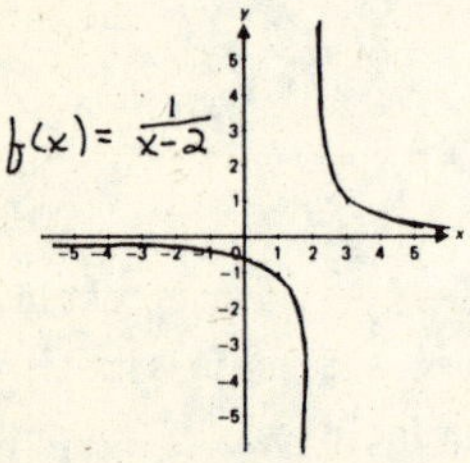

As x approaches 2 from the left, the outputs get smaller and smaller without bound. As x approaches 2 from the right, the outputs get larger and larger without bound. We say that

$\lim_{x\to 2} \frac{1}{x - 2}$ does not exist.

Remember: For a limit to exist, the limits from the left and right must both exist and be the same.

19. Find $\lim_{x\to 2} \frac{3x^2 - 4x + 2}{7x^2 - 5x + 3}$.

The rational function $\frac{3x^2 - 4x + 2}{7x^2 - 5x + 3}$ is continuous at all real numbers except those for which $7x^2 - 5x + 3 = 0$. Since $7x^2 - 5x + 3 \neq 0$ when $x = 2$, the function is continuous at $x = 2$. Thus we can substitute to find the limit.

$\lim_{x\to 2} \frac{3x^2 - 4x + 2}{7x^2 - 5x + 3}$

$= \frac{3\cdot 2^2 - 4\cdot 2 + 2}{7\cdot 2^2 - 5\cdot 2 + 3}$ (Substituting)

$= \frac{12 - 8 + 2}{28 - 10 + 3}$

$= \frac{6}{21}$

$= \frac{2}{7}$ (Simplifying)

21. Find $\lim_{x\to 2} \frac{x^2 + x - 6}{x^2 - 4}$.

The function $\frac{x^2 + x - 6}{x^2 - 4}$ is not continuous at $x = 2$ ($x^2 - 4 = 0$ when $x = 2$). We first use some algebraic simplification.

$\lim_{x\to 2} \frac{x^2 + x - 6}{x^2 - 4}$

$= \lim_{x\to 2} \frac{(x - 2)(x + 3)}{(x - 2)(x + 2)}$ (Factoring numerator and denominator)

$= \lim_{x\to 2} \frac{x + 3}{x + 2}$ (Simplifying, assuming $x \neq 2$)

Using input-output tables, we see that as inputs x approach 2 from the left, outputs $\frac{x + 3}{x + 2}$ approach $\frac{5}{4}$. As inputs x approach 2 from the right, outputs $\frac{x + 3}{x + 2}$ approach $\frac{5}{4}$. Since the limit from the left, $\frac{5}{4}$, is the same as the limit from the right, $\frac{5}{4}$, we have

$$\lim_{x\to 2} \frac{x + 3}{x + 2} = \frac{5}{4}.$$

Thus, $\lim_{x\to 2} \frac{x^2 + x - 6}{x^2 - 4}$ is $\frac{5}{4}$.

Note that limit principles could have been used as an alternative to input-output tables in finding the limit.

23. Find $\lim_{x\to 1} \frac{1 - \sqrt{x}}{1 - x}$.

The function $\frac{1 - \sqrt{x}}{1 - x}$ is not continuous at $x = 1$ and, assuming we see no further algebraic simplification for $\frac{1 - \sqrt{x}}{1 - x}$, direct substitution is not possible. We can use an input-output table or a graph.

Here we use an input-output table.

x	$\frac{1 - \sqrt{x}}{1 - x}$
0	1
0.5	0.586
0.7	0.544
0.9	0.513
0.95	0.506
0.99	0.501
0.999	0.500

These inputs approach 1 from the left.

x	$\frac{1 - \sqrt{x}}{1 - x}$
2	0.414
1.5	0.449
1.3	0.467
1.1	0.488
1.05	0.494
1.01	0.499
1.001	0.500

These inputs approach 1 from the right.

From the tables we see that as inputs x approach 1 from the left, outputs $\frac{1 - \sqrt{x}}{1 - x}$ approach 0.5. As inputs x approach 1 from the right, outputs $\frac{1 - \sqrt{x}}{1 - x}$ approach 0.5. Since the limit from the left is the same as the limit from the right, we have

$$\lim_{x \to 1} \frac{1 - \sqrt{x}}{1 - x} = 0.5, \text{ or } \frac{1}{2}.$$

Note that algebraic simplification of $\frac{1 - \sqrt{x}}{1 - x}$, although perhaps not obvious, is possible. A second method for finding this limit uses such simplification and some limit principles.

$$\lim_{x \to 1} \frac{1 - \sqrt{x}}{1 - x}$$

$$= \lim_{x \to 1} \frac{1 - \sqrt{x}}{1 - (\sqrt{x})^2} \qquad (x = (\sqrt{x})^2)$$

$$= \lim_{x \to 1} \frac{1 - \sqrt{x}}{(1 - \sqrt{x})(1 + \sqrt{x})} \qquad \text{(Factoring the denominator)}$$

$$= \lim_{x \to 1} \frac{1}{1 + \sqrt{x}} \qquad \text{(Simplifying, assuming } x \neq 1\text{)}$$

$$= \frac{1}{\lim_{x \to 1} (1 + \sqrt{x})} \qquad \text{(By L4)}$$

$$= \frac{1}{\lim_{x \to 1} 1 + \lim_{x \to 1} \sqrt{x}} \qquad \text{(By L3)}$$

$$= \frac{1}{1 + 1} \qquad \text{(By L1 and L2)}$$

$$= \frac{1}{2}$$

Thus, $\lim_{x \to 1} \frac{1 - \sqrt{x}}{1 - x} = \frac{1}{2}$, or 0.5.

25. Find $\lim_{h \to 0} (6x^2 + 6xh + 2h^2)$.

We treat x as a constant since we are interested only in the way in which the expression varies when $h \to 0$. Since $6x^2 + 6xh + 2h^2$ is continuous at $h = 0$, we can substitute to find the limit.

$$\lim_{h \to 0} (6x^2 + 6xh + 2h^2)$$

$$= 6x^2 + 6x \cdot 0 + 2 \cdot 0^2 \qquad \text{(Substituting)}$$

$$= 6x^2 + 0 + 0$$

$$= 6x^2$$

27. Find $\lim_{h \to 0} \frac{-2x - h}{x^2(x + h)^2}$.

We treat x as a constant since we are interested only in the way in which the expression varies when $h \to 0$. Since $\frac{-2x - h}{x^2(x + h)^2}$ is continuous at $h = 0$, we can substitute to find the limit.

$$\lim_{h \to 0} \frac{-2x - h}{x^2(x + h)^2} = \frac{-2x - 0}{x^2(x + 0)^2} \qquad \text{(Substituting 0 for h)}$$

$$= \frac{-2x}{x^2 \cdot x^2}$$

$$= \frac{-2x}{x^4}$$

$$= \frac{-2}{x^3} \qquad \text{(Simplifying)}$$

29. Find $\lim_{x \to \infty} \frac{2x - 4}{5x}$.

We will use some algebra and the fact that as $x \to \infty$, $\frac{b}{ax^n} \to 0$, for any positive integer n.

$$\lim_{x \to \infty} \frac{2x - 4}{5x}$$

$$= \lim_{x \to \infty} \frac{2x - 4}{5x} \cdot \frac{(1/x)}{(1/x)} \qquad \text{(Multiplying by a form of 1)}$$

$$= \lim_{x \to \infty} \frac{2x \cdot \frac{1}{x} - 4 \cdot \frac{1}{x}}{5x \cdot \frac{1}{x}}$$

$$= \lim_{x \to \infty} \frac{2 - \frac{4}{x}}{5}$$

$$= \frac{2 - 0}{5} \qquad \left(\text{As } x \to \infty, \frac{4}{x} \to 0\right)$$

$$= \frac{2}{5}$$

31. Find $\lim_{x \to \infty} (5 - \frac{2}{x})$.

We will use the fact that as $x \to \infty$, $\frac{b}{ax^n} \to 0$, for any positive integer n.

$$\lim_{x \to \infty} \left(5 - \frac{2}{x}\right)$$

$$= 5 - 0 \qquad \left(\text{As } x \to \infty, \frac{2}{x} \to 0\right)$$

$$= 5$$

33. Find $\lim_{x\to\infty} \frac{2x - 5}{4x + 3}$.

We will use some algebra and the fact that as $x \to \infty$, $\frac{b}{ax^n} \to 0$, for any positive integer n.

$$\lim_{x\to\infty} \frac{2x - 5}{4x + 3}$$

$$= \lim_{x\to\infty} \frac{2x - 5}{4x + 3} \cdot \frac{(1/x)}{(1/x)} \quad \text{(Multiplying by a form of 1)}$$

$$= \lim_{x\to\infty} \frac{2x \cdot \frac{1}{x} - 5 \cdot \frac{1}{x}}{4x \cdot \frac{1}{x} + 3 \cdot \frac{1}{x}}$$

$$= \lim_{x\to\infty} \frac{2 - \frac{5}{x}}{4 + \frac{3}{x}}$$

$$= \frac{2 - 0}{4 + 0} \quad \left(\text{As } x \to \infty, \frac{5}{x} \to 0 \text{ and } \frac{3}{x} \to 0\right)$$

$$= \frac{2}{4}$$

$$= \frac{1}{2}$$

35. Find $\lim_{x\to\infty} \frac{2x^2 - 5}{3x^2 - x + 7}$.

We will use some algebra and the fact that as $x \to \infty$, $\frac{b}{ax^n} \to 0$, for any positive integer n.

$$\lim_{x\to\infty} \frac{2x^2 - 5}{3x^2 - x + 7}$$

$$= \lim_{x\to\infty} \frac{2x^2 - 5}{3x^2 - x + 7} \cdot \frac{(1/x^2)}{(1/x^2)} \quad \text{(Multiplying by a form of 1)}$$

$$= \lim_{x\to\infty} \frac{2x^2 \cdot \frac{1}{x^2} - 5 \cdot \frac{1}{x^2}}{3x^2 \cdot \frac{1}{x^2} - x \cdot \frac{1}{x^2} + 7 \cdot \frac{1}{x^2}}$$

$$= \lim_{x\to\infty} \frac{2 - \frac{5}{x^2}}{3 - \frac{1}{x} + \frac{7}{x^2}}$$

$$= \frac{2 - 0}{3 - 0 + 0} \quad \left(\text{As } x \to \infty, \frac{5}{x^2} \to 0, \frac{1}{x} \to 0, \text{ and } \frac{7}{x^2} \to 0\right)$$

$$= \frac{2}{3}$$

37. Find $\lim_{x\to\infty} \frac{4 - 3x}{5 - 2x^2}$.

We divide the numerator and the denominator by x^2, the highest power of x in the denominator.

$$\lim_{x\to\infty} \frac{4 - 3x}{5 - 2x^2} = \lim_{x\to\infty} \frac{\frac{4}{x^2} - \frac{3}{x}}{\frac{5}{x^2} - 2}$$

$$= \frac{0 - 0}{0 - 2}$$

$$= 0$$

39. Find $\lim_{x\to\infty} \frac{8x^4 - 3x^2}{5x^2 + 6x}$.

We divide the numerator and the denominator by x^2, the highest power of x in the denominator.

$$\lim_{x\to\infty} \frac{8x^4 - 3x^2}{5x^2 + 6x} = \lim_{x\to\infty} \frac{8x^2 - 3}{5 + \frac{6}{x}}$$

$$= \frac{\lim_{x\to\infty} 8x^2 - 3}{5 + 0} = \infty$$

41. Find $\lim_{x\to\infty} \frac{6x^4 - 5x^2 + 7}{8x^6 + 4x^3 - 8x}$.

We divide the numerator and the denominator by x^6, the highest power of x in the denominator.

$$\lim_{x\to\infty} \frac{6x^4 - 5x^2 + 7}{8x^6 + 4x^3 - 8x} = \lim_{x\to\infty} \frac{\frac{6}{x^2} - \frac{5}{x^4} + \frac{7}{x^6}}{8 + \frac{4}{x^3} - \frac{8}{x^5}}$$

$$= \frac{0 - 0 + 0}{8 + 0 - 0}$$

$$= 0$$

43. Find $\lim_{x\to\infty} \frac{11x^5 + 4x^3 - 6x + 2}{6x^3 + 5x^2 + 3x - 1}$.

We divide the numerator and the denominator by x^3, the highest power of x in the denominator.

$$\lim_{x\to\infty} \frac{11x^5 + 4x^3 - 6x + 2}{6x^3 + 5x^2 + 3x - 1} = \lim_{x\to\infty} \frac{11x^2 + 4 - \frac{6}{x^2} + \frac{2}{x^3}}{6 + \frac{5}{x} + \frac{3}{x^2} - \frac{1}{x^3}}$$

$$= \frac{\lim_{x\to\infty} 11x^2 + 4 - 0 + 0}{6 + 0 + 0 - 0}$$

$$= \infty$$

45. It is helpful to graph the function.

$$f(x) = \begin{cases} 1, & \text{for } x \neq 2 \\ -1, & \text{for } x = 2 \end{cases}$$

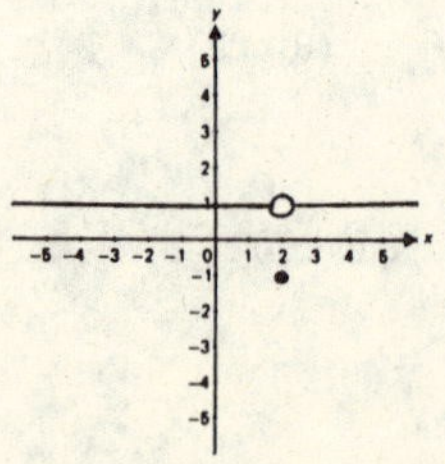

a) Find $\lim_{x\to 0} f(x)$.

As inputs x approach 0 from the left, outputs f(x) approach 1. Thus, the limit from the left is 1. As inputs x approach 0 from the right, outputs f(x) approach 1, so the limit from the right is 1. Since the limit from the left, 1, is the same as the limit from the right, 1, we have

$$\lim_{x\to 0} f(x) = 1.$$

b) Find $\lim_{x\to 2^-} f(x)$.

As inputs x approach 2 from the left, outputs, f(x), approach 1. Thus the limit from the left is 1.

$\lim_{x\to 2^-} f(x) = 1$

c) Find $\lim_{x\to 2^+} f(x)$.

As inputs x approach 2 from the right, outputs, f(x), approach 1. Thus the limit from the right is 1.

$\lim_{x\to 2^+} f(x) = 1$

d) Since the limit from the left, 1, is the same as the limit from the right, 1, we have

$\lim_{x\to 2} f(x) = 1.$

e) The function f(x) <u>is</u> <u>continuous</u> at x = 0 because

1) f(0) exists, f(0) = 1,

2) $\lim_{x\to 0} f(x)$ exists, $\lim_{x\to 0} f(x) = 1$,

and

3) $\lim_{x\to 0} f(x) = f(0) = 1.$

The function f(x) <u>is</u> <u>not</u> <u>continuous</u> at x = 2 because the $\lim_{x\to 2} f(x)$ is not the same as the function value at x = 2.

$\lim_{x\to 2} f(x) = 1$

$f(2) = -1$

$\lim_{x\to 2} f(x) \neq f(2)$

<u>47</u>. a) The value at the beginning of the first year is $10,000.

The depreciation during the first year is 8%·10,000, or $800.

The value at the beginning of the second year is 10,000 - 800, or $9200.

The depreciation during the second year is 8%·$9200, or $736.

The value at the beginning of the third year is 9200 - 736, or $8464.

The depreciation during the third year is 8%·8464, or $677.12.

The value at the beginning of the fourth year is 8464 - 677.12, or $7786.88.

The depreciation during the fourth year is 8%·7786.88, or $622.95.

The value at the beginning of the fifth year is 7786.88 - 622.95, or $7163.93.

The depreciation during the fifth year is 8%·7163.93, or $573.11.

b) The value at the beginning of the sixth year is 7163.93 - 573.11, or $6590.82.

The depreciation during the sixth year is 8%·6590.82, or $527.27.

The value at the beginning of the seventh year is 6590.82 - 527.27, or $6063.55.

The depreciation during the seventh year is 8%·6063.55, or $485.08.

The value at the beginning of the eighth year is 6063.55 - 485.08, or $5578.47.

The depreciation during the eighth year is 8%·5578.47, or $446.28.

The value at the beginning of the ninth year is 5578.47 - 446.28, or $5132.19.

The depreciation during the ninth year is 8%·5132.19, or $410.58.

The value at the beginning of the tenth year is 5132.19 - 410.58, or $4721.61.

The depreciation during the tenth year is 8%·4721.61, or $377.73.

The total depreciation at the end of ten years is $800 + $736 + $677.12 + $622.95 + $573.11 + $527.27 + $485.08 + $446.28 + $410.58 + $377.73, or $5656.12.

c) The sum of the annual depreciation costs cannot exceed the new cost which was $10,000. The limit of the sum is $10,000.

<u>49</u>. We divide the numerator and the denominator by x, the highest power of x in the denominator..

$$\lim_{x\to\infty} \frac{13x + 100}{x} = \lim_{x\to\infty} \frac{13 + \frac{100}{x}}{1}$$

$$= \frac{13 + 0}{1}$$

$$= 13$$

The limit is $13 per unit. This means that as more and more units are produced, the average production cost gets closer and closer to $13 per unit.

<u>51</u>.

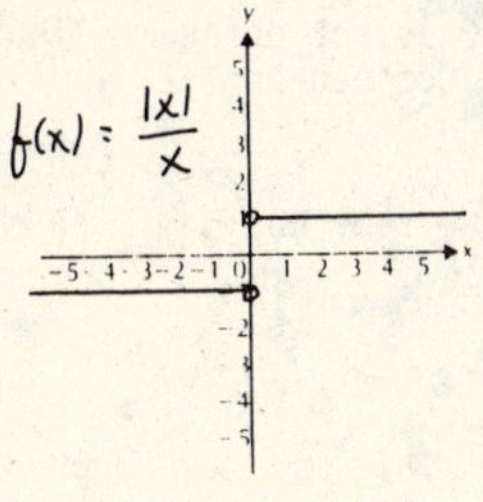

$\lim_{x\to 0^-} \frac{|x|}{x} = -1$, $\lim_{x\to 0^+} \frac{|x|}{x} = 1$, so $\lim_{x\to 0} \frac{|x|}{x}$ does not exist.

53. $\lim_{x\to 1} \frac{x^3 - 1}{x^2 - 1} = \lim_{x\to 1} \frac{(x - 1)(x^2 + x + 1)}{(x - 1)(x + 1)}$

$= \lim_{x\to 1} \frac{x^2 + x + 1}{x + 1}$ (Assuming $x \neq 1$)

$= \frac{1^2 + 1 + 1}{1 + 1}$

$= \frac{3}{2}$

55. We divide the numerator and the denominator by x^2, the highest power of x in the denominator.

$$\lim_{x\to\infty} \frac{-6x^3 + 7x}{2x^2 - 3x - 10} = \lim_{x\to\infty} \frac{-6x + \frac{7}{x}}{2 - \frac{3}{x} - \frac{10}{x^2}}$$

$$= \frac{\lim_{x\to\infty} -6x + 0}{2 - 0 - 0}$$

$$= -\infty$$

(The numerator increases without bound negatively while the denominator approaches 2.)

57. a)

x	$\frac{\sqrt{x^2 + 3x + 2}}{x - 3}$
10	1.6413
100	1.0464
10,000	1.0005
1,000,000	1.0000

These inputs approach ∞.

From the table we see that $\lim_{x\to\infty} f(x) = 1$.

x	$\frac{\sqrt{x^2 + 3x + 2}}{x - 3}$
-10	-0.6527
-100	-0.9563
-10,000	-0.9996
-1,000,000	-1.0000

These inputs approach $-\infty$.

From the table we see that $\lim_{x\to\infty} f(x) = -1$.

b) For values of x in (-2,-1), $x^2 + 3x + 2 < 0$ so $\sqrt{x^2 + 3x + 2}$ is not defined. Thus, the outputs are not defined on (-2,-1).

c) From the graph we see that $\lim_{x\to -2^-} f(x) = 0$ and $\lim_{x\to -1+} f(x) = 0$.

Exercise Set 2.3

1. a) $f(x) = 7x^2$

so

$f(x + h) = 7(x + h)^2$ (Substituting x + h for x)

$= 7(x^2 + 2xh + h^2)$

$= 7x^2 + 14xh + 7h^2$

Then

$\frac{f(x + h) - f(x)}{h}$ (Difference quotient)

$= \frac{(7x^2 + 14xh + 7h^2) - 7x^2}{h}$ (Substituting)

$= \frac{14xh + 7h^2}{h}$

$= \frac{h(14x + 7h)}{h}$ (Factoring the numerator)

$= \frac{h}{h} \cdot (14x + 7h)$

$= 14x + 7h$, (Simplified difference quotient)

or $7(2x + h)$

b) The difference quotient column in the table can be easily completed using the simplified difference quotient.

$14x + 7h$ (Simplified difference quotient)

$= 14(4) + 7(2) = 56 + 14 = 70$

(Substituting 4 for x and 2 for h)

$= 14(4) + 7(1) = 56 + 7 = 63$

(Substituting 4 for x and 1 for h)

$= 14(4) + 7(0.1) = 56 + 0.7 = 56.7$

(Substituting 4 for x and 0.1 for h)

$= 14(4) + 7(0.01) = 56 + 0.07 = 56.07$

(Substituting 4 for x and 0.01 for h)

The values in the table are 70, 63, 56.7, and 56.07.

3. a) $f(x) = -7x^2$

so

$f(x + h) = -7(x + h)^2$ (Substituting x + h for x)

$= -7(x^2 + 2xh + h^2)$

$= -7x^2 - 14xh - 7h^2$

Then

$\frac{f(x + h) - f(x)}{h}$ (Difference quotient)

$= \frac{(-7x^2 - 14xh - 7h^2) - (-7x^2)}{h}$ (Substituting)

$= \frac{-7x^2 - 14xh - 7h^2 + 7x^2}{h}$

$= \frac{-14xh - 7h^2}{h}$

$= \frac{h(-14x - 7h)}{h}$ (Factoring the numerator)

$= \frac{h}{h} \cdot (-14x - 7h)$

$= -14x - 7h$ (Simplified difference quotient)

or $-7(2x + h)$

b) The difference quotient column in the table can be easily completed using the simplified difference quotient.

$-14x - 7h$ (Simplified difference quotient)

$= -14(4) - 7(2) = -56 - 14 = -70$

(Substituting 4 for x and 2 for h)

$= -14(4) - 7(1) = -56 - 7 = -63$

(Substituting 4 for x and 1 for h)

$= -14(4) - 7(0.1) = -56 - 0.7 = -56.7$

(Substituting 4 for x and 0.1 for h)

$= -14(4) - 7(0.01) = -56 - 0.07 = -56.07$

(Substituting 4 for x and 0.01 for h)

The values in the table are -70, -63, -56.7, and -56.07.

5. a) $f(x) = 7x^3$

so

$f(x + h) = 7(x + h)^3$

(Substituting x + h for x)

$= 7(x^3 + 3x^2h + 3xh^2 + h^3)$

$= 7x^3 + 21x^2h + 21xh^2 + 7h^3$

Then

$\frac{f(x + h) - f(x)}{h}$ (Difference quotient)

$= \frac{(7x^3 + 21x^2h + 21xh^2 + 7h^3) - 7x^3}{h}$

(Substituting)

$= \frac{21x^2h + 21xh^2 + 7h^3}{h}$

$= \frac{h(21x^2 + 21xh + 7h^2)}{h}$

(Factoring the numerator)

$= \frac{h}{h} \cdot (21x^2 + 21xh + 7h^2)$

$= 21x^2 + 21xh + 7h^2$

(Simplified difference quotient)

or $7(3x^2 + 3xh + h^2)$

b) The difference quotient column in the table can be easily completed using the simplified difference quotient.

$21x^2 + 21xh + 7h^2$

(Simplified difference quotient)

$= 21 \cdot 4^2 + 21 \cdot 4 \cdot 2 + 7 \cdot 2^2$

(Substituting 4 for x and 2 for h)

$= 21 \cdot 16 + 21 \cdot 8 + 7 \cdot 4$

$= 336 + 168 + 28$

$= 532$

$= 21 \cdot 4^2 + 21 \cdot 4 \cdot 1 + 7 \cdot 1^2$

(Substituting 4 for x and 1 for h)

$= 21 \cdot 16 + 21 \cdot 4 + 7 \cdot 1$

$= 336 + 84 + 7$

$= 427$

$= 21(4)^2 + 21(4)(0.1) + 7(0.1)^2$

(Substituting 4 for x and 0.1 for h)

$= 21(16) + 21(0.4) + 7(0.01)$

$= 336 + 8.4 + 0.07$

$= 344.47$

$= 21(4)^2 + 21(4)(0.01) + 7(0.01)^2$

(Substituting 4 for x and 0.01 for h)

$= 21(16) + 21(0.04) + 7(0.0001)$

$= 336 + 0.84 + 0.0007$

$= 336.8407$

The values in the table are 532, 427, 344.47, and 336.8407.

7. a) $f(x) = \frac{5}{x}$

so

$f(x + h) = \frac{5}{x + h}$ (Substituting x + h for x)

Then

$\frac{f(x + h) - f(x)}{h}$ (Difference quotient)

$= \frac{\frac{5}{x + h} - \frac{5}{x}}{h}$ (Substituting)

$= \frac{\frac{5}{x + h} \cdot \frac{x}{x} - \frac{5}{x} \cdot \frac{x + h}{x + h}}{h}$

(Multiplying by 1)

$= \frac{\frac{5x}{x(x + h)} - \frac{5(x + h)}{x(x + h)}}{h}$

$= \frac{\frac{5x - 5x - 5h}{x(x + h)}}{h}$

$= \frac{\frac{-5h}{x(x + h)}}{\frac{h}{1}}$ $\left(h = \frac{h}{1}\right)$

$= \frac{-5h}{x(x + h)} \cdot \frac{1}{h}$ (Multiplying by the reciprocal)

$= \frac{h}{h} \cdot \frac{-5}{x(x + h)}$

$= \frac{-5}{x(x + h)}$ (Simplified difference quotient)

b) The difference quotient column in the table can be easily completed using the simplified difference quotient.

$$\frac{-5}{x(x + h)}$$ (Simplified difference quotient)

$$= \frac{-5}{4(4 + 2)} = \frac{-5}{24} \approx -0.2083$$
(Substituting 4 for x and 2 for h)

$$= \frac{-5}{4(4 + 1)} = \frac{-5}{20} = -0.25$$
(Substituting 4 for x and 1 for h)

$$= \frac{-5}{4(4 + 0.1)} = \frac{-5}{16.4} \approx -0.3049$$
(Substituting 4 for x and 0.1 for h)

$$= \frac{-5}{4(4 + 0.01)} = \frac{-5}{16.04} \approx -0.3117$$
(Substituting 4 for x and 0.01 for h)

The values in the table are -0.2083, -0.25, -0.3049, and -0.3117.

9. a) $f(x) = -2x + 5$

so

$f(x + h) = -2(x + h) + 5$ (Substituting x + h for x)

$= -2x - 2h + 5$

Then

$$\frac{f(x + h) - f(x)}{h}$$ (Difference quotient)

$$= \frac{(-2x - 2h + 5) - (-2x + 5)}{h}$$ (Substituting)

$$= \frac{-2x - 2h + 5 + 2x - 5}{h}$$

$$= \frac{-2h}{h}$$

$= -2$ (Simplified difference quotient)

b) For a linear function average rates of change are the same, for any choice of x_1 and x_2, being equal to the slope of the line.

$f(x) = -2x + 5$ is a linear function whose slope is -2. Thus, the average rate of change is also -2 for <u>any</u> choice of x and h. All values in the difference quotient column of the table are -2.

11. a) $f(x) = x^2 - x$

so

$f(x + h) = (x + h)^2 - (x + h)$ (Substituting x + h for x)

$= x^2 + 2xh + h^2 - x - h$

Then

$$\frac{f(x + h) - f(x)}{h}$$ (Difference quotient)

$$= \frac{(x^2 + 2xh + h^2 - x - h) - (x^2 - x)}{h}$$ (Substituting)

$$= \frac{x^2 + 2xh + h^2 - x - h - x^2 + x}{h}$$

$$= \frac{2xh + h^2 - h}{h}$$

$$= \frac{h(2x + h - 1)}{h}$$

$= 2x + h - 1$ (Simplified difference quotient)

b) The difference quotient column in the table can be easily completed using the simplified difference quotient.

$2x + h - 1$ (Simplified difference quotient)

$= 2(4) + 2 - 1 = 9$
(Substituting 4 for x and 2 for h)

$= 2(4) + 1 - 1 = 8$
(Substituting 4 for x and 1 for h)

$= 2(4) + 0.1 - 1 = 7.1$
(Substituting 4 for x and 0.1 for h)

$= 2(4) + 0.01 - 1 = 7.01$
(Substituting 4 for x and 0.01 for h)

The values in the table are 9, 8, 7.1, and 7.01.

13. a) When the input is 0, the output, U(0), is 0. When the input is 1, the output, U(1) = 70.

Let $(x_1,y_1) = (0,0)$ and $(x_2,y_2) = (1,70)$.

The average rate of change is

$$\frac{y_2 - y_1}{x_2 - x_1} = \frac{70 - 0}{1 - 0} = \frac{70}{1} = 70 \frac{\text{pleasure units}}{\text{unit of product}}.$$

When the input is 1, the output, U(1), is 70. When the input is 2, the output, U(2), is 109.

Let $(x_1,y_1) = (1,70)$ and $(x_2,y_2) = (2,109)$.

The average rate of change is

$$\frac{y_2 - y_1}{x_2 - x_1} = \frac{109 - 70}{2 - 1} = \frac{39}{1} = 39 \frac{\text{pleasure units}}{\text{unit of product}}.$$

When the input is 2, the output, U(2), is 109. When the input is 3, the output, U(3), is 138.

Let $(x_1,y_1) = (2,109)$ and $(x_2,y_2) = (3,138)$.

The average rate of change is

$$\frac{y_2 - y_1}{x_2 - x_1} = \frac{138 - 109}{3 - 2} = \frac{29}{1} = 29 \frac{\text{pleasure units}}{\text{unit of product}}.$$

When the input is 3, the output, U(3), is 138. When the input is 4, the output, U(4), is 161.

Let $(x_1,y_1) = (3,138)$ and $(x_2,y_2) = (4,161)$.

The average rate of change is

$$\frac{y_2 - y_1}{x_2 - x_1} = \frac{161 - 138}{4 - 3} = \frac{23}{1} = 23 \frac{\text{pleasure units}}{\text{unit of product}}.$$

b) The average rates of change are decreasing because the more you get the less pleasure you get from each additional unit.

15. a) $R(x) = -0.01x^2 + 1000x$ (Total revenue)

$R(301) = -0.01(301)^2 + 1000(301)$ (Substituting 301 for x)

$= -0.01(90{,}601) + 1000(301)$

$= -906.01 + 301{,}000$

$= 300{,}093.99$

The total revenue from the sale of 301 units is $300,093.99.

b) $R(x) = -0.01x^2 + 1000x$ (Total revenue)

$R(300) = -0.01(300)^2 + 1000(300)$ (Substituting 300 for x)

$= -0.01(90{,}000) + 1000(300)$

$= -900 + 300{,}000$

$= 299{,}100$

The total revenue from the sale of 300 units is $299,100.

c) $R(301) = 300{,}093.99$ (Part a)

$R(300) = 299{,}100$ (Part b)

$R(301) - R(300) = 300{,}093.99 - 299{,}100$

$= 993.99$

The change in total revenue from the sale of 300 units to 301 units is $993.99.

d) The average rate of change of the total revenue with respect to the number of units sold, as the number of units sold changes from 300 to 301, is

$$\frac{R(301) - R(300)}{301 - 300}$$

$$= \frac{300{,}093.99 - 299{,}100}{301 - 300}$$

$$= \frac{993.99}{1}$$

$= \$993.99$

17. a) We will find the average rate of change in millions of dollars per year. Let $(x_1,y_1) = (1986, 930)$ and $(x_2,y_2) = (1989, 15{,}242)$.

$$\frac{y_2 - y_1}{x_2 - x_1} = \frac{15{,}242 - 930}{1989 - 1986} = \frac{14{,}312}{3} \approx 4770.67$$

The average rate of change of total revenue from 1986 to 1989 is about $4770.67 million/year.

b) We will find the average rate of change in millions of dollars per year.

Let $(x_1,y_1) = (1988, 13{,}007)$ and $(x_2,y_2) = (1989, 15{,}242)$.

$$\frac{y_2 - y_1}{x_2 - x_1} = \frac{15{,}242 - 13{,}007}{1989 - 1988} = \frac{2235}{1} = 2235$$

The average rate of change of total revenue from 1988 to 1989 is $2235 million/year.

19. a) As t changes from 0 to 8, M(t) changes from 0 to 10. The average rate of change is

$\frac{10 - 0}{8 - 0} = \frac{10}{8} = \frac{5}{4}$, or 1.25 words per minute.

As t changes from 8 to 16, M(t) changes from 10 to 20. The average rate of change is

$\frac{20 - 10}{16 - 8} = \frac{10}{8} = \frac{5}{4}$, or 1.25 words per minute.

As t changes from 16 to 24, M(t) changes from 20 to 25. The average rate of change is

$\frac{25 - 20}{24 - 16} = \frac{5}{8}$, or 0.625 words per minute.

As t changes from 24 to 32, M(t) changes from 25 to 25. The average rate of change is

$\frac{25 - 25}{32 - 24} = \frac{0}{8}$, or 0 words per minute.

As t changes from 32 to 36, M(t) changes from 25 to 25. The average rate of change is

$\frac{25 - 25}{36 - 32} = \frac{0}{4}$, or 0 words per minute.

b) You have reached a saturation point; you cannot memorize any more words.

21. a) $s(t) = 16t^2$ (s is distance in feet; t is time in seconds)

$s(3) = 16\cdot 3^2$ (Substituting 3 for t)

$= 16\cdot 9$

$= 144$

The object will fall 144 feet in 3 seconds.

b) $s(t) = 16t^2$

$s(5) = 16\cdot 5^2$ (Substituting 5 for t)

$= 16\cdot 25$

$= 400$

The object will fall 400 feet in 5 seconds.

c) The average rate of change of distance with respect to time during the time from 3 to 5 seconds is

$$\frac{s(5) - s(3)}{5 - 3}$$

$$= \frac{400 \text{ feet} - 144 \text{ feet}}{5 \text{ seconds} - 3 \text{ seconds}}$$

$$= \frac{256 \text{ feet}}{2 \text{ seconds}}$$

$$= 128 \frac{\text{ft}}{\text{sec}}$$

The average velocity is $128 \frac{\text{ft}}{\text{sec}}$.

23. a) For each curve, as t changes from 0 to 4, P(t) changes from 0 to 500. Thus, the average growth rate for each is

$\frac{500 - 0}{4 - 0} = \frac{500}{4} = 125 \frac{\text{million people}}{\text{yr}}$.

b) No

c) Country A:

As t changes from 0 to 1, P(t) changes from 0 to 290. The average growth rate is

$$\frac{290 - 0}{1 - 0} = \frac{290}{1} = 290\ \frac{\text{million people}}{\text{yr}}.$$

As t changes from 1 to 2, P(t) changes from 290 to 250. The average growth rate is

$$\frac{250 - 290}{2 - 1} = \frac{-40}{1} = -40\ \frac{\text{million people}}{\text{yr}}.$$

As t changes from 2 to 3, P(t) changes from 250 to 200. The average growth rate is

$$\frac{200 - 250}{3 - 2} = \frac{-50}{1} = -50\ \frac{\text{million people}}{\text{yr}}.$$

As t changes from 3 to 4, P(t) changes from 200 to 500. The average growth rate is

$$\frac{500 - 200}{4 - 3} = \frac{300}{1} = 300\ \frac{\text{million people}}{\text{yr}}.$$

Country B:

As t changes from 0 to 1, P(t) changes from 0 to 125. The average growth rate is

$$\frac{125 - 0}{1 - 0} = \frac{125}{1} = 125\ \frac{\text{million people}}{\text{yr}}.$$

As t changes from 1 to 2, P(t) changes from 125 to 250. The average growth rate is

$$\frac{250 - 125}{2 - 1} = \frac{125}{1} = 125\ \frac{\text{million people}}{\text{yr}}.$$

As t changes from 2 to 3, P(t) changes from 250 to 375. The average growth rate is

$$\frac{375 - 250}{3 - 2} = \frac{125}{1} = 125\ \frac{\text{million people}}{\text{yr}}.$$

As t changes from 3 to 4, P(t) changes from 375 to 500. The average growth rate is

$$\frac{500 - 375}{4 - 3} = \frac{125}{1} = 125\ \frac{\text{million people}}{\text{yr}}.$$

d) A

25. Let $(x_1, y_1) = (1960, 1{,}523{,}000)$ and $(x_2, y_2) = (1982, 2{,}495{,}000)$.

The average rate of change is

$$\frac{y_2 - y_1}{x_2 - x_1} = \frac{2{,}495{,}000 - 1{,}523{,}000}{1982 - 1960}$$

$$= \frac{972{,}000}{22}$$

$$\approx 44{,}182\ \frac{\text{marriages}}{\text{yr}}.$$

27. $f(x) = ax^2 + bx + c$

so

$f(x + h) = a(x + h)^2 + b(x + h) + c$ (Substituting x + h for x)

$= a(x^2 + 2xh + h^2) + b(x + h) + c$

$= ax^2 + 2axh + ah^2 + bx + bh + c$

Then

$\frac{f(x + h) - f(x)}{h}$ (Difference quotient)

$= \frac{(ax^2 + 2axh + ah^2 + bx + bh + c) - (ax^2 + bx + c)}{h}$ (Substituting)

$= \frac{2axh + ah^2 + bh}{h}$

$= \frac{h(2ax + ah + b)}{h}$ (Factoring the numerator)

$= 2ax + ah + b$ (Simplified difference quotient)

29. $f(x) = \sqrt{x}$

so

$f(x + h) = \sqrt{x + h}$ (Substituting x + h for x)

Then

$\frac{f(x + h) - f(x)}{h}$ (Difference quotient)

$= \frac{\sqrt{x + h} - \sqrt{x}}{h}$ (Substituting)

$= \frac{\sqrt{x + h} - \sqrt{x}}{h} \cdot \frac{\sqrt{x + h} + \sqrt{x}}{\sqrt{x + h} + \sqrt{x}}$ (Multiplying by 1)

$= \frac{(x + h) - x}{h(\sqrt{x + h} + \sqrt{x})}$

$= \frac{h}{h(\sqrt{x + h} + \sqrt{x})}$

$= \frac{1}{\sqrt{x + h} + \sqrt{x}}$ (Simplified difference quotient)

31. $f(x) = \frac{1}{x^2}$

so

$f(x + h) = \frac{1}{(x + h)^2}$ (Substituting x + h for x)

Then

$\frac{f(x + h) - f(x)}{h}$ (Difference quotient)

$= \frac{\frac{1}{(x + h)^2} - \frac{1}{x^2}}{h}$ (Substituting)

$= \frac{\frac{1}{(x + h)^2} \cdot \frac{x^2}{x^2} - \frac{1}{x^2} \cdot \frac{(x + h)^2}{(x + h)^2}}{h}$

$= \frac{\frac{x^2 - (x + h)^2}{x^2(x + h)^2}}{\frac{h}{1}}$

$= \frac{x^2 - x^2 - 2xh - h^2}{x^2(x + h)^2} \cdot \frac{1}{h}$

$= \frac{-2xh - h^2}{x^2(x + h)^2} \cdot \frac{1}{h}$

$= \frac{h(-2x - h)}{x^2(x + h)^2} \cdot \frac{1}{h}$

$= \frac{-2x - h}{x^2(x + h)^2}$ (Simplified difference quotient)

33. $f(x) = \frac{x}{1 + x}$

so

$f(x + h) = \frac{x + h}{1 + x + h}$ (Substituting x + h for x)

Then

$\frac{f(x + h) - f(x)}{h}$ (Difference quotient)

$= \frac{\frac{x + h}{1 + x + h} - \frac{x}{1 + x}}{h}$ (Substituting)

$= \frac{\frac{x + h}{1 + x + h} \cdot \frac{1 + x}{1 + x} - \frac{x}{1 + x} \cdot \frac{1 + x + h}{1 + x + h}}{\frac{h}{1}}$

$= \frac{(x^2 + xh + x + h) - (x + x^2 + xh)}{(1 + x)(1 + x + h)} \cdot \frac{1}{h}$

$= \frac{h}{(1 + x)(1 + x + h)} \cdot \frac{1}{h}$

$= \frac{1}{(1 + x)(1 + x + h)}$ (Simplified difference quotient)

Exercise Set 2.4

1. Write down and simplify the difference quotient.

$f(x) = 5x^2$

so

$f(x + h) = 5(x + h)^2$

$= 5(x^2 + 2xh + h^2)$

$= 5x^2 + 10xh + 5h^2$

Then

$\frac{f(x + h) - f(x)}{h}$

$= \frac{(5x^2 + 10xh + 5h^2) - 5x^2}{h}$ (Substituting)

$= \frac{10xh + 5h^2}{h}$

$= \frac{h(10x + 5h)}{h}$

$= 10x + 5h$ (Simplified difference quotient)

Find the limit of the difference quotient as $h \to 0$.

$\lim_{h\to 0} \frac{f(x + h) - f(x)}{h}$

$= \lim_{h\to 0} (10x + 5h)$

$= 10x$ (As $h \to 0$, $10x + 5h \to 10x$)

Thus, $f'(x) = 10x$.

$f'(-2) = 10(-2) = -20$ (Substituting -2 for x)

At x = -2 the curve has a tangent line whose slope is -20.

$f'(-1) = 10(-1) = -10$ (Substituting -1 for x)

At x = -1 the curve has a tangent line whose slope is -10.

$f'(0) = 10 \cdot 0 = 0$ (Substituting 0 for x)

$f'(1) = 10 \cdot 1 = 10$ (Substituting 1 for x)

$f'(2) = 10 \cdot 2 = 20$ (Substituting 2 for x)

3. Write down and simplify the difference quotient.

$f(x) = -5x^2$

so

$f(x + h) = -5(x + h)^2$

$= -5(x^2 + 2xh + h^2)$

$= -5x^2 - 10xh - 5h^2$

Then

$\frac{f(x + h) - f(x)}{h}$

$= \frac{(-5x^2 - 10xh - 5h^2) - (-5x^2)}{h}$ (Substituting)

$= \frac{-5x^2 - 10xh - 5h^2 + 5x^2}{h}$

$= \frac{-10xh - 5h^2}{h}$

$= \frac{h(-10x - 5h)}{h}$

$= -10x - 5h$ (Simplified difference quotient)

Find the limit of the difference quotient as $h \to 0$.

$\lim_{h\to 0} \frac{f(x + h) - f(x)}{h}$

$= \lim_{h\to 0} (-10x - 5h)$

$= -10x$ (As $h \to 0$, $-10x - 5h \to -10x$)

Thus, $f'(x) = -10x$.

$f'(-2) = -10(-2) = 20$ (Substituting -2 for x)

At x = -2 the curve has a tangent line whose slope is 20.

$f'(-1) = -10(-1) = 10$ (Substituting -1 for x)

At x = -1 the curve has a tangent line whose slope is 10.

$f'(0) = -10 \cdot 0 = 0$ (Substituting 0 for x)

$f'(1) = -10 \cdot 1 = -10$ (Substituting 1 for x)

$f'(2) = -10 \cdot 2 = -20$ (Substituting 2 for x)

5. Write down and simplify the difference quotient.

$f(x) = 5x^3$

so

$$f(x + h) = 5(x + h)^3$$
$$= 5(x^3 + 3x^2h + 3xh^2 + h^3)$$
$$= 5x^3 + 15x^2h + 15xh^2 + 5h^3$$

Then

$$\frac{f(x + h) - f(x)}{h}$$

$$= \frac{(5x^3 + 15x^2h + 15xh^2 + 5h^3) - (5x^3)}{h}$$ (Substituting)

$$= \frac{15x^2h + 15xh^2 + 5h^3}{h}$$

$$= \frac{h(15x^2 + 15xh + 5h^2)}{h}$$

$$= 15x^2 + 15xh + 5h^2$$ (Simplified difference quotient)

Find the limit of the difference quotient as $h \to 0$.

$$\lim_{h \to 0} \frac{f(x + h) - f(x)}{h}$$

$$= \lim_{h \to 0} (15x^2 + 15xh + 5h^2)$$

$$= 15x^2$$ (As $h \to 0$, $15x^2 + 15xh + 5h^2 \to 15x^2$)

Thus, $f'(x) = 15x^2$.

$f'(-2) = 15(-2)^2 = 60$ (Substituting -2 for x)

At $x = -2$ the curve has a tangent line whose slope is 60.

$f'(-1) = 15(-1)^2 = 15$ (Substituting -1 for x)

At $x = -1$ the curve has a tangent line whose slope is 15.

$f'(0) = 15 \cdot 0^2 = 0$ (Substituting 0 for x)

$f'(1) = 15 \cdot 1^2 = 15$ (Substituting 1 for x)

$f'(2) = 15 \cdot 2^2 = 60$ (Substituting 2 for x)

7. Write down and simplify the difference quotient.

$f(x) = 2x + 3$

so

$$f(x + h) = 2(x + h) + 3$$
$$= 2x + 2h + 3$$

Then

$$\frac{f(x + h) - f(x)}{h}$$

$$= \frac{(2x + 2h + 3) - (2x + 3)}{h}$$ (Substituting)

$$= \frac{2x + 2h + 3 - 2x - 3}{h}$$

$$= \frac{2h}{h}$$

$$= 2$$ (Simplified difference quotient)

Find the limit of the difference quotient as $h \to 0$.

$$\lim_{h \to 0} \frac{f(x + h) - f(x)}{h}$$

$$= \lim_{h \to 0} 2$$

$$= 2$$ (As $h \to 0$, $2 \to 2$)

Thus, $f'(x) = 2$.

Note that $f'(x) = 2$ is a constant function. For all inputs, x, the output is 2.

$f'(-2) = 2$

$f'(-1) = 2$

$f'(0) = 2$

$f'(1) = 2$

$f'(2) = 2$

We should observe that $f(x) = 2x + 3$ is a linear function. Every nonvertical line has a constant slope. The slope of this line is 2.

9. Write down and simplify the difference quotient.

$f(x) = -4x$

so

$$f(x + h) = -4(x + h)$$
$$= -4x - 4h$$

Then

$$\frac{f(x + h) - f(x)}{h}$$

$$= \frac{(-4x - 4h) - (-4x)}{h}$$ (Substituting)

$$= \frac{-4x - 4h + 4x}{h}$$

$$= \frac{-4h}{h}$$

$$= -4$$ (Simplified difference quotient)

Find the limit of the difference quotient as $h \to 0$.

$$\lim_{h \to 0} \frac{f(x + h) - f(x)}{h}$$

$$= \lim_{h \to 0} -4$$

$$= -4$$ (As $h \to 0$, $-4 \to -4$)

Thus, $f'(x) = -4$.

Note that $f'(x) = -4$ is a constant function. For all inputs, x, the output is -4.

$f'(-2) = -4$

$f'(-1) = -4$

$f'(0) = -4$

$f'(1) = -4$

$f'(2) = -4$

We should observe that $f(x) = -4x$ is a linear function. Every nonvertical line has a constant slope. The slope of the line is -4.

11. Write down and simplify the difference quotient.

$f(x) = x^2 + x$

so

$f(x + h) = (x + h)^2 + (x + h)$

$= x^2 + 2xh + h^2 + x + h$

Then

$\frac{f(x + h) - f(x)}{h}$

$= \frac{(x^2 + 2xh + h^2 + x + h) - (x^2 + x)}{h}$ (Substituting)

$= \frac{x^2 + 2xh + h^2 + x + h - x^2 - x}{h}$

$= \frac{2xh + h^2 + h}{h}$

$= \frac{h(2x + h + 1)}{h}$

$= 2x + h + 1$ (Simplified difference quotient)

Find the limit of the difference quotient as $h \to 0$.

$\lim_{h\to 0} \frac{f(x + h) - f(x)}{h}$

$= \lim_{h\to 0} (2x + h + 1)$

$= 2x + 1$ (As $h \to 0$, $2x + h + 1 \to 2x + 1$)

Thus, $f'(x) = 2x + 1$.

$f'(-2) = 2(-2) + 1 = -3$ (Substituting -2 for x)

At $x = -2$ the curve has a tangent line whose slope is -3.

$f'(-1) = 2(-1) + 1 = -1$ (Substituting -1 for x)

At $x = -1$ the curve has a tangent line whose slope is -1.

$f'(0) = 2\cdot 0 + 1 = 1$ (Substituting 0 for x)

$f'(1) = 2\cdot 1 + 1 = 3$ (Substituting 1 for x)

$f'(2) = 2\cdot 2 + 1 = 5$ (Substituting 2 for x)

13. Write down and simplify the difference quotient.

$f(x) = \frac{4}{x}$

so

$f(x + h) = \frac{4}{x + h}$

Then

$\frac{f(x + h) - f(x)}{h}$

$= \frac{\frac{4}{x + h} - \frac{4}{x}}{h}$ (Substituting)

$= \frac{\frac{4}{x + h} \cdot \frac{x}{x} - \frac{4}{x} \cdot \frac{x + h}{x + h}}{h}$

$= \frac{\frac{4x - 4x - 4h}{x(x + h)}}{h}$

$= \frac{\frac{-4h}{x(x + h)}}{\frac{h}{1}}$

$= \frac{-4h}{x(x + h)} \cdot \frac{1}{h}$

$= \frac{-4}{x(x + h)}$ (Simplified difference quotient)

Find the limit of the difference quotient as $h \to 0$.

$\lim_{h\to 0} \frac{f(x + h) - f(x)}{h}$

$= \lim_{h\to 0} \frac{-4}{x(x + h)}$

$= \frac{-4}{x\cdot x}$ (As $h \to 0$, $x + h \to x$)

$= \frac{-4}{x^2}$

Thus, $f'(x) = \frac{-4}{x^2}$.

$f'(-2) = \frac{-4}{(-2)^2} = \frac{-4}{4} = -1$

At $x = -2$, the curve has a tangent line whose slope is -1.

$f'(-1) = \frac{-4}{(-1)^2} = \frac{-4}{1} = -4$

$f'(0)$ does not exist because $f(0)$ does not exist. We say "f is not differentiable at 0." When a function is not defined at a point, it is not differentiable at that point.

$f'(1) = \frac{-4}{1^2} = \frac{-4}{1} = -4$

$f'(2) = \frac{-4}{2^2} = \frac{-4}{4} = -1$

15. Write down and simplify the difference quotient.

$f(x) = mx$

so

$f(x + h) = m(x + h)$

$= mx + mh$

Then

$\frac{f(x + h) - f(x)}{h}$

$= \frac{(mx + mh) - mx}{h}$ (Substituting)

$= \frac{mh}{h}$

$= m$ (Simplified difference quotient)

Find the limit of the difference quotient as $h \to 0$.

$\lim_{h\to 0} \frac{f(x + h) - f(x)}{h}$

$= \lim_{h\to 0} m$

$= m$ (As $h \to 0$, $m \to m$)

Thus, $f'(x) = m$.

Note that $f'(x) = m$ is a constant function. For <u>all</u> inputs, x, the output is m.

$f'(-2) = m$

$f'(-1) = m$

$f'(0) = m$

$f'(1) = m$

$f'(2) = m$

We should observe that $f(x) = mx$ is a linear function. Every nonvertical line has a constant slope.

17. From Example 3 we know that $f'(x) = 2x$.

$f'(3) = 2\cdot 3 = 6$, so the slope of the line tangent to the curve at (3,9) is 6. We substitute the point (3,9) and the slope 6 in the point-slope equation to find the equation of the tangent line:

$$y - y_1 = m(x - x_1)$$
$$y - 9 = 6(x - 3)$$
$$y - 9 = 6x - 18$$
$$y = 6x - 9$$

$f'(-1) = 2(-1) = -2$, so the slope of the line tangent to the curve at (-1,1) is -2. We substitute the point (-1,1) and the slope -2 in the point-slope equation to find the equation of the tangent line:

$$y - y_1 = m(x - x_1)$$
$$y - 1 = -2[x - (-1)]$$
$$y - 1 = -2(x + 1)$$
$$y - 1 = -2x - 2$$
$$y = -2x - 1$$

$f'(10) = 2\cdot 10 = 20$, so the slope of the line tangent to the curve at (10, 100) is 20. We substitute the point (10, 100) and the slope 20 in the point-slope equation to find the equation of the tangent line:

$$y - y_1 = m(x - x_1)$$
$$y - 100 = 20(x - 10)$$
$$y - 100 = 20x - 200$$
$$y = 20x - 100$$

19. Write down and simplify the difference quotient.

$$f(x) = \frac{5}{x}$$

$$f(x + h) = \frac{5}{x + h}$$

Then

$$\frac{f(x + h) - f(x)}{h} = \frac{\frac{5}{x + h} - \frac{5}{x}}{h}$$

$$= \frac{\frac{5}{x + h}\cdot\frac{x}{x} - \frac{5}{x}\cdot\frac{x + h}{x + h}}{h}$$

$$= \frac{\frac{5x - 5x - 5h}{x(x + h)}}{h}$$

$$= \frac{-5h}{x(x + h)}\cdot\frac{1}{h}$$

$$= \frac{-5}{x(x + h)} \quad \text{(Simplified difference quotient)}$$

Find the limit of the difference quotient as $h \to 0$.

$$\lim_{h\to 0} \frac{f(x + h) - f(x)}{h} = \lim_{h\to 0} \frac{-5}{x(x + h)}$$

$$= \frac{-5}{x\cdot x}$$

$$= \frac{-5}{x^2}$$

Thus, $f'(x) = \frac{-5}{x^2}$.

$f'(1) = \frac{-5}{1^2} = -5$, so the slope of the line tangent to the curve at (1,5) is -5. We substitute the point (1,5) and the slope -5 in the point-slope equation to find the equation of the tangent line:

$$y - y_1 = m(x - x_1)$$
$$y - 5 = -5(x - 1)$$
$$y - 5 = -5x + 5$$
$$y = -5x + 10$$

$f'(-1) = \frac{-5}{(-1)^2} = -5$, so the slope of the line tangent to the curve at (-1,-5) is -5. We substitute the point (-1,-5) and the slope -5 in the point-slope equation to find the equation of the tangent line:

$$y - y_1 = m(x - x_1)$$
$$y - (-5) = -5[x - (-1)]$$
$$y + 5 = -5(x + 1)$$
$$y + 5 = -5x - 5$$
$$y = -5x - 10$$

$f'(100) = \frac{-5}{100^2} = -0.0005$, so the slope of the line tangent to the curve at (100, 0.05) is -0.0005. We substitute the point (100, 0.05) and the slope -0.0005 in the point-slope equation to find the equation of the tangent line:

$$y - 0.05 = -0.0005(x - 100)$$
$$y - 0.05 = -0.0005x + 0.05$$
$$y = -0.0005x + 0.1$$

21. First we find $f'(x)$:

$$\frac{f(x + h) - f(x)}{h} = \frac{4 - (x + h)^2 - (4 - x^2)}{h}$$

$$= \frac{4 - (x^2 + 2xh + h^2) - 4 + x^2}{h}$$

$$= \frac{4 - x^2 - 2xh - h^2 - 4 + x^2}{h}$$

$$= \frac{-2xh - h^2}{h}$$

$$= \frac{h(-2x - h)}{h}$$

$$= -2x - h$$

$$f'(x) = \lim_{h\to 0} (-2x - h) = -2x$$

$f'(-1) = -2(-1) = 2$, so the slope of the line tangent to the curve at $(-1,3)$ is 2. We substitute the point $(-1,3)$ and the slope 2 in the point-slope equation to find the equation of the tangent line:

$$y - y_1 = m(x - x_1)$$
$$y - 3 = 2[x - (-1)]$$
$$y - 3 = 2(x + 1)$$
$$y - 3 = 2x + 2$$
$$y = 2x + 5$$

$f'(0) = -2\cdot 0 = 0$, so the slope of the line tangent to the curve at $(0,4)$ is 0. We substitute the point $(0,4)$ and the slope 0 in the point-slope equation to find the equation of the tangent line:

$$y - y_1 = m(x - x_1)$$
$$y - 4 = 0(x - 0)$$
$$y - 4 = 0$$
$$y = 4$$

$f'(5) = -2\cdot 5 = -10$, so the slope of the line tangent to the curve at $(5,-21)$ is -10. We substitute the point $(5,-21)$ and the slope -10 in the point-slope equation to find the equation of the tangent line:

$$y - y_1 = m(x - x_1)$$
$$y - (-21) = -10(x - 5)$$
$$y + 21 = -10x + 50$$
$$y = -10x + 29$$

23. If a function has a "sharp point" or "corner," it will not have a derivative at that point. Thus, the function is not differentiable at x_3, x_4, and x_6. The function has a vertical tangent at x_{12}. Vertical lines have no slope, hence there is no derivative at the point x_{12}. Also, if a function is discontinuous at some point a, then it is not differentiable at a. The function is discontinuous at point x_0. Thus, it is also not differentiable at x_0.

25. The postage function does not have any "sharp points" or "corners" nor does it have any vertical tangents. However, it is discontinous at all natural numbers. Therefore, the postage function is not differentiable for 1, 2, 3, 4, and so on.

27. We found the simplified difference quotient in Exercise 31, Exercise Set 2.3. We now find the limit of the difference quotient as $h \to 0$.

$$\lim_{h\to 0} \frac{-2x - h}{x^2(x + h)^2}$$

$$= \frac{-2x}{x^2\cdot x^2} \qquad [\text{As } h \to 0,\ (-2x - h) \to -2x \text{ and } (x + h) \to x]$$

$$= \frac{-2x}{x^4}$$

$$= \frac{-2}{x^3}$$

Thus, $f'(x) = \frac{-2}{x^3}$.

29. We found the simplified difference quotient in Exercise 33, Exercise Set 2.3. We now find the limit of the difference quotient as $h \to 0$.

$$\lim_{h\to 0} \frac{1}{(1 + x)(1 + x + h)}$$

$$= \frac{1}{(1 + x)(1 + x)}$$

[As $h \to 0$, $(1 + x + h) \to (1 + x)$]

$$= \frac{1}{(1 + x)^2}, \text{ or } \frac{1}{1 + 2x + x^2}$$

Thus, $f'(x) = \frac{1}{(1 + x)^2}$.

31. a)

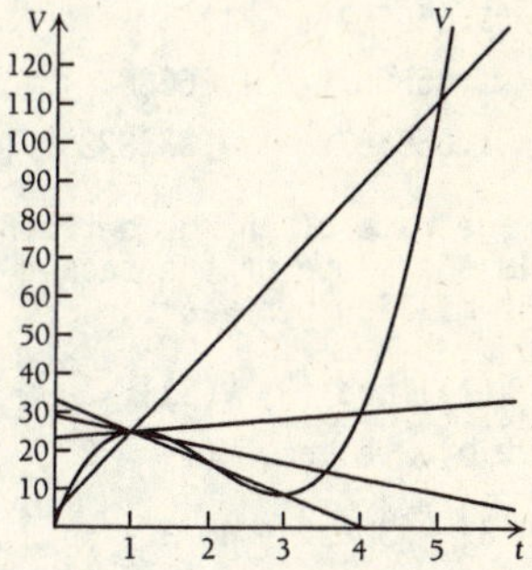

b) $V(1) = 5\cdot 1^3 - 30\cdot 1^2 + 45\cdot 1 + 5\sqrt{1}$
$= 5 - 30 + 45 + 5$
$= 25$

$V(5) = 5\cdot 5^3 - 30\cdot 5^2 + 45\cdot 5 + 5\sqrt{5}$
≈ 111.180

Find the equation of the secant line:

$$m = \frac{V(5) - V(1)}{5 - 1}$$
$$= \frac{111.180 - 25}{5 - 1}$$
$$= \frac{86.180}{4}$$
$$= 21.545$$

$$V - V_1 = m(t - t_1)$$
$$V - 25 = 21.545(t - 1)$$
$$V - 25 = 21.545t - 21.545$$
$$V = 21.545t + 3.455$$

c) The average rate of change between year 1 and year 5 is the slope of the secant line, \$21.545 million per year.

d) For (1, V(1)) and (4, V(4)):

From part b), we know V(1) = 25.

$V(4) = 5\cdot 4^3 - 30\cdot 4^2 + 45\cdot 4 + 5\sqrt{4}$

$= 30$

$m = \frac{V(4) - V(1)}{4 - 1}$

$= \frac{30 - 25}{4 - 1}$

$= \frac{5}{3}$

≈ 1.666667

$V - V_1 = m(t - t_1)$

$V - 25 = 1.666667(t - 1)$

$V - 25 = 1.666667t - 1.666667$

$V = 1.666667t + 23.333333$ (Secant line)

The average rate of change between year 1 and year 4 is the slope of the secant line, \$1.666667 million per year.

For (1, V(1)) and (3, V(3)):

From part b), we know V(1) = 25.

$V(3) = 5\cdot 3^3 - 30\cdot 3^2 + 45\cdot 3 + 5\sqrt{3}$

≈ 8.660254

$m = \frac{V(3) - V(1)}{3 - 1}$

$= \frac{8.660254 - 25}{3 - 1}$

$= \frac{-16.339746}{2}$

$= -8.169873$

$V - V_1 = m(t - t_1)$

$V - 25 = -8.169873(t - 1)$

$V - 25 = -8.169873t + 8.169873$

$V = -8.169873t + 33.169873$ (Secant line)

The average rate of change between year 1 and year 3 is the slope of the secant line, -\$8.169873 million per year.

For (1, V(1)) and (1.5, V(1.5)):

From part b), we know V(1) = 25.

$V(1.5) = 5(1.5)^3 - 30(1.5)^2 + 45(1.5) + 5\sqrt{1.5}$

≈ 22.998724

$m = \frac{V(1.5) - V(1)}{1.5 - 1}$

$= \frac{22.998724 - 25}{1.5 - 1}$

$= \frac{-2.001276}{0.5}$

$= -4.002551$

$V - V_1 = m(t - t_1)$

$V - 25 = -4.002551(t - 1)$

$V - 25 = -4.002551t + 4.002551$

$V = -4.002551t + 29.002551$

(Secant line)

The average rate of change between year 1 and year 1.5 is the slope of the secant line, -\$4.002551 million per year.

e) At the point (1, V(1)) the slope of the tangent line to the graph appears to be 0.

Exercise Set 2.5

1. $y = x^7$

$\frac{dy}{dx} = \frac{d}{dx} x^7$

$= 7x^{7-1}$ (Theorem 1)

$= 7x^6$

3. $y = 15$ (Constant function)

$\frac{dy}{dx} = \frac{d}{dx} 15$

$= 0$ (Theorem 2)

5. $y = 4x^{150}$

$\frac{dy}{dx} = \frac{d}{dx} 4x^{150}$

$= 4 \cdot \frac{d}{dx} x^{150}$ (Theorem 3)

$= 4\cdot 150x^{150-1}$ (Theorem 1)

$= 600x^{149}$

7. $y = x^3 + 3x^2$

$\frac{dy}{dx} = \frac{d}{dx} (x^3 + 3x^2)$

$= \frac{d}{dx} x^3 + \frac{d}{dx} 3x^2$ (Theorem 4)

$= \frac{d}{dx} x^3 + 3 \cdot \frac{d}{dx} x^2$ (Theorem 3)

$= 3x^{3-1} + 3\cdot 2x^{2-1}$ (Theorem 1)

$= 3x^2 + 6x$

9. $y = 8\sqrt{x}$

$y = 8x^{1/2}$

$\frac{dy}{dx} = \frac{d}{dx} 8x^{1/2}$

$= 8 \cdot \frac{d}{dx} x^{1/2}$ (Theorem 3)

$= 8 \cdot \frac{1}{2} x^{1/2-1}$ (Theorem 1)

$= 4x^{-1/2}$, or $\frac{4}{x^{1/2}}$, or $\frac{4}{\sqrt{x}}$

11. $y = x^{0.07}$

$\frac{dy}{dx} = \frac{d}{dx} x^{0.07}$

$= 0.07x^{0.07-1}$ (Theorem 1)

$= 0.07x^{-0.93}$, or $\frac{0.07}{x^{0.93}}$

13. $y = \frac{1}{2} x^{4/5}$

$\frac{dy}{dx} = \frac{d}{dx} \frac{1}{2} x^{4/5}$

$= \frac{1}{2} \cdot \frac{d}{dx} x^{4/5}$ (Theorem 3)

$= \frac{1}{2} \cdot \frac{4}{5} x^{4/5-1}$ (Theorem 1)

$= \frac{2}{5} x^{-1/5}$, or $\frac{2}{5x^{1/5}}$, or $\frac{2}{5\sqrt[5]{x}}$

15. $y = x^{-3}$

$\frac{dy}{dx} = \frac{d}{dx} x^{-3}$

$= -3x^{-3-1}$ (Theorem 1)

$= -3x^{-4}$, or $-\frac{3}{x^4}$

17. $y = 3x^2 - 8x + 7$

$\frac{dy}{dx} = \frac{d}{dx} (3x^2 - 8x + 7)$

$= \frac{d}{dx} 3x^2 - \frac{d}{dx} 8x + \frac{d}{dx} 7$ (Theorem 4)

$= 3 \cdot \frac{d}{dx} x^2 - 8 \cdot \frac{d}{dx} x + \frac{d}{dx} 7$ (Theorem 3)

$= 3 \cdot 2x^{2-1} - 8 \cdot 1x^{1-1} + 0$ (Theorem 1) (Theorem 2)

$= 6x - 8$ ($x^{1-1} = x^0 = 1$)

19. $y = \sqrt[4]{x} - \frac{1}{x}$

$y = x^{1/4} - x^{-1}$

$\frac{dy}{dx} = \frac{d}{dx} (x^{1/4} - x^{-1})$

$= \frac{d}{dx} x^{1/4} - \frac{d}{dx} x^{-1}$ (Theorem 4)

$= \frac{1}{4} x^{1/4-1} - (-1)x^{-1-1}$ (Theorem 1)

$= \frac{1}{4} x^{-3/4} + x^{-2}$, or $\frac{1}{4\sqrt[4]{x^3}} + \frac{1}{x^2}$

21. $f(x) = 0.64x^{2.5}$

$\frac{d}{dx} f(x) = \frac{d}{dx} 0.64x^{2.5}$

$= 0.64 \cdot \frac{d}{dx} x^{2.5}$ (Theorem 3)

$= 0.64(2.5x^{2.5-1})$ (Theorem 1)

$= 1.6x^{1.5}$

$f'(x) = 1.6x^{1.5}$

23. $f(x) = \frac{5}{x} - x$

$\frac{d}{dx} f(x) = \frac{d}{dx} (5x^{-1} - x)$

$= \frac{d}{dx} 5x^{-1} - \frac{d}{dx} x$ (Theorem 4)

$= 5 \cdot \frac{d}{dx} x^{-1} - \frac{d}{dx} x$ (Theorem 3)

$= 5 \cdot (-1)x^{-1-1} - 1x^{1-1}$ (Theorem 1)

$= -5x^{-2} - 1$ ($x^{1-1} = x^0 = 1$)

$f'(x) = -5x^{-2} - 1$, or $\frac{-5}{x^2} - 1$

25. $f(x) = 4x - 7$

$\frac{d}{dx} f(x) = \frac{d}{dx} (4x - 7)$

$= \frac{d}{dx} 4x - \frac{d}{dx} 7$ (Theorem 4)

$= 4 \cdot \frac{d}{dx} x - \frac{d}{dx} 7$ (Theorem 3)

$= 4 \cdot 1 - 0$

(Theorem 1, $\frac{d}{dx} x = 1 \cdot x^{1-1} = x^0 = 1$)

(Theorem 2, $\frac{d}{dx} 7 = 0$)

$= 4$

$f'(x) = 4$

27. $f(x) = 4x + 9$

$\frac{d}{dx} f(x) = \frac{d}{dx} (4x + 9)$

$= \frac{d}{dx} 4x + \frac{d}{dx} 9$ (Theorem 4)

$= 4 \cdot \frac{d}{dx} x + \frac{d}{dx} 9$ (Theorem 3)

$= 4 \cdot 1 + 0$ (Theorems 1 and 2)

$= 4$

$f'(x) = 4$

29. $f(x) = \frac{x^4}{4} = \frac{1}{4} x^4$

$\frac{d}{dx} f(x) = \frac{d}{dx} \frac{1}{4} x^4$

$= \frac{1}{4} \cdot \frac{d}{dx} x^4$ (Theorem 3)

$= \frac{1}{4} \cdot 4x^{4-1}$ (Theorem 1)

$= x^3$

$f'(x) = x^3$

31. $f(x) = -0.01x^2 - 0.5x + 70$

$$\frac{d}{dx} f(x) = \frac{d}{dx}(-0.01x^2 - 0.5x + 70)$$

$$= \frac{d}{dx}(-0.01x^2) - \frac{d}{dx} 0.5x + \frac{d}{dx} 70$$

(Theorem 4)

$$= -0.01 \cdot \frac{d}{dx} x^2 - 0.5 \cdot \frac{d}{dx} x + \frac{d}{dx} 70$$

(Theorem 3)

$$= -0.01 \cdot 2x - 0.5 \cdot 1 + 0$$

(Theorems 1 and 2)

$$= -0.02x - 0.5$$

$$f'(x) = -0.02x - 0.5$$

33. $f(x) = 3x^{-2/3} + x^{3/4} + x^{6/5} + \frac{8}{x^3}$

$$= 3x^{-2/3} + x^{3/4} + x^{6/5} + 8x^{-3}$$

$$f'(x) = 3 \cdot (-\tfrac{2}{3}) \cdot x^{-2/3-1} + \tfrac{3}{4} \cdot x^{3/4-1} + \tfrac{6}{5} \cdot x^{6/5-1} + 8 \cdot (-3) \cdot x^{-3-1}$$

(Theorems 1 and 3)

$$= -2x^{-5/3} + \tfrac{3}{4} x^{-1/4} + \tfrac{6}{5} x^{1/5} - 24x^{-4}$$

35. $f(x) = \frac{2}{x} - \frac{x}{2} = 2x^{-1} - \frac{1}{2}x$

$$f'(x) = 2(-1)x^{-1-1} - \tfrac{1}{2} \cdot 1$$

$$= -2x^{-2} - \tfrac{1}{2}, \text{ or } -\frac{2}{x^2} - \frac{1}{2}$$

37. $f(x) = \frac{16}{x} - \frac{8}{x^3} + \frac{1}{x^4}$

$$= 16x^{-1} - 8x^{-3} + x^{-4}$$

$$f'(x) = 16(-1)x^{-1-1} - 8(-3)x^{-3-1} + (-4)x^{-4-1}$$

$$= -16x^{-2} + 24x^{-4} - 4x^{-5}$$

39. $f(x) = \sqrt{x} + \sqrt[3]{x} - \sqrt[4]{x} + \sqrt[5]{x}$

$$= x^{1/2} + x^{1/3} - x^{1/4} + x^{1/5}$$

$$f'(x) = \tfrac{1}{2}x^{1/2-1} + \tfrac{1}{3}x^{1/3-1} - \tfrac{1}{4}x^{1/4-1} + \tfrac{1}{5}x^{1/5-1}$$

$$= \tfrac{1}{2}x^{-1/2} + \tfrac{1}{3}x^{-2/3} - \tfrac{1}{4}x^{-3/4} + \tfrac{1}{5}x^{-4/5}$$

41. $f(x) = x^3 - 2x + 1$

$$f'(x) = 3x^2 - 2 \cdot 1 + 0 = 3x^2 - 2$$

At (2,5), $f'(x) = f'(2) = 3 \cdot 2^2 - 2 = 3 \cdot 4 - 2 = 12 - 2 = 10$, so we have

$$y - y_1 = m(x - x_1)$$
$$y - 5 = 10(x - 2)$$
$$y - 5 = 10x - 20$$
$$y = 10x - 15$$

At (-1,2), $f'(x) = f'(-1) = 3(-1)^2 - 2 = 3 \cdot 1 - 2 = 3 - 2 = 1$, so we have

$$y - y_1 = m(x - x_1)$$
$$y - 2 = 1[x - (-1)]$$
$$y - 2 = x + 1$$
$$y = x + 3$$

At (0,1), $f'(x) = f'(0) = 3 \cdot 0^2 - 2 = 0 - 2 = -2$, so we have

$$y - y_1 = m(x - x_1)$$
$$y - 1 = -2(x - 0)$$
$$y - 1 = -2x$$
$$y = -2x + 1$$

43. $y = x^2$ so $\frac{dy}{dx} = 2x$

A horizontal line has slope 0. We first find the values of x for which dy/dx = 0. That is, we want to find x such that 2x = 0.

Solve: $2x = 0$

$x = 0$ (Multiplying by $\frac{1}{2}$)

Next we find the point on the graph. We determine the second coordinate from the original equation $y = x^2$.

For $x = 0$, $y = 0^2 = 0$.

At the point (0,0), there is a horizontal tangent.

45. $y = -x^3$ so $\frac{dy}{dx} = -3x^2$

A horizontal tangent has slope 0. We first find the values of x for which dy/dx = 0. That is, we want to find x such that $-3x^2 = 0$.

Solve: $-3x^2 = 0$

$x^2 = 0$ (Multiplying by $-\frac{1}{3}$)

$x = 0$ (Taking square root)

Next we find the point on the graph. We determine the second coordinate from the original equation $y = -x^3$.

For $x = 0$, $y = -0^3 = 0$.

At the point (0,0), there is a horizontal tangent.

47. $y = 3x^2 - 5x + 4$

$$\frac{dy}{dx} = 6x - 5$$

A horizontal tangent has slope 0. We first find the values of x for which dy/dx = 0. That is, we want to find x such that 6x - 5 = 0.

Solve: $6x - 5 = 0$

$6x = 5$

$x = \frac{5}{6}$

Next we find the point on the graph. We determine the second coordinate from the original equation $y = 3x^2 - 5x + 4$.

For $x = \frac{5}{6}$, $y = 3(\frac{5}{6})^2 - 5(\frac{5}{6}) + 4$

$= 3(\frac{25}{36}) - (\frac{25}{6}) + 4$

$= \frac{75}{36} - \frac{25}{6} \cdot \frac{6}{6} + 4 \cdot \frac{36}{36}$

$= \frac{75}{36} - \frac{150}{36} + \frac{144}{36}$

$= \frac{69}{36}$

$= \frac{23}{12}$

At the point $(\frac{5}{6}, \frac{23}{12})$, there is a horizontal tangent.

The result is illustrated in the following graph.

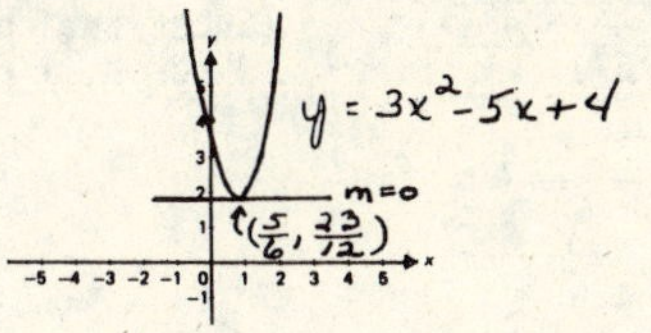

49. $y = -0.01x^2 - 0.5x + 70$

$\frac{dy}{dx} = -0.02x - 0.5$

First find the values of x for which $dy/dx = 0$. That is, we want to find x such that $-0.02x - 0.5 = 0$.

Solve: $-0.02x - 0.5 = 0$

$-0.02x = 0.5$

$x = \frac{0.5}{-0.02}$

$x = -25$

Next we find the point on the graph. We determine the second coordinate from the original equation $y = -0.01x^2 - 0.5x + 70$.

For $x = -25$, $y = -0.01(-25)^2 - 0.5(-25) + 70$

$= -0.01(625) + 12.5 + 70$

$= -6.25 + 12.5 + 70$

$= 76.25$

At the point (-25, 76.25), there is a horizontal tangent.

51. $y = 2x + 4$ (Linear function)

$\frac{dy}{dx} = 2$ (Slope is 2)

For all values of x, $dy/dx = 2$. There are no values of x for which $dy/dx = 0$. Thus, there are no points on the graph at which there is a horizontal tangent.

53. $y = 4$ (Horizontal line)

$\frac{dy}{dx} = 0$ (Slope is 0)

For all values of x, $dy/dx = 0$. The tangent is horizontal at all points on the graph.

55. $y = -x^3 + x^2 + 5x - 1$

$\frac{dy}{dx} = -3x^2 + 2x + 5$

First find the values of x for which $dy/dx = 0$. That is, we want to find x such that $-3x^2 + 2x + 5 = 0$. We can simplify our work by multiplying by -1. We get $3x^2 - 2x - 5 = 0$.

We factor and solve:

$3x^2 - 2x - 5 = 0$

$(3x - 5)(x + 1) = 0$ (Factoring)

$3x - 5 = 0$ or $x + 1 = 0$ (Principle of zero products)

$3x = 5$ or $x = -1$

$x = \frac{5}{3}$ or $x = -1$

Next we find the points on the graph. We determine the second coordinate from the original equation $y = -x^3 + x^2 + 5x - 1$.

For $x = \frac{5}{3}$, $y = -(\frac{5}{3})^3 + (\frac{5}{3})^2 + 5 \cdot \frac{5}{3} - 1$

$= -\frac{125}{27} + \frac{25}{9} + \frac{25}{3} - 1$

$= -\frac{125}{27} + \frac{75}{27} + \frac{225}{27} - \frac{27}{27}$

$= \frac{148}{27}$

For $x = -1$, $y = -(-1)^3 + (-1)^2 + 5(-1) - 1$

$= 1 + 1 - 5 - 1$

$= -4$

At the points $(\frac{5}{3}, \frac{148}{27})$ and (-1,-4), there are horizontal tangents.

The result is illustrated in the following graph.

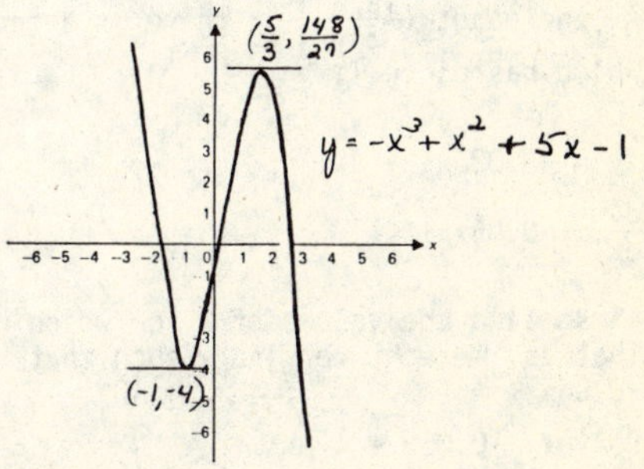

57. $y = \frac{1}{3}x^3 - 3x + 2$

$\frac{dy}{dx} = x^2 - 3$

First find the values of x for which $dy/dx = 0$. That is, we want to find x such that $x^2 - 3 = 0$.

Solve: $x^2 - 3 = 0$

$x^2 = 3$

$x = \pm\sqrt{3}$

Next we find the points on the graph. We determine the second coordinate from the original equation $y = \frac{1}{3}x^3 - 3x + 2$.

For $x = \sqrt{3}$, $y = \frac{1}{3}(\sqrt{3})^3 - 3\sqrt{3} + 2$

$= \frac{1}{3}(3\sqrt{3}) - 3\sqrt{3} + 2$

$= \sqrt{3} - 3\sqrt{3} + 2$

$= 2 - 2\sqrt{3}$

For $x = -\sqrt{3}$, $y = \frac{1}{3}(-\sqrt{3})^3 - 3(-\sqrt{3}) + 2$

$= \frac{1}{3}(-3\sqrt{3}) - 3(-\sqrt{3}) + 2$

$= -\sqrt{3} + 3\sqrt{3} + 2$

$= 2 + 2\sqrt{3}$

At the points $(\sqrt{3}, 2 - 2\sqrt{3})$ and $(-\sqrt{3}, 2 + 2\sqrt{3})$, there are horizontal tangents.

59. $y = 20x - x^2$

$\frac{dy}{dx} = 20 - 2x$

First find the values of x for which dy/dx = 1. That is, we want to find x such that 20 - 2x = 1.

Solve: $20 - 2x = 1$

$-2x = -19$

$x = \frac{19}{2}$

Next we find the point on the graph. We determine the second coordinate from the original equation $y = 20x - x^2$.

For $x = \frac{19}{2}$, $y = 20(\frac{19}{2}) - (\frac{19}{2})^2$

$= 190 - \frac{361}{4}$

$= \frac{399}{4}$

At the point $(\frac{19}{2}, \frac{399}{4})$, there is a tangent line which has slope 1.

61. $y = -0.025x^2 + 4x$

$\frac{dy}{dx} = -0.05x + 4$

First find the values of x for which dy/dx = 1. That is, we want to find x such that $-0.05x + 4 = 1$.

Solve: $-0.05x + 4 = 1$

$-0.05x = -3$

$x = 60$

Next we find the point on the graph. We determine the second coordinate from the original equation $y = -0.025x^2 + 4x$.

For $x = 60$, $y = -0.025\cdot 60^2 + 4\cdot 60$

$= -0.025\cdot 3600 + 240$

$= -90 + 240$

$= 150$

At the point (60,150), there is a tangent line which has slope 1.

63. $y = \frac{1}{3}x^3 + 2x^2 + 2x$

$\frac{dy}{dx} = x^2 + 4x + 2$

First find the values of x for which dy/dx = 1. That is, we want to find x such that $x^2 + 4x + 2 = 1$.

Solve: $x^2 + 4x + 2 = 1$

$x^2 + 4x + 1 = 0$ (Adding -1)

This is a quadratic equation, not readily factorable, so we use the quadratic formula where a = 1, b = 4, and c = 1.

$x = \frac{-b \pm \sqrt{b^2 - 4ac}}{2a}$

$x = \frac{-4 \pm \sqrt{4^2 - 4\cdot 1\cdot 1}}{2\cdot 1}$ (Substituting 1 for a, 4 for b, 1 for c)

$= \frac{-4 \pm \sqrt{16 - 4}}{2}$

$= \frac{-4 \pm \sqrt{12}}{2}$

$= \frac{-4 \pm 2\sqrt{3}}{2}$ $(\sqrt{12} = \sqrt{4\cdot 3} = 2\sqrt{3})$

$= \frac{2(-2 \pm \sqrt{3})}{2}$

$= -2 \pm \sqrt{3}$

Next we find the points on the graph. We determine the second coordinate from the original equation $y = \frac{1}{3}x^3 + 2x^2 + 2x$.

For $x = -2 + \sqrt{3}$,

$y = \frac{1}{3}(-2 + \sqrt{3})^3 + 2(-2 + \sqrt{3})^2 + 2(-2 + \sqrt{3})$

$= \frac{1}{3}(-26 + 15\sqrt{3}) + 2(7 - 4\sqrt{3}) - 4 + 2\sqrt{3}$

$= -\frac{26}{3} + 5\sqrt{3} + 14 - 8\sqrt{3} - 4 + 2\sqrt{3}$

$= \frac{4}{3} - \sqrt{3}$

For $x = -2 - \sqrt{3}$,

$y = \frac{1}{3}(-2 - \sqrt{3})^3 + 2(-2 - \sqrt{3})^2 + 2(-2 - \sqrt{3})$

$= \frac{1}{3}(-26 - 15\sqrt{3}) + 2(7 + 4\sqrt{3}) - 4 - 2\sqrt{3}$

$= -\frac{26}{3} - 5\sqrt{3} + 14 + 8\sqrt{3} - 4 - 2\sqrt{3}$

$= \frac{4}{3} + \sqrt{3}$

At the points $(-2 + \sqrt{3}, \frac{4}{3} - \sqrt{3})$ and $(-2 - \sqrt{3}, \frac{4}{3} + \sqrt{3})$, there are tangent lines which have slope 1.

65. $y = x^4 - \frac{4}{3}x^2 - 4$

$\frac{dy}{dx} = 4x^3 - \frac{8}{3}x$

First find the values of x for which dy/dx = 0. That is, we want to find x such that $4x^3 - \frac{8}{3}x = 0$.

Solve: $4x^3 - \frac{8}{3}x = 0$

$12x^3 - 8x = 0$

$4x(3x^2 - 2) = 0$

$4x = 0$ or $3x^2 - 2 = 0$

$x = 0$ or $x^2 = \frac{2}{3}$

$x = 0$ or $x = \pm\sqrt{\frac{2}{3}}$

Next we find the points on the graph. We determine the second coordinate from the original equation $y = x^4 - \frac{4}{3}x^2 - 4$.

For $x = 0$, $y = 0^4 - \frac{4}{3}\cdot 0^2 - 4 = -4$

For $x = \pm\sqrt{\frac{2}{3}}$, $y = \left(\pm\sqrt{\frac{2}{3}}\right)^4 - \frac{4}{3}\left(\pm\sqrt{\frac{2}{3}}\right)^2 - 4$

$= \frac{4}{9} - \frac{4}{3}\cdot\frac{2}{3} - 4$

$= \frac{4}{9} - \frac{8}{9} - \frac{36}{9}$

$= -\frac{40}{9}$

At the points $(0, -4)$, $\left(\sqrt{\frac{2}{3}}, -\frac{40}{9}\right)$, and $\left(-\sqrt{\frac{2}{3}}, -\frac{40}{9}\right)$, there are horizontal tangents.

67. $y = x(x - 1) = x^2 - x$

$\frac{dy}{dx} = 2x - 1$

69. $y = (x - 2)(x + 3) = x^2 + x - 6$

$\frac{dy}{dx} = 2x + 1$

71. $y = \frac{x^5 + x}{x^2} = \frac{x^5}{x^2} + \frac{x}{x^2} = x^3 + x^{-1}$

$\frac{dy}{dx} = 3x^2 - x^{-2}$, or $3x^2 - \frac{1}{x^2}$

73. $y = (-4x)^3 = -64x^3$

$\frac{dy}{dx} = -192x^2$

75. $y = \sqrt[3]{8x} = \sqrt[3]{8}\,\sqrt[3]{x} = 2\sqrt[3]{x} = 2x^{1/3}$

$\frac{dy}{dx} = \frac{2}{3}x^{-2/3}$, or $\frac{2}{3x^{2/3}}$, or $\frac{2}{3\sqrt[3]{x^2}}$

77. $y = (x + 1)^3$

$= x^3 + 3\cdot x^2\cdot 1 + 3\cdot x\cdot 1^2 + 1^3$

$= x^3 + 3x^2 + 3x + 1$

$\frac{dy}{dx} = 3x^2 + 6x + 3$

79. See the answer section in the text.

81. a) $f(x) = 5x^3 - 30x^2 + 45x + 5\sqrt{x}$

$= 5x^3 - 30x^2 + 45x + 5x^{1/2}$

$f'(x) = 15x^2 - 60x + 45 + 5\cdot\frac{1}{2}x^{-1/2}$

$= 15x^2 - 60x + 45 + \frac{5}{2\sqrt{x}}$

The graphs are shown in the answer section in the text.

b) See the answer section in the text.

Exercise Set 2.6

1. $s(t) = t^3 + t$ (Distance function, s in feet and t in seconds)

a) $v(t) = s'(t) = 3t^2 + 1$

b) $a(t) = v'(t) = 6t$

c) $v(4) = 3\cdot 4^2 + 1$

$= 3\cdot 16 + 1$

$= 48 + 1$

$= 49$

The velocity when t = 4 sec is 49 $\frac{ft}{sec}$.

$a(4) = 6\cdot 4 = 24$

The acceleration when t = 4 sec is 24 $\frac{ft}{sec^2}$.

3. $R(x) = 5x$

$C(x) = 0.001x^2 + 1.2x + 60$

a) $P(x) = R(x) - C(x)$

$= 5x - (0.001x^2 + 1.2x + 60)$

$= 5x - 0.001x^2 - 1.2x - 60$

$= -0.001x^2 + 3.8x - 60$

b) $R(x) = 5x$

$R(100) = 5\cdot 100 = 500$ (Substituting 100)

The total revenue from the sale of the first 100 units is $500.

$C(x) = 0.001x^2 + 1.2x + 60$

$C(100) = 0.001(100)^2 + 1.2(100) + 60$ (Substituting 100)

$= 0.001(10{,}000) + 1.2(100) + 60$

$= 10 + 120 + 60$

$= 190$

The total cost of producing the first 100 units is $190.

$P(x) = R(100) - C(100)$

$P(100) = 500 - 190$

$= 310$

The total profit from the production and sale of the first 100 units is \$310.

c) $R(x) = 5x$

so

$R'(x) = 5$

$C(x) = 0.001x^2 + 1.2x + 60$

so

$C'(x) = 0.002x + 1.2$

$P(x) = -0.001x^2 + 3.8x - 60$

so

$P'(x) = -0.002x + 3.8$

d) $R'(x) = 5$

$R'(100) = \$5$ per unit (Substituting 100)

$C'(x) = 0.002x + 1.2$

$C'(100) = 0.002(100) + 1.2$

$= 0.2 + 1.2$

$= \$1.4$, or \$1.40 per unit

$P'(100) = R'(100) - C'(100)$

$= \$5.00 - \$1.40$

$= \$3.60$ per unit

5. $N(a) = -a^2 + 300a + 6$

a) The rate of change is given by $N'(a)$.

$N'(a) = -2a + 300$

b) Since a is measured in thousands of dollars, we find $N(10)$.

$N(10) = -(10)^2 + 300(10) + 6$

$= -100 + 3000 + 6$

$= 2906$

After spending \$10,000 on advertising, 2906 units will be sold.

c) $N'(10) = -2\cdot 10 + 300$

$= -20 + 300$

$= 280$

The rate of change is 280 units/thousand dollars.

7. $M(t) = -2t^2 + 100t + 180$

a) $M(5) = -2\cdot 5^2 + 100\cdot 5 + 180$

$= -2\cdot 25 + 500 + 180$

$= -50 + 500 + 180$

$= 630$

The productivity is 630 units per month after 5 years of service.

$M(10) = -2\cdot 10^2 + 100\cdot 10 + 180$

$= -2\cdot 100 + 1000 + 180$

$= -200 + 1000 + 180$

$= 980$

The productivity is 980 units per month after 10 years of service.

$M(25) = -2\cdot 25^2 + 100\cdot 25 + 180$

$= -2\cdot 625 + 2500 + 180$

$= -1250 + 2500 + 180$

$= 1430$

The productivity is 1430 units per month after 25 years of service.

$M(45) = -2\cdot 45^2 + 100\cdot 45 + 180$

$= -2\cdot 2025 + 4500 + 180$

$= -4050 + 4500 + 180$

$= 630$

The productivity is 630 units per month after 45 years of service.

b) The marginal productivity is given by $M'(t)$.

$M'(t) = -4t + 100$

c) $M'(5) = -4\cdot 5 + 100 = -20 + 100 = 80$

At $t = 5$ the monthly marginal productivity is 80 units per year.

$M'(10) = -4\cdot 10 + 100 = -40 + 100 = 60$

At $t = 10$ the monthly marginal productivity is 60 units per year.

$M'(25) = -4\cdot 25 + 100 = -100 + 100 = 0$

At $t = 25$ the monthly marginal productivity is 0 units per year.

$M'(45) = -4\cdot 45 + 100 = -180 + 100 = -80$

At $t = 45$ the monthly marginal productivity is -80 units per year.

9. $D(p) = 100 - \sqrt{p} = 100 - p^{1/2}$

a) $\frac{dD}{dp} = -\frac{1}{2}p^{-1/2}$, or $-\frac{1}{2p^{1/2}}$, or $-\frac{1}{2\sqrt{p}}$

b) $D(25) = 100 - \sqrt{25} = 100 - 5 = 95$

The consumer will want to buy 95 units when the price is \$25 per unit.

c) At $p = 25$, $\frac{dD}{dp} = -\frac{1}{2\sqrt{25}} = -\frac{1}{2\cdot 5} = -\frac{1}{10}$ units per dollar

11. $C(r) = 2\pi r$ (Linear function, 2π is a constant)

The rate of change of the circumference with respect to the radius is given by $C'(r)$.

$C'(r) = 2\pi$

13. $T(t) = -0.1t^2 + 1.2t + 98.6$

[T is temperature (°F), and t is time (days)]

a) $T'(t) = -0.1 \cdot 2t + 1.2 + 0$

$= -0.2t + 1.2$

b) $T(t) = -0.1t^2 + 1.2t + 98.6$

$T(1.5) = -0.1(1.5)^2 + 1.2(1.5) + 98.6$ (Substituting 1.5 for t)

$= -0.1(2.25) + 1.2(1.5) + 98.6$

$= -0.225 + 1.8 + 98.6$

$= 100.175$

At 1.5 days, the temperature is 100.175°F.

c) $T'(t) = -0.2t + 1.2$

$T'(1.5) = -0.2(1.5) + 1.2$

$= -0.3 + 1.2$

$= 0.9$

The rate of change of the temperature with respect to time at 1.5 days is 0.9°F per day.

15. $T = W^{1.31}$

$\frac{dT}{dW} = 1.31W^{1.31-1}$

$= 1.31W^{0.31}$

17. $R(Q) = Q^2\left[\frac{k}{2} - \frac{Q}{3}\right] = \frac{k}{2}Q^2 - \frac{Q^3}{3} = \frac{k}{2}Q^2 - \frac{1}{3}Q^3$

$\frac{dR}{dQ} = \frac{k}{2} \cdot 2Q - \frac{1}{3} \cdot 3Q^2 = kQ - Q^2$

19. $A(t) = 0.08t + 19.7$

The rate of change of the median age A with respect to time t is given by $A'(t)$.

$A'(t) = 0.08$

Exercise Set 2.7

1. $y = x^3 \cdot x^8$, or x^{11}

Differentiate $y = x^3 \cdot x^8$ using the Product Rule, (Theorem 5).

$\frac{d}{dx} x^3 \cdot x^8 = x^3 \cdot 8x^7 + 3x^2 \cdot x^8$

$= 8x^{10} + 3x^{10}$

$= 11x^{10}$

Differentiate $y = x^{11}$ using the Power Rule (Theorem 1).

$\frac{d}{dx} x^{11} = 11x^{10}$

3. $y = \frac{-1}{x}$, or $-1 \cdot x^{-1}$

Differentiate $y = \frac{-1}{x}$ using the Quotient Rule (Theorem 6).

$\frac{d}{dx} \frac{-1}{x} = \frac{x \cdot 0 - 1 \cdot (-1)}{x^2}$

$= \frac{0 + 1}{x^2}$

$= \frac{1}{x^2}$, or x^{-2}

Differentiate $y = -1 \cdot x^{-1}$ using the Power Rule (Theorem 1) and Theorem 3.

$\frac{d}{dx} -1 \cdot x^{-1} = -1 \cdot \frac{d}{dx} x^{-1}$ (Theorem 3)

$= -1 \cdot (-1)x^{-1-1}$ (Theorem 1)

$= 1 \cdot x^{-2}$

$= x^{-2}$, or $\frac{1}{x^2}$

5. $y = \frac{x^8}{x^5}$, or x^3

Differentiate $y = \frac{x^8}{x^5}$ using the Quotient Rule (Theorem 6).

$\frac{d}{dx} \frac{x^8}{x^5} = \frac{x^5 \cdot 8x^7 - 5x^4 \cdot x^8}{(x^5)^2}$

$= \frac{8x^{12} - 5x^{12}}{x^{10}}$

$= \frac{3x^{12}}{x^{10}}$

$= 3x^2$

Differentiate $y = x^3$ using the Power Rule (Theorem 1).

$\frac{d}{dx} x^3 = 3x^2$

7. $y = (8x^5 - 3x^2 + 20)(8x^4 - 3\sqrt{x})$

$= (8x^5 - 3x^2 + 20)(8x^4 - 3x^{1/2})$

Using the Product Rule, we have

$$\frac{dy}{dx} = \underset{①}{(8x^5 - 3x^2 + 20)}\underset{②}{(32x^3 - \frac{3}{2}x^{-1/2})} \underset{⑤}{+} \underset{③}{(40x^4 - 6x)}\underset{④}{(8x^4 - 3x^{1/2})}$$

Steps to follow:

① Write down the first factor.

② Multiply it by the derivative of the second factor.

③ Write down the derivative of the first factor.

④ Multiply it by the second factor.

⑤ Add the result of 1 and 2 to the result of 3 and 4 .

The derivative could also be expressed using radical symbols.

$$\frac{dy}{dx} = (8x^5 - 3x^2 + 20)(32x^3 - \frac{3}{2\sqrt{x}}) +$$
$$(40x^4 - 6x)(8x^4 - 3\sqrt{x})$$

9. $f(x) = x(300 - x)$

Using the Product Rule, we have

① ②⑤③ ④

$$f'(x) = x\cdot(-1) + 1\cdot(300 - x)$$

(Follow steps listed below)

$$= -x + 300 - x$$
$$= 300 - 2x$$

Write down:

① First factor

② Derivative of second factor

③ Derivative of first factor

④ Second factor

⑤ Plus sign

We could have multiplied the factors and then differentiated, avoiding the Product Rule.

$$f(x) = x(300 - x)$$
$$= 300x - x^2$$
$$f'(x) = 300 - 2x$$

11. $f(x) = (4\sqrt{x} - 6)(x^3 - 2x + 4)$, or

$(4x^{1/2} - 6)(x^3 - 2x + 4)$

$f'(x) = (4x^{1/2} - 6)(3x^2 - 2) +$
$(2x^{-1/2})(x^3 - 2x + 4)$, or

$(4\sqrt{x} - 6)(3x^2 - 2) + \frac{2}{\sqrt{x}}(x^3 - 2x + 4)$

13. $f(x) = (x + 3)^2 = (x + 3)(x + 3)$

$$f'(x) = (x + 3)\cdot 1 + 1\cdot(x + 3)$$
$$= x + 3 + x + 3$$
$$= 2x + 6$$

15. $f(x) = (x^3 - 4x)^2 = (x^3 - 4x)(x^3 - 4x)$

$$f'(x) = (x^3 - 4x)(3x^2 - 4) + (3x^2 - 4)(x^3 - 4x)$$
$$= 3x^5 - 16x^3 + 16x + 3x^5 - 16x^3 + 16x$$
$$= 6x^5 - 32x^3 + 32x$$

17. $f(x) = 5x^{-3}(x^4 - 5x^3 + 10x - 2)$

$$f'(x) = 5x^{-3}(4x^3 - 15x^2 + 10) -$$
$$15x^{-4}(x^4 - 5x^3 + 10x - 2)$$

Simplifying, we get

$$f'(x) = 20 - 75x^{-1} + 50x^{-3} - 15 + 75x^{-1} - 150x^{-3} +$$
$$30x^{-4}$$
$$= 5 - 100x^{-3} + 30x^{-4}.$$

19. $f(x) = (x + \frac{2}{x})(x^2 - 3)$, or $(x + 2x^{-1})(x^2 - 3)$

$f'(x) = (x + 2x^{-1})(2x) + (1 - 2x^{-2})(x^2 - 3)$, or

$(x + \frac{2}{x})(2x) + (1 - \frac{2}{x^2})(x^2 - 3)$

Simplifying, we get

$$f'(x) = 2x^2 + 4 + x^2 - 5 + \frac{6}{x^2}$$
$$= 3x^2 + \frac{6}{x^2} - 1.$$

21. $f(x) = \frac{x}{300 - x}$

Using the Quotient Rule, we have

② ③④ ⑤⑥

$$f'(x) = \frac{(300 - x)\cdot 1 - (-1)\cdot x}{(300 - x)^2}$$

①

(Follow steps listed below)

$$= \frac{300 - x + x}{(300 - x)^2}$$
$$= \frac{300}{(300 - x)^2}$$

Steps to follow:

① Square the denominator.

② Write down the denominator.

③ Multiply the denominator by the derivative of the numerator.

④ Write a minus sign.

⑤ Write down the derivative of the denominator.

⑥ Multiply it by the numerator.

23. $f(x) = \frac{3x - 1}{2x + 5}$

Using the Quotient Rule, we have

② ③④⑤ ⑥

$$f'(x) = \frac{(2x + 5)\cdot 3 - 2\cdot(3x - 1)}{(2x + 5)^2}$$

①

(Follow steps listed below)

$$= \frac{6x + 15 - 6x + 2}{(2x + 5)^2}$$
$$= \frac{17}{(2x + 5)^2}$$

Steps to follow:

① Square the denominator.

② Write down the denominator.

③ Multiply the denominator by the derivative of the numerator.

④ Write a minus sign.

(5) Write down the derivative of the denominator.

(6) Multiply it by the numerator.

25. $y = \frac{x^2 + 1}{x^3 - 1}$

Using the Quotient Rule, we have

$$\frac{dy}{dx} = \frac{\overset{2}{(x^3 - 1)}\cdot\overset{3}{2x} \overset{4}{-} \overset{5}{3x^2}\cdot\overset{6}{(x^2 + 1)}}{\underset{1}{(x^3 - 1)^2}}$$

(Follow numbered steps listed in Exercise 23)

$$= \frac{2x^4 - 2x - 3x^4 - 3x^2}{(x^3 - 1)^2}$$

$$= \frac{-x^4 - 3x^2 - 2x}{(x^3 - 1)^2}$$

27. $y = \frac{x}{1 - x}$

Using the Quotient Rule, we have

$$\frac{dy}{dx} = \frac{(1 - x)\cdot 1 - (-1)\cdot x}{(1 - x)^2}$$

$$= \frac{1 - x + x}{(1 - x)^2}$$

$$= \frac{1}{(1 - x)^2}$$

29. $y = \frac{x - 1}{x + 1}$

Using the Quotient Rule, we have

$$\frac{dy}{dx} = \frac{(x + 1)\cdot 1 - 1\cdot(x - 1)}{(x + 1)^2}$$

$$= \frac{x + 1 - x + 1}{(x + 1)^2}$$

$$= \frac{2}{(x + 1)^2}$$

31. $f(x) = \frac{1}{x - 3}$

Using the Quotient Rule, we have

$$f'(x) = \frac{(x - 3)\cdot 0 - 1\cdot 1}{(x - 3)^2}$$

$$= \frac{-1}{(x - 3)^2}$$

33. $f(x) = \frac{3x^2 + 2x}{x^2 + 1}$

$$f'(x) = \frac{(x^2 + 1)\cdot(6x + 2) - 2x\cdot(3x^2 + 2x)}{(x^2 + 1)^2}$$

(Using the Quotient Rule)

$$= \frac{(6x^3 + 2x^2 + 6x + 2) - (6x^3 + 4x^2)}{(x^2 + 1)^2}$$

$$= \frac{6x^3 + 2x^2 + 6x + 2 - 6x^3 - 4x^2}{(x^2 + 1)^2}$$

$$= \frac{-2x^2 + 6x + 2}{(x^2 + 1)^2}$$

35. $f(x) = \frac{3x^2 - 5x}{x^8}$

Using the Quotient Rule:

$$f'(x) = \frac{x^8(6x - 5) - 8x^7(3x^2 - 5x)}{(x^8)^2}$$

$$= \frac{6x^9 - 5x^8 - 24x^9 + 40x^8}{x^{16}} \quad [(x^8)^2 = x^{16}]$$

$$= \frac{-18x^9 + 35x^8}{x^{16}}$$

$$= \frac{x^8(-18x + 35)}{x^8\cdot x^8} \quad \text{(Factoring both numerator and denominator)}$$

$$= \frac{-18x + 35}{x^8} \quad \text{(Simplifying)}$$

Dividing first:

$$f(x) = \frac{3x^2 - 5x}{x^8} = 3x^{-6} - 5x^{-7}$$

$$f'(x) = -18x^{-7} + 35x^{-8}$$

This is equivalent to the result we found using the Quotient Rule.

37. $g(x) = \frac{4x + 3}{\sqrt{x}}$, or $\frac{4x + 3}{x^{1/2}}$

$$g'(x) = \frac{\sqrt{x}(4) - \frac{1}{2}x^{-1/2}(4x + 3)}{(\sqrt{x})^2}$$

$$= \frac{4\sqrt{x} - \frac{1}{2}x^{-1/2}(4x + 3)}{x}$$

Simplifying, we get

$$g'(x) = \frac{4\sqrt{x} - \frac{4x + 3}{2\sqrt{x}}}{x}$$

$$= \frac{4\sqrt{x}\cdot\frac{2\sqrt{x}}{2\sqrt{x}} - \frac{4x + 3}{2\sqrt{x}}}{x} \quad \text{(Subtracting in the numerator)}$$

$$= \frac{\frac{8x - (4x + 3)}{2\sqrt{x}}}{x} = \frac{\frac{8x - 4x - 3}{2\sqrt{x}}}{x}$$

$$= \frac{\frac{4x - 3}{2\sqrt{x}}}{x} = \frac{4x - 3}{2\sqrt{x}}\cdot\frac{1}{x}$$

$$= \frac{4x - 3}{2x\sqrt{x}}, \text{ or } \frac{4x - 3}{2x^{3/2}}.$$

39. $y = \frac{8}{x^2 + 4}$

$$\frac{dy}{dx} = \frac{(x^2 + 4)(0) - 2x(8)}{(x^2 + 4)^2}$$

$$= \frac{-16x}{(x^2 + 4)^2}$$

When $x = 0$, $\frac{dy}{dx} = \frac{-16\cdot 0}{(0^2 + 4)^2} = 0$, so the slope of the tangent line at (0,2) is 0. We use the point-slope equation.

$$y - y_1 = m(x - x_1)$$

$$y - 2 = 0(x - 0)$$

$$y - 2 = 0$$

$$y = 2$$

When $x = -2$, $\frac{dy}{dx} = \frac{-16(-2)}{[(-2)^2 + 4]^2} = \frac{32}{64} = \frac{1}{2}$, so the slope of the tangent line at (-2,1) is $\frac{1}{2}$. We use the point-slope equation.

$$y - y_1 = m(x - x_1)$$
$$y - 1 = \frac{1}{2}[x - (-2)]$$
$$y - 1 = \frac{1}{2}(x + 2)$$
$$y - 1 = \frac{1}{2}x + 1$$
$$y = \frac{1}{2}x + 2$$

41. $D(p) = \frac{2p + 300}{10p + 11}$

a) $D'(p) = \frac{(10p + 11)(2) - 10(2p + 300)}{(10p + 11)^2}$

$= \frac{20p + 22 - 20p - 3000}{(10p + 11)^2}$

$= \frac{-2978}{(10p + 11)^2}$

b) $D'(4) = \frac{-2978}{(10 \cdot 4 + 11)^2} = \frac{-2978}{51^2} = -\frac{2978}{2601}$

43. $D(p) = 400 - p$

a) $R(p) = p \cdot D(p) = p(400 - p) = 400p - p^2$

b) $R'(p) = 400 - 2p$

45. $D(p) = \frac{4000}{p} + 3$

a) $R(p) = p \cdot D(p) = p\left[\frac{4000}{p} + 3\right] = 4000 + 3p$

b) $R'(p) = 3$

47. $A(x) = \frac{C(x)}{x}$ (Average cost)

$A'(x) = \frac{x \cdot C'(x) - 1 \cdot C(x)}{x^2}$ (Marginal average cost)

$= \frac{x\,C'(x) - C(x)}{x^2}$

49. $f(x) = \frac{x^3}{\sqrt{x} - 5}$, or $\frac{x^3}{x^{1/2} - 5}$

$f'(x) = \frac{(\sqrt{x} - 5) \cdot 3x^2 - \frac{1}{2}x^{-1/2} \cdot x^3}{(\sqrt{x} - 5)^2}$

(Using the Quotient Rule)

$= \frac{(\sqrt{x} - 5) \cdot 3x^2 - \frac{x^3}{2\sqrt{x}}}{(\sqrt{x} - 5)^2} \cdot \frac{2\sqrt{x}}{2\sqrt{x}}$

$= \frac{2\sqrt{x}(\sqrt{x} - 5) \cdot 3x^2 - x^3}{2\sqrt{x}(\sqrt{x} - 5)^2}$

$= \frac{(2x - 10\sqrt{x})3x^2 - x^3}{2\sqrt{x}(\sqrt{x} - 5)^2}$

$= \frac{6x^3 - 30x^2\sqrt{x} - x^3}{2\sqrt{x}(\sqrt{x} - 5)^2}$

$= \frac{5x^3 - 30x^2\sqrt{x}}{2\sqrt{x}(\sqrt{x} - 5)^2}$

51. $f(v) = \frac{3}{1 + v + v^2}$

$f'(v) = \frac{(1 + v + v^2) \cdot 0 - (1 + 2v) \cdot 3}{(1 + v + v^2)^2}$

(Using the Quotient Rule)

$= \frac{-3(1 + 2v)}{(1 + v + v^2)^2}$

53. $p(t) = \frac{t}{1 - t + t^2 - t^3}$

$p'(t) = \frac{(1 - t + t^2 - t^3) \cdot 1 - (-1 + 2t - 3t^2) \cdot t}{(1 - t + t^2 - t^3)^2}$

(Using the Quotient Rule)

$= \frac{1 - t + t^2 - t^3 + t - 2t^2 + 3t^3}{(1 - t + t^2 - t^3)^2}$

$= \frac{1 - t^2 + 2t^3}{(1 - t + t^2 - t^3)^2}$

55. $h(x) = \frac{x^3 + 5x^2 - 2}{\sqrt{x}}$, or $\frac{x^3 + 5x^2 - 2}{x^{1/2}}$

$h'(x) = \frac{\sqrt{x}\,(3x^2 + 10x) - \frac{1}{2}x^{-1/2}(x^3 + 5x^2 - 2)}{(\sqrt{x})^2}$

(Using the Quotient Rule)

$= \frac{\sqrt{x}\,(3x^2 + 10x) - \frac{x^3 + 5x^2 - 2}{2\sqrt{x}}}{x} \cdot \frac{2\sqrt{x}}{2\sqrt{x}}$

$= \frac{2x(3x^2 + 10x) - (x^3 + 5x^2 - 2)}{2x\sqrt{x}}$

$= \frac{6x^3 + 20x^2 - x^3 - 5x^2 + 2}{2x\sqrt{x}}$

$= \frac{5x^3 + 15x^2 + 2}{2x\sqrt{x}}$

57. $f(x) = x(3x^3 + 6x - 2)(3x^4 + 7)$

$= [x(3x^3 + 6x - 2)](3x^4 + 7)$ (Grouping the factors)

$f'(x) = [x(3x^3 + 6x - 2)] \cdot \frac{d}{dx}(3x^4 + 7) + \left\{\frac{d}{dx}[x(3x^3 + 6x - 2)]\right\} \cdot (3x^4 + 7)$

(Using the Product Rule on the entire expression)

$= [x(3x^3 + 6x - 2)](12x^3) + [x(9x^2 + 6) + 1 \cdot (3x^3 + 6x - 2)](3x^4 + 7)$

(Finding $\frac{d}{dx}(3x^4 + 7)$ and using the Product Rule on the expression in the brackets)

$= 12x^4(3x^3 + 6x - 2) + [x(9x^2 + 6) + (3x^3 + 6x - 2)](3x^4 + 7)$

59. $f(t) = (t^5 + 3) \cdot \dfrac{t^3 - 1}{t^3 + 1}$

$$f'(t) = (t^5 + 3) \cdot \frac{(t^3 + 1)\cdot 3t^2 - 3t^2(t^3 - 1)}{(t^3 + 1)^2} + 5t^4 \cdot \frac{t^3 - 1}{t^3 + 1}$$

$$= (t^5 + 3) \cdot \frac{3t^5 + 3t^2 - 3t^5 + 3t^2}{(t^3 + 1)^2} + 5t^4 \cdot \frac{t^3 - 1}{t^3 + 1}$$

$$= \frac{6t^2(t^5 + 3)}{(t^3 + 1)^2} + \frac{5t^4(t^3 - 1)}{t^3 + 1}$$

61. $f'(x) = \dfrac{(2x^2 + 3)(4x^3 - 7x + 2)}{x^7 - 2x^6 + 9}$

We will use the Quotient Rule. We must also use the Product Rule to find the derivative of the numerator.

$$f'(x) = \{(x^7 - 2x^6 + 9)[(2x^2 + 3)(12x^2 - 7) + 4x(4x^3 - 7x + 2)] - (7x^6-12x^5)(2x^2+3)(4x^3-7x+2)\}/(x^7-2x^6+9)^2$$

63. $f(x) = x(x + 2)(x - 2)$

$= x(x^2 - 4)$

$= x^3 - 4x$

$f'(x) = 3x^2 - 4$

See the answer section in the text for the result.

65. $f(x) = \left[x + \dfrac{2}{x}\right](x^2 - 3)$

$= x^3 - x - \dfrac{6}{x}$, or $x^3 - x - 6x^{-1}$

$f'(x) = 3x^2 - 1 + 6x^{-2}$, or $3x^2 - 1 + \dfrac{6}{x^2}$

See the answer section in the text for the result.

67. $f(x) = \dfrac{0.3x}{0.04 + x^2}$

$$f'(x) = \frac{(0.04 + x^2)(0.3) - 2x(0.3x)}{(0.04 + x^2)^2}$$

$$= \frac{0.012 + 0.3x^2 - 0.6x^2}{(0.04 + x^2)^2}$$

$$= \frac{0.012 - 0.3x^2}{(0.04 + x^2)^2}$$

See the answer section in the text for the result.

Exercise Set 2.8

1. $y = (1 - x)^{55}$

$\dfrac{dy}{dx} = 55(1 - x)^{54}\cdot(-1)$ (Using the Extended Power Rule)

$= -55(1 - x)^{54}$ [Don't forget: $\dfrac{d}{dx}(1 - x) = -1$]

3. $y = \sqrt{1 + 8x} = (1 + 8x)^{1/2}$

$\dfrac{dy}{dx} = \dfrac{1}{2}(1 + 8x)^{1/2-1} \cdot 8$ (Using the Extended Power Rule)

[Don't forget: $\dfrac{d}{dx}(1 + 8x) = 8$]

$= 4(1 + 8x)^{-1/2}$, or $\dfrac{4}{\sqrt{1 + 8x}}$

5. $y = \sqrt{3x^2 - 4} = (3x^2 - 4)^{1/2}$

$\dfrac{dy}{dx} = \dfrac{1}{2}(3x^2 - 4)^{1/2-1} \cdot 6x$ (Using the Extended Power Rule)

[Don't forget: $\dfrac{d}{dx}(3x^2 - 4) = 6x$]

$= 3x(3x^2 - 4)^{-1/2}$, or $\dfrac{3x}{\sqrt{3x^2 - 4}}$

7. $y = (3x^2 - 6)^{-40}$

$\dfrac{dy}{dx} = -40(3x^2 - 6)^{-40-1} \cdot (6x)$ (Using the Extended Power Rule)

[Don't forget: $\dfrac{d}{dx}(3x^2 - 6) = 6x$]

$= -240x(3x^2 - 6)^{-41}$, or $-\dfrac{240x}{(3x^2 - 6)^{41}}$

9. $y = x\sqrt{2x + 3} = x(2x + 3)^{1/2}$

$\dfrac{dy}{dx} = x \cdot \left[\dfrac{1}{2}(2x + 3)^{1/2-1} \cdot 2\right] + 1\cdot(2x + 3)^{1/2}$

(Using the Product Rule and the Extended Power Rule)

$= x(2x + 3)^{-1/2} + (2x + 3)^{1/2}$, or

$\dfrac{x}{\sqrt{2x + 3}} + \sqrt{2x + 3}$

The above answers are acceptable, but further simplification can be done.

$= \dfrac{x}{\sqrt{2x + 3}} + \sqrt{2x + 3} \cdot \dfrac{\sqrt{2x + 3}}{\sqrt{2x + 3}}$

(Multiplying the second term by a form of 1)

$= \dfrac{x}{\sqrt{2x + 3}} + \dfrac{2x + 3}{\sqrt{2x + 3}}$

$= \dfrac{3x + 3}{\sqrt{2x + 3}}$, or $\dfrac{3(x + 1)}{\sqrt{2x + 3}}$

11. $y = x^2\sqrt{x - 1} = x^2(x - 1)^{1/2}$

$\dfrac{dy}{dx} = x^2 \cdot \dfrac{1}{2}(x - 1)^{1/2-1} \cdot 1 + 2x(x - 1)^{1/2}$

(Using the Product Rule and the Extended Power Rule)

$= \dfrac{1}{2}x^2(x - 1)^{-1/2} + 2x(x - 1)^{1/2}$

$\dfrac{x^2}{2\sqrt{x - 1}} + 2x\sqrt{x - 1}$

The above answers are acceptable, but further simplification can be done.

$$= \frac{x^2}{2\sqrt{x - 1}} + 2x\sqrt{x - 1} \cdot \frac{2\sqrt{x - 1}}{2\sqrt{x - 1}}$$

(Multiplying the second term by a form of 1)

$$= \frac{x^2}{2\sqrt{x - 1}} + \frac{4x(x - 1)}{2\sqrt{x - 1}}$$

$$= \frac{x^2 + 4x^2 - 4x}{2\sqrt{x - 1}}$$

$$= \frac{5x^2 - 4x}{2\sqrt{x - 1}}$$

13. $y = \frac{1}{(3x + 8)^2} = (3x + 8)^{-2}$

$$\frac{dy}{dx} = -2(3x + 8)^{-2-1} \cdot 3$$

(Using the Extended Power Rule)

$$= -6(3x + 8)^{-3}, \text{ or } \frac{-6}{(3x + 8)^3}$$

15. $f(x) = (1 + x^3)^3 - (1 + x^3)^4$

$$f'(x) = 3(1 + x^3)^2 \cdot 3x^2 - 4(1 + x^3)^3 \cdot 3x^2$$

(Using the Difference Rule and the Extended Power Rule)

$$= 3x^2(1 + x^3)^2[3 - 4(1 + x^3)]$$

[Factoring out $3x^2(1 + x^3)^2$]

$$= 3x^2(1 + x^3)^2(3 - 4 - 4x^3)$$

$$= 3x^2(1 + x^3)^2(-1 - 4x^3), \text{ or}$$

$$-3x^2(1 + x^3)^2(1 + 4x^3)$$

17. $f(x) = x^2 + (200 - x)^2$

$$f'(x) = 2x + 2(200 - x)^1 \cdot (-1)$$

(Using the Sum Rule, the Power Rule, and the Extended Power Rule)

$$= 2x - 400 + 2x$$

$$= 4x - 400$$

19. $f(x) = (x + 6)^{10}(x - 5)^4$

$$f'(x) = (x+ 6)^{10} \cdot 4(x- 5)^3 \cdot 1 + 10(x+ 6)^9 \cdot 1 \cdot (x- 5)^4$$

(Using the Product Rule and the Extended Power Rule)

$$= 4(x + 6)^{10}(x - 5)^3 + 10(x + 6)^9(x - 5)^4$$

$$= 2(x + 6)^9(x - 5)^3[2(x + 6) + 5(x - 5)]$$

[Factoring out $2(x + 6)^9(x - 5)^3$]

$$= 2(x + 6)^9(x - 5)^3(2x + 12 + 5x - 25)$$

$$= 2(x + 6)^9(x - 5)^3(7x - 13)$$

21. $f(x) = (x - 4)^8(3 - x)^4$

$$f'(x) = (x - 4)^8 \cdot 4(3 - x)^3 \cdot (-1) + 8(x - 4)^7 \cdot 1 \cdot (3 - x)^4$$

(Using the Product Rule and the Extended Power Rule)

$$= -4(x - 4)^8(3 - x)^3 + 8(x - 4)^7(3 - x)^4$$

$$= 4(x - 4)^7(3 - x)^3[-(x - 4) + 2(3 - x)]$$

[Factoring out $4(x - 4)^7(3 - x)^3$]

$$= 4(x - 4)^7(3 - x)^3(-x + 4 + 6 - 2x)$$

$$= 4(x - 4)^7(3 - x)^3(10 - 3x)$$

23. $f(x) = -4x(2x - 3)^3$

$$f'(x) = -4x \cdot 3(2x - 3)^2 \cdot 2 + (-4) \cdot (2x - 3)^3$$

(Using the Product Rule and the Extended Power Rule)

$$= -24x(2x - 3)^2 - 4(2x - 3)^3$$

$$= -4(2x - 3)^2[6x + (2x - 3)]$$

[Factoring out $-4(2x - 3)^2$]

$$= -4(2x - 3)^2(8x - 3), \text{ or}$$

$$4(2x - 3)^2(3 - 8x)$$

25. $f(x) = \sqrt{\frac{1 - x}{1 + x}} = \left[\frac{1 - x}{1 + x}\right]^{1/2}$

$$f'(x) = \frac{1}{2}\left[\frac{1 - x}{1 + x}\right]^{1/2-1} \cdot \frac{(1 + x)(-1) - (1)(1 - x)}{(1 + x)^2}$$

(Using the Extended Power Rule and the Quotient Rule)

$$= \frac{1}{2}\left[\frac{1 - x}{1 + x}\right]^{-1/2} \cdot \frac{-1 - x - 1 + x}{(1 + x)^2}$$

$$= \frac{1}{2}\left[\frac{1 - x}{1 + x}\right]^{-1/2} \cdot \frac{-2}{(1 + x)^2}$$

$$= \left[\frac{1 - x}{1 + x}\right]^{-1/2} \cdot \frac{-1}{(1 + x)^2}$$

27. $f(x) = \left[\frac{3x - 1}{5x + 2}\right]^4$

$$f'(x) = 4\left[\frac{3x - 1}{5x + 2}\right]^3 \cdot \frac{(5x + 2)(3) - 5(3x - 1)}{(5x + 2)^2}$$

(Using the Extended Power Rule and Quotient Rule)

$$= 4\left[\frac{3x - 1}{5x + 2}\right]^3 \cdot \frac{15x + 6 - 15x + 5}{(5x + 2)^2}$$

$$= 4\left[\frac{3x - 1}{5x + 2}\right]^3 \cdot \frac{11}{(5x + 2)^2}$$

$$= \left[\frac{3x - 1}{5x + 2}\right]^3 \cdot \frac{44}{(5x + 2)^2}$$

29. $f(x) = \sqrt[3]{x^4 + 3x^2} = (x^4 + 3x^2)^{1/3}$

$$f'(x) = \frac{1}{3}(x^4 + 3x^2)^{1/3-1}(4x^3 + 6x)$$

(Using the Extended Power Rule)

$$= \frac{1}{3}(x^4 + 3x^2)^{-2/3}(4x^3 + 6x)$$

31. $f(x) = (2x^3 - 3x^2 + 4x + 1)^{100}$

$f'(x) = 100(2x^3 - 3x^2 + 4x + 1)^{99}(6x^2 - 6x + 4)$

(Using the Extended Power Rule)

33. $g(x) = \left[\frac{2x + 3}{5x - 1}\right]^{-4}$

$g'(x) = -4\left[\frac{2x + 3}{5x - 1}\right]^{-5} \cdot \frac{(5x - 1)(2) - 5(2x + 3)}{(5x - 1)^2}$

(Using the Extended Power Rule and the Quotient Rule)

$= -4\left[\frac{2x + 3}{5x - 1}\right]^{-5} \cdot \frac{10x - 2 - 10x - 15}{(5x - 1)^2}$

$= -4\left[\frac{2x + 3}{5x - 1}\right]^{-5} \cdot \frac{-17}{(5x - 1)^2}$

$= \left[\frac{2x + 3}{5x - 1}\right]^{-5} \cdot \frac{68}{(5x - 1)^2}$

35. $f(x) = \sqrt{\frac{x^2 + 1}{x^2 - 1}} = \left[\frac{x^2 + 1}{x^2 - 1}\right]^{1/2}$

$f'(x) = \frac{1}{2}\left[\frac{x^2 + 1}{x^2 - 1}\right]^{1/2-1} \cdot \frac{(x^2 - 1)(2x) - 2x(x^2 + 1)}{(x^2 - 1)^2}$

(Using the Extended Power Rule and the Quotient Rule)

$= \frac{1}{2}\left[\frac{x^2 + 1}{x^2 - 1}\right]^{-1/2} \cdot \frac{2x^3 - 2x - 2x^3 - 2x}{(x^2 - 1)^2}$

$= \frac{1}{2}\left[\frac{x^2 + 1}{x^2 - 1}\right]^{-1/2} \cdot \frac{-4x}{(x^2 - 1)^2}$

$= \left[\frac{x^2 + 1}{x^2 - 1}\right]^{-1/2} \cdot \frac{-2x}{(x^2 - 1)^2}$

37. $f(x) = \frac{(2x + 3)^4}{(3x - 2)^5}$

$f'(x) = \frac{(3x-2)^5 \cdot 4(2x+3)^3(2) - 5(3x-2)^4(3)(2x+3)^4}{[(3x - 2)^5]^2}$

(Using the Quotient Rule and the Extended Power Rule)

$= \frac{(3x - 2)^4(2x + 3)^3[8(3x - 2) - 15(2x + 3)]}{(3x - 2)^{10}}$

(Factoring)

$= \frac{(3x - 2)^4(2x + 3)^3(24x - 16 - 30x - 45)}{(3x - 2)^{10}}$

$= \frac{(3x - 2)^4}{(3x - 2)^4} \cdot \frac{(2x + 3)^3(-6x - 61)}{(3x - 2)^6}$

$= \frac{(2x + 3)^3(-6x - 61)}{(3x - 2)^6}$

39. $f(x) = 12(2x + 1)^{2/3}(3x - 4)^{5/4}$

$f'(x) = 12(2x + 1)^{2/3} \cdot \frac{5}{4}(3x - 4)^{5/4-1}(3) +$

$12 \cdot \frac{2}{3}(2x + 1)^{2/3-1}(2)(3x - 4)^{5/4}$

(Using the Product Rule and the Extended Power Rule)

$= 12 \cdot \frac{5}{4} \cdot 3(2x + 1)^{2/3}(3x - 4)^{1/4} +$

$12 \cdot \frac{2}{3} \cdot 2(2x + 1)^{-1/3}(3x - 4)^{5/4}$

$= 45(2x + 1)^{2/3}(3x - 4)^{1/4} +$

$16(2x + 1)^{-1/3}(3x - 4)^{5/4}$

We can also simplify by factoring:

$f'(x) = (2x+1)^{-1/3}(3x-4)^{1/4}[45(2x+1) + 16(3x-4)]$

$= (2x+1)^{-1/3}(3x-4)^{1/4}(90x+45+48x-64)$

$= (2x + 1)^{-1/3}(3x - 4)^{1/4}(138x - 19)$

41. $y = \sqrt{u} = u^{1/2}$, $u = x^2 - 1$

$\frac{dy}{du} = \frac{1}{2} u^{1/2-1} = \frac{1}{2} u^{-1/2} = \frac{1}{2\sqrt{u}}$

$\frac{du}{dx} = 2x$

$\frac{dy}{dx} = \frac{dy}{du} \cdot \frac{du}{dx} = \frac{1}{2\sqrt{u}} \cdot 2x$

$= \frac{2x}{2\sqrt{x^2 - 1}}$ (Substituting $x^2 - 1$ for u)

$= \frac{x}{\sqrt{x^2 - 1}}$ (Simplifying)

43. $y = u^{50}$, $u = 4x^3 - 2x^2$

$\frac{dy}{du} = 50u^{49}$

$\frac{du}{dx} = 12x^2 - 4x$

$\frac{dy}{dx} = \frac{dy}{du} \cdot \frac{du}{dx}$

$= 50u^{49}(12x^2 - 4x)$

$= 50(4x^3 - 2x^2)^{49}(12x^2 - 4x)$ (Substituting for u)

45. $y = u(u + 1)$, $u = x^3 - 2x$

$\frac{dy}{du} = u \cdot 1 + 1 \cdot (u + 1)$ (Product Rule)

$= u + u + 1$

$= 2u + 1$

$\frac{du}{dx} = 3x^2 - 2$

$\frac{dy}{dx} = \frac{dy}{du} \cdot \frac{du}{dx} = (2u + 1)(3x^2 - 2)$

$= [2(x^3 - 2x) + 1](3x^2 - 2)$

(Substituting for u)

$= (2x^3 - 4x + 1)(3x^2 - 2)$

47. $y = \sqrt{x^2 + 3x} = (x^2 + 3x)^{1/2}$

$\frac{dy}{dx} = \frac{1}{2}(x^2 + 3x)^{-1/2}(2x + 3)$

$= \frac{2x + 3}{2\sqrt{x^2 + 3x}}$

When $x = 1$, $\frac{dy}{dx} = \frac{2 \cdot 1 + 3}{2\sqrt{1^2 + 3 \cdot 1}}$

$= \frac{2 + 3}{2\sqrt{4}}$

$= \frac{5}{2 \cdot 2}$

$= \frac{5}{4}$

Thus, the slope of the tangent line at (1,2) is $\frac{5}{4}$. We use the point-slope equation.

$$y - y_1 = m(x - x_1)$$
$$y - 2 = \frac{5}{4}(x - 1)$$
$$y - 2 = \frac{5}{4}x - \frac{5}{4}$$
$$y = \frac{5}{4}x + \frac{3}{4}$$

49. $f(x) = \frac{x^2}{(1 + x)^5}$, or $x^2(1 + x)^{-5}$

a) $f'(x) = \frac{(1 + x)^5 \cdot 2x - 5(1 + x)^4 \cdot 1 \cdot x^2}{[(1 + x)^5]^2}$

(Using the Quotient Rule and the Extended Power Rule)

$$= \frac{2x(1 + x)^5 - 5x^2(1 + x)^4}{(1 + x)^{10}}$$

$$= \frac{x(1 + x)^4[2(1 + x) - 5x]}{(1 + x)^{10}}$$

[Factoring out $x(1 + x)^4$ in the numerator]

$$= \frac{x(2 - 3x)}{(1 + x)^6}, \text{ or } \frac{2x - 3x^2}{(1 + x)^6} \quad \text{(Simplifying)}$$

b) $f'(x) = x^2 \cdot (-5)(1 + x)^{-6} \cdot 1 + 2x(1 + x)^{-5}$

(Using the Product Rule and the Extended Power Rule)

$$= -5x^2(1 + x)^{-6} + 2x(1 + x)^{-5}$$
$$= x(1 + x)^{-5}[-5x(1 + x)^{-1} + 2]$$
$$= \frac{x}{(1 + x)^5} \cdot \left[\frac{-5x}{1 + x} + 2 \cdot \frac{1 + x}{1 + x}\right]$$
$$= \frac{x}{(1 + x)^5} \cdot \frac{2 - 3x}{1 + x}$$
$$= \frac{2x - 3x^2}{(1 + x)^6}$$

c) Same

51. $f(x) = 3x^2 + 2, \quad g(x) = 2x - 1$

$$f \circ g(x) = f(g(x)) = f(2x - 1)$$
$$= 3(2x - 1)^2 + 2$$
$$= 3(4x^2 - 4x + 1) + 2$$
$$= 12x^2 - 12x + 3 + 2$$
$$= 12x^2 - 12x + 5$$

$$g \circ f(x) = g(f(x)) = g(3x^2 + 2)$$
$$= 2(3x^2 + 2) - 1$$
$$= 6x^2 + 4 - 1$$
$$= 6x^2 + 3$$

53. $f(x) = 4x^2 - 1, \quad g(x) = \frac{2}{x}$

$$f \circ g(x) = f(g(x)) = f\left(\frac{2}{x}\right)$$
$$= 4\left(\frac{2}{x}\right)^2 - 1$$
$$= 4 \cdot \frac{4}{x^2} - 1$$
$$= \frac{16}{x^2} - 1$$

$$g \circ f(x) = g(f(x)) = g(4x^2 - 1)$$
$$= \frac{2}{4x^2 - 1}$$

55. $f(x) = x^2 + 1, \quad g(x) = x^2 - 1$

$$f \circ g(x) = f(g(x)) = f(x^2 - 1)$$
$$= (x^2 - 1)^2 + 1$$
$$= x^4 - 2x^2 + 1 + 1$$
$$= x^4 - 2x^2 + 2$$

$$g \circ f(x) = g(f(x)) = g(x^2 + 1)$$
$$= (x^2 + 1)^2 - 1$$
$$= x^4 + 2x^2 + 1 - 1$$
$$= x^4 + 2x^2$$

57. $h(x) = (3x^2 - 7)^5$

Let $f(x) = x^5$ and $g(x) = 3x^2 - 7$.

$f \circ g(x) = f(g(x)) = f(3x^2 - 7) = (3x^2 - 7)^5$

Thus, $h(x) = f \circ g(x)$.

Answers may vary.

59. $h(x) = \frac{x^3 + 1}{x^3 - 1}$

Let $f(x) = \frac{x + 1}{x - 1}$ and $g(x) = x^3$.

$f \circ g(x) = f(g(x)) = f(x^3) = \frac{x^3 + 1}{x^3 - 1}$

Thus, $h(x) = f \circ g(x)$.

Answers may vary.

61. $C(x) = 1000\sqrt{x^3 + 2}$

$$= 1000(x^3 + 2)^{1/2}$$

$C'(x) = 1000 \cdot \frac{1}{2}(x^3 + 2)^{-1/2} \cdot 3x^2$

(Using Theorem 3 and the Extended Power Rule)

$$= 1500x^2(x^3 + 2)^{-1/2}, \text{ or } \frac{1500x^2}{\sqrt{x^3 + 2}}$$

63. The marginal profit $P'(x)$ is given by $R'(x) - C'(x)$:

$$P'(x) = R'(x) - C'(x)$$
$$= \frac{2000x}{\sqrt{x^2 + 3}} - \frac{1500x^2}{\sqrt{x^2 + 3}}, \text{ or}$$
$$\frac{2000x - 1500x^2}{\sqrt{x^2 + 3}}$$

65. $A = \$1000(1 + i)^3$

$\frac{dA}{di} = \$1000 \cdot 3(1 + i)^2 \cdot 1$

(Using Theorem 3 and the Extended Power Rule)

$$= \$3000(1 + i)^2$$

67. $D(p) = \frac{80,000}{p}$, $p = 1.6t + 9$

a) We substitute 1.6t + 9 for p in D(p):

$$D(t) = \frac{80,000}{1.6t + 9}$$

b) $D'(t) = \frac{(1.6t + 9)(0) - 1.6(80,000)}{(1.6t + 9)^2}$ (Quotient Rule)

$$= \frac{-128,000}{(1.6t + 9)^2}$$

c) $D'(100) = \frac{-128,000}{[1.6(100) + 9]^2}$

$$= \frac{-128,000}{169^2}$$

$$= \frac{-128,000}{28,561}$$

$$\approx -4.48 \text{ units/day}$$

69. $y = \sqrt[3]{x^3 - 6x + 1} = (x^3 - 6x + 1)^{1/3}$

$$\frac{dy}{dx} = \frac{1}{3}(x^3 - 6x + 1)^{-2/3}(3x^2 - 6)$$

(Using the Extended Power Rule)

$$= \frac{x^2 - 2}{\sqrt[3]{(x^3 - 6x + 1)^2}}$$ (Simplifying)

71. $y = \frac{x}{\sqrt{x - 1}} = \frac{x}{(x - 1)^{1/2}}$

$$\frac{dy}{dx} = \frac{(x - 1)^{1/2}\cdot 1 - \frac{1}{2}(x - 1)^{-1/2}\cdot 1\cdot x}{[(x - 1)^{1/2}]^2}$$

(Using the Quotient Rule and the Extended Power Rule)

$$= \frac{\sqrt{x - 1} - \frac{x}{2\sqrt{x - 1}}}{x - 1} \cdot \frac{2\sqrt{x - 1}}{2\sqrt{x - 1}}$$

(Multiplying by a form of 1)

$$= \frac{2(x - 1) - x}{2(x - 1)\sqrt{x - 1}}$$

$$= \frac{x - 2}{2(x - 1)^{3/2}}$$

$[(x - 1)\sqrt{x - 1} = (x - 1)^{2/2}(x - 1)^{1/2} = (x - 1)^{3/2}]$

73. $u = \frac{(1 + 2v)^4}{v^4}$

$$\frac{du}{dv} = \frac{v^4\cdot 4(1 + 2v)^3\cdot 2 - 4v^3(1 + 2v)^4}{(v^4)^2}$$

(Using the Quotient Rule and the Extended Power Rule)

$$= \frac{8v^4(1 + 2v)^3 - 4v^3(1 + 2v)^4}{v^8}$$

$$= \frac{4v^3(1 + 2v)^3[2v - (1 + 2v)]}{v^8}$$

$$= \frac{4v^3(1 + 2v)^3(-1)}{v^8}$$

$$= \frac{-4(1 + 2v)^3}{v^5}$$ (Simplifying)

75. $y = \frac{\sqrt{1 - x^2}}{1 - x} = \frac{(1 - x^2)^{1/2}}{1 - x}$

$$\frac{dy}{dx} = \frac{(1 - x)\frac{1}{2}(1 - x^2)^{-1/2}(-2x) - (-1)(1 - x^2)^{1/2}}{(1 - x)^2}$$

(Using the Quotient Rule and the Extended Power Rule)

$$= \frac{\frac{-x(1 - x)}{\sqrt{1 - x^2}} + \sqrt{1 - x^2}}{(1 - x)^2} \cdot \frac{\sqrt{1 - x^2}}{\sqrt{1 - x^2}}$$

(Multiplying by a form of 1)

$$= \frac{-x(1 - x) + (1 - x^2)}{(1 - x)^2\sqrt{1 - x^2}}$$

$$= \frac{-x + x^2 + 1 - x^2}{(1 - x)^2\sqrt{1 - x^2}}$$

$$= \frac{1 - x}{(1 - x)^2\sqrt{1 - x^2}}$$

$$= \frac{1}{(1 - x)\sqrt{1 - x^2}}$$

Rationalizing the denominator, we get

$$\frac{dy}{dx} = \frac{1}{(1 - x)\sqrt{1 - x^2}} \cdot \frac{\sqrt{1 - x^2}}{\sqrt{1 - x^2}}$$

$$= \frac{\sqrt{1 - x^2}}{(1 - x)(1 - x^2)}.$$

77. $y = \left[\frac{x^2 - x - 1}{x^2 + 1}\right]^3$

$$\frac{dy}{dx} = 3\left[\frac{x^2 - x - 1}{x^2 + 1}\right]^2 \cdot \frac{(x^2+1)(2x-1) - 2x(x^2-x-1)}{(x^2 + 1)^2}$$

(Using the Extended Power Rule and the Quotient Rule)

$$= \frac{3(x^2-x-1)^2}{(x^2 + 1)^2} \cdot \frac{2x^3-x^2+2x-1-2x^3+2x^2+2x}{(x^2 + 1)^2}$$

$$= \frac{3(x^2 - x - 1)^2(x^2 + 4x - 1)}{(x^2 + 1)^4}$$

79. $s = \frac{\sqrt{t} - 1}{\sqrt{t} + 1} = \frac{t^{1/2} - 1}{t^{1/2} + 1}$

$$\frac{ds}{dt} = \frac{(t^{1/2} + 1)\left[\frac{1}{2} t^{-1/2}\right] - \left[\frac{1}{2} t^{-1/2}\right](t^{1/2} - 1)}{(t^{1/2} + 1)^2}$$

(Using the Quotient Rule)

$$= \frac{\frac{1}{2} t^0 + \frac{1}{2} t^{-1/2} - \frac{1}{2} t^0 + \frac{1}{2} t^{-1/2}}{(t^{1/2} + 1)^2}$$

$$= \frac{t^{-1/2}}{(t^{1/2} + 1)^2}$$

$$= \frac{1}{\sqrt{t}(\sqrt{t} + 1)^2}$$

81. $f(x) = 1.68x\sqrt{9.2 - x^2} = 1.68x(9.2 - x^2)^{1/2}$

$f'(x) = 1.68x\left[\frac{1}{2}\right](9.2 - x^2)^{-1/2}(-2x) + 1.68(9.2 - x^2)^{1/2}$

$= \frac{-1.68x^2}{(9.2 - x^2)^{1/2}} + 1.68(9.2 - x^2)^{1/2}$

$= \frac{-1.68x^2 + 1.68(9.2 - x^2)}{(9.2 - x^2)^{1/2}}$

$= \frac{-1.68x^2 + 15.456 - 1.68x^2}{(9.2 - x^2)^{1/2}}$

$= \frac{-3.36x^2 + 15.456}{(9.2 - x^2)^{1/2}}$

See the answer section in the text for the result.

Exercise Set 2.9

1. $y = 3x + 5$

$\frac{dy}{dx} = 3$ (First derivative)

$\frac{d^2y}{dx^2} = 0$ (Second derivative)

3. $y = -\frac{1}{x} = -1 \cdot x^{-1}$

$\frac{dy}{dx} = -1 \cdot (-1)x^{-1-1}$

$= x^{-2}$ (First derivative)

$\frac{d^2y}{dx^2} = -2x^{-2-1}$

$= -2x^{-3}$, or $-\frac{2}{x^3}$ (Second derivative)

5. $y = x^{1/4}$

$\frac{dy}{dx} = \frac{1}{4}x^{1/4-1}$

$= \frac{1}{4}x^{-3/4}$ (First derivative)

$\frac{d^2y}{dx^2} = \frac{1}{4} \cdot (-\frac{3}{4})x^{-3/4-1}$

$= -\frac{3}{16}x^{-7/4}$, or $-\frac{3}{16x^{7/4}}$ (Second derivative)

7. $y = x^4 + \frac{4}{x} = x^4 + 4x^{-1}$

$\frac{dy}{dx} = 4x^{4-1} + 4 \cdot (-1) \cdot x^{-1-1}$

$= 4x^3 - 4x^{-2}$ (First derivative)

$\frac{d^2y}{dx^2} = 4 \cdot 3x^{3-1} - 4 \cdot (-2) \cdot x^{-2-1}$

$= 12x^2 + 8x^{-3}$, or $12x^2 + \frac{8}{x^3}$

(Second derivative)

9. $y = x^{-3}$

$\frac{dy}{dx} = -3x^{-3-1}$

$= -3x^{-4}$ (First derivative)

$\frac{d^2y}{dx^2} = -3 \cdot (-4) \cdot x^{-4-1}$

$= 12x^{-5}$, or $\frac{12}{x^5}$ (Second derivative)

11. $y = x^n$

$\frac{dy}{dx} = nx^{n-1}$ (First derivative)

$\frac{d^2y}{dx^2} = n \cdot (n - 1) \cdot x^{(n-1)-1}$

$= n(n - 1)x^{n-2}$ (Second derivative)

13. $y = x^4 - x^2$

$\frac{dy}{dx} = 4x^{4-1} - 2x^{2-1}$

$= 4x^3 - 2x$

$\frac{d^2y}{dx^2} = 4 \cdot 3x^{3-1} - 2$

$= 12x^2 - 2$

15. $y = \sqrt{x - 1} = (x - 1)^{1/2}$

$\frac{dy}{dx} = \frac{1}{2}(x - 1)^{1/2-1} \cdot 1$

$= \frac{1}{2}(x - 1)^{-1/2}$ (First derivative)

$\frac{d^2y}{dx^2} = \frac{1}{2} \cdot \left[-\frac{1}{2}\right](x - 1)^{-1/2-1} \cdot 1$

$= -\frac{1}{4}(x - 1)^{-3/2}$, or $-\frac{1}{4(x - 1)^{3/2}}$, or

$-\frac{1}{4\sqrt{(x - 1)^3}}$ (Second derivative)

17. $y = ax^2 + bx + c$ (x and y are variables; a, b, and c are constants)

$\frac{dy}{dx} = a \cdot 2x + b + 0$

$= 2ax + b$ (First derivative) (2a and b are constants)

$\frac{d^2y}{dx^2} = 2a + 0$

$= 2a$ (Second derivative)

19. $y = (x^2 - 8x)^{43}$

$\frac{dy}{dx} = 43(x^2 - 8x)^{42}(2x - 8)$ (First derivative)

$\frac{d^2y}{dx^2} = 43(x^2 - 8x)^{42}(2) + 43 \cdot 42(x^2 - 8x)^{41}(2x - 8)(2x - 8)$

$= 2 \cdot 43(x^2 - 8x)^{41}[x^2 - 8x + 21(2x - 8)^2]$ (Factoring)

$= 86(x^2 - 8x)^{41}[x^2 - 8x + 21(4x^2 - 32x + 64)]$

$= 86(x^2 - 8x)^{41}(x^2 - 8x + 84x^2 - 672x + 1344)$

$= 86(x^2 - 8x)^{41}(85x^2 - 680x + 1344)$

(Second derivative)

21. $y = (x^4 - 4x^2)^{50}$

$\frac{dy}{dx} = 50(x^4 - 4x^2)^{49}(4x^3 - 8x)$ (First derivative)

$\frac{d^2y}{dx^2} = 50(x^4 - 4x^2)^{49}(12x^2 - 8) +$

$50\cdot 49(x^4 - 4x^2)^{48}(4x^3 - 8x)(4x^3 - 8x)$

$= 50\cdot 4(x^4 - 4x^2)^{49}(3x^2 - 2) +$

$50\cdot 49\cdot 4\cdot 4(x^4 - 4x^2)^{48}(x^3 - 2x)(x^3 - 2x)$

(Factoring the terms)

$= 200(x^4 - 4x^2)^{48}[(x^4 - 4x^2)(3x^2 - 2) +$

$49\cdot 4(x^3 - 2x)^2]$

(Factoring the expression)

$= 200x^2(x^4 - 4x^2)^{48}[(x^2 - 4)(3x^2 - 2) +$

$196(x^2 - 2)^2]$ (Factoring out x^2)

$= 200x^2(x^4 - 4x^2)^{48}[3x^4 - 14x^2 + 8 +$

$196(x^4 - 4x^2 + 4)]$

$= 200x^2(x^4 - 4x^2)^{48}(3x^4 - 14x^2 + 8 + 196x^4 -$

$784x^2 + 784)$

$= 200x^2(x^4 - 4x^2)^{48}(199x^4 - 798x^2 + 792)$

(Second derivative)

23. $y = x^{2/3} + 4x$

$\frac{dy}{dx} = \frac{2}{3}x^{-1/3} + 4$ (First derivative)

$\frac{d^2y}{dx^2} = \frac{2}{3}\left(-\frac{1}{3}\right)x^{-1/3-1}$

$= -\frac{2}{9}x^{-4/3}$ (Second derivative)

25. $y = (x - 8)^{3/4}$

$\frac{dy}{dx} = \frac{3}{4}(x - 8)^{3/4-1}\cdot 1$

$= \frac{3}{4}(x - 8)^{-1/4}$ (First derivative)

$\frac{d^2y}{dx^2} = \frac{3}{4}\left(-\frac{1}{4}\right)(x - 8)^{-1/4-1}\cdot 1$

$= -\frac{3}{16}(x - 8)^{-5/4}$ (Second derivative)

27. $y = \frac{1}{x^2} + \frac{2}{x^3} = x^{-2} + 2x^{-3}$

$\frac{dy}{dx} = -2x^{-3} - 6x^{-4}$ (First derivative)

$\frac{d^2y}{dx^2} = 6x^{-4} + 24x^{-5}$ (Second derivative)

29. $y = x^4$

$\frac{dy}{dx} = 4x^3$ (First derivative)

$\frac{d^2y}{dx^2} = 4\cdot 3x^2$

$= 12x^2$ (Second derivative)

$\frac{d^3y}{dx^3} = 12\cdot 2x$

$= 24x$ (Third derivative)

$\frac{d^4y}{dy^4} = 24$ (Fourth derivative)

31. $y = x^6 - x^3 + 2x$

$\frac{dy}{dx} = 6x^5 - 3x^2 + 2$ (First derivative)

$\frac{d^2y}{dx^2} = 30x^4 - 6x$ (Second derivative)

$\frac{d^3y}{dx^3} = 120x^3 - 6$ (Third derivative)

$\frac{d^4y}{dx^4} = 360x^2$ (Fourth derviative)

$\frac{d^5y}{dx^5} = 720x$ (Fifth derivative)

33. $y = (x^2 - 5)^{10}$

$\frac{dy}{dx} = 10(x^2 - 5)^9\cdot 2x$

$= 20x(x^2 - 5)^9$

$\frac{d^2y}{dx^2} = 20x\cdot 9(x^2 - 5)^8\cdot 2x + 20(x^2 - 5)^9$

$= 360x^2(x^2 - 5)^8 + 20(x^2 - 5)^9$

$= 20(x^2 - 5)^8[18x^2 + (x^2 - 5)]$

$= 20(x^2 - 5)^8(19x^2 - 5)$

35. $s(t) = t^3 + t^2 + 2t$

$v(t) = s'(t) = 3t^2 + 2t + 2$

$a(t) = s''(t) = 6t + 2$

37. $P(t) = 100{,}000(1 + 0.6t + t^2)$

$= 100{,}000 + 60{,}000t + 100{,}000t^2$

(This function gives the number of people in a population at time t)

Whenever a quantity is a function of time, the first derivative gives the rate of change with respect to time and the second derivative gives the acceleration.

$P'(t) = 60{,}000 + 200{,}000t$

(This function gives how fast the size of a population is changing.)

$P''(t) = 200{,}000$ (Acceleration)

(This function gives the acceleration in the size of the population.)

39. $y = x^{-1} + x^{-2}$

$y' = -1\cdot x^{-1-1} + (-2)\cdot x^{-2-1}$

$= -x^{-2} - 2x^{-3}$

$y'' = -1\cdot(-2)\cdot x^{-2-1} - 2\cdot(-3)\cdot x^{-3-1}$

$= 2x^{-3} + 6x^{-4}$

$y''' = 2\cdot(-3)\cdot x^{-3-1} + 6\cdot(-4)\cdot x^{-4-1}$

$= -6x^{-4} - 24x^{-5}$

41. $y = x\sqrt{1 + x^2} = x(1 + x^2)^{1/2}$

$y' = x \cdot \frac{1}{2}(1 + x^2)^{-1/2} \cdot 2x + 1\cdot(1 + x^2)^{1/2}$

$= \frac{x^2}{(1 + x^2)^{1/2}} + (1 + x^2)^{1/2} \cdot \frac{(1 + x^2)^{1/2}}{(1 + x^2)^{1/2}}$

(Multiplying the second term by a form of 1)

$= \frac{x^2 + 1 + x^2}{(1 + x^2)^{1/2}}$

$= \frac{2x^2 + 1}{(1 + x^2)^{1/2}}$

$y'' = \frac{(1 + x^2)^{1/2}\cdot 4x - \frac{1}{2}(1 + x^2)^{-1/2}\cdot 2x\cdot(2x^2 + 1)}{[(1 + x^2)^{1/2}]^2}$

$= \frac{4x(1+x^2)^{1/2} - x(2x^2+1)(1+x^2)^{-1/2}}{1 + x^2} \cdot \frac{(1+x^2)^{1/2}}{(1+x^2)^{1/2}}$

(Multiplying by a form of 1)

$= \frac{4x(1 + x^2) - x(2x^2 + 1)}{(1 + x^2)^{3/2}}$

$= \frac{4x + 4x^3 - 2x^3 - x}{(1 + x^2)^{3/2}}$

$= \frac{2x^3 + 3x}{(1 + x^2)^{3/2}}$

$y''' = \frac{(1+x^2)^{3/2}(6x^2+3) - \frac{3}{2}(1+x^2)^{1/2}\cdot 2x\cdot(2x^3+3x)}{[(1 + x^2)^{3/2}]^2}$

$= \frac{(1+x^2)^{3/2}(6x^2+3) - 3x(1+x^2)^{1/2}(2x^3+3x)}{(1 + x^2)^{6/2}}$

$= \frac{(1 + x^2)^{1/2}[(1+x^2)(6x^2+3) - 3x(2x^3 + 3x)]}{(1 + x^2)^{1/2}(1 + x^2)^{5/2}}$

$= \frac{6x^2 + 3 + 6x^4 + 3x^2 - 6x^4 - 9x^2}{(1 + x^2)^{5/2}}$

$= \frac{3}{(1 + x^2)^{5/2}}$

43. $y = \frac{3x - 1}{2x + 3}$

$y' = \frac{(2x + 3)\cdot 3 - 2\cdot(3x - 1)}{(2x + 3)^2}$

$= \frac{6x + 9 - 6x + 2}{(2x + 3)^2}$

$= \frac{11}{(2x + 3)^2}$, or $11(2x + 3)^{-2}$

$y'' = 11\cdot(-2)(2x + 3)^{-3}\cdot 2$

$= -44(2x + 3)^{-3}$, or $\frac{-44}{(2x + 3)^3}$

$y''' = -44\cdot(-3)\cdot(2x + 3)^{-4}\cdot 2$

$= 264(2x + 3)^{-4}$, or $\frac{264}{(2x + 3)^4}$

45. $y = \frac{x}{\sqrt{x - 1}}$, or $x(x - 1)^{-1/2}$

$y' = x\cdot\left[-\frac{1}{2}\right](x - 1)^{-3/2}\cdot 1 + 1\cdot(x - 1)^{-1/2}$

$= \frac{-x}{2(x - 1)^{3/2}} + \frac{1}{(x - 1)^{1/2}} \cdot \frac{2(x - 1)}{2(x - 1)}$

(Multiplying the second term by a form of 1)

$= \frac{-x + 2x - 2}{2(x - 1)^{3/2}}$

$= \frac{x - 2}{2(x - 1)^{3/2}}$

$y'' = \frac{2(x - 1)^{3/2}\cdot 1 - 2 \cdot \frac{3}{2}(x - 1)^{1/2}\cdot 1\cdot(x - 2)}{[2(x - 1)^{3/2}]^2}$

$= \frac{2(x - 1)^{3/2} - 3(x - 2)(x - 1)^{1/2}}{4(x - 1)^3} \cdot \frac{(x-1)^{-1/2}}{(x-1)^{-1/2}}$

(Multiplying by a form of 1)

$= \frac{2(x - 1) - 3(x - 2)}{4(x - 1)^{5/2}}$

$= \frac{2x - 2 - 3x + 6}{4(x - 1)^{5/2}}$

$= \frac{4 - x}{4(x - 1)^{5/2}}$

$y''' = \frac{4(x - 1)^{5/2}(-1) - 4 \cdot \frac{5}{2}(x - 1)^{3/2}\cdot 1(4 - x)}{[4(x - 1)^{5/2}]^2}$

$= \frac{-4(x-1)^{5/2} - 10(4-x)(x-1)^{3/2}}{16(x - 1)^5} \cdot \frac{(x - 1)^{-1/2}}{(x - 1)^{-1/2}}$

(Multiplying by a form of 1)

$= \frac{-4(x - 1)^2 - 10(4 - x)(x - 1)}{16(x - 1)^{9/2}}$

$= \frac{-4(x^2 - 2x + 1) - 10(-x^2 + 5x - 4)}{16(x - 1)^{9/2}}$

$= \frac{-4x^2 + 8x - 4 + 10x^2 - 50x + 40}{16(x - 1)^{9/2}}$

$= \frac{6x^2 - 42x + 36}{16(x - 1)^{9/2}}$

$= \frac{6(x^2 - 7x + 6)}{16(x - 1)^{9/2}}$

$= \frac{6(x - 6)(x - 1)}{16(x - 1)^{9/2}}$

$= \frac{3(x - 6)}{8(x - 1)^{7/2}}$

47. $f(x) = \frac{x}{x - 1}$

$f'(x) = \frac{(x - 1)\cdot 1 - 1\cdot x}{(x - 1)^2}$

$= \frac{x - 1 - x}{(x - 1)^2}$

$= \frac{-1}{(x - 1)^2}$

$f''(x) = \frac{(x - 1)^2\cdot 0 - 2(x - 1)\cdot 1\cdot(-1)}{[(x - 1)^2]^2}$

$= \frac{2(x - 1)}{(x - 1)4}$

$= \frac{2}{(x - 1)^3}$

49. $f(x) = \frac{x - 1}{x + 2}$

$f'(x) = \frac{(x + 2)\cdot 1 - 1\cdot(x - 1)}{(x + 2)^2}$

$= \frac{x + 2 - x + 1}{(x + 2)^2}$

$= \frac{3}{(x + 2)^2}$ (First derivative)

Express $f'(x)$ as $3(x + 2)^{-2}$. Then we have:

$f''(x) = -6(x + 2)^{-3} \cdot 1$

$= \frac{-6}{(x + 2)^3}$ (Second derivative)

Express $f''(x)$ as $-6(x + 2)^{-3}$. Then we have:

$f'''(x) = 18(x + 2)^{-4} \cdot 1$

$= \frac{18}{(x + 2)^4}$ (Third derivative)

Express $f'''(x)$ as $18(x + 2)^{-4}$. Then we have:

$f^{(4)}(x) = -72(x + 2)^{-5} \cdot 1$

$= \frac{-72}{(x + 2)^5}$ (Fourth derivative)

Express $f^{(4)}(x)$ as $-72(x + 2)^{-5}$. Then we have:

$f^{(5)}(x) = 360(x + 2)^{-6} \cdot 1$

$= \frac{360}{(x + 2)^6}$ (Fifth derivative)

51. $f(x) = 0.1x^4 - x^2 + 0.4$

$f'(x) = 0.4x^3 - 2x$

$f''(x) = 1.2x^2 - 2$

See the answer section in the text for the result.

53. $f(x) = x^4 + x^3 - 4x^2 - 2x + 4$

$f'(x) = 4x^3 + 3x^2 - 8x - 2$

$f''(x) = 12x^2 + 6x - 8$

See the answer section in the text for the result.

Exercise Set 3.1

1. $f(x) = x^2 - 4x + 5$

First, find the critical points.

$f'(x) = 2x - 4$

$f'(x)$ exists for all real numbers. We solve $f'(x) = 0$:

$$2x - 4 = 0$$
$$2x = 4$$
$$x = 2$$

The only critical point is 2. We use 2 to divide the real number line into two intervals, A: $(-\infty,2)$ and B: $(2,\infty)$:

A | B, divided at 2

We use a test value in each interval to determine the sign of the derivative in each interval.

A: Test 0, $f'(0) = 2\cdot0 - 4 = -4 < 0$

B: Test 3, $f'(3) = 2\cdot3 - 4 = 2 > 0$

We see that $f(x)$ is decreasing on $(-\infty,2)$ and increasing on $(2,\infty)$, and the change from decreasing to increasing indicates that a relative minimum occurs at $x = 2$. We substitute into the original equation to find $f(2)$:

$$f(2) = 2^2 - 4\cdot2 + 5 = 1$$

Thus, there is a relative minimum at (2,1). We use the information obtained to sketch the graph. Other function values are listed below.

x	f(x)
-2	17
-1	10
0	5
1	2
2	1
3	2
4	5
5	10

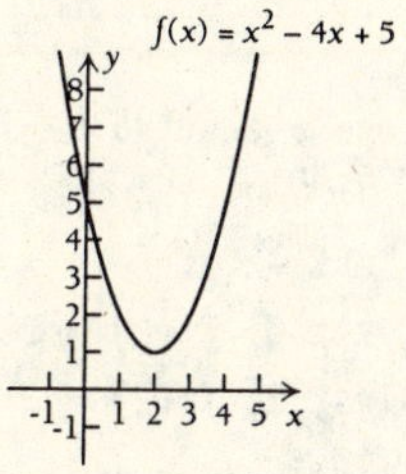

3. $f(x) = 5 + x - x^2$

First, find the critical points.

$f'(x) = 1 - 2x$

$f'(x)$ exists for all real numbers. We solve $f'(x) = 0$:

$$1 - 2x = 0$$
$$1 = 2x$$
$$\frac{1}{2} = x$$

The only critical point is $\frac{1}{2}$. We use $\frac{1}{2}$ to divide the real number line into two intervals, A: $\left(-\infty,\frac{1}{2}\right)$ and B: $\left(\frac{1}{2},\infty\right)$:

A | B, divided at $\frac{1}{2}$

We use a test value in each interval to determine the sign of the derivative in each interval.

A: Test 0, $f'(0) = 1 - 2\cdot0 = 1 > 0$

B: Test 1, $f'(1) = 1 - 2\cdot1 = -1 < 0$

We see that $f(x)$ is increasing on $\left(-\infty,\frac{1}{2}\right)$ and decreasing on $\left(\frac{1}{2},\infty\right)$, so there is a relative maximum at $x = \frac{1}{2}$. We find $f\left(\frac{1}{2}\right)$:

$$f\left(\frac{1}{2}\right) = 5 + \frac{1}{2} - \left(\frac{1}{2}\right)^2 = \frac{21}{4}$$

Thus, there is a relative maximum at $\left(\frac{1}{2},\frac{21}{4}\right)$. We use the information obtained to sketch the graph. Other function values are listed below.

x	f(x)
-3	-7
-2	-1
-1	3
0	5
$\frac{1}{2}$	$\frac{21}{4}$
1	5
2	3
3	-1
4	-7

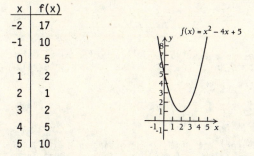

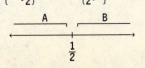

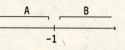

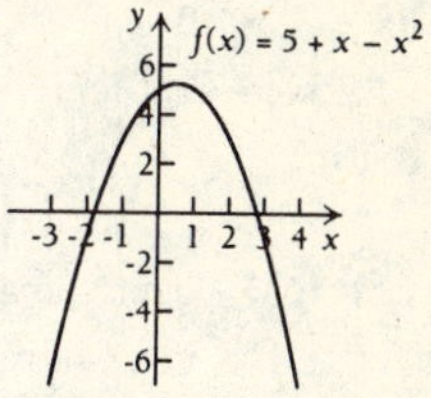

5. $f(x) = 1 + 6x + 3x^2$

First, find the critical points.

$f'(x) = 6 + 6x$

$f'(x)$ exists for all real numbers. We solve $f'(x) = 0$:

$$6 + 6x = 0$$
$$6x = -6$$
$$x = -1$$

The only critical point is -1. We use -1 to divide the real number line into two intervals: A: $(-\infty,-1)$ and B: $(-1,\infty)$.

A | B, divided at -1

We use a test value in each interval to determine the sign of the derivative in each interval.

A: Test -2, $f'(-2) = 6 + 6(-2) = -6 < 0$

B: Test 0, $f'(0) = 6 + 6\cdot0 = 6 > 0$

We see that $f(x)$ is decreasing on $(-\infty,-1)$ and increasing on $(-1,\infty)$, so there is a relative minimum at $x = -1$. We find $f(-1)$:

$$f(-1) = 1 + 6(-1) + 3(-1)^2 = -2$$

Thus, there is a relative minimum at (-1,-2). We use the information obtained to sketch the graph. Other function values are listed below.

x	f(x)
-3	10
-2	1
-1	-2
0	1
1	10
2	25

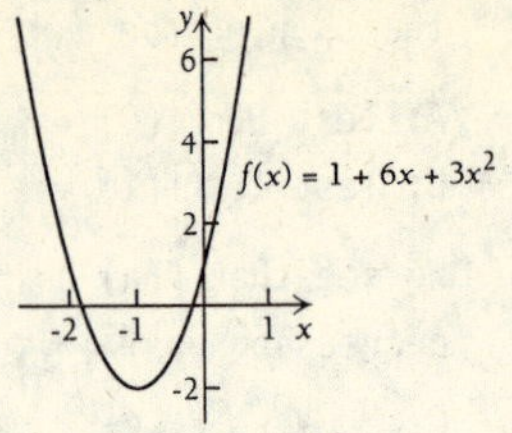

7. $f(x) = x^3 - x^2 - x + 2$

First, find the critical points.

$f'(x) = 3x^2 - 2x - 1$

$f'(x)$ exists for all real numbers. We solve $f'(x) = 0$:

$$3x^2 - 2x - 1 = 0$$
$$(3x + 1)(x - 1) = 0$$
$$3x + 1 = 0 \quad \text{or} \quad x - 1 = 0$$
$$3x = -1 \quad \text{or} \quad x = 1$$
$$x = -\frac{1}{3} \quad \text{or} \quad x = 1$$

The critical points are $-\frac{1}{3}$ and 1. We use them to divide the real number line into three intervals, A: $\left(-\infty, -\frac{1}{3}\right)$, B: $\left(-\frac{1}{3}, 1\right)$, and C: $(1, \infty)$.

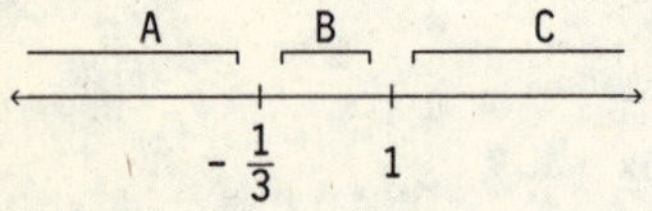

We use a test value in each interval to determine the sign of the derivative in each interval.

A: Test -1, $f'(-1) = 3(-1)^2 - 2(-1) - 1 = 3 + 2 - 1 = 4 > 0$

B: Test 0, $f'(0) = 3(0)^2 - 2(0) - 1 = -1 < 0$

C: Test 2, $f'(2) = 3(2)^2 - 2(2) - 1 = 12 - 4 - 1 = 7 > 0$

We see that $f(x)$ is increasing on $\left(-\infty, -\frac{1}{3}\right)$, decreasing on $\left(-\frac{1}{3}, 1\right)$, and increasing again on $(1, \infty)$, so there is a relative maximum at $x = -\frac{1}{3}$ and a relative minimum at $x = 1$. We find $f\left(-\frac{1}{3}\right)$:

$$f\left(-\frac{1}{3}\right) = \left(-\frac{1}{3}\right)^3 - \left(-\frac{1}{3}\right)^2 - \left(-\frac{1}{3}\right) + 2$$
$$= -\frac{1}{27} - \frac{1}{9} + \frac{1}{3} + 2 \qquad -1-3+9+54$$
$$= \frac{59}{27}$$

Then we find $f(1)$:

$$f(1) = 1^3 - 1^2 - 1 + 2$$
$$= 1 - 1 - 1 + 2$$
$$= 1$$

There is a relative maximum at $\left(-\frac{1}{3}, \frac{59}{27}\right)$, and there is a relative minimum at $(1,1)$. We use the information obtained to sketch the graph. Other function values are listed below.

x	f(x)
-2	-8
-1	1
0	2
2	4
3	17

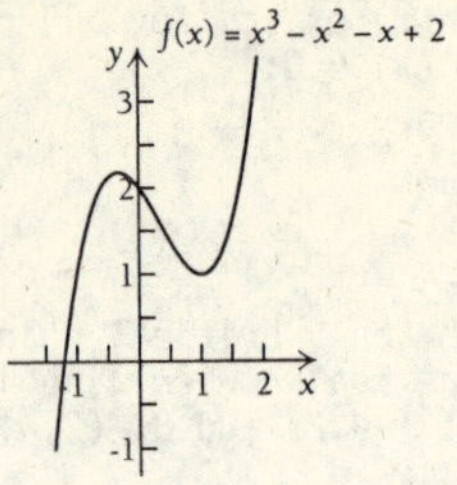

9. $f(x) = x^3 - 3x + 6$

First, find the critical points.

$f'(x) = 3x^2 - 3$

$f'(x)$ exists for all real numbers. We solve $f'(x) = 0$:

$$3x^2 - 3 = 0$$
$$x^2 - 1 = 0 \quad \text{Dividing by 3}$$
$$(x + 1)(x - 1) = 0$$
$$x + 1 = 0 \quad \text{or} \quad x - 1 = 0$$
$$x = -1 \quad \text{or} \quad x = 1$$

The critical points are -1 and 1. We use them to divide the real number line into three intervals, A: $(-\infty, -1)$, B: $(-1, 1)$, and C: $(1, \infty)$.

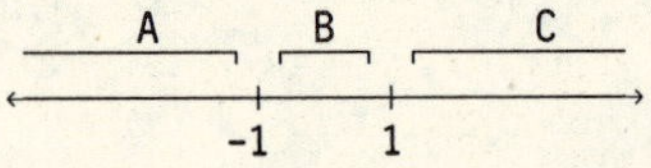

We use a test value in each interval to determine the sign of the derivative in each interval.

A: Test -2, $f'(-2) = 3(-2)^2 - 3 = 12 - 3 = 9 > 0$

B: Test 0, $f'(0) = 3 \cdot 0^2 - 3 = 0 - 3 = -3 < 0$

C: Test 2, $f'(2) = 3 \cdot 2^2 - 3 = 9 > 0$

We see that $f(x)$ is increasing on $(-\infty, -1)$, decreasing on $(-1, 1)$, and increasing again on $(1, \infty)$, so there is a relative maximum at $x = -1$ and a relative minimum at $x = 1$. We find $f(-1)$:

$$f(-1) = (-1)^3 - 3(-1) + 6 = -1 + 3 + 6 = 8$$

Then we find $f(1)$:

$$f(1) = 1^3 - 3 \cdot 1 + 6 = 1 - 3 + 6 = 4$$

There is a relative maximum at $(-1,8)$, and there is a relative minimum at $(1,4)$. We use the information obtained to sketch the graph. Other function values are listed below.

x	f(x)
-3	-12
-2	4
0	6
2	8
3	24

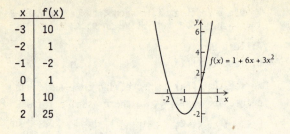

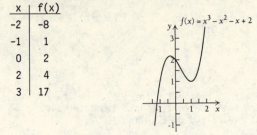

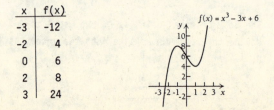

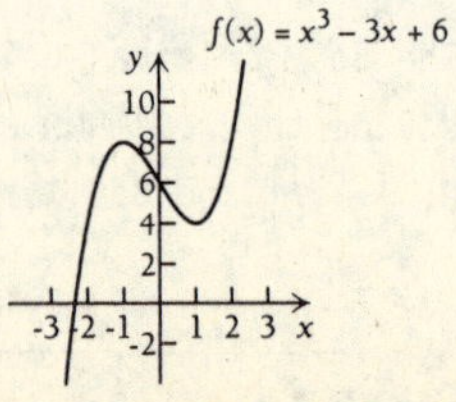

11. $f(x) = 3x^2 - 2x^3$

First, find the critical points.

$f'(x) = 6x - 6x^2$

$f'(x)$ exists for all real numbers. We solve $f'(x) = 0$:

$6x - 6x^2 = 0$

$x - x^2 = 0$ Dividing by 6

$x(1 - x) = 0$

$x = 0$ or $1 - x = 0$

$x = 0$ or $1 = x$

The critical points are 0 and 1. We use them to divide the real number line into three intervals, A: $(-\infty,0)$, B: $(0,1)$, and C: $(1,\infty)$.

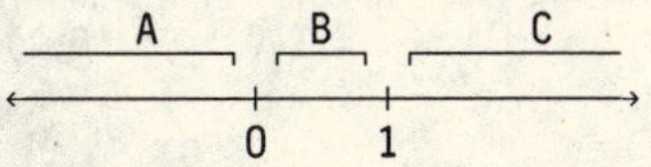

We use a test value in each interval to determine the sign of the derivative in each interval.

A: Test -1, $f'(-1) = 6(-1) - 6(-1)^2 = -6 - 6 = -12 < 0$

B: Test $\frac{1}{2}$, $f'\left(\frac{1}{2}\right) = 6\left(\frac{1}{2}\right) - 6\left(\frac{1}{2}\right)^2 = 3 - \frac{3}{2} = \frac{3}{2} > 0$

C: Test 2, $f'(2) = 6\cdot2 - 6\cdot2^2 = 12 - 24 = -12 < 0$

We see that $f(x)$ is decreasing on $(-\infty,0)$, increasing on $(0,1)$, and decreasing again on $(1,\infty)$, so there is a relative minimum at $x = 0$ and a relative maximum at $x = 1$. We find $f(0)$:

$f(0) = 3\cdot0^2 - 2\cdot0^3 = 0 - 0 = 0$

Then we find $f(1)$:

$f(1) = 3\cdot1^2 - 2\cdot1^3 = 3 - 2 = 1$

There is a relative minimum at (0,0), and there is a relative maximum at (1,1). We use the information obtained to sketch the graph. Other function values are listed below.

x	f(x)
-2	28
-1	5
2	-4
3	-27

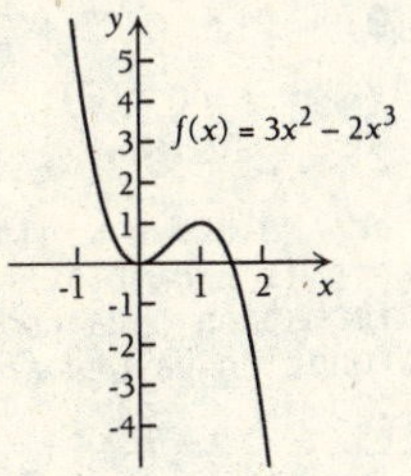

13. $f(x) = 2x^3$

First, find the critical points.

$f'(x) = 6x^2$

$f'(x)$ exists for all real numbers. We solve $f'(x) = 0$:

$6x^2 = 0$

$x^2 = 0$ Dividing by 6

$x = 0$

The only critical point is 0. We use 0 to divide the real number line into two intervals, A: $(-\infty,0)$, and B: $(0,\infty)$:

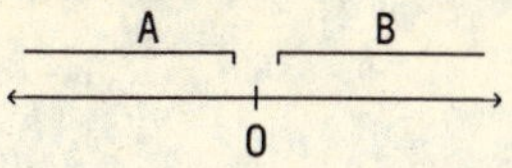

We use a test value in each interval to determine the sign of the derivative in each interval.

A: Test -1, $f'(-1) = 6(-1)^2 = 6\cdot1 = 6 > 0$

B: Test 1, $f'(1) = 6(1)^2 = 6\cdot1 = 6 > 0$

We see that $f(x)$ is increasing on both intervals, so the function has no relative extrema. We use the information obtained to sketch the graph. Some function values are listed below.

x	f(x)
-2	-16
-1	-2
0	0
1	2
2	16

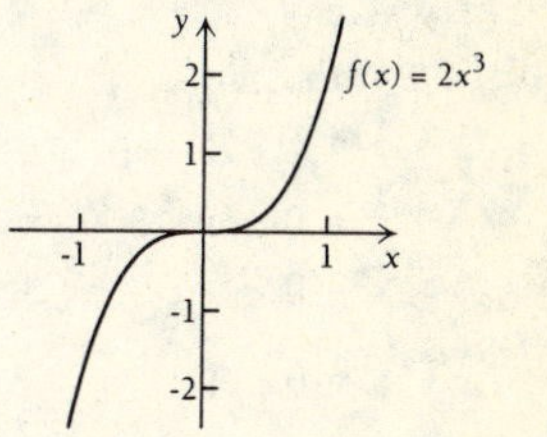

15. $f(x) = x^3 - 6x^2 + 10$

First, find the critical points.

$f'(x) = 3x^2 - 12x$

$f'(x)$ exists for all real numbers. We solve $f'(x) = 0$:

$3x^2 - 12x = 0$

$x^2 - 4x = 0$ Dividing by 3

$x(x - 4) = 0$

$x = 0$ or $x - 4 = 0$

$x = 0$ or $x = 4$

The critical points are 0 and 4. We use them to divide the real number line into three intervals, A: $(-\infty,0)$, B: $(0,4)$, and C: $(4,\infty)$.

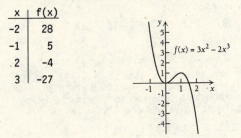

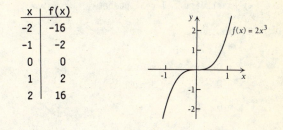

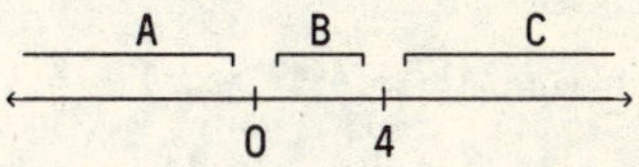

We use a test value in each interval to determine the sign of the derivative in each interval.

A: Test -1, $f'(-1) = 3(-1)^2 - 12(-1) = 3 + 12 = 15 > 0$

B: Test 1, $f'(1) = 3\cdot1^2 - 12\cdot1 = 3 - 12 = -9 < 0$

C: Test 5, $f'(5) = 3\cdot5^2 - 12\cdot5 = 75 - 60 = 15 > 0$

We see that $f(x)$ is increaseing on $(-\infty,0)$, decreasing on $(0,4)$, and increasing again on $(4,\infty)$, so there is a relative maximum at $x = 0$ and a relative minimum at $x = 4$. We find $f(0)$:

$f(0) = 0^3 - 6\cdot0^2 + 10 = 0 - 0 + 10 = 10$

Then we find $f(4)$:

$f(4) = 4^3 - 6\cdot4^2 + 10 = 64 - 96 + 10 = -22$

There is a relative maximum at (0,10), and there is a relative minimum at (4,-22). We use the information obtained to sketch the graph. Other function values are listed below.

x	f(x)
-2	-22
-1	3
1	5
2	-6
3	-17

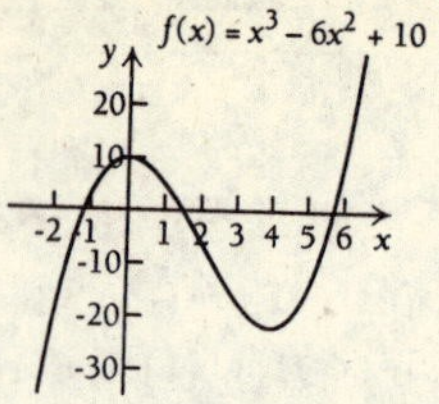

17. $f(x) = x^3 - x^4$

First, find the critical points.

$f'(x) = 3x^2 - 4x^3$

$f'(x)$ exists for all real numbers. We solve $f'(x) = 0$:

$$3x^2 - 4x^3 = 0$$
$$x^2(3 - 4x) = 0$$
$$x^2 = 0 \text{ or } 3 - 4x = 0$$
$$x = 0 \text{ or } 3 = 4x$$
$$x = 0 \text{ or } \frac{3}{4} = x$$

The critical points are 0 and $\frac{3}{4}$. We use them to divide the real number line into three intervals, A: $(-\infty,0)$, B: $\left[0,\frac{3}{4}\right]$, and C: $\left[\frac{3}{4},\infty\right]$.

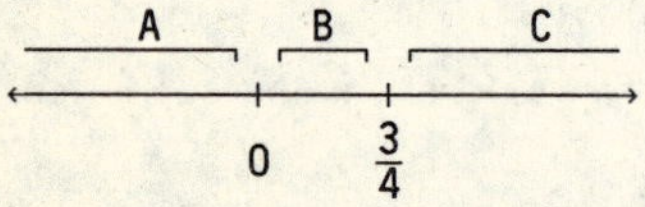

We use a test value in each interval to determine the sign of the derivative in each interval.

A: Test -1, $f'(-1) = 3(-1)^2 - 4(-1)^3 = 3\cdot1 - 4(-1) = 7 > 0$

B: Test $\frac{1}{2}$, $f'\left(\frac{1}{2}\right) = 3\left(\frac{1}{2}\right)^2 - 4\left(\frac{1}{2}\right)^3 = 3\cdot\frac{1}{4} - 4\cdot\frac{1}{8} = \frac{1}{4} > 0$

C: Test 1, $f'(1) = 3\cdot1^2 - 4\cdot1^3 = 3\cdot1 - 4\cdot1 = -1 < 0$

We see that $f(x)$ is increasing on both $(-\infty,0)$ and $\left[0,\frac{3}{4}\right]$ and is decreasing on $\left[\frac{3}{4},\infty\right]$, so there is no relative extremum at $x = 0$ but there is a relative maximum at $x = \frac{3}{4}$. We find $f\left(\frac{3}{4}\right)$:

$$f\left(\frac{3}{4}\right) = \left(\frac{3}{4}\right)^3 - \left(\frac{3}{4}\right)^4 = \frac{27}{64} - \frac{81}{256} = \frac{27}{256}$$

There is a relative maximum at $\left(\frac{3}{4},\frac{27}{256}\right)$. We use the information obtained to sketch the graph. Other function values are listed below.

x	f(x)
-2	-24
-1	-2
0	0
$\frac{1}{2}$	$\frac{1}{16}$
1	0
2	-8

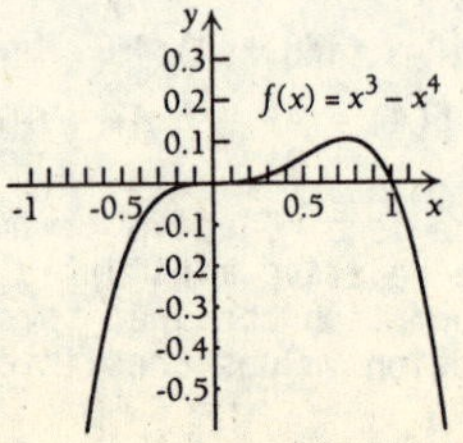

19. $f(x) = x^4 - 8x^2 + 3$

First, find the critical points.

$f'(x) = 4x^3 - 16x$

$f'(x)$ exists for all real numbers. We solve $f'(x) = 0$.

$$4x^3 - 16x = 0$$
$$x^3 - 4x = 0 \quad \text{(Dividing by 4)}$$
$$x(x^2 - 4) = 0$$
$$x(x + 2)(x - 2) = 0$$
$$x = 0 \text{ or } x + 2 = 0 \text{ or } x - 2 = 0$$
$$x = 0 \text{ or } x = -2 \text{ or } x = 2$$

The critical points are -2, 0, and 2. We use them to divide the real number line into four intervals, A: $(-\infty,-2)$, B: $(-2,0)$: C: $(0,2)$, and D: $(2,\infty)$.

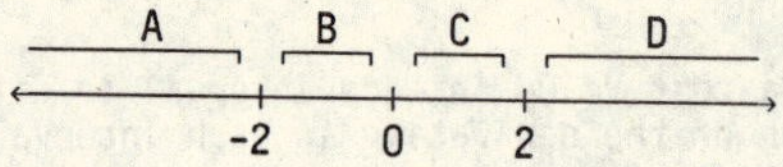

We use a test value in each interval to determine the sign of the derivative in each interval.

A: Test -3, $f'(-3) = 4(-3)^3 - 16(-3) = -108 + 48 = -60 < 0$

B: Test -1, $f'(-1) = 4(-1)^3 - 16(-1) = -4 + 16 = 12 > 0$

C: Test 1, $f'(1) = 4\cdot1^3 - 16\cdot1 = 4 - 16 = -12 < 0$

D: Test 3, $f'(3) = 4\cdot3^3 - 16\cdot3 = 108 - 48 = 60 > 0$

We see that $f(x)$ is decreasing on $(-\infty,-2)$, increasing on $(-2,0)$, decreasing again on $(0,2)$, and increasing again on $(2,\infty)$. Thus, there is a relative minimum at $x = -2$, a relative maximum at $x = 0$, and another relative minimum at $x = 2$.

We find $f(-2)$:

$$f(-2) = (-2)^4 - 8(-2)^2 + 3 = 16 - 32 + 3 = -13$$

Then we find $f(0)$:

$$f(0) = 0^4 - 8\cdot0^2 + 3 = 0 - 0 + 3 = 3$$

Finally, we find $f(2)$:

$$f(2) = 2^4 - 8\cdot2^2 + 3 = 16 - 32 + 3 = -13$$

There are relative minima at $(-2,-13)$ and $(2,-13)$, and there is a relative maximum at $(0,3)$. We use the information obtained to sketch the graph. Other function values are listed below.

x	f(x)
-3	12
-1	-4
1	-4
3	12

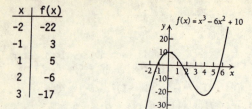

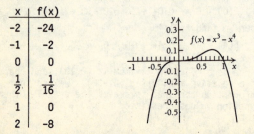

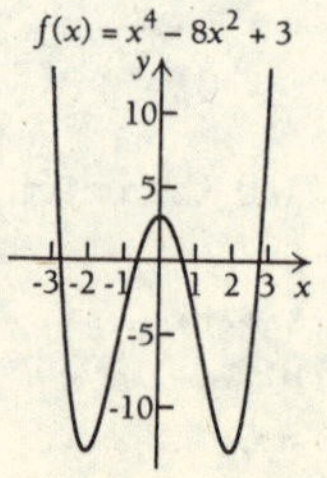

21. $f(x) = 1 - x^{2/3}$

First, find the critical points.

$$f'(x) = -\frac{2}{3}x^{-1/3} = -\frac{2}{3\sqrt[3]{x}}$$

$f'(x)$ does not exist for $x = 0$. The equation $f'(x) = 0$ has no solution, so the only critical point is 0. We use it to divide the real number line into two intervals, A: $(-\infty,0)$ and B: $(0,\infty)$.

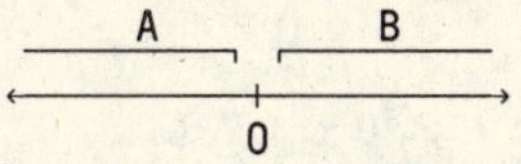

We use a test value in each interval to determine the sign of the derivative in each interval.

A: Test -1, $f'(-1) = -\frac{2}{3\sqrt[3]{-1}} = -\frac{2}{3(-1)} = \frac{2}{3} > 0$

B: Test 1, $f'(1) = -\frac{2}{3\sqrt[3]{1}} = -\frac{2}{3\cdot 1} = -\frac{2}{3} < 0$

We see that $f(x)$ is increasing on $(-\infty,0)$ and decreasing on $(0,\infty)$, so there is a relative maximum at $x = 0$.

We find $f(0)$:

$$f(0) = 1 - 0^{2/3} = 1 - 0 = 1$$

There is a relative maximum at (0,1). We use the information obtained to sketch the graph. Other function values are listed below.

x	f(x)
-27	-8
-8	-3
-1	0
1	0
8	-3
27	8

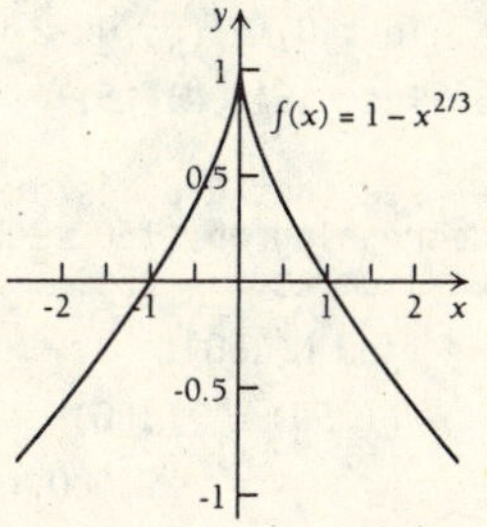

23. $f(x) = \frac{-8}{x^2 + 1} = -8(x^2 + 1)^{-1}$

First, find the critical points.

$$f'(x) = -8(-1)(x^2 + 1)^{-2}(2x)$$
$$= 16x(x^2 + 1)^{-2}$$
$$= \frac{16x}{(x^2 + 1)^2}$$

$f'(x)$ exists for all real numbers. We solve $f'(x) = 0$:

$$\frac{16x}{(x^2 + 1)^2} = 0$$

$16x = 0$ (Multiplying by $(x^2 + 1)^2$)

$x = 0$

The only critical point is 0. We use it to divide the real number line into two intervals, A: $(-\infty,0)$ and B: $(0,\infty)$.

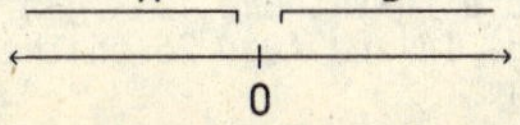

We use a test value in each interval to determine the sign of the derivative in each interval.

A: Test -1, $f'(-1) = \frac{16(-1)}{[(-1)^2 + 1]^2} = \frac{-16}{4} = -4 < 0$

B: Test 1, $f'(1) = \frac{16\cdot 1}{(1^2 + 1)^2} = \frac{16}{4} = 4 > 0$

We see that $f(x)$ is decreasing on $(-\infty,0)$ and increasing on $(0,\infty)$, so there is a relative minimum at $x = 0$.

We find $f(0)$:

$$f(0) = \frac{-8}{0^2 + 1} = \frac{-8}{1} = -8$$

There is a relative minimum at (0,-8). We use the information obtained to sketch the graph. Other function values are listed below.

x	f(x)
-4	$-\frac{8}{17}$
-3	$-\frac{4}{5}$
-2	$-\frac{8}{5}$
-1	-4
1	4
2	$-\frac{8}{5}$
3	$-\frac{4}{5}$
4	$-\frac{8}{17}$

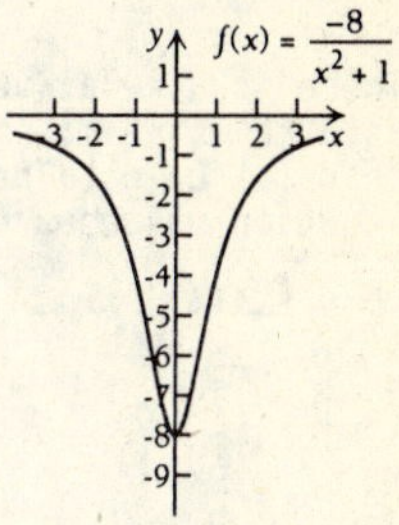

25. $f(x) = \frac{4x}{x^2 + 1}$

First, find the critical points.

$$f'(x) = \frac{(x^2 + 1)(4) - 2x(4x)}{(x^2 + 1)^2} \quad \text{(Quotient Rule)}$$
$$= \frac{4x^2 + 4 - 8x^2}{(x^2 + 1)^2}$$
$$= \frac{4 - 4x^2}{(x^2 + 1)^2}$$

$f'(x)$ exists for all real numbers. We solve $f'(x) = 0$:

$$\frac{4 - 4x^2}{(x^2 + 1)^2} = 0$$

$4 - 4x^2 = 0$ (Multiplying by $(x^2 + 1)^2$)

$1 - x^2 = 0$ (Dividing by 4)

$(1 - x)(1 + x) = 0$

$1 - x = 0$ or $1 + x = 0$

$1 = x$ or $x = -1$

The critical points are -1 and 1. We use them to divide the real number line into three intervals, A: $(-\infty,-1)$, B: $(-1,1)$, and C: $(1,\infty)$.

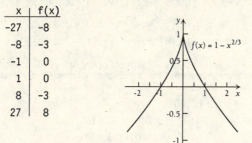

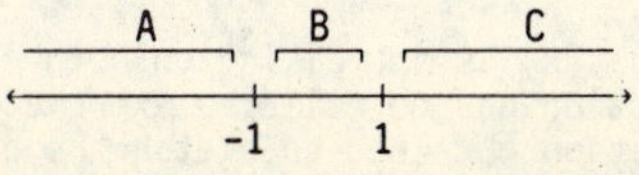

We use a test value in each interval to determine the sign of the derivative in each interval.

A: Test (-2), $f'(-2) = \frac{4 - 4(-2)^2}{[(-2)^2 + 1]^2} = \frac{-12}{25} < 0$

B: Test 0, $f'(0) = \frac{4 - 4\cdot 0^2}{(0^2 + 1)^2} = \frac{4}{1} = 4 > 0$

C: Test 2, $f'(2) = \frac{4 - 4\cdot 2^2}{(2^2 + 1)^2} = \frac{-12}{25} < 0$

We see that f(x) is decreasing on $(-\infty,-1)$, increasing on (-1,1), and decreasing again on $(1,\infty)$, so there is a relative minimum at x = -1 and a relative maximum at x = 1.

We find f(-1):

$$f(-1) = \frac{4(-1)}{(-1)^2 + 1} = \frac{-4}{2} = -2$$

Then we find f(1):

$$f(1) = \frac{4\cdot 1}{1^2 + 1} = \frac{4}{2} = 2$$

There is a relative minimum at (-1,-2), and there is a relative maximum at (1,2). We use the information obtained to sketch the graph. Other function values are listed below.

x	f(x)
-4	$-\frac{16}{17}$
-3	$-\frac{6}{5}$
-2	$-\frac{8}{5}$
0	0
2	$\frac{8}{5}$
3	$\frac{6}{5}$
4	$\frac{16}{17}$

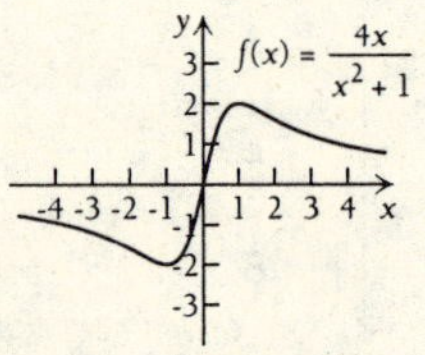

27. $f(x) = \sqrt[3]{x} = x^{1/3}$

First, find the critical points.

$f'(x) = \frac{1}{3}x^{-2/3} = \frac{1}{3\sqrt[3]{x^2}}$

f'(x) does not exist for x = 0. The equation f'(x) = 0 has no solution, so the only critical point is 0. We use it to divide the real number line into two intervals, A: $(-\infty,0)$ and B: $(0,\infty)$.

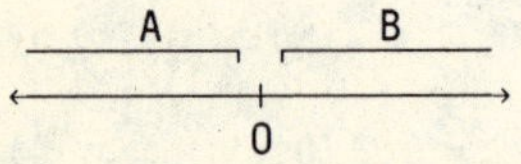

We use a test value in each interval to determine the sign of the derivative in each interval.

A: Test -1, $f'(-1) = \frac{1}{3\sqrt[3]{(-1)^2}} = \frac{1}{3\cdot 1} = \frac{1}{3} > 0$

B: Test 1, $f'(1) = \frac{1}{3\sqrt[3]{1^2}} = \frac{1}{3\cdot 1} = \frac{1}{3} > 0$

We see that f(x) is increasing on both intervals, so the function has no relative extrema. We use the information obtained to sketch the graph. Some function values are listed below.

x	f(x)
-27	-3
-8	-2
-1	-1
0	0
1	1
8	2
27	3

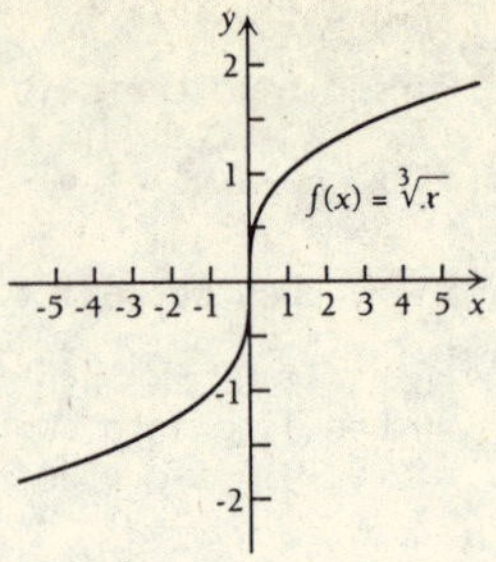

29. $N(a) = -a^2 + 300a + 6,\ 0 \leqslant a \leqslant 300$

First, find the critical points.

$N'(a) = -2a + 300$

N'(a) exists for all real numbers. We solve N'(a) = 0:

$$-2a + 300 = 0$$
$$-2a = -300$$
$$a = 150$$

The only critical point is 150. We use it to divide the interval [0,300] (the domain of N(a)) into two intervals, A: [0,150) and B: (150, 300].

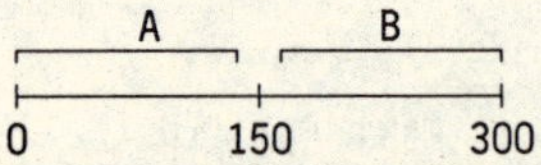

We use a test value in each interval to determine the sign of the derivative in each interval.

A: Test 0, $N'(0) = -2\cdot 0 + 300 = 300 > 0$

B: Test 151, $N'(151) = -2\cdot 151 + 300 = -2 < 0$

We see that N(a) is increasing on [0,150) and decreasing on (150, 300], so there is a relative maximum at x = 150.

We find N(150):

$$N(150) = -(150)^2 + 300(150) + 6$$
$$= -22{,}500 + 45{,}000 + 6$$
$$= 22{,}506$$

There is a relative maximum at (150, 22,506). We use the information obtained to sketch the graph. Other function values are listed below.

x	f(x)
0	6
50	12,506
100	20,006
200	20,006
250	12,506
300	6

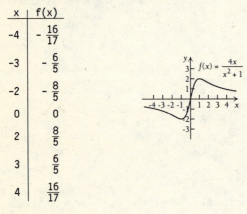

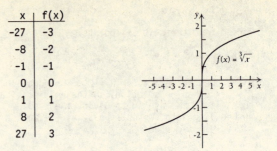

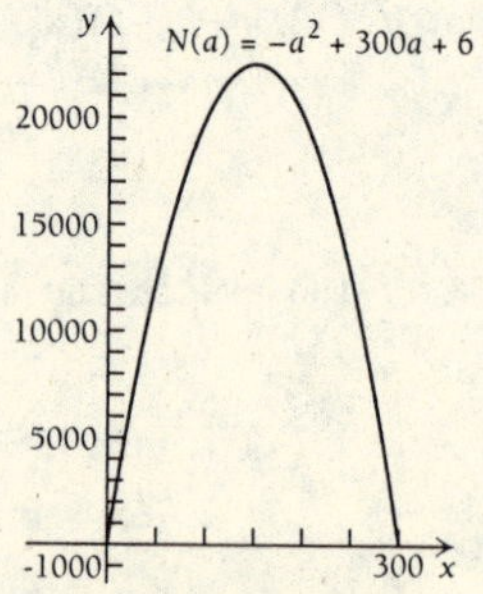

31. We look for points c in an interval (a,b) for which the graph is increasing on (a,c) and decreasing on (c,b). There are 5 such points. They occur on Aug. 5, Aug. 9, Aug. 13, Aug. 28, and Aug. 31. (Note that there is a gap in the dates between Aug. 14 and Aug. 26 on the horizontal axis.)

33. Using the graph shown in the answer section in the text, we estimate that relative minima occur at (-5, 425) and (4, -304) and that a relative maximum occurs at (-2, 560).

Exercise Set 3.2

1. $f(x) = 2 - x^2$

a) Find f'(x) and f''(x).

$f'(x) = -2x$

$f''(x) = -2$

b) Find the critical points of f.

Since f'(x) exists for all values of x, the only critical points are where $-2x = 0$.

$-2x = 0$

$x = 0$ (Critical point)

Find the function value at x = 0.

$f(0) = 2 - 0^2 = 2$

This gives the point (0,2) on the graph.

c) Use the Second-Derivative Test:

$f''(x) = -2$

$f''(0) = -2 < 0$

This tells us that (0,2) is a relative maximum. Then we can deduce that f(x) is increasing on $(-\infty,0)$ and decreasing on $(0,\infty)$.

d) Find possible inflection points.

The second derivative, f''(x), exists and is -2 for all real numbers. Note that f''(x) is never 0. Thus, there are no possible inflection points.

e) Since f''(x) is always negative (f''(x) = -2), f is concave down on the interval $(-\infty,\infty)$.

f) Sketch the graph using the preceding information. By solving $2 - x^2 = 0$ we can easily find the x-intercepts. They are $(-\sqrt{2},0)$ and $(\sqrt{2},0)$. Other function values can also be calculated.

x	f(x)
-2	-2
-1	1
1	1
2	-2

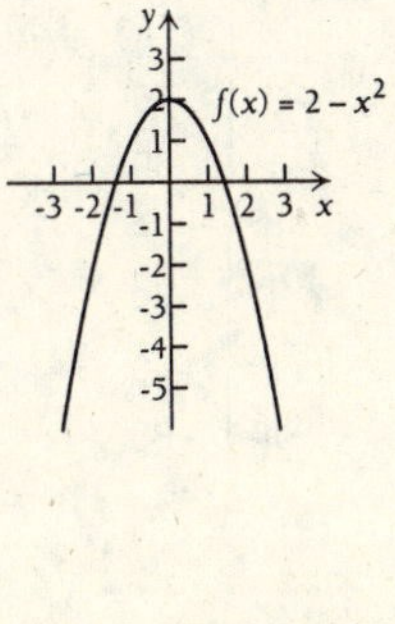

3. $f(x) = x^2 + x - 1$

a) Find f'(x) and f''(x).

$f'(x) = 2x + 1$

$f''(x) = 2$

b) Find the critical points of f.

Since $f'(x) = 2x + 1$ exists for all values of x, the only critical points are where $2x + 1 = 0$.

$2x + 1 = 0$

$2x = -1$

$x = -\frac{1}{2}$ (Critical point)

Find the function value at $x = -\frac{1}{2}$.

$f\left(-\frac{1}{2}\right) = \left(-\frac{1}{2}\right)^2 + \left(-\frac{1}{2}\right) - 1$

$= \frac{1}{4} - \frac{2}{4} - \frac{4}{4}$

$= -\frac{5}{4}$

This gives the point $\left(-\frac{1}{2}, -\frac{5}{4}\right)$ on the graph.

c) Use the Second-Derivative Test:

$f''(x) = 2$

$f''\left(-\frac{1}{2}\right) = 2 > 0$

This tells us that $\left(-\frac{1}{2}, -\frac{5}{4}\right)$ is a relative minimum. Then we can deduce that f(x) is decreasing on $\left(-\infty, -\frac{1}{2}\right)$ and increasing on $\left(-\frac{1}{2}, \infty\right)$.

d) Find the possible inflection points. The second derivative, f''(x), exists and is 2 for all real numbers. Note that f''(x) is never 0. Thus, there are no possible inflection points.

e) Note that f''(x) is always positive, f''(x) = 2. Thus, f is concave up on the interval $(-\infty,\infty)$.

f) Sketch the graph using the preceding information. Other function values can be calculated.

x	f(x)
-3	5
-2	1
-1	-1
0	-1
1	1
2	5

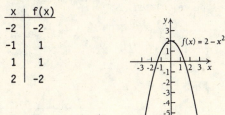

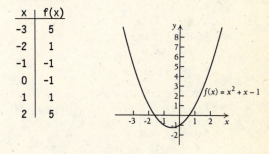

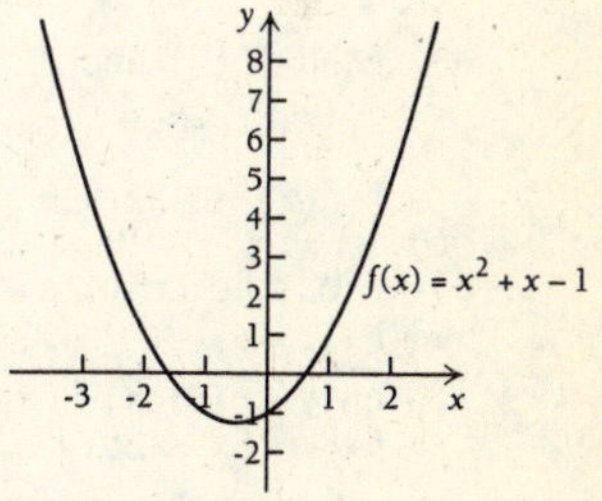

5. $f(x) = -4x^2 + 3x - 1$

a) Find f'(x) and f''(x).

$f'(x) = -8x + 3$

$f''(x) = -8$

b) Find the critical points of f.

Since f'(x) exists for all values of x, the only critical points are where $-8x + 3 = 0$.

$-8x + 3 = 0$

$-8x = -3$

$x = \frac{3}{8}$ (Critical point)

Then $f\left(\frac{3}{8}\right) = -4\left(\frac{3}{8}\right)^2 + 3\left(\frac{3}{8}\right) - 1$

$= -\frac{9}{16} + \frac{9}{8} - 1$

$= -\frac{7}{16}.$

This gives the point $\left(\frac{3}{8}, -\frac{7}{16}\right)$ on the graph.

c) Use the Second-Derivative Test:

$f''(x) = -8$

$f''\left(\frac{3}{8}\right) = -8 < 0$

This tells us that $\left(\frac{3}{8}, -\frac{7}{16}\right)$ is a relative maximum. Then we can deduce that f(x) is increasing on $\left(-\infty, \frac{3}{8}\right)$ and decreasing on $\left(\frac{3}{8}, \infty\right)$.

d) Find the possible inflection points. f''(x) exists and is -8 for all real numbers. Note that f''(x) is never 0. Thus, there are no possible inflection points.

e) Since f''(x) is always negative (f''(x) = -8), f is concave down on $(-\infty,\infty)$.

f) Sketch the graph using the preceding information. Other function values can also be calculated.

x	f(x)
-1	-8
0	-1
1	-2
2	-11

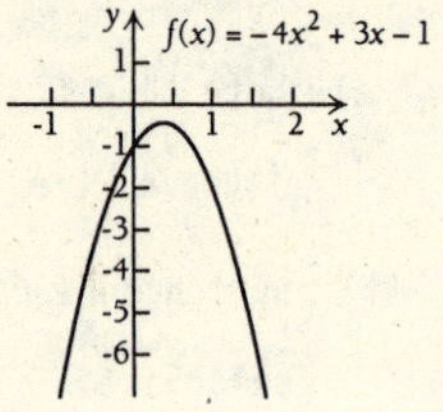

7. $f(x) = 2x^3 - 3x^2 - 36x + 28$

a) Find f'(x) and f''(x).

$f'(x) = 6x^2 - 6x - 36$

$f''(x) = 12x - 6$

b) Find the critical points of f.

Since f'(x) exists for all values of x, the only critical points are where $6x^2 - 6x - 36 = 0$.

$6x^2 - 6x - 36 = 0$

$6(x + 2)(x - 3) = 0$

$x + 2 = 0$ or $x - 3 = 0$

$x = -2$ or $x = 3$ (Critical points)

Then $f(-2) = 2(-2)^3 - 3(-2)^2 - 36(-2) + 28$

$= -16 - 12 + 72 + 28$

$= 72,$

and $f(3) = 2\cdot3^3 - 3\cdot3^2 - 36\cdot3 + 28$

$= 54 - 27 - 108 + 28$

$= -53.$

These give the points (-2,72) and (3,-53) on the graph.

c) Use the Second-Derivative Test:

$f''(-2) = 12(-2) - 6 = -30 < 0$, so (-2,72) is a relative maximum.

$f''(3) = 12\cdot3 - 6 = 30 > 0$, so (3,-53) is a relative minimum.

Then if we use the points -2 and 3 to divide the real number line into three intervals, $(-\infty,-2)$, $(-2,3)$, and $(3,\infty)$, we know that f is increasing on $(-\infty,-2)$, decreasing on $(-2,3)$, and increasing again on $(3,\infty)$.

d) Find the possible inflection points.

f''(x) exists for all values of x, so we solve f''(x) = 0.

$12x - 6 = 0$

$12x = 6$

$x = \frac{1}{2}$ (Possible inflection point)

Then $f\left(\frac{1}{2}\right) = 2\left(\frac{1}{2}\right)^3 - 3\left(\frac{1}{2}\right)^2 - 36\left(\frac{1}{2}\right) + 28$

$= \frac{1}{4} - \frac{3}{4} - 18 + 28$

$= \frac{19}{2}.$

This gives the point $\left(\frac{1}{2}, \frac{19}{2}\right)$ on the graph.

e) To determine the concavity we use the possible inflection point, $\frac{1}{2}$, to divide the real number line into two intervals, A: $\left(-\infty, \frac{1}{2}\right)$ and B: $\left(\frac{1}{2}, \infty\right)$. Test a point in each interval.

A: Test 0, $f''(0) = 12\cdot0 - 6 = -6 < 0$

B: Test 1, $f''(1) = 12\cdot1 - 6 = 6 > 0$

Then f is concave down on $\left(-\infty, \frac{1}{2}\right)$ and concave up on $\left(\frac{1}{2}, \infty\right)$, so $\left(\frac{1}{2}, \frac{19}{2}\right)$ is an inflection point.

f) Sketch the graph using the preceding information. Other function values can also be calculated.

x	f(x)
-3	55
-1	59
0	28
1	-9
2	-40
4	-36

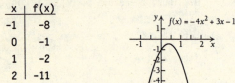

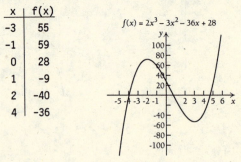

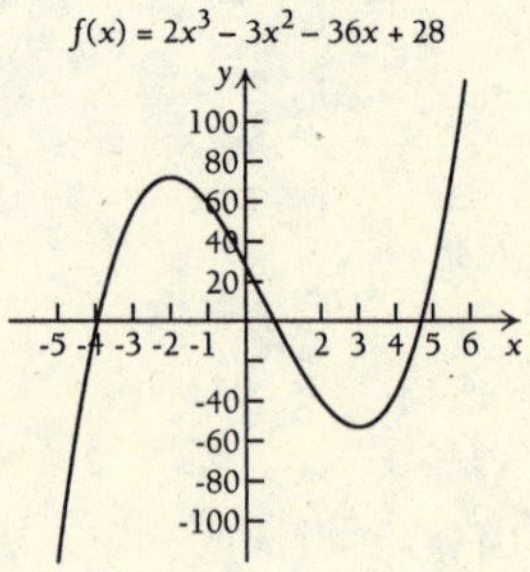

9. $f(x) = \frac{8}{3}x^3 - 2x + \frac{1}{3}$

a) Find f'(x) and f''(x).

$f'(x) = 8x^2 - 2$

$f''(x) = 16x$

b) Find the critical points of f.

Now $f'(x) = 8x^2 - 2$ exists for all values of x, so the only critical points of f are where $8x^2 - 2 = 0$.

$$8x^2 - 2 = 0$$
$$8x^2 = 2$$
$$x^2 = \frac{2}{8}$$
$$x^2 = \frac{1}{4}$$
$$x = \pm\frac{1}{2} \quad \text{(Critical points)}$$

Then $f\left(-\frac{1}{2}\right) = \frac{8}{3}\left(-\frac{1}{2}\right)^3 - 2\left(-\frac{1}{2}\right) + \frac{1}{3}$

$= -\frac{1}{3} + 1 + \frac{1}{3}$

$= 1,$

and $f\left(\frac{1}{2}\right) = \frac{8}{3}\left(\frac{1}{2}\right)^3 - 2\left(\frac{1}{2}\right) + \frac{1}{3}$

$= \frac{1}{3} - 1 + \frac{1}{3}$

$= -\frac{1}{3}$

These give the points $\left(-\frac{1}{2}, 1\right)$ and $\left(\frac{1}{2}, -\frac{1}{3}\right)$ on the graph.

c) Use the Second-Derivative Test:

$f''\left(-\frac{1}{2}\right) = 16\left(-\frac{1}{2}\right) = -8 < 0$, so $\left(-\frac{1}{2}, 1\right)$ is a relative maximum.

$f''\left(\frac{1}{2}\right) = 16 \cdot \frac{1}{2} = 8 > 0$, so $\left(\frac{1}{2}, -\frac{1}{3}\right)$ is a relative minimum.

Then if we use the points $-\frac{1}{2}$ and $\frac{1}{2}$ to divide the real number line into three intervals, A: $\left(-\infty, -\frac{1}{2}\right)$, B: $\left(-\frac{1}{2}, \frac{1}{2}\right)$, and C: $\left(\frac{1}{2}, \infty\right)$, we know that f is increasing on $\left(-\infty, -\frac{1}{2}\right)$, decreasing on $\left(-\frac{1}{2}, \frac{1}{2}\right)$, and increasing again on $\left(\frac{1}{2}, \infty\right)$.

d) Find the possible inflection points.

Now $f''(x) = 16x$ exists for all values of x, so the only critical points of f' are where $16x = 0$.

$$16x = 0$$
$$x = 0 \quad \text{(Possible inflection point)}$$

Then $f(0) = \frac{8}{3} \cdot 0^3 - 2\cdot 0 + \frac{1}{3} = \frac{1}{3}$, so $\left(0, \frac{1}{3}\right)$ is another point on the graph.

e) To determine the concavity we use the possible inflection point, 0, to divide the real number line into two intervals, A: $(-\infty,0)$ and B: $(0,\infty)$. Test a point in each interval.

A: Test -1, $f''(-1) = 16(-1) = -16 < 0$

B: Test 1, $f''(1) = 16\cdot 1 = 16 > 0$

Then f'(x) is concave down on $(-\infty,0)$ and concave up on $(0,\infty)$, so $\left(0,\frac{1}{3}\right)$ is an inflection point.

f) Sketch the graph using the preceding information. Other function values can also be calculated.

x	f(x)
-2	-17
-1	$-\frac{1}{3}$
0	$\frac{1}{3}$
1	1
2	$\frac{53}{3}$

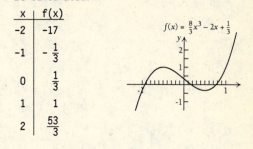

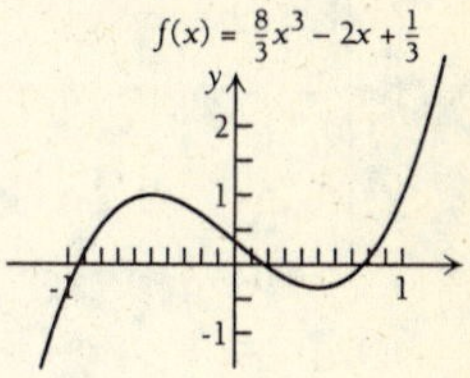

11. $f(x) = -x^3 + 3x^2 - 4$

a) Find f'(x) and f''(x).

$f'(x) = -3x^2 + 6x$

$f''(x) = -6x + 6$

b) Find the critical point of f.

Since f'(x) exists for all values of x, the only critical points are where $-3x^2 + 6x = 0$.

$$-3x^2 + 6x = 0$$
$$-3x(x - 2) = 0$$
$$-3x = 0 \text{ or } x - 2 = 0$$
$$x = 0 \text{ or } x = 2 \quad \text{(Critical points)}$$

Then $f(0) = -0^3 + 3\cdot 0^2 - 4 = -4$

and $f(2) = -2^3 + 3\cdot 2^2 - 4 = -8 + 12 - 4 = 0.$

These give the points (0,-4) and (2,0) on the graph.

c) Use the Second-Derivative Test:

$f''(0) = -6\cdot 0 + 6 = 6 > 0$, so (0,-4) is a relative minimum.

$f''(2) = -6\cdot 2 + 6 = -12 + 6 = -6 < 0$, so (2,0) is a relative maximum.

Then if we use the points 0 and 2 to divide the real number line into three intervals, $(-\infty,0)$, $(0,2)$, and $(2,\infty)$, we know that f is decreasing on $(-\infty,0)$, increasing on (0,2), and decreasing again on $(2,\infty)$.

d) Find the possible inflection points.

f''(x) exists for all values of x, so we solve $f''(x) = 0$.

$$-6x + 6 = 0$$
$$-6x = -6$$
$$x = 1 \quad \text{(Possible inflection point)}$$

Then $f(1) = -1^3 + 3\cdot 1^2 - 4 = -1 + 3 - 4 = -2$, so (1,-2) is another point on the graph.

e) To determine the concavity we use the possible inflection point, 1, to divide the real number line into two intervals, A: $(-\infty,1)$ and B: $(1,\infty)$. Test a point in each interval.

A: Test 0, $f''(0) = -6\cdot0 + 6 = 6 > 0$

B: Test 2, $f''(2) = -6\cdot2 + 6 = -6 < 0$

Then f is concave up on $(-\infty,0)$ and concave down on $(0,\infty)$, so $(1,-2)$ is an inflection point.

f) Sketch the graph. Other function values can also be calculated.

x	f(x)
-2	16
-1	0
3	-4
4	-20

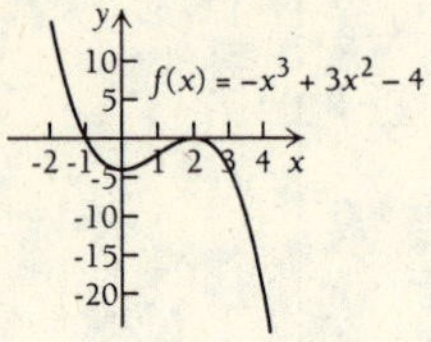

13. $f(x) = 3x^4 - 16x^3 + 18x^2$

a) $f'(x) = 12x^3 - 48x^2 + 36x$

$f''(x) = 36x^2 - 96x + 36$

b) Since $f'(x)$ exists for all values of x, the only critical points are where $f'(x) = 0$.

$$12x^3 - 48x^2 + 36x = 0$$
$$12x(x^2 - 4x + 3) = 0$$
$$12x(x - 1)(x - 3) = 0$$

$$12x = 0 \text{ or } x - 1 = 0 \text{ or } x - 3 = 0$$
$$x = 0 \text{ or } x = 1 \text{ or } x = 3$$

Then $f(0) = 3\cdot0^4 - 16\cdot0^3 + 18\cdot0^2 = 0$,

$f(1) = 3\cdot1^4 - 16\cdot1^3 + 18\cdot1^2 = 5$,

and $f(3) = 3\cdot3^4 - 16\cdot3^3 + 18\cdot3^2 = -27$.

These give the points (0,0), (1,5), and (3,-27) on the graph.

c) Use the Second-Derivative Test:

$f''(0) = 36\cdot0^2 - 96\cdot0 + 36 = 36 > 0$, so (0,0) is a relative minimum.

$f''(1) = 36\cdot1^2 - 96\cdot1 + 36 = -24 < 0$, so (1,5) is a relative maximum.

$f''(3) = 36\cdot3^2 - 96\cdot3 + 36 = 72 > 0$, so (3,-27) is a relative minimum.

Then if we use the points 0, 1, and 3 to divide the real number line into four intervals, $(-\infty,0)$, $(0,1)$, $(1,3)$, and $(3,\infty)$, we know that f is decreasing on $(-\infty,0)$ and on $(1,3)$ and is increasing on $(0,1)$ and $(3,\infty)$.

d) $f''(x)$ exists for all values of x, so the only possible inflection points are where $f''(x) = 0$.

$$36x^2 - 96x + 36 = 0$$
$$12(3x^2 - 8x + 3) = 0$$
$$3x^2 - 8x + 3 = 0$$

Using the quadratic formula, we find

$x = \dfrac{4 \pm \sqrt{7}}{3}$, so $x \approx 0.45$ or $x \approx 2.22$ are possible inflection points.

Then $f(0.45) \approx 2.31$ and $f(2.22) \approx -13.48$, so (0.45, 2.31) and (2.22, -13.48) are two more points on the graph.

e) To determine the concavity we use the points 0.45 and 2.22 to divide the real number line into three intervals, A: $(-\infty, 0.45)$, B: $(0.45, 2.22)$, and C: $(2.22, \infty)$. Test a point in each interval.

A: Test 0, $f''(0) = 36\cdot0^2 - 96\cdot0 + 36 = 36 > 0$

B: Test 1, $f''(1) = 36\cdot1^2 - 96\cdot1 + 36 = -24 < 0$

C: Test 3, $f''(3) = 36\cdot3^2 - 96\cdot3 + 36 = 72 > 0$

Then f is concave up on $(-\infty, 0.45)$, concave down on $(0.45, 2.22)$, and concave up on $(2.22, \infty)$, so (0.45, 2.31) and (2.22, -13.48) are inflection points.

f) Sketch the graph. Other function values can be calculated.

x	f(x)
-1	37
2	-8
4	32

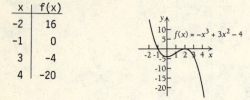

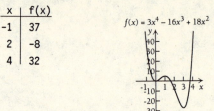

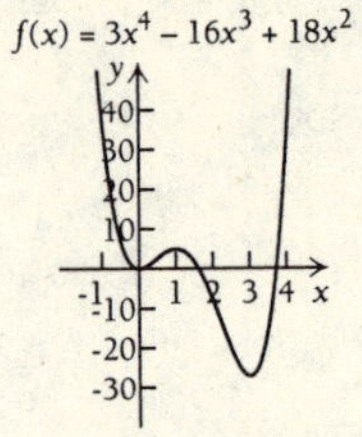

15. $f(x) = (x + 1)^{2/3}$

a) $f'(x) = \frac{2}{3}(x + 1)^{-1/3} = \dfrac{2}{3\sqrt[3]{x + 1}}$

$f''(x) = -\frac{2}{9}(x + 1)^{-4/3} = -\dfrac{2}{9\sqrt[3]{(x + 1)^4}}$

b) Since $f'(-1)$ does not exist, -1 is a critical point. The equation $f'(x) = 0$ has no solution, so the only critical point is -1.

Now $f(-1) = (-1 + 1)^{2/3} = 0^{2/3} = 0$.

This gives the point (-1,0) on the graph.

c) We cannot use the Second-Derivative Test, because $f''(-1)$ is not defined. We will use the First-Derivative Test. Use -1 to divide the real number line into two intervals, A: $(-\infty,-1)$ and B: $(-1,\infty)$. Test a point in each interval.

A: Test -2, $f'(-2) = \dfrac{2}{3\sqrt[3]{-2 + 1}} = -\dfrac{2}{3} < 0$

B: Test 0, $f'(0) = \dfrac{2}{3\sqrt[3]{0 + 1}} = \dfrac{2}{3} > 0$

Since $f(x)$ is decreasing on $(-\infty,-1)$ and increasing on $(-1,\infty)$, there is a relative minimum at (-1,0).

d) As $f''(-1)$ does not exist, -1 is a possible inflection point. The equation $f''(x) = 0$ has no solution, so the only possible inflection point is -1. We have already found $f(-1)$ in step b).

e) To determine the concavity we use -1 to divide the real number line into two intervals as in step c). Test a point in each interval.

A: Test -2, $f''(-2) = -\dfrac{2}{9\sqrt[3]{(-2+1)^4}} = -\dfrac{2}{9} < 0$

B: Test 0, $f''(0) = -\dfrac{2}{9\sqrt[3]{(0+1)^4}} = -\dfrac{2}{9} < 0$

Then f is concave down on both intervals, so there is no inflection point.

f) Sketch the graph using the preceding information. Other function values can be calculated.

x	f(x)
-9	4
-2	1
0	1
7	4

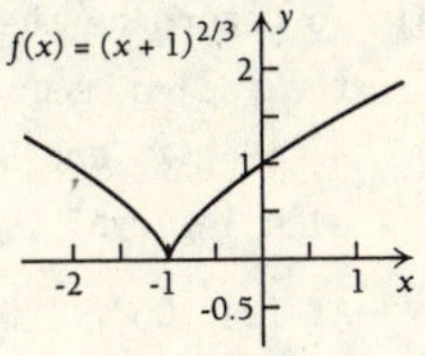

17. $f(x) = x^4 - 6x^2$

a) $f'(x) = 4x^3 - 12x$

$f''(x) = 12x^2 - 12$

b) Since $f'(x)$ exists for all values of x, the only critical points are where $4x^3 - 12x = 0$.

$4x^3 - 12x = 0$

$4x(x^2 - 3) = 0$

$4x = 0$ or $x^2 - 3 = 0$

$x = 0$ or $x^2 = 3$

$x = 0$ or $x = \pm\sqrt{3}$

The critical points are $-\sqrt{3}$, 0, and $\sqrt{3}$.

$f(-\sqrt{3}) = (-\sqrt{3})^4 - 6(-\sqrt{3})^2$
$= 9 - 6\cdot3 = 9 - 18 = -9$

$f(0) = 0^4 - 6\cdot0^2 = 0 - 0 = 0$

$f(\sqrt{3}) = (\sqrt{3})^4 - 6(\sqrt{3})^2$
$= 9 - 6\cdot3 = 9 - 18 = -9$

These give the points $(-\sqrt{3},-9)$, (0,0), and $(\sqrt{3},-9)$ on the graph.

c) Use the Second-Derivative Test:

$f''(-\sqrt{3}) = 12(-\sqrt{3})^2 - 12 = 12\cdot3 - 12 =$ 24 > 0, so $(-\sqrt{3}, -9)$ is a relative minimum.

$f''(0) = 12\cdot0^2 - 12 = -12 < 0$, so (0,0) is a relative maximum.

$f''(\sqrt{3}) = 12(\sqrt{3})^2 - 12 = 12\cdot3 - 12 = 24 > 0$, so $(\sqrt{3}, -9)$ is a relative minimum.

Then if we use the points $-\sqrt{3}$, 0, and $\sqrt{3}$ to divide the real number line into four intervals, $(-\infty, -\sqrt{3})$, $(-\sqrt{3}, 0)$, $(0, \sqrt{3})$, and $(\sqrt{3}, \infty)$, we know that f is decreasing on $(-\infty, -\sqrt{3})$ and on $(0, \sqrt{3})$ and is increasing on $(-\sqrt{3}, 0)$ and on $(\sqrt{3}, \infty)$.

d) Since $f''(x)$ exists for all values of x, the only possible inflection points are where $12x^2 - 12 = 0$.

$12x^2 - 12 = 0$

$x^2 - 1 = 0$

$(x + 1)(x - 1) = 0$

$x + 1 = 0$ or $x - 1 = 0$

$x = -1$ or $x = 1$

The possible inflection points are -1 and 1.

$f(-1) = (-1)^4 - 6(-1)^2 = 1 - 6\cdot1 = -5$

$f(1) = 1^4 - 6\cdot1^2 = 1 - 6\cdot1 = -5$

These give the points (-1,-5) and (1,-5) on the graph.

e) To determine the concavity we use the points -1 and 1 to divide the real number line into three intervals, A: $(-\infty,-1)$, B: $(-1,1)$, and $(1,\infty)$. Test a point in each interval.

A: Test -2, $f''(-2) = 12(-2)^2 - 12 = 36 > 0$

B: Test 0, $f''(0) = 12\cdot0^2 - 12 = -12 < 0$

C: Test 2, $f''(2) = 12\cdot2^2 - 12 = 36 > 0$

We see that f is concave up on the intervals $(-\infty,-1)$ and $(1,\infty)$ and concave down on the interval (-1,1), so (-1,-5) and (1,-5) are inflection points.

f) Sketch the graph using the preceding information. By solving $x^4 - 6x^2 = 0$ we can find the x-intercepts. They are helpful in graphing.

$x^4 - 6x^2 = 0$

$x^2(x^2 - 6) = 0$

$x^2 = 0$ or $x^2 - 6 = 0$

$x = 0$ or $x^2 = 6$

$x = 0$ or $x = \pm\sqrt{6}$

The x-intercepts are (0,0), $(-\sqrt{6},0)$, and $(\sqrt{6},0)$.

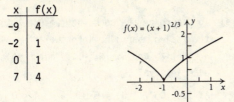

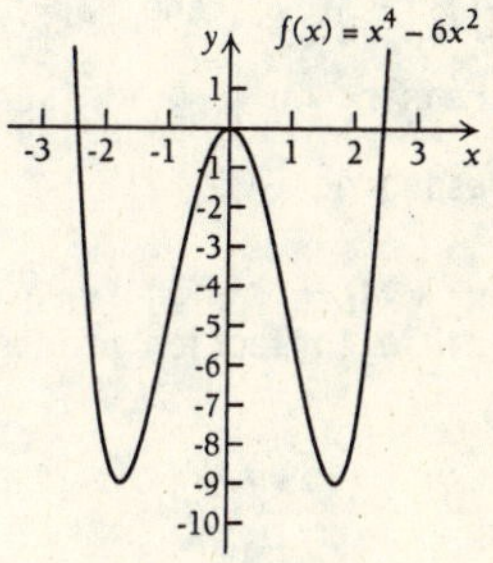

19. $f(x) = x^3 - 2x^2 - 4x + 3$

a) $f'(x) = 3x^2 - 4x - 4$

$f''(x) = 6x - 4$

b) Now $f'(x)$ exists for all values of x, so the only critical points of f are where $3x^2 - 4x - 4 = 0$.

$3x^2 - 4x - 4 = 0$

$(3x + 2)(x - 2) = 0$

$3x + 2 = 0$ or $x - 2 = 0$

$x = -\frac{2}{3}$ or $x = 2$

The critical points are $-\frac{2}{3}$ and 2.

$f\left(-\frac{2}{3}\right) = \left(-\frac{2}{3}\right)^3 - 2\left(-\frac{2}{3}\right)^2 - 4\left(-\frac{2}{3}\right) + 3$

$= -\frac{8}{27} - \frac{8}{9} + \frac{8}{3} + 3$

$= -\frac{8}{27} - \frac{24}{27} + \frac{72}{27} + \frac{81}{27}$

$= \frac{121}{27}$

$f(2) = 2^3 - 2\cdot2^2 - 4\cdot2 + 3$

$= 8 - 8 - 8 + 3$

$= -5$

These give the points $\left(-\frac{2}{3}, \frac{121}{27}\right)$ and (2,-5) on the graph.

c) Use the Second-Derivative Test:

$f''\left(-\frac{2}{3}\right) = 6\left(-\frac{2}{3}\right) - 4 = -4 - 4 = -8 < 0$, so $\left(-\frac{2}{3}, \frac{121}{27}\right)$ is a relative maximum.

$f''(2) = 6\cdot2 - 4 = 12 - 4 = 8 > 0$, so (2,-5) is a relative minimum.

Then if we use the points $-\frac{2}{3}$ and 2 to divide the real number line into three intervals, $\left(-\infty, -\frac{2}{3}\right)$, $\left(-\frac{2}{3}, 2\right)$, and $(2,\infty)$, we know that f is increasing on $\left(-\infty, -\frac{2}{3}\right)$ and on $(2,\infty)$ and is decreasing on $\left(-\frac{2}{3}, 2\right)$.

d) Now $f''(x)$ exists for all values of x, so the only possible inflection points are where

$6x - 4 = 0$.

$6x - 4 = 0$

$6x = 4$

$x = \frac{4}{6}$

$x = \frac{2}{3}$ (Possible inflection point)

$f\left(\frac{2}{3}\right) = \left(\frac{2}{3}\right)^3 - 2\left(\frac{2}{3}\right)^2 - 4\left(\frac{2}{3}\right) + 3$

$= \frac{8}{27} - \frac{8}{9} - \frac{8}{3} + 3$

$= \frac{8}{27} - \frac{24}{27} - \frac{72}{27} + \frac{81}{27}$

$= -\frac{7}{27}$

This gives another point $\left(\frac{2}{3}, -\frac{7}{27}\right)$ on the graph.

e) To determine the concavity we use $\frac{2}{3}$ to divide the real number line into two intervals, A: $\left(-\infty, \frac{2}{3}\right)$ and B: $\left(\frac{2}{3}, \infty\right)$. Test a point in each interval.

A: Test 0, $f''(0) = 6\cdot0 - 4 = -4 < 0$

B: Test 1, $f''(1) = 6\cdot1 - 4 = 2 > 0$

We see that f is concave down on $\left(-\infty, \frac{2}{3}\right)$ and concave up on $\left(\frac{2}{3}, \infty\right)$, so $\left(\frac{2}{3}, -\frac{7}{27}\right)$ is an inflection point.

f) Sketch the graph using the preceding information. Other function values can also be calculated.

x	f(x)
-2	-5
-1	4
0	3
1	-2
3	0
4	19

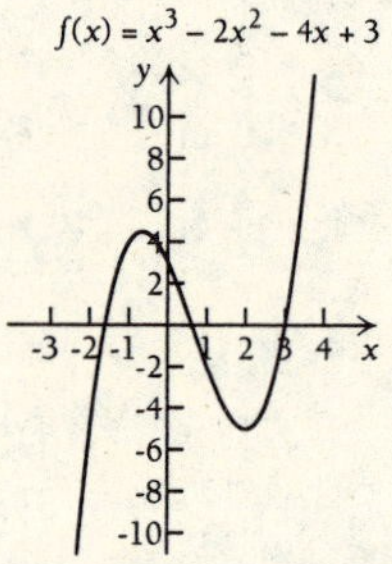

21. $f(x) = 3x^4 + 4x^3$

a) $f'(x) = 12x^3 + 12x^2$

$f''(x) = 36x^2 + 24x$

b) Since $f'(x)$ exists for all values of x, the only critical points of f are where $12x^3 + 12x^2 = 0$.

$12x^3 + 12x^2 = 0$

$12x^2(x + 1) = 0$

$12x^2 = 0$ or $x + 1 = 0$

$x = 0$ or $x = -1$

The critical points are 0 and -1.

$f(0) = 3\cdot0^4 + 4\cdot0^3 = 0 + 0 = 0$

$f(-1) = 3(-1)^4 + 4(-1)^3 = 3\cdot1 + 4(-1)$

$= 3 - 4$

$= -1$

These give the points (0,0) and (-1,-1) on the graph.

c) Use the Second-Derivative Test:

$f''(-1) = 36(-1)^2 + 24(-1) = 36 - 24 = 12 > 0$, so $(-1,-1)$ is a relative minimum.

$f''(0) = 36\cdot 0^2 + 24\cdot 0 = 0$, so this test fails. We will use the First-Derivative Test. Use 0 to divide the interval $(-1,\infty)$ into two intervals: A: $(-1,0)$ and $(0,\infty)$. Test a point in each interval.

A: Test $-\frac{1}{2}$, $f'\left(-\frac{1}{2}\right) = 12\left(-\frac{1}{2}\right)^3 + 12\left(\frac{1}{2}\right)^2 = \frac{3}{2} > 0$

B: Test 2, $f'(2) = 12\cdot 2^3 + 12\cdot 2^2 = 144 > 0$

Since f is increasing on both intervals, (0,0) is not a relative extremum. Since $(-1,-1)$ is a relative minimum, we also know that f is decreasing on $(-\infty,-1)$.

d) Now $f''(x)$ exists for all values of x, so the only possible inflection points are where $36x^2 + 24x = 0$.

$36x^2 + 24x = 0$

$12x(3x + 2) = 0$

$12x = 0$ or $3x + 2 = 0$

$x = 0$ or $x = -\frac{2}{3}$

The possible inflection points are 0 and $-\frac{2}{3}$.

$f(0) = 3\cdot 0^4 + 4\cdot 0^3 = 0$ [Already found in step b)]

$f\left(-\frac{2}{3}\right) = 3\left(-\frac{2}{3}\right)^4 + 4\left(-\frac{2}{3}\right)^3$

$= 3\cdot\frac{16}{81} + 4\cdot\left(-\frac{8}{27}\right)$

$= \frac{16}{27} - \frac{32}{27}$

$= -\frac{16}{27}$

This gives one additional point $\left(-\frac{2}{3}, -\frac{16}{27}\right)$ on the graph.

e) To determine the concavity we use $-\frac{2}{3}$ and 0 to divide the real number line into three intervals, A: $\left(-\infty, -\frac{2}{3}\right)$, B: $\left(-\frac{2}{3}, 0\right)$, and C: $(0,\infty)$. Test a point in each interval.

A: Test -1, $f''(-1) = 36(-1)^2 + 24(-1) = 12 > 0$

B: Test $-\frac{1}{2}$, $f''\left(-\frac{1}{2}\right) = 36\left(-\frac{1}{2}\right)^2 + 24\left(-\frac{1}{2}\right) = -3 < 0$

C: Test 1, $f''(1) = 36\cdot 1^2 + 24\cdot 1 = 60 > 0$

We see that f is concave up on the intervals $\left(-\infty, -\frac{2}{3}\right)$ and $(0,\infty)$ and concave down on the interval $\left(-\frac{2}{3}, 0\right)$, so $\left(-\frac{2}{3}, -\frac{16}{27}\right)$ and (0,0) are both inflection points.

f) Sketch the graph using the preceding information. By solving $3x^4 + 4x^3 = 0$ we can find the x-intercepts. They are helpful in graphing.

$3x^4 + 4x^3 = 0$

$x^3(3x + 4) = 0$

$x^3 = 0$ or $3x + 4 = 0$

$x = 0$ or $x = -\frac{4}{3}$

The intercepts are (0,0) and $\left(-\frac{4}{3}, 0\right)$.

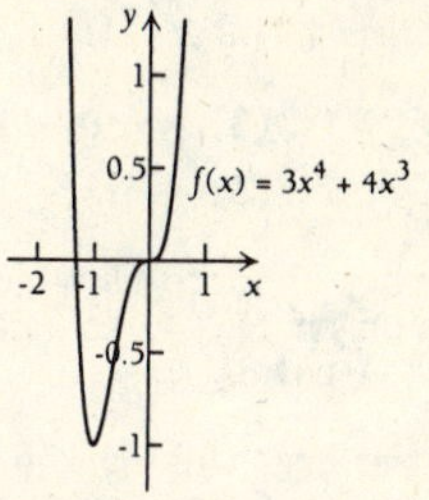

23. $f(x) = x^3 - 6x^2 - 135x$

a) $f'(x) = 3x^2 - 12x - 135$

$f''(x) = 6x - 12$

b) Since $f'(x)$ exists for all values of x, the only critical points of f are where $3x^2 - 12x - 135 = 0$.

$3x^2 - 12x - 135 = 0$

$x^2 - 4x - 45 = 0$

$(x - 9)(x + 5) = 0$

$x - 9 = 0$ or $x + 5 = 0$

$x = 9$ or $x = -5$

The critical points are 9 and -5.

$f(9) = 9^3 - 6\cdot 9^2 - 135\cdot 9$

$= 729 - 486 - 1215$

$= -972$

$f(-5) = (-5)^3 - 6(-5)^2 - 135(-5)$

$= -125 - 150 + 675$

$= 400$

These give the points (9,-972) and (-5,400) on the graph.

c) Use the Second-Derivative Test:

$f''(-5) = 6(-5) - 12 = -30 - 12 = -42 < 0$, so (-5,400) is a relative maximum.

$f''(9) = 6\cdot 9 - 12 = 54 - 12 = 42 > 0$, so (9,-972) is a relative minimum.

Then if we use the points -5 and 9 to divide the real number line into three intervals, $(-\infty,-5)$, $(-5,9)$, and $(9,\infty)$, we know that f is increasing on $(-\infty,-5)$ and on $(9,\infty)$ and is decreasing on $(-5,9)$.

d) Now $f''(x)$ exists for all values of x, so the only possible inflection points are where $6x - 12 = 0$.

$6x - 12 = 0$

$6x = 12$

$x = 2$ (Possible inflection point)

$f(2) = 2^3 - 6\cdot 2^2 - 135\cdot 2$
$= 8 - 24 - 270$
$= -286$

This gives another point (2,-286) on the graph.

e) To determine the concavity we use 2 to divide the real number line into two intervals, A: $(-\infty,2)$ and B: $(2,\infty)$. Test a point in each interval.

A: Test 0, $f''(0) = 6\cdot 0 - 12 = -12 < 0$

B: Test 3, $f''(3) = 6\cdot 3 - 12 = 6 > 0$

We see that f is concave down on $(-\infty,2)$ and concave up on $(2,\infty)$, so (2,-286) is an inflection point.

f) Sketch the graph using the preceding information. Other function values can also be calculated.

x	f(x)
-11	-572
-10	-250
-9	0
-3	324
0	0
2	-286
5	-700
15	0
16	400

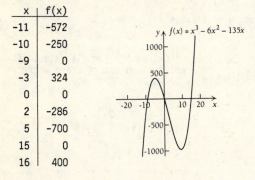

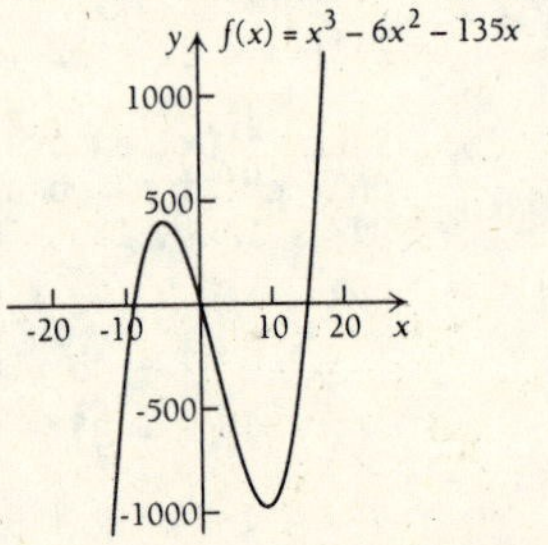

25. $f(x) = \dfrac{x}{x^2 + 1}$

a) $f'(x) = \dfrac{(x^2 + 1)(1) - 2x\cdot x}{(x^2 + 1)^2}$ (Quotient Rule)

$= \dfrac{x^2 + 1 - 2x^2}{(x^2 + 1)^2}$

$= \dfrac{1 - x^2}{(x^2 + 1)^2}$

$f''(x) = \dfrac{(x^2 + 1)^2(-2x)-2(x^2 + 1)(2x)(1 - x^2)}{[(x^2 + 1)^2]^2}$

(Quotient Rule)

$= \dfrac{(x^2 + 1)[(x^2 + 1)(-2x) - 4x(1 - x^2)]}{(x^2 + 1)^4}$

$= \dfrac{-2x^3 - 2x - 4x + 4x^3}{(x^2 + 1)^3}$

$= \dfrac{2x^3 - 6x}{(x^2 + 1)^3}$

b) Since $f'(x)$ exists for all real numbers, the only critical points are where $f'(x) = 0$.

$\dfrac{1 - x^2}{(x^2 + 1)^2} = 0$

$1 - x^2 = 0$ (Multiplying by $(x^2 + 1)^2$)

$(1 + x)(1 - x) = 0$

$1 + x = 0$ or $1 - x = 0$

$x = -1$ or $1 = x$ (Critical points)

Then $f(-1) = \dfrac{-1}{(-1)^2 + 1} = -\dfrac{1}{2}$ and

$f(1) = \dfrac{1}{1^2 + 1} = \dfrac{1}{2}$, so $\left(-1, -\dfrac{1}{2}\right)$ and $\left(1, \dfrac{1}{2}\right)$ are on the graph.

c) Use the Second-Derivative Test:

$f''(-1) = \dfrac{2(-1)^3 - 6(-1)}{[(-1)^2 + 1]^3}$

$= \dfrac{-2 + 6}{2^3}$

$= \dfrac{4}{8} = \dfrac{1}{2} > 0$, so $\left(-1, -\dfrac{1}{2}\right)$ is a relative minimum.

$f''(1) = \dfrac{2\cdot 1^3 - 6\cdot 1}{[(1)^2 + 1]^3}$

$= \dfrac{2 - 6}{2^3}$

$= \dfrac{-4}{8} = -\dfrac{1}{2} < 0$, so $\left(1, \dfrac{1}{2}\right)$ is a relative maximum.

Then if we use -1 and 1 to divide the real number line into three intervals, $(-\infty,-1)$, $(-1,1)$, and $(1,\infty)$, we know that f is decreasing on $(-\infty,-1)$ and on $(1,\infty)$ and is increasing on $(-1,1)$.

d) $f''(x)$ exists for all real numbers, so the only possible inflection points are where $f''(x) = 0$.

$\dfrac{2x^3 - 6x}{(x^2 + 1)^3} = 0$

$2x(x^2 - 3) = 0$

$2x = 0$ or $x^2 - 3 = 0$

$x = 0$ or $x^2 = 3$

$x = 0$ or $x = \pm\sqrt{3}$

(Possible inflection points)

$f(-\sqrt{3}) = \dfrac{-\sqrt{3}}{(-\sqrt{3})^2 + 1} = -\dfrac{\sqrt{3}}{4}$

$f(0) = \dfrac{0}{0^2 + 1} = 0$

$f(\sqrt{3}) = \dfrac{\sqrt{3}}{(\sqrt{3})^2 + 1} = \dfrac{\sqrt{3}}{4}$

These give the points $\left(-\sqrt{3}, -\dfrac{\sqrt{3}}{4}\right)$, (0,0), and $\left(\sqrt{3}, \dfrac{\sqrt{3}}{4}\right)$ on the graph.

e) To determine the concavity we use $-\sqrt{3}$, 0, and $\sqrt{3}$ to divide the real number line into four intervals, A: $(-\infty, -\sqrt{3})$, B: $(-\sqrt{3}, 0)$, C: $(0, \sqrt{3})$, and D: $(\sqrt{3}, \infty)$. Test a point in each interval.

A: Test -2, $f''(-2) = \dfrac{-4}{125} < 0$

B: Test -1, $f''(-1) = \dfrac{1}{2} > 0$

C: Test 1, $f''(1) = \dfrac{-1}{2} < 0$

D: Test 2, $f''(2) = \dfrac{4}{125} > 0$

Then f is concave down on $(-\infty, -\sqrt{3})$ and on $(0, \sqrt{3})$ and is concave up on $(-\sqrt{3}, 0)$ and on $(\sqrt{3}, \infty)$, so $\left(-\sqrt{3}, -\frac{\sqrt{3}}{4}\right)$, $(0,0)$, and $\left(\sqrt{3}, \frac{\sqrt{3}}{4}\right)$ are all inflection points.

f) Sketch the graph using the preceding information. Other function values can also be calculated.

x	f(x)
-3	$-\frac{3}{10}$
-2	$-\frac{2}{5}$
2	$\frac{2}{5}$
3	$\frac{3}{10}$

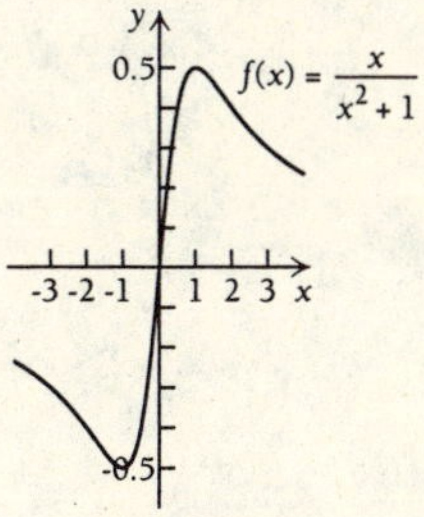

27. $f(x) = \frac{3}{x^2 + 1} = 3(x^2 + 1)^{-1}$

a) $f'(x) = 3(-1)(x^2 + 1)^{-2}(2x)$

$= -6x(x^2 + 1)^{-2}$, or $\frac{-6x}{(x^2 + 1)^2}$

$$f''(x) = \frac{(x^2 + 1)^2(-6) - 2(x^2 + 1)(2x)(-6x)}{[(x^2 + 1)^2]^2}$$

$$= \frac{(x^2 + 1)[(x^2 + 1)(-6) - 2(2x)(-6x)]}{(x^2 + 1)^4}$$

$$= \frac{-6x^2 - 6 + 24x^2}{(x^2 + 1)^3}$$

$$= \frac{18x^2 - 6}{(x^2 + 1)^3}$$

b) Since $f'(x)$ exists for all real numbers, the only critical points are where $f'(x) = 0$.

$\frac{-6x}{(x^2 + 1)^2} = 0$

$-6x = 0$ (Multiplying by $(x^2 + 1)^2$)

$x = 0$ (Critical point)

Then $f(0) = \frac{3}{0^2 + 1} = 3$, so $(0,3)$ is on the graph.

c) Use the Second-Derivative Test:

$f''(0) = \frac{18\cdot 0^2 - 6}{(0^2 + 1)^3} = -6 < 0$, so $(0,3)$ is a relative maximum.

Then if we use 0 to divide the real number line into two intervals, $(-\infty,0)$ and $(0,\infty)$, we know that f is increasing on $(-\infty,0)$ and decreasing on $(0,\infty)$.

d) $f''(x)$ exists for all real numbers, so the only possible inflection points are where $f''(x) = 0$.

$\frac{18x^2 - 6}{(x^2 + 1)^3} = 0$

$18x^2 - 6 = 0$ (Multiplying by $(x^2 + 1)^3$)

$18x^2 = 6$

$x^2 = \frac{1}{3}$

$x = \pm \frac{1}{\sqrt{3}}$ (Possible inflection points)

$$f\left(-\frac{1}{\sqrt{3}}\right) = \frac{3}{\left(-\frac{1}{\sqrt{3}}\right)^2 + 1} = \frac{3}{\frac{1}{3} + 1} = \frac{3}{\frac{4}{3}} = \frac{3}{1} \cdot \frac{3}{4} = \frac{9}{4}$$

$$f\left(\frac{1}{\sqrt{3}}\right) = \frac{3}{\left(\frac{1}{\sqrt{3}}\right)^2 + 1} = \frac{3}{\frac{1}{3} + 1} = \frac{3}{\frac{4}{3}} = \frac{3}{1} \cdot \frac{3}{4} = \frac{9}{4}$$

These give the points $\left(-\frac{1}{\sqrt{3}}, \frac{9}{4}\right)$ and $\left(\frac{1}{\sqrt{3}}, \frac{9}{4}\right)$ on the graph.

e) To determine the concavity we use $-\frac{1}{\sqrt{3}}$ and $\frac{1}{\sqrt{3}}$ to divide the real number line into three intervals, A: $\left(-\infty, -\frac{1}{\sqrt{3}}\right)$, B: $\left(-\frac{1}{\sqrt{3}}, \frac{1}{\sqrt{3}}\right)$, and C: $\left(\frac{1}{\sqrt{3}}, \infty\right)$. Test a point in each interval.

A: Test -1, $f''(-1) = \frac{18(-1)^2 - 6}{[(-1)^2 + 1]^3} = \frac{12}{8} = \frac{3}{2} > 0$

B: Test 0, $f''(0) = \frac{18\cdot 0^2 - 6}{(0^2 + 1)^3} = -6 < 0$

C: Test 1, $f''(1) = \frac{18\cdot 1^2 - 6}{(1^2 + 1)^3} = \frac{12}{8} = \frac{3}{2} > 0$

Then f is concave up on $\left(-\infty, -\frac{1}{\sqrt{3}}\right)$ and on $\left(\frac{1}{\sqrt{3}}, \infty\right)$ and is concave down on $\left(-\frac{1}{\sqrt{3}}, \frac{1}{\sqrt{3}}\right)$, so $\left(-\frac{1}{\sqrt{3}}, \frac{9}{4}\right)$ and $\left(\frac{1}{\sqrt{3}}, \frac{9}{4}\right)$ are both inflection points.

f) Sketch the graph using the preceding information. Other function values can also be calculated.

x	f(x)
-3	$\frac{3}{10}$
-1	$\frac{3}{2}$
1	$\frac{3}{2}$
3	$\frac{3}{10}$

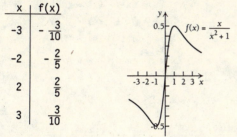

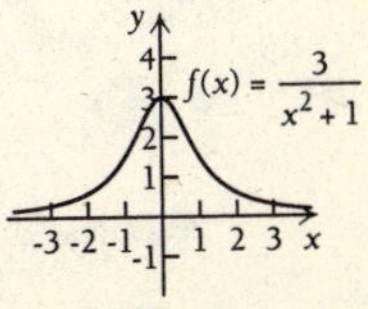

29. $f(x) = (x - 1)^3$

a) $f'(x) = 3(x - 1)^2(1) = 3(x - 1)^2$

$f''(x) = 3 \cdot 2(x - 1)(1) = 6(x - 1)$

b) Since $f'(x)$ exists for all real numbers, the only critical points are where $f'(x) = 0$.

$3(x - 1)^2 = 0$

$(x - 1)^2 = 0$

$x - 1 = 0$

$x = 1$ (Critical point)

Then $f(1) = (1 - 1)^3 = 0^3 = 0$, so $(1,0)$ is on the graph.

c) The Second-Derivative Test fails since $f''(1) = 0$, so we use the First-Derivative Test. Use 1 to divide the real number line into two intervals, A: $(-\infty,1)$ and B: $(1,\infty)$. Test a point in each interval.

A: Test 0, $f'(0) = 3(0 - 1)^2 = 3 > 0$

B: Test 2, $f'(2) = 3(2 - 1)^2 = 3 > 0$

Since f is increasing on both intervals, $(1,0)$ is not a relative extremum.

d) $f''(x)$ exists for all real numbers, so the only possible inflection points are where $f''(x) = 0$.

$6(x - 1) = 0$

$x - 1 = 0$

$x = 1$ (Possible inflection point)

From step b), we know that $(1,0)$ is on the graph.

e) To determine the concavity we use 1 to divide the real number line as in step c).

A: Test 0, $f''(0) = 6(0 - 1) = -6 < 0$

B: Test 2, $f''(2) = 6(2 - 1) = 6 > 0$

Then f is concave down on $(-\infty, 1)$ and concave up on $(1,\infty)$, so $(1,0)$ is an inflection point.

f) Sketch the graph using the preceding information. Other function values can also be calculated.

x	f(x)
-3	-64
-2	-27
-1	-8
0	-1
2	1
3	8
4	27

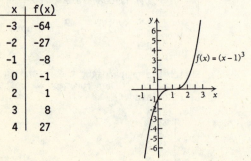

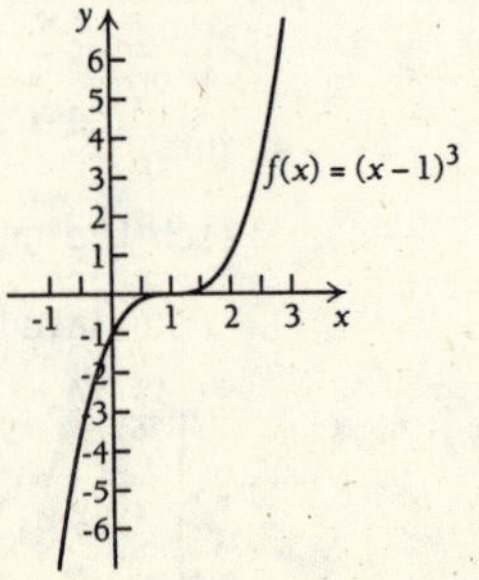

31. $f(x) = x^2(1 - x)^2$

$= x^2(1 - 2x + x^2)$

$= x^2 - 2x^3 + x^4$

a) $f'(x) = 2x - 6x^2 + 4x^3$

$f''(x) = 2 - 12x + 12x^2$

b) Since $f'(x)$ exists for all real numbers, the only critical points are where $f'(x) = 0$.

$2x - 6x^2 + 4x^3 = 0$

$2x(1 - 3x + 2x^2) = 0$

$2x(1 - x)(1 - 2x) = 0$

$2x = 0$ or $1 - x = 0$ or $1 - 2x = 0$

$x = 0$ or $1 = x$ or $1 = 2x$

$x = 0$ or $1 = x$ or $\frac{1}{2} = x$

(Critical points)

$f(0) = 0^2(1 - 0)^2 = 0$

$f(1) = 1^2(1 - 1)^2 = 0$

$f\left(\frac{1}{2}\right) = \left(\frac{1}{2}\right)^2\left(1 - \frac{1}{2}\right)^2 = \frac{1}{4} \cdot \frac{1}{4} = \frac{1}{16}$

Thus, $(0,0)$, $(1,0)$, and $\left(\frac{1}{2},\frac{1}{16}\right)$ are on the graph.

c) Use the Second-Derivative Test:

$f''(0) = 2 - 12 \cdot 0 + 12 \cdot 0^2 = 2 > 0$, so $(0,0)$ is a relative minimum.

$f''\left(\frac{1}{2}\right) = 2 - 12 \cdot \frac{1}{2} + 12\left(\frac{1}{2}\right)^2 = 2 - 6 + 3 =$

$-1 < 0$, so $\left(\frac{1}{2},\frac{1}{16}\right)$ is a relative maximum.

$f''(1) = 2 - 12 \cdot 1 + 12 \cdot 1^2 = 2 > 0$, so $(1,0)$ is a relative minimum.

Then if we use the points 0, $\frac{1}{2}$, and 1 to divide the real number line into four intervals, $(-\infty,0)$, $\left(0,\frac{1}{2}\right)$, $\left(\frac{1}{2},1\right)$ and $(1,\infty)$, we know that f is decreasing on $(-\infty,0)$ and on $\left(\frac{1}{2},1\right)$ and is increasing on $\left(0,\frac{1}{2}\right)$ and on $(1,\infty)$.

d) $f''(x)$ exists for all real numbers, so the only possible inflection points are where $f''(x) = 0$.

$2 - 12x + 12x^2 = 0$

$2(1 - 6x + 6x^2) = 0$

Using the quadratic formula we find

$x = \frac{3 \pm \sqrt{3}}{6}$

$x \approx 0.21$ or $x \approx 0.79$ (Possible inflection points)

$f(0.21) \approx 0.03$ and $f(0.79) \approx 0.03$, so $(0.21, 0.03)$ and $(0.79, 0.03)$ are on the graph.

e) To determine the concavity we use 0.21 and 0.79 to divide the real number line into three intervals, A: $(-\infty, 0.21)$, B: $(0.21, 0.79)$, and C: $(0.79, \infty)$.

A: Test 0, $f''(0) = 2 - 12\cdot0 + 12\cdot0^2 = 2 > 0$

B: Test 0.5, $f''(0.5) = 2 - 12(0.5) + 12(0.5)^2 = -1 < 0$

C: Test 1, $f''(1) = 2 - 12\cdot1 + 12\cdot1^2 = 2 > 0$

Then f is concave up on $(-\infty, 0.21)$ and on $(0.79, \infty)$ and is concave down on $(0.21, 0.79)$, so $(0.21, 0.03)$ and $(0.79, 0.03)$ are both inflection points.

f) Sketch the graph using the preceding information. Other function values can also be calculated.

x	f(x)
-2	36
-1	4
2	4
3	36

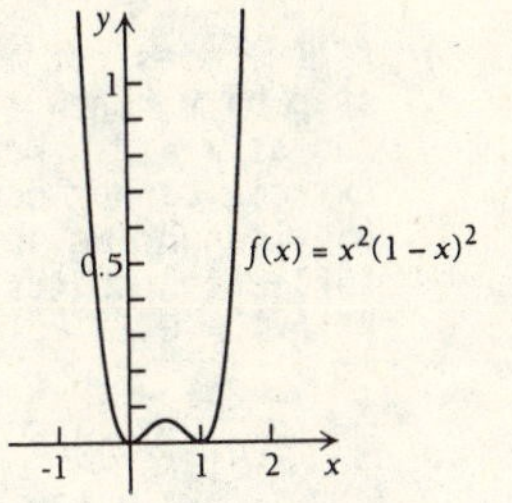

33. $f(x) = 20x^3 - 3x^5$

a) $f'(x) = 60x^2 - 15x^4$

$f''(x) = 120x - 60x^3$

b) Since $f'(x)$ exists for all real numbers, the only critical points are where $f'(x) = 0$.

$$60x^2 - 15x^4 = 0$$
$$15x^2(4 - x^2) = 0$$
$$15x^2(2 + x)(2 - x) = 0$$
$$15x^2 = 0 \text{ or } 2 + x = 0 \text{ or } 2 - x = 0$$
$$x = 0 \text{ or } x = -2 \text{ or } 2 = x$$

(Critical points)

$f(0) = 20\cdot0^3 - 3\cdot0^5 = 0$

$f(-2) = 20(-2)^3 - 3(-2)^5 = -160 + 96 = -64$

$f(2) = 20\cdot2^3 - 3\cdot2^5 = 160 - 96 = 64$

Thus, (0,0), (-2,-64), and (2,64) are on the graph.

c) Use the Second-Derivative Test:

$f''(-2) = 120(-2) - 60(-2)^3 = -240 + 480 = 240 > 0$, so (-2,-64) is a relative minimum.

$f''(2) = 120\cdot2 - 60\cdot2^3 = 240 - 480 = -240 < 0$, so (2,64) is a relative maximum.

$f''(0) = 120\cdot0 - 60\cdot0^3 = 0$, so we will use the First-Derivative Test on $x = 0$. Use 0 to divide the interval (-2,2) into two intervals, A: (-2,0), and B: (0,2). Test a point in each interval.

A: Test -1, $f'(-1) = 60(-1)^2 - 15(-1)^4 = 60 - 15 = 45 > 0$

B: Test 1, $f'(1) = 60\cdot1^2 - 15\cdot1^4 = 60 - 15 = 45 > 0$

Then f is increasing on both intervals, so (0,0) is not a relative extremum.

If we use the points -2 and 2 to divide the real number line into three intervals, $(-\infty,-2)$, $(-2,2)$, and $(2,\infty)$, we know that f is decreasing on $(-\infty,-2)$ and on $(2,\infty)$ and is increasing on $(-2,2)$.

d) $f''(x)$ exists for all real numbers, so the only possible inflection points are where $f''(x) = 0$.

$$120x - 60x^3 = 0$$
$$60x(2 - x^2) = 0$$
$$60x = 0 \text{ or } 2 - x^2 = 0$$
$$x = 0 \text{ or } 2 = x^2$$
$$x = 0 \text{ or } \pm\sqrt{2} = x$$

(Possible inflection points)

$f(-\sqrt{2}) = 20(-\sqrt{2})^3 - 3(-\sqrt{2})^5 = -40\sqrt{2} + 12\sqrt{2} = -28\sqrt{2}$

$f(0) = 0$ (from step b))

$f(\sqrt{2}) = 20(\sqrt{2})^3 - 3(\sqrt{2})^5 = 40\sqrt{2} - 12\sqrt{2} = 28\sqrt{2}$

Thus, $(-\sqrt{2}, -28\sqrt{2})$ and $(\sqrt{2}, 28\sqrt{2})$ are also on the graph.

e) To determine the concavity we use $-\sqrt{2}$, 0, and $\sqrt{2}$ to divide the real number line into four intervals, A: $(-\infty, -\sqrt{2})$, B: $(-\sqrt{2}, 0)$, C: $(0, \sqrt{2})$, and D: $(\sqrt{2}, \infty)$.

A: Test -2, $f''(-2) = 120(-2) - 60(-2)^3 = 240 > 0$

B: Test -1, $f''(-1) = 120(-1) - 60(-1)^3 = -60 < 0$

C: Test 1, $f''(1) = 120\cdot1 - 60\cdot1^3 = 60 > 0$

D: Test 2, $f''(2) = 120\cdot2 - 60\cdot2^3 = -240 < 0$

Then f is concave up on $(-\infty, -\sqrt{2})$ and on $(0, \sqrt{2})$ and is concave down on $(-\sqrt{2}, 0)$ and on $(\sqrt{2}, \infty)$, so $(-\sqrt{2}, -28\sqrt{2})$, (0,0), and $(\sqrt{2}, 28\sqrt{2})$ are all inflection points.

f) Sketch the graph using the preceding information. Other function values can also be calculated.

x	f(x)
-3	189
-2	-64
-1	-17
1	17
2	64
3	-189

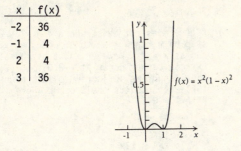

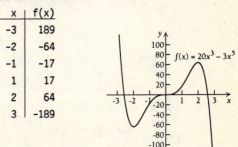

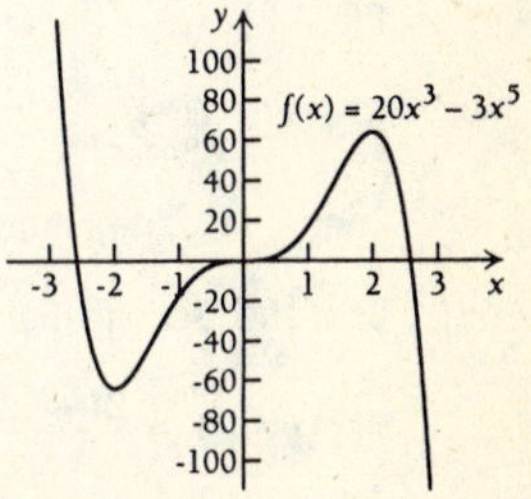

35. $f(x) = x\sqrt{4 - x^2} = x(4 - x^2)^{1/2}$

a) $f'(x) = x \cdot \frac{1}{2}(4 - x^2)^{-1/2}(-2x) + 1\cdot(4 - x^2)^{1/2}$

$= \frac{-x^2}{\sqrt{4 - x^2}} + \sqrt{4 - x^2}$

$= \frac{-x^2 + 4 - x^2}{\sqrt{4 - x^2}}$

$= \frac{4 - 2x^2}{\sqrt{4 - x^2}}$, or $(4 - 2x^2)(4 - x^2)^{-1/2}$

$f''(x) = (4 - 2x^2)\left[-\frac{1}{2}\right](4 - x^2)^{-3/2}(-2x) - 4x(4 - x^2)^{-1/2}$

$= \frac{x(4 - 2x^2)}{(4 - x^2)^{3/2}} - \frac{4x}{(4 - x^2)^{1/2}}$

$= \frac{x(4 - 2x^2) - 4x(4 - x^2)}{(4 - x^2)^{3/2}}$

$= \frac{4x - 2x^3 - 16x + 4x^3}{(4 - x^2)^{3/2}}$

$= \frac{2x^3 - 12x}{(4 - x^2)^{3/2}}$

b) $f'(x)$ does not exist where $4 - x^2 = 0$. Solve:

$4 - x^2 = 0$

$(2 + x)(2 - x) = 0$

$2 + x = 0$ or $2 - x = 0$

$x = -2$ or $2 = x$

Note that $f(x)$ is not defined for $x < -2$ or $x > 2$. (For these values $4 - x^2 < 0$.) Therefore, relative extrema cannot occur at $x = -2$ or $x = 2$, because there is no open interval containing -2 or 2 on which the function is defined. For this reason, we do not consider -2 and 2 further in our discussion of relative extrema.

Critical points occur where $f'(x) = 0$. Solve:

$\frac{4 - 2x^2}{\sqrt{4 - x^2}} = 0$

$4 - 2x^2 = 0$

$4 = 2x^2$

$2 = x^2$

$\pm\sqrt{2} = x$ (Critical points)

$f(-\sqrt{2}) = -\sqrt{2}\sqrt{4 - (-\sqrt{2})^2} = -\sqrt{2}\cdot\sqrt{2} = -2$

$f(\sqrt{2}) = \sqrt{2}\sqrt{4 - (\sqrt{2})^2} = \sqrt{2}\cdot\sqrt{2} = 2$

Then $(-\sqrt{2}, -2)$ and $(\sqrt{2}, 2)$ are on the graph.

c) Use the Second-Derivative Test:

$f''(-\sqrt{2}) = \frac{2(-\sqrt{2})^3 - 12(-\sqrt{2})}{[4 - (-\sqrt{2})^2]^{3/2}} =$

$\frac{-4\sqrt{2} + 12\sqrt{2}}{2^{3/2}} = \frac{8\sqrt{2}}{2\sqrt{2}} = 4 > 0$, so $(-\sqrt{2}, -2)$ is a relative minimum.

$f''(\sqrt{2}) = \frac{2(\sqrt{2})^3 - 12\sqrt{2}}{[4 - (\sqrt{2})^2]^{3/2}} = \frac{4\sqrt{2} - 12\sqrt{2}}{2^{3/2}} =$

$\frac{-8\sqrt{2}}{2\sqrt{2}} = -4 < 0$, so $(\sqrt{2}, 2)$ is a relative maximum.

If we use the points $-\sqrt{2}$ and $\sqrt{2}$ to divide the interval $(-2,2)$ into three intervals, $(-2, -\sqrt{2})$, $(-\sqrt{2}, \sqrt{2})$, and $(\sqrt{2}, 2)$, we know that f is decreasing on $(-2, -\sqrt{2})$ and on $(\sqrt{2}, 2)$ and is increasing on $(-\sqrt{2}, \sqrt{2})$.

d) $f''(x)$ does not exist where $4 - x^2 = 0$. From step b) we know that this occurs at $x = -2$ and at $x = 2$. However, just as relative extrema cannot occur at $(-2,0)$ and $(2,0)$, they cannot be inflection points either. Inflection points could occur where $f''(x) = 0$.

$\frac{2x^3 - 12x}{(4 - x^2)^{3/2}} = 0$

$2x^3 - 12x = 0$

$2x(x^2 - 6) = 0$

$2x = 0$ or $x^2 - 6 = 0$

$x = 0$ or $x^2 = 6$

$x = 0$ or $x = \pm\sqrt{6}$

Note that $f(x)$ is not defined for $x = \pm\sqrt{6}$. Therefore, the only possible inflection point is $x = 0$.

$f(0) = 0\sqrt{4 - 0^2} = 0\cdot 2 = 0$

Then $(0,0)$ is on the graph.

e) To determine the concavity we use 0 to divide the interval $(-2,2)$ into two intervals, A: $(-2,0)$ and B: $(0,2)$.

A: Test -1, $f''(-1) = \frac{2(-1)^3 - 12(-1)}{[4 - (-1)^2]^{3/2}} =$

$\frac{10}{3^{3/2}} > 0$

B: Test 1, $f''(1) = \frac{2\cdot 1^3 - 12\cdot 1}{(4 - 1^2)^{3/2}} = \frac{-10}{3^{3/2}} < 0$

Then f is concave up on $(-2,0)$ and concave down on $(0,2)$, so $(0,0)$ is an inflection point.

f) Sketch the graph using the preceding information. Other function values can also be calculated.

x	f(x)
$-\sqrt{3}$	$-\sqrt{3}$
-1	$-\sqrt{3}$
1	$\sqrt{3}$
$\sqrt{3}$	$\sqrt{3}$

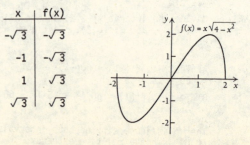

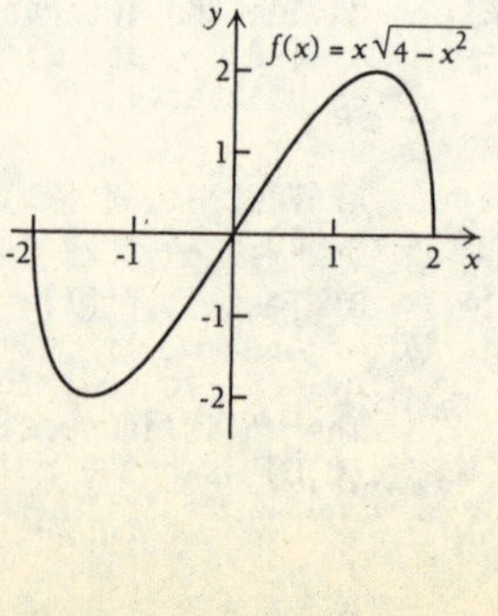

37. $f(x) = (x - 1)^{1/3} - 1$

a) $f'(x) = \frac{1}{3}(x - 1)^{-2/3}$, or $\frac{1}{3(x - 1)^{2/3}}$

$f''(x) = \frac{1}{3}\left(-\frac{2}{3}\right)(x - 1)^{-5/3}$

$= -\frac{2}{9}(x - 1)^{-5/3}$, or $-\frac{2}{9(x - 1)^{5/3}}$

b) $f'(x)$ does not exist for $x = 1$. The equation $f'(x) = 0$ has no solution, so $x = 1$ is the only critical point. $f(1) = (1 - 1)^{1/3} - 1 = 0 - 1 = -1$, so $(1,-1)$ is on the graph.

c) Use the First-Derivative Test: Use 1 to divide the real number line into two intervals, A: $(-\infty,1)$ and B: $(1,\infty)$. Test a point in each interval.

A: Test 0, $f'(0) = \frac{1}{3(0 - 1)^{2/3}} = \frac{1}{3\cdot 1} = \frac{1}{3} > 0$

B: Test 2, $f'(2) = \frac{1}{3(2 - 1)^{2/3}} = \frac{1}{3\cdot 1} = \frac{1}{3} > 0$

Then f is increasing on both intervals, so $(1,-1)$ is not a relative extremum.

d) $f''(1)$ does not exist for $x = 1$. The equation $f''(x) = 0$ has no solution, so $x = 1$ is the only possible inflection point. From step b) we know $(1,-1)$ is on the graph.

e) To determine the concavity we use 1 to divide the real number line as in step c). Test a point in each interval.

A: Test 0, $f''(0) = -\frac{2}{9(0 - 1)^{5/3}} = -\frac{2}{9(-1)} = \frac{2}{9} > 0$

B: Test 2, $f''(2) = -\frac{2}{9(2 - 1)^{5/3}} = -\frac{2}{9\cdot 1} = -\frac{2}{9} < 0$

Then f is concave up on $(-\infty,1)$ and concave down on $(1,\infty)$, so $(1,-1)$ is an inflection point.

f) Sketch the graph using the preceding information. Other function values can also be calculated.

x	f(x)
-7	-3
0	-2
2	0
9	1

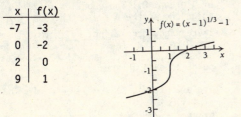

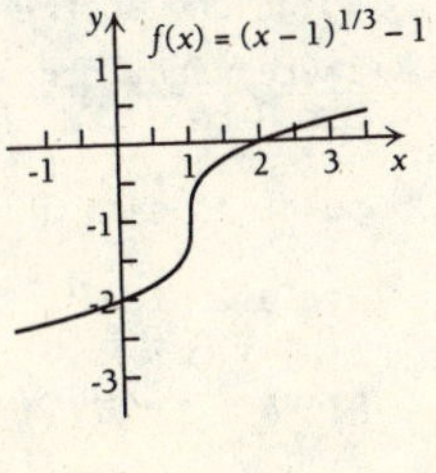

39. $R(x) = 50x - 0.5x^2$

$C(x) = 4x + 10$

$P(x) = R(x) - C(x)$

$= (50x - 0.5x^2) - (4x + 10)$

$= -0.5x^2 + 46x - 10$

We will restrict the domains of all three functions to $x \geqslant 0$ since a negative number of units cannot be produced and sold.

First graph $R(x) = 50x - 0.5x^2$

a) $R'(x) = 50 - x$

$R''(x) = -1$

b) Since $R'(x)$ exists for all $x \geqslant 0$, the only critical points are where $50 - x = 0$.

$50 - x = 0$

$50 = x$ (Critical point)

Find the function value at $x = 50$.

$R(50) = 50\cdot 50 - 0.5(50)^2$

$= 2500 - 1250$

$= 1250$

This gives the point (50,1250) on the graph.

c) Use the Second-Derivative Test:

$R''(50) = -1 < 0$, so (50,1250) is a relative maximum.

Then if we use 50 to divide the interval $(0,\infty)$ into two intervals, (0,50) and $(50,\infty)$, we know that R is increasing on (0,50) and decreasing on $(50,\infty)$.

d) Since $R''(x)$ exists for all $x \geqslant 0$, and is always negative ($R''(x) = -1$), the equation $R''(x) = 0$ has no solution. Thus there are no possible inflection points.

e) Since $R''(x) < 0$ for all $x \geqslant 0$, R is concave down on the interval $(0,\infty)$.

f) Sketch the graph using the preceding information. The x-intercepts of R are easily found by solving $50x - 0.5x^2 = 0$.

$50x - 0.5x^2 = 0$

$500x - 5x^2 = 0$

$5x(100 - x) = 0$

$5x = 0$ or $100 - x = 0$

$x = 0$ or $100 = x$

The x-intercepts are (0,0) and (100,0).

Next we graph $C(x) = 4x + 10$. This is a linear function with slope 4 and y-intercept (0,10).

Finally we graph $P(x) = -0.5x^2 + 46x - 10$.

a) $P'(x) = -x + 46$

$P''(x) = -1$

b) Since $P'(x)$ exists for all $x \geqslant 0$, the only critical points are where $-x + 46 = 0$.

$-x + 46 = 0$

$46 = x$ (Critical point)

Find the function value at x = 46.

$P(46) = -0.5(46)^2 + 46 \cdot 46 - 10$

$= -1058 + 2116 - 10$

$= 1048$

This gives the point (46,1048) on the graph.

c) Use the Second-Derivative Test:

$P''(46) = -1 < 0$, so (46,1048) is a relative maximum.

Then if we use 46 to divide the interval $(0,\infty)$ into two intervals, (0,46) and $(46,\infty)$, we know that P is increasing on (0,46) and decreasing on $(46,\infty)$

d) Since $P''(x)$ exists for all $x \geqslant 0$, and is always negative ($P''(x) = -1$), the equation $P''(x) = 0$ has no solution. Thus there are no possible inflection points.

e) Since $P''(x) < 0$ for all $x \geqslant 0$, P is concave down on the interval $(0,\infty)$.

f) Sketch the graph using the preceding information.

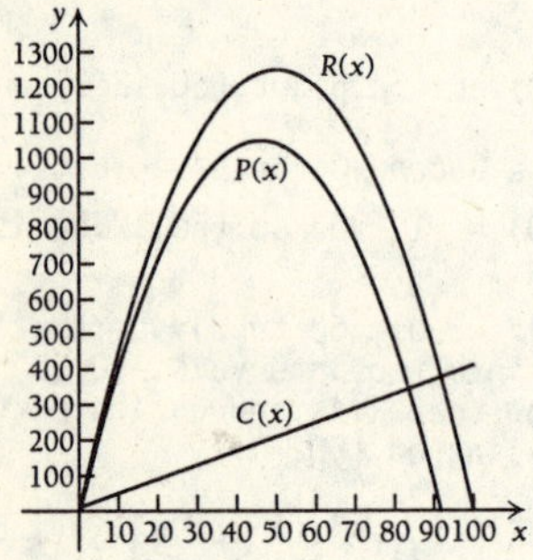

41. $V(r) = k(20r^2 - r^3),\ 0 \leqslant r \leqslant 20$

$V'(r) = k(40r - 3r^2)$

$V''(r) = k(40 - 6r)$

$V'(r)$ exists for all r in [0,20], so the only critical points occur where $V'(r) = 0$.

$k(40r - 3r^2) = 0$

$40r - 3r^2 = 0$

$r(40 - 3r) = 0$

$r = 0$ or $40 - 3r = 0$

$r = 0$ or $\frac{40}{3} = r$ (Critical points)

Use the Second-Derivative Test:

$V''(0) = k(40 - 6 \cdot 0) = 40k > 0$

$V''\left(\frac{40}{3}\right) = k\left(40 - 6 \cdot \frac{40}{3}\right) = k(40 - 80) = -40k < 0$

Since $V''\left(\frac{40}{3}\right) < 0$, we know that there is a relative maximum at $x = \frac{40}{3}$. Thus the maximum velocity is required to remove an object whose radius is $\frac{40}{3}$, or $13\frac{1}{3}$ mm.

43. Observe that h is increasing for all values of x for which g is positive and h is decreasing for all values of x for which g is negative. Furthermore, for all values of x for which g = 0, h has a relative extremum. Then, g = h'.

45. Using the graph shown in the answer section in the text, we estimate that there is a relative minimum at (0,0) and a relative maximum at (1,1).

47. Using the graph shown in the answer section in the text, we estimate that there is a relative minimum at (1,-1).

49. Using the graph shown in the answer section in the text, we estimate that there is a relative minimum at $\left(\frac{1}{4}, -\frac{1}{4}\right)$.

Exercise Set 3.3

1. $f(x) = \frac{4}{x}$, or $4x^{-1}$

a) Intercepts. Since the numerator is the constant 4, there are no x-intercepts. The number 0 is not in the domain of the function, so there are no y-intercepts.

b) Asymptotes.

Vertical. The denominator is 0 for x = 0, so the line x = 0 is a vertical asymptote.

Horizontal. The degree of the numerator is less than the degree of the denominator, so y = 0 is a horizontal asymptote.

Oblique. There is no oblique asymptote since the degree of the numerator is not one more than the degree of the denominator.

c) Derivatives.

$f'(x) = -4x^{-2} = -\frac{4}{x^2}$

$f''(x) = 8x^{-3} = \frac{8}{x^3}$

d) Critical points. The number 0 is not in the domain of f. Now $f'(x)$ exists for all values of x except 0. The equation $f'(x) = 0$ has no solution, so there are no critical points.

e) Increasing, decreasing, relative extrema. Use 0 to divide the real number line into two intervals, A: $(-\infty,0)$ and B: $(0,\infty)$. Test a point in each interval.

A: Test -1, $f'(-1) = -\frac{4}{(-1)^2} = -4 < 0$

B: Test 1, $f'(1) = -\frac{4}{1^2} = -4 < 0$

Then f is decreasing on both intervals. Since there are no critical points, there are no relative extrema.

f) Inflection points. $f''(0)$ does not exist, but because f(0) does not exist there cannot be an inflection point at 0. The equation $f''(x) = 0$ has no solution, so there are no inflection points.

g) Concavity. Use 0 to divide the real number line as in step e). Note that for any $x < 0$, $x^3 < 0$, so

$$f''(x) = \frac{8}{x^3} < 0$$

and for any $x > 0$, $x^3 > 0$, so

$$f''(x) = \frac{8}{x^3} > 0.$$

Then f is concave down on $(-\infty,0)$ and concave up on $(0,\infty)$.

h) Sketch. Use the preceding information to sketch the graph. Compute function values as needed.

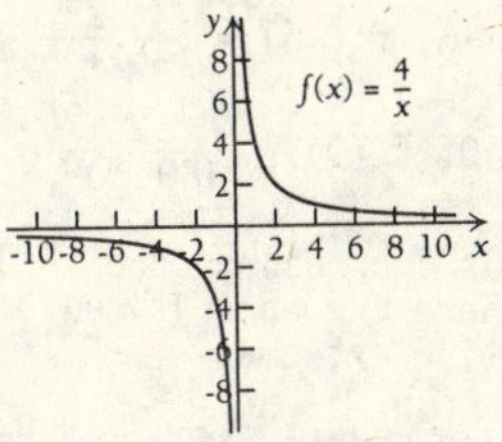

3. $f(x) = \frac{-2}{x - 5}$

a) Intercepts. Since the numerator is the constant -2, there are no x-intercepts. To find the y-intercepts we compute f(0):

$$f(0) = \frac{-2}{0 - 5} = \frac{-2}{-5} = \frac{2}{5}$$

Then $\left(0,\frac{2}{5}\right)$ is the y-intercept.

b) Asymptotes.

Vertical. The denominator is 0 for $x = 5$, so the line $x = 5$ is a vertical asymptote.

Horizontal. The degree of the numerator is less than the degree of the denominator, so $y = 0$ is a horizontal asymptote.

Oblique. There is no oblique asymptote since the degree of the numerator is not one more than the degree of the denominator.

c) Derivatives.

$$f'(x) = 2(x - 5)^{-2} = \frac{2}{(x - 5)^2}$$

$$f''(x) = -4(x - 5)^{-3} = -\frac{4}{(x - 5)^3}$$

d) Critical points. f'(5) does not exist, but because f(5) does not exist, $x = 5$ is not a critical point. The equation $f'(x) = 0$ has no solution, so there are no critical points.

e) Increasing, decreasing, relative extrema. Use 5 to divide the real number line into two intervals, A: $(-\infty,5)$ and B: $(5,\infty)$. Test a point in each interval.

A: Test 0, $f'(0) = \frac{2}{(0 - 5)^2} = \frac{2}{25} > 0$

B: Test 6, $f'(6) = \frac{2}{(6 - 5)^2} = 2 > 0$

Then f is increasing on both intervals. Since there are no critical points, there are no relative extrema.

f) Inflection points. f''(5) does not exist, but because f(5) does not exist there cannot be an inflection point at 5. The equation $f''(x) = 0$ has no solution, so there are no inflection points.

g) Concavity. Use 5 to divide the real number line as in step e). Note that for any $x < 5$, $(x - 5)^3 < 0$, so

$$f''(x) = -\frac{4}{(x - 5)^3} > 0$$

and for any $x > 5$, $(x - 5)^3 > 0$, so

$$f''(x) = -\frac{4}{(x - 5)^3} < 0.$$

Then f is concave up on $(-\infty,5)$ and concave down $(5,\infty)$.

h) Sketch. Use the preceding information to sketch the graph. Compute function values as needed.

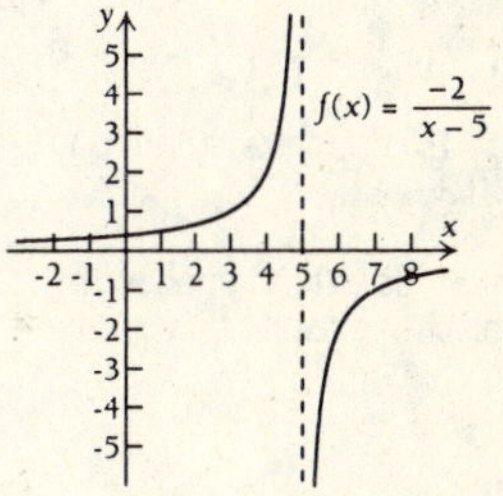

5. $f(x) = \frac{1}{x - 3}$

a) Intercepts. Since the numerator is the constant 1, there are no x-intercepts. $f(0) = \frac{1}{0 - 3} = -\frac{1}{3}$, so $\left(0,-\frac{1}{3}\right)$ is the y-intercept.

b) Asymptotes.

Vertical. The denominator is 0 for $x = 3$, so the line $x = 3$ is a vertical asymptote.

Horizontal. The degree of the numerator is less than the degree of the denominator, so $y = 0$ is a horizontal asymptote.

Oblique. There is no oblique asymptote since the degree of the numerator is not one more than the degree of the denominator.

c) Derivatives.

$$f'(x) = -(x - 3)^{-2} = -\frac{1}{(x - 3)^2}$$

$$f''(x) = 2(x - 3)^{-3} = \frac{2}{(x - 3)^3}$$

d) Critical points. f'(3) does not exist, but because f(3) does not exist, $x = 3$ is not a critical point. The equation $f'(x) = 0$ has no solution, so there are no critical points.

e) Increasing, decreasing, relative extrema. Use 3 to divide the real number line into two intervals, A: $(-\infty,3)$ and B: $(3,\infty)$. Test a point in each interval.

A: Test 0, $f'(0) = -\frac{1}{(0-3)^2} = -\frac{1}{9} < 0$

B: Test 4, $f'(4) = -\frac{1}{(4-3)^2} = -1 < 0$

Then f is decreasing on both intervals. Since there are no critical points, there are no relative extrema.

f) Inflection points. $f''(3)$ does not exist, but because $f(3)$ does not exist there cannot be an inflection point at 3. The equation $f''(x) = 0$ has no solution, so there are no inflection points.

g) Concavity. Use 3 to divide the real number line as in step e). Note that for any $x < 3$, $(x-3)^3 < 0$, so

$$f''(x) = \frac{2}{(x-3)^3} < 0$$

and for any $x > 3$, $(x-3)^3 > 0$, so

$$f''(x) = \frac{2}{(x-3)^3} > 0.$$

Then f is concave down on $(-\infty,3)$ and concave up on $(3,\infty)$

h) Sketch. Use the preceding information to sketch the graph. Compute function values as needed.

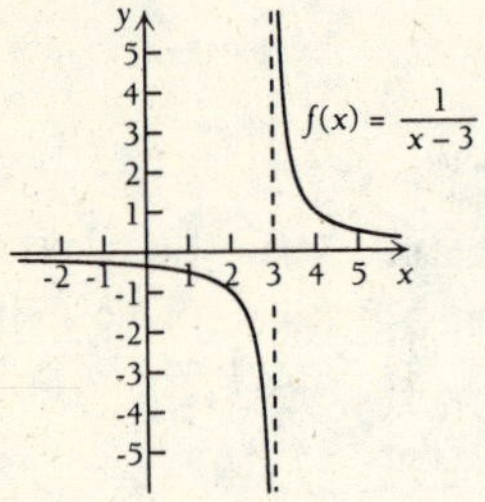

7. $f(x) = \frac{-2}{x+5}$

a) Intercepts. Since the numerator is the constant -2, there are no x-intercepts. $f(0) = \frac{-2}{0+5} = -\frac{2}{5}$, so $\left(0,-\frac{2}{5}\right)$ is the y-intercept.

b) Asymptotes.

Vertical. The denominator is 0 for $x = -5$, so the line $x = -5$ is a vertical asymptote.

Horizontal. The degree of the numerator is less than the degree of the denominator, so $y = 0$ is a horizontal asymptote.

Oblique. There is no oblique asymptote since the degree of the numerator is not one more than the degree of the denominator.

c) Derivatives.

$$f'(x) = 2(x+5)^{-2} = \frac{2}{(x+5)^2}$$

$$f''(x) = -4(x+5)^{-3} = \frac{-4}{(x+5)^3}$$

d) Critical points. $f'(-5)$ does not exist, but because $f(-5)$ does not exist, $x = -5$ is not a critical point. The equation $f'(x) = 0$ has no solution, so there are no critical points.

e) Increasing, decreasing, relative extrema. Use -5 to divide the real number line into two intervals, A: $(-\infty,-5)$ and B: $(-5,\infty)$. Test a point in each interval.

A: Test -6, $f'(-6) = \frac{2}{(-6+5)^2} = 2 > 0$

B: Test 0, $f'(0) = \frac{2}{(0+5)^2} = \frac{2}{25} > 0$

Then f is increasing on both intervals. Since there are no critical points there are no relative extrema.

f) Inflection points. $f''(-5)$ does not exist, but because $f(-5)$ does not exist there cannot be an inflection point at -5. The equation $f''(x) = 0$ has no solution, so there are no inflection points.

g) Concavity. Use -5 to divide the real number line as in step e). Test a point in each interval.

A: Test -6, $f''(-6) = \frac{-4}{(-6+5)^3} = 4 > 0$

B: Test 0, $f''(0) = \frac{-4}{(0+5)^3} = -\frac{4}{125} < 0$

Then f is concave up on $(-\infty,-5)$ and concave down on $(-5,\infty)$.

h) Sketch. Use the preceding information to sketch the graph. Compute function values as needed.

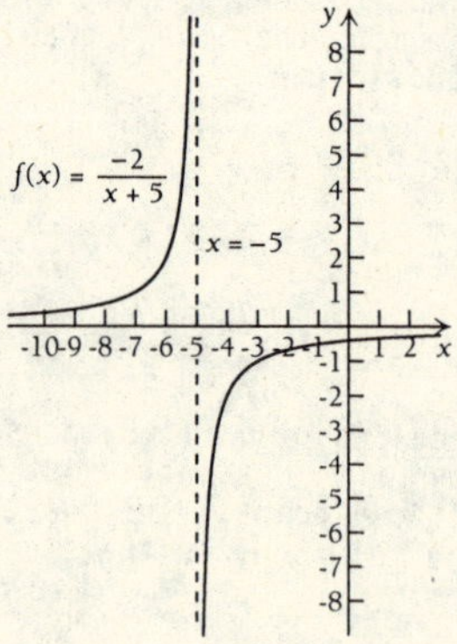

9. $f(x) = \frac{2x + 1}{x}$

a) Intercepts. To find the x-intercepts, solve $f(x) = 0$.

$$\frac{2x + 1}{x} = 0$$

$$2x + 1 = 0$$

$$2x = -1$$

$$x = -\frac{1}{2}$$

Since $x = -\frac{1}{2}$ does not make the denominator 0, the x-intercept is $\left(-\frac{1}{2},0\right)$. The number 0 is not in the domain of f, so there are no y-intercepts.

b) Asymptotes.

Vertical. The denominator is 0 for $x = 0$, so the line $x = 0$ is a vertical asymptote.

Horizontal. The numerator and denominator have the same degree, so $y = \frac{2}{1}$, or $y = 2$, is a horizontal asymptote.

Oblique. There is no oblique asymptote since the degree of the numerator is not one more than the degree of the denominator.

c) Derivatives.

$$f'(x) = -\frac{1}{x^2}$$

$$f''(x) = 2x^{-3}, \text{ or } \frac{2}{x^3}$$

d) Critical points. $f'(0)$ does not exist, but because $f(0)$ does not exist $x = 0$ is not a critical point. The equation $f'(x) = 0$ has no solution, so there are no critical points.

e) Increasing, decreasing, relative extrema. Use 0 to divide the real number line into two intervals: A: $(-\infty,0)$ and B: $(0,\infty)$. Test a point in each interval.

A: Test -1, $f'(-1) = -\frac{1}{(-1)^2} = -1 < 0$

B: Test 1, $f'(1) = -\frac{1}{1^2} = -1 < 0$

Then f is decreasing on both intervals. Since there are no critical points, there are no relative extrema.

f) Inflection points. $f''(0)$ does not exist, but because $f(0)$ does not exist there cannot be an inflection point at 0. The equation $f''(x) = 0$ has no solution, so there are no inflection points.

g) Concavity. Use 0 to divide the real number line as in step e). Test a point in each interval.

A: Test -1, $f''(-1) = \frac{2}{(-1)^3} = -2 < 0$

B: Test 1, $f''(1) = \frac{2}{1^3} = 2 > 0$

Then f is concave down on $(-\infty,0)$ and concave up on $(0,\infty)$.

h) Sketch. Use the preceding information to sketch the graph. Compute function values as needed.

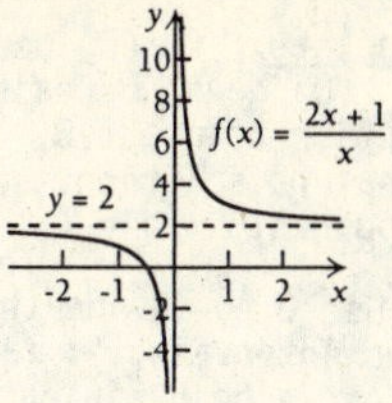

11. $f(x) = x + \frac{9}{x} = \frac{x^2 + 9}{x}$

a) Intercepts. The equation $f(x) = 0$ has no real number solution, so there are no x-intercepts. The number 0 is not in the domain of the function, so there are no y-intercepts.

b) Asymptotes.

Vertical. The denominator is 0 for $x = 0$, so the line $x = 0$ is a vertical asymptote.

Horizontal. The degree of the numerator is greater than the degree of the denominator, so there are no horizontal asymptotes.

Oblique. As $|x|$ gets very large, $f(x) = x + \frac{9}{x}$ approaches x, so $y = x$ is an oblique asymptote.

c) Derivatives.

$$f'(x) = 1 - 9x^{-2} = 1 - \frac{9}{x^2}$$

$$f''(x) = 18x^{-3} = \frac{18}{x^3}$$

d) Critical points. $f'(0)$ does not exist, but because $f(0)$ does not exist $x = 0$ is not a critical point. Solve $f'(x) = 0$.

$$1 - \frac{9}{x^2} = 1$$

$$1 = \frac{9}{x^2}$$

$$x^2 = 9$$

$$x = \pm 3$$

Thus, -3 and 3 are critical points. $f(-3) = -6$ and $f(3) = 6$, so $(-3,-6)$ and $(3,6)$ are on the graph.

e) Increasing, decreasing, relative extrema. Use -3, 0, and 3 to divide the real number line into four intervals, A: $(-\infty,-3)$, B: $(-3,0)$, C: $(0,3)$, and D: $(3,\infty)$. Test a point in each interval.

A: Test -4, $f'(-4) = 1 - \frac{9}{(-4)^2} = \frac{7}{16} > 0$

B: Test -1, $f'(-1) = 1 - \frac{9}{(-1)^2} = -8 < 0$

C: Test 1, $f'(1) = 1 - \frac{9}{1^2} = -8 < 0$

D: Test 4, $f'(4) = 1 - \frac{9}{4^2} = \frac{7}{16} > 0$

Then f is increasing on $(-\infty,-3)$ and on $(3,\infty)$ and is decreasing on $(-3,0)$ and on $(0,3)$. Thus, there is a relative maximum at $(-3,-6)$ and a relative minimum at $(3,6)$.

f) Inflection points. $f''(0)$ does not exist, but because $f(0)$ does not exist there cannot be an inflection point at 0. The equation $f''(x) = 0$ has no solution, so there are no inflection points.

g) Concavity. Use 0 to divide the real number line into two intervals, A: $(-\infty,0)$ and B: $(0,\infty)$. Test a point in each interval.

A: Test -1, $f''(-1) = \frac{18}{(-1)^3} = -18 < 0$

B: Test 1, $f''(1) = \frac{18}{1^3} = 18 > 0$

Then f is concave down on $(-\infty,0)$ and concave up on $(0,\infty)$.

h) Sketch. Use the preceding information to sketch the graph. Compute other function values as needed.

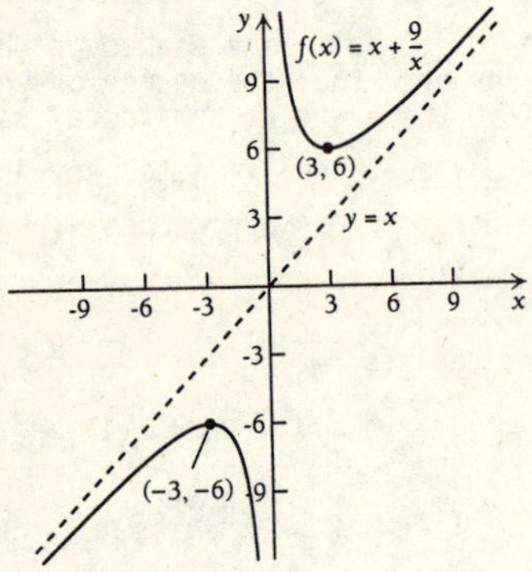

13. $f(x) = \frac{2}{x^2}$

a) Intercepts. Since the numerator is the constant 2, there are no x-intercepts. The number 0 is not in the domain of the function, so there are no y-intercepts.

b) Asymptotes.

Vertical. The denominator is 0 for $x = 0$, so the line $x = 0$ is a vertical asymptote.

Horizontal. The degree of the numerator is less than the degree of the denominator, so $y = 0$ is a horizontal asymptote.

Oblique. There is no oblique asymptote since the degree of the numerator is not one more than the degree of the denominator.

c) Derivatives.

$$f'(x) = -4x^{-3} = -\frac{4}{x^3}$$

$$f''(x) = 12x^{-4} = \frac{12}{x^4}$$

d) Critical points. $f'(0)$ does not exist, but because $f(0)$ does not exist $x = 0$ is not a critical point. The equation $f'(x) = 0$ has no solution, so there are no critical points.

e) Increasing, decreasing, relative extrema. Use 0 to divide the real number line into two intervals, A: $(-\infty,0)$ and B: $(0,\infty)$. Test a point in each interval.

A: Test -1, $f'(-1) = -\frac{4}{(-1)^3} = 4 > 0$

B: Test 1, $f'(1) = -\frac{4}{1^3} = -4 < 0$

Then f is increasing on $(-\infty,0)$ and is decreasing on $(0,\infty)$. Since there are no critical points, there are no relative extrema.

f) Inflection points. $f''(0)$ does not exist, but because $f(0)$ does not exist there cannot be an inflection point at 0. The equation $f''(x) = 0$ has no solution, so there are no inflection points.

g) Concavity. Use 0 to divide the real number line as in step e). Test a point in each interval.

A: Test -1, $f''(-1) = \frac{12}{(-1)^4} = 12 > 0$

B: Test 1, $f''(1) = \frac{12}{1^4} = 12 > 0$

Then f is concave down on both intervals.

h) Sketch. Use the preceding information to sketch the graph. Compute function values as needed.

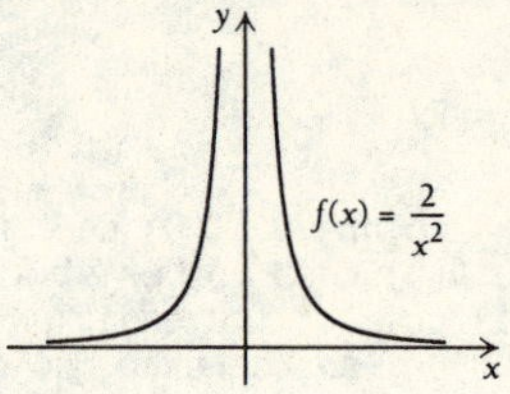

15. $f(x) = \frac{x}{x - 3}$

a) Intercepts. The numerator is 0 for $x = 0$ and this value of x does not make the denominator 0, so $(0,0)$ is the x-intercept. $f(0) = \frac{0}{0 - 3} = 0$, so the y-intercept is the x-intercept $(0,0)$.

b) Asymptotes.

Vertical. The denominator is 0 for $x = 3$, so the line $x = 3$ is a vertical asymptote.

Horizontal. The numerator and denominator have the same degree, so $y = \frac{1}{1}$, or $y = 1$, is a horizontal asymptote.

Oblique. There is no oblique asymptote since the degree of the numerator is not one more than the degree of the denominator.

c) Derivatives.

$$f'(x) = -\frac{3}{(x - 3)^2}$$

$$f''(x) = 6(x - 3)^{-3} = \frac{6}{(x - 3)^3}$$

d) Critical points. $f'(3)$ does not exist, but because $f(3)$ does not exist $x = 3$ is not a critical point. The equation $f'(x) = 0$ has no solution, so there are no critical points.

e) Increasing, decreasing, relative extrema. Use 3 to divide the real number line into two intervals, A: $(-\infty,3)$ and B: $(3,\infty)$. Test a point in each interval.

A: Test 0, $f'(0) = -\frac{1}{3} < 0$

B: Test 4, $f'(4) = -3 < 0$

Then f is decreasing on both intervals. Since there are no critical points, there are no relative extrema.

f) Inflection points. $f''(3)$ does not exist, but because $f(3)$ does not exist there cannot be an inflection point at 3. The equation $f''(x) = 0$ has no solution, so there are no inflection points.

g) Concavity. Use 3 to divide the real number line as in step e).

A: Test 0, $f''(0) = -\frac{2}{9} < 0$

B: Test 4, $f''(4) = 6 > 0$

Then f is concave down on $(-\infty,3)$ and concave up on $(3,\infty)$.

h) Sketch. Use the preceding information to sketch the graph. Compute function values as needed.

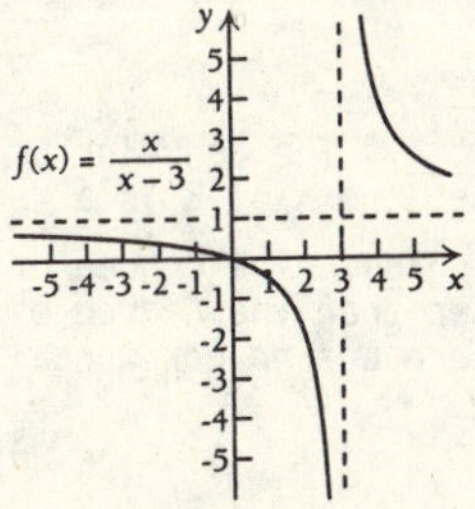

17. $f(x) = \frac{1}{x^2 + 3}$

a) Intercepts. Since the numerator is the constant 1, there are no x-intercepts.

$f(0) = \frac{1}{0^2 + 3} = \frac{1}{3}$, so $\left(0,\frac{1}{3}\right)$ is the y-intercept.

b) Asymptotes.

Vertical. $x^2 + 3 = 0$ has no real number solutions, so there are no vertical asymptotes.

Horizontal. The degree of the numerator is less than the degree of the denominator, so $y = 0$ is a horizontal asymptote.

Oblique. There is no oblique asymptote since the degree of the numerator is not one more than the degree of the denominator.

c) Derivatives.

$$f'(x) = -\frac{2x}{(x^2 + 3)^2}$$

$$f''(x) = \frac{6x^2 - 6}{(x^2 + 3)^3}$$

d) Critical points. $f'(x)$ exists for all real numbers. Solve $f'(x) = 0$.

$$-\frac{2x}{(x^2 + 3)^2} = 0$$

$$-2x = 0$$

$$x = 0 \quad \text{(Critical point)}$$

From step a) we already know $\left(0,\frac{1}{3}\right)$ is on the graph.

e) Increasing, decreasing, relative extrema. Use 0 to divide the real number line into two intervals, A: $(-\infty,0)$ and B: $(0,\infty)$. Test a point in each interval.

A: Test -1, $f'(-1) = \frac{1}{8} > 0$

B: Test 1, $f'(1) = -\frac{1}{8} < 0$

Then f is increasing on $(-\infty,0)$ and decreasing on $(0,\infty)$. Thus, $\left(0,\frac{1}{3}\right)$ is a relative maximum.

f) Inflection points. $f''(x)$ exists for all real numbers. Solve $f''(x) = 0$.

$$\frac{6x^2 - 6}{(x^2 + 3)^3} = 0$$

$$6x^2 - 6 = 0$$

$$6(x + 1)(x - 1) = 0$$

$x = -1$ or $x = 1$ (Possible inflection points)

$f(-1) = \frac{1}{4}$ and $f(1) = \frac{1}{4}$, so $\left(-1,\frac{1}{4}\right)$ and $\left(1,\frac{1}{4}\right)$ are on the graph.

g) Concavity. Use -1 and 1 to divide the real number line into three intervals, A: $(-\infty,-1)$, B: $(-1,1)$, and C: $(1,\infty)$. Test a point in each interval.

A: Test -2, $f''(-2) = \frac{18}{343} > 0$

B: Test 0, $f''(0) = -\frac{2}{9} < 0$

C: Test 2, $f''(2) = \frac{18}{343} > 0$

Then f is concave up on $(-\infty,-1)$ and on $(1,\infty)$ and is concave down on $(-1,1)$. Thus, $\left(-1,\frac{1}{4}\right)$ and $\left(1,\frac{1}{4}\right)$ are both inflection points.

h) Sketch. Use the preceding information to sketch the graph. Compute other function values as needed.

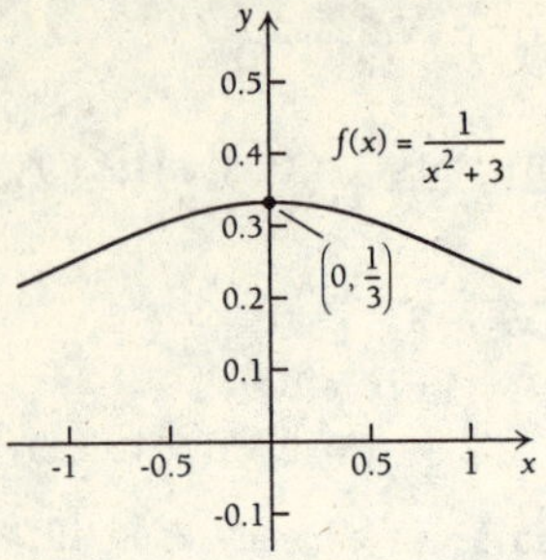

19. $f(x) = \frac{x - 1}{x + 2}$

a) Intercepts. The numerator is 0 for x = 1 and this value of x does not make the denominator 0, so (1,0) is the x-intercept.

$f(0) = \frac{0 - 1}{0 + 2} = -\frac{1}{2}$, so $\left(0, -\frac{1}{2}\right)$ is the y-intercept.

b) Asymptotes.

Vertical. The denominator is 0 for x = -2, so the line x = -2 is a vertical asymptote.

Horizontal. The numerator and the denominator have the same degree, so $y = \frac{1}{1}$, or y = 1, is a horizontal asymptote.

Oblique. There is no oblique asymptote since the degree of the numerator is not one more than the degree of the denominator.

c) Derivatives.

$$f'(x) = \frac{3}{(x + 2)^2}$$

$$f''(x) = \frac{-6}{(x + 2)^3}$$

d) Critical points. f'(-2) does not exist, but because f(-2) does not exist x = -2 is not a critical point. The equation f'(x) = 0 has no solution, so there are no critical points.

e) Increasing, decreasing, relative extrema. Use -2 to divide the real number into two intervals, A: $(-\infty,-2)$ and B: $(-2,\infty)$. Test a point in each interval.

A: Test -3, f'(-3) = 3 > 0

B: Test -1, f'(-1) = 3 > 0

Then f is increasing on both intervals. Since there are no critical points, there are no relative extrema.

f) Inflection points. f''(-2) does not exist, but because f(-2) does not exist there cannot be an inflection point at -2. The equation f''(x) = 0 has no solution, so there are no inflection points.

g) Concavity. Use -2 to divide the real number line as in step e). Test a point in each interval.

A: Test -3, f''(-3) = 6 > 0

B: Test -1, f''(-1) = -6 < 0

Then f is concave up on $(-\infty,-2)$ and concave down on $(-2,\infty)$.

h) Sketch. Use the preceding information to sketch the graph. Compute function values as needed.

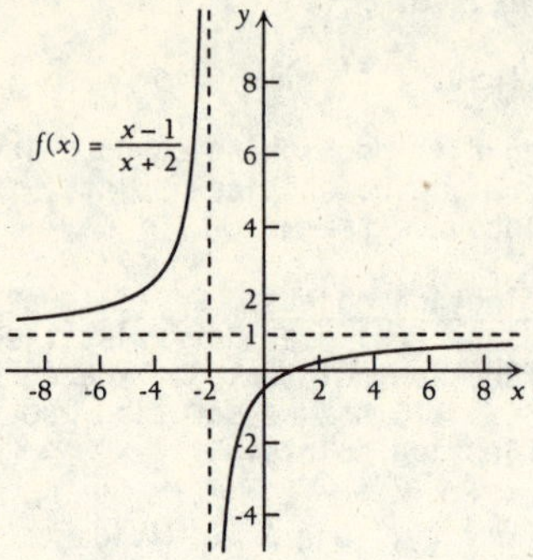

21. $f(x) = \frac{x^2 - 4}{x + 3}$

a) Intercepts. The numerator $x^2 - 4 = (x + 2)(x - 2)$ is 0 for x = -2 or x = 2, and neither of these values makes the denominator 0. Thus, the x-intercepts are (-2,0) and (2,0).

$f(0) = \frac{0^2 - 4}{0 + 3} = -\frac{4}{3}$, so $\left(0, -\frac{4}{3}\right)$ is the y-intercept.

b) Asymptotes.

Vertical. The denominator is 0 for x = -3, so the line x = -3 is a vertical asymptote.

Horizontal. The degree of the numerator is greater than the degree of the denominator, so there are no horizontal asymptotes.

Oblique.

$$f(x) = x - 3 + \frac{5}{x + 3}$$

$$\begin{array}{r} x - 3 \\ x + 3 \overline{) x^2 \quad\quad - 4} \\ \underline{x^2 + 3x} \\ -3x - 4 \\ \underline{-3x - 9} \\ 5 \end{array}$$

As |x| gets very large, f(x) approaches x - 3, so y = x - 3 is an oblique asymptote.

c) Derivatives.

$$f'(x) = \frac{x^2 + 6x + 4}{(x + 3)^2}$$

$$f''(x) = \frac{10}{(x + 3)^3}$$

d) Critical points. f'(-3) does not exist, but because f(-3) does not exist x = -3 is not a critical point. Solve f'(x) = 0.

$$\frac{x^2 + 6x + 4}{(x + 3)^2} = 0$$

$$x^2 + 6x + 4 = 0$$

$$x = -3 \pm \sqrt{5} \quad \text{(Using the quadratic formula)}$$

$$x \approx -5.24 \quad \text{or} \quad x \approx -0.76 \quad \text{(Critical points)}$$

f(-5.24) ≈ -10.47 and f(-0.76) ≈ -1.53, so (-5.24, -10.47) and (-0.76, -1.53) are on the graph.

e) Increasing, decreasing, relative extrema. Use -5.24, -3, and -0.76 to divide the real number line into four intervals, A: (-∞, -5.24), B: (-5.24, -3), C: (-3, -0.76), and D: (-0.76, ∞). Test a point in each interval.

A: Test -6, $f'(-6) = \frac{4}{9} > 0$

B: Test -4, $f'(-4) = -4 < 0$

C: Test -2, $f'(-2) \quad -4 < 0$

D: Test 0, $f'(0) = \frac{4}{9} > 0$

Then f is increasing on (-∞, -5.24) and on (-0.76, ∞) and is decreasing on (-5.24, -3) and on (-3, -0.76). Thus, (-5.24, -10.47) is a relative maximum and (-0.76, -1.53) is a relative minimum.

f) Inflection points. f''(-3) does not exist, but because f(-3) does not exist there cannot be an inflection point at -3. The equation f''(x) = 0 has no solution, so there are no inflection points.

g) Concavity. Use -3 to divide the real number line into two intervals, A: (-∞,-3) and B: (-3,∞). Test a point in each interval.

A: Test -4, $f''(-4) = -10 < 0$

B: Test -2, $f''(-2) = 10 > 0$

Then f is concave down on (-∞,-3) and concave up on (-3,∞).

h) Sketch. Use the preceding information to sketch the graph. Compute other function values as needed.

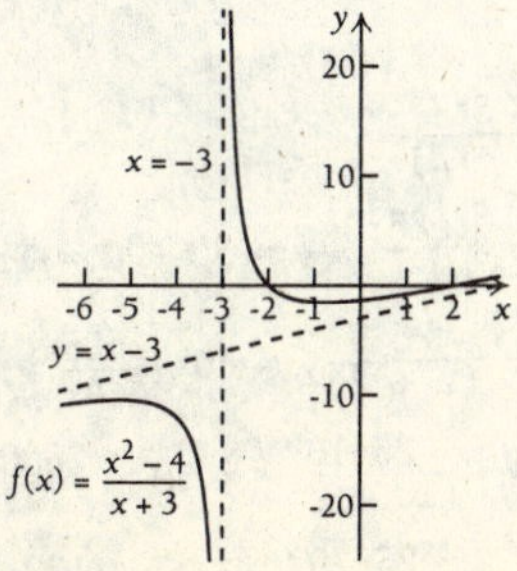

23. $f(x) = \frac{x - 1}{x^2 - 2x - 3}$

a) Intercepts. The numerator is 0 for x = 1, and this value of x does not make the denominator 0. Then (1,0) is the x-intercept.

$f(0) = \frac{0 - 1}{0^2 - 2 \cdot 0 - 3} = \frac{1}{3}$, so $\left(0,\frac{1}{3}\right)$ is the y-intercept.

b) Asymptotes.

Vertical. The denominator $x^2 - 2x - 3 = (x + 1)(x - 3)$ is 0 for x = -1 or x = 3. Then the lines x = -1 and x = 3 are vertical asymptotes.

Horizontal. The degree of the numerator is less than the degree of the denominator, so y = 0 is a horizontal asymptote.

Oblique. There is no oblique asymptote since the degree of the numerator is not one more than the degree of the denominator.

c) Derivatives.

$$f'(x) = \frac{-x^2 + 2x - 5}{(x^2 - 2x - 3)^2}$$

$$f''(x) = \frac{2x^3 - 6x^2 + 30x - 26}{(x^2 - 2x - 3)^3}$$

d) Critical points. f'(-1) and f'(3) do not exist, but because f(-1) and f(3) do not exist x = -1 and x = 3 are not critical points. The equation f'(x) = 0 has no real number solution, so there are no critical points.

e) Increasing, decreasing, relative extrema. Use -1 and 3 to divide the real number line into three intervals, A: (-∞,-1), B: (-1,3), and C: (3,∞). Test a point in each interval.

A: Test -2, $f'(-2) = -\frac{13}{25} < 0$

B: Test 0, $f'(0) = -\frac{5}{9} < 0$

C: Test 4, $f'(4) = -\frac{13}{25} < 0$

Then f is decreasing on all three intervals. Since there are no critical points, there are no relative extrema.

f) Inflection points. f''(-1) and f''(3) do not exist, but because f(-1) and f(3) do not exist there cannot be an inflection point at -1 or at 3. Solve f''(x) = 0.

$$\frac{2x^3 - 6x^2 + 30x - 26}{(x^2 - 2x - 3)^3} = 0$$

$$2x^3 - 6x^2 + 30x - 26 = 0$$

$$(x - 1)(2x^2 - 4x + 26) = 0$$

$$x - 1 = 0 \quad \text{or} \quad 2x^2 - 4x + 26 = 0$$

$$x = 1 \qquad \text{No real number solution}$$

f(1) = 0, so (1,0) is on the graph and is a possible inflection point.

g) Concavity. Use -1, 1, and 3 to divide the real number line into four intervals, A: (-∞,-1), B: (-1,1), C: (1,3), and D: (3,∞). Test a point in each interval.

A: Test -2, $f''(-2) = -\frac{126}{125} < 0$

B: Test 0, $f''(0) = \frac{26}{27} > 0$

C: Test 2, $f''(2) = -\frac{26}{27} < 0$

D: Test 4, $f''(4) = \frac{126}{125} > 0$

Then f is concave down on $(-\infty,-1)$ and on $(1,3)$ and is concave up on $(-1,1)$ and on $(3,\infty)$. Thus, $(1,0)$ is an inflection point.

h) Sketch. Use the preceding information to sketch the graph. Compute other function values as needed.

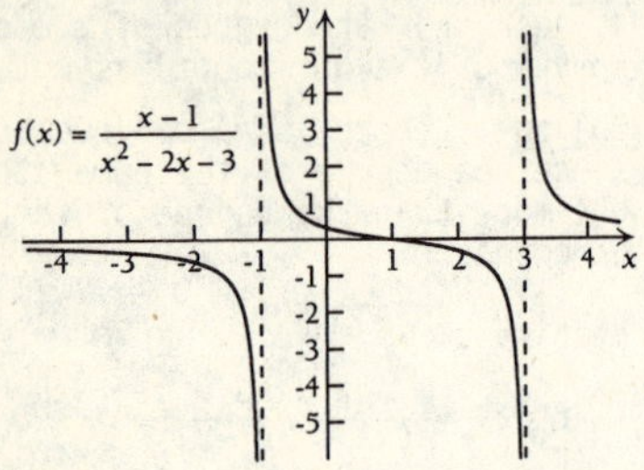

25. $f(x) = \frac{2x^2}{x^2 - 16}$

a) Intercepts. The numerator is 0 for x = 0, and this value of x does not make the denominator 0, so (0,0) is the x-intercept.

f(0) = 0, so the y-intercept is the x-intercept (0,0).

b) Asymptotes.

Vertical. The denominator $x^2 - 16 = (x + 4)(x - 4)$ is 0 for x = -4 or x = 4, so the lines x = -4 and x = 4 are vertical asymptotes.

Horizontal. The numerator and denominator have the same degree, so $y = \frac{2}{1}$, or y = 2, is a horizontal asymptote.

Oblique. There is no oblique asymptote since the degree of the numerator is not one more than the degree of the denominator.

c) Derivatives.

$$f'(x) = \frac{-64x}{(x^2 - 16)^2}$$

$$f''(x) = \frac{192x^2 + 1024}{(x^2 - 16)^3}$$

d) Critical points. f'(-4) and f'(4) do not exist, but because f(-4) and f(4) do not exist x = -4 and x = 4 are not critical points. Solve f'(x) = 0.

$$\frac{-64x}{(x^2 - 16)^2} = 0$$

$$-64x = 0$$

$$x = 0 \quad \text{(Critical point)}$$

From step a) we already know that (0,0) is on the graph.

e) Increasing, decreasing, relative extrema. Use -4, 0, and 4 to divide the real number line into four intervals, A: $(-\infty,-4)$, B: $(-4,0)$, C: $(0,4)$, and D: $(4,\infty)$. Test a point in each interval.

A: Test -5, $f'(-5) = \frac{320}{81} > 0$

B: Test -1, $f'(-1) = \frac{64}{225} > 0$

C: Test 1, $f'(1) = -\frac{64}{225} < 0$

D: Test 5, $f'(5) = -\frac{320}{81} < 0$

Then f is increasing on $(-\infty,-4)$ and on $(-4,0)$ and is decreasing on $(0,4)$ and on $(4,\infty)$. Thus, there is a relative maximum at (0,0).

f) Inflection points. f''(-4) and f''(4) do not exist, but because f(-4) and f(4) do not exist there cannot be an inflection point at -4 or at 4. The equation f''(x) = 0 has no solution, so there are no inflection points.

g) Concavity. Use -4 and 4 to divide the real number line into three intervals, A: $(-\infty,-4)$, B: $(-4,4)$, and C: $(4,\infty)$. Test a point in each interval.

A: Test -5, $f''(-5) = \frac{5824}{729} > 0$

B: Test 0, $f''(0) = -\frac{1}{4} < 0$

C: Test 5, $f''(5) = \frac{5824}{729} > 0$

Then f is concave up on $(-\infty,-4)$ and on $(4,\infty)$ and is concave down on $(-4,4)$.

h) Sketch. Use the preceding information to sketch the graph. Compute other function values as needed.

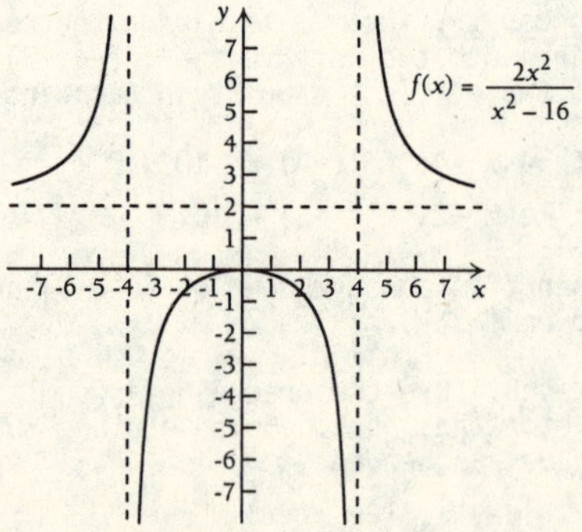

27. $V(t) = 50 - \frac{25t^2}{(t + 2)^2}$

a) $V(0) = 50 - \frac{25 \cdot 0^2}{(0 + 2)^2} = 50 - 0 = \$50$

$V(5) = 50 - \frac{25 \cdot 5^2}{(5 + 2)^2} = 50 - \frac{625}{49} \approx \$37.24$

$V(10) = 50 - \frac{25 \cdot 10^2}{(10 + 2)^2} = 50 - \frac{2500}{144} \approx \$32.64$

$V(70) = 50 - \frac{25 \cdot 70^2}{(70 + 2)^2} = 50 - \frac{122,500}{5184} \approx \$26.37$

b) $f'(x) = -\frac{(t + 2)^2(50t) - 2(t + 2)(25t^2)}{(t + 2)^4}$

$= -\frac{(t + 2)[(t + 2)(50t) - 2(25t^2)]}{(t + 2)^4}$

$= -\frac{50t^2 + 100t - 50t^2}{(t + 2)^3}$

$= -\frac{100t}{(t + 2)^3}$

$f''(x) = -\frac{(t + 2)^3(100) - 3(t + 2)^2(100t)}{(t + 2)^6}$

$= -\frac{(t + 2)^2[(t + 2)(100) - 3(100t)]}{(t + 2)^6}$

$= -\frac{100t + 200 - 300t}{(t + 2)^4}$

$= -\frac{200 - 200t}{(t + 2)^4}$, or $\frac{200t - 200}{(t + 2)^4}$

$f'(x)$ exists for all values of x in $[0,\infty)$.

Solve $f'(x) = 0$.

$-\frac{100t}{(t + 2)^3} = 0$

$-100t = 0$

$t = 0$ (Critical point)

Use the Second-Derivative Test:

$f''(0) = \frac{200 \cdot 0 - 200}{(0 + 2)^4} = -\frac{200}{16} < 0$

Thus, there is a relative maximum at $t = 0$.

$V(0) = \$50$, so the maximum value of the product is \$50.

c) Using the techniques of this section we find the following additional information:

Intercepts. No x-intercepts in $[0,\infty)$; y-intercept is (0,50).

Asymptotes. There are no vertical asymptotes in $[0,\infty)$. The line $y = 25$ is a horizontal asymptote. There is no oblique asymptote.

Increasing, decreasing, relative extrema. $V(t)$ is decreasing on $[0,\infty)$. The only relative extremum is at (0,50).

Inflection points, concavity. $V(t)$ is concave down on [0,1) and concave up on $(1,\infty)$, and (1, 47.22) is an inflection point.

We use this information and compute other function values as needed to sketch the graph.

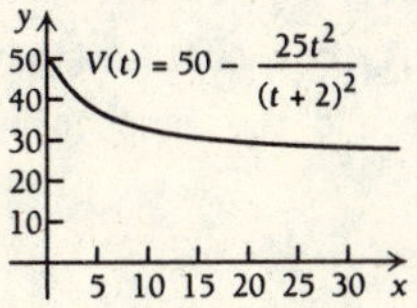

d) $V(t) = 50 - \frac{25t^2}{(t + 2)^2}$

$= \frac{50(t + 2)^2 - 25t^2}{(t + 2)^2}$

$= \frac{50t^2 + 200t + 200 - 25t^2}{(t + 2)^2}$

$= \frac{25t^2 + 200t + 200}{t^2 + 4t + 4}$

$\lim_{t\to\infty} V(t) = \lim_{t\to\infty} \frac{25 + \frac{200}{t} + \frac{200}{t^2}}{1 + \frac{4}{t} + \frac{4}{t^2}}$

$= 25$

e) Yes; the value below which V will never fall is $\lim_{t\to\infty} V(t)$, or \$25.

29. $C(p) = \frac{\$48,000}{100 - p}$

We will only consider the interval [0,100) since it is not possible to remove less than 0% or more than 100% of the pollutants and C(p) is not defined for p = 100.

a) $C(0) = \frac{\$48,000}{100 - 0} = \$480$

$C(20) = \frac{\$48,000}{100 - 20} = \$600$

$C(80) = \frac{\$48,000}{100 - 80} = \$2400$

$C(90) = \frac{\$48,000}{100 - 90} = \$4800$

b) $\lim_{p\to 100^-} C(p) = \lim_{p\to 100^-} \frac{\$48,000}{100 - p} = \infty$

c) Using the techniques of this section we find the following additional information.

Intercepts. No x-intercept; (0,480) in the y-intercept.

Asymptotes. Vertical. p = 100

Horiziontal. y = 0

Oblique. None

Increasing, decreasing, relative extrema.

C(p) is increasing on [0,100). There are no relative extrema.

Inflection points, concavity. C(p) is concave up on [0,100). There is no inflection point.

We use this information and compute other function values as needed to sketch the graph.

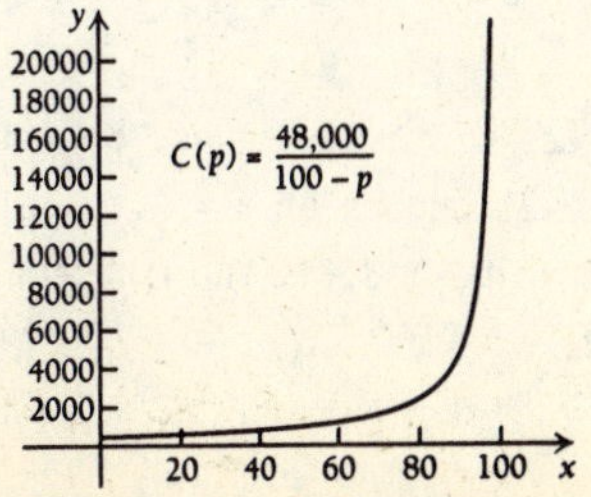

d) From the result in part b), we see that the company cannot afford to remove 100% of the pollutants.

31. See the answer section in the text.

33. See the answer section in the text.

35. See the answer section in the text.

Exercise Set 3.4

1. a) Over the interval [20,80] the function has a maximum value when the speed is 41 mph.

b) Over the interval [20,80] the function has a minimum value when the speed is 80 mph.

c) When the speed is 70 mph, the function value, or miles per gallon, is 13.5 mpg.

d) When the speed is 55 mph, the function value, or miles per gallon, is 16.5 mpg.

e) The mileage at 70 mph is 13.5 mpg.
The mileage at 55 mph is 16.5 mpg.

The increase in mileage is 16.5 - 13.5, or 3 mpg.

The percent of increase is

$\frac{3}{13.5}$ $\frac{\text{(Increase)}}{\text{(Original mpg)}}$, or $\approx 0.22 = 22\%$.

3. $f(x) = 5 + x - x^2$; $[0,2]$

a) Find $f'(x)$.

$f'(x) = 1 - 2x$

b) Find the critical points. The derivative exists for all real numbers. Thus we merely solve $f'(x) = 0$.

$1 - 2x = 0$

$1 = 2x$

$\frac{1}{2} = x$

c) Determine the critical point and the endpoints. These points are

$0, \frac{1}{2}$, and 2.

d) Find the function values at the points in part (c):

$f(0) = 5 + 0 - 0^2 = 5$

$f\left(\frac{1}{2}\right) = 5 + \frac{1}{2} - \left(\frac{1}{2}\right)^2 = 5 + \frac{1}{2} - \frac{1}{4} = 5\frac{1}{4}$ (Max.)

$f(2) = 5 + 2 - 2^2 = 5 + 2 - 4 = 3$ (Min.)

The largest of these values, $5\frac{1}{4}$, is the maximum. It occurs at $x = \frac{1}{2}$. The smallest of these values, 3, is the minimum. It occurs at $x = 2$.

Thus, on the interval [0,2], the

absolute maximum = $5\frac{1}{4}$ at $x = \frac{1}{2}$

and the

absolute minimum = 3 at $x = 2$.

5. $f(x) = x^3 - x^2 - x + 2$; $[0,2]$

a) Find $f'(x)$.

$f'(x) = 3x^2 - 2x - 1$

b) Find the critical points. The derivative exists for all real numbers. Thus we merely solve $f'(x) = 0$.

$3x^2 - 2x - 1 = 0$

$(3x + 1)(x - 1) = 0$ (Factoring)

$3x + 1 = 0$ or $x - 1 = 0$ (Principle of zero products)

$x = -\frac{1}{3}$ or $x = 1$

Note that $-\frac{1}{3}$ is not in the interval [0,2], so 1 is the only critical point.

c) Determine the critical point and the endpoints. These points are 0, 1, and 2.

d) Find the function values at the points in part (c):

$f(0) = 0^3 - 0^2 - 0 + 2 = 2$

$f(1) = 1^3 - 1^2 - 1 + 2 = 1 - 1 - 1 + 2 = 1$ (Minimum)

$f(2) = 2^3 - 2^2 - 2 + 2 = 8 - 4 - 2 + 2 = 4$ (Maximum)

The largest of these values, 4, is the maximum. It occurs at $x = 2$. The smallest of these values, 1, is the minimum. It occurs at $x = 1$.

Thus, on the interval [0,2], the

absolute maximum = 4 at $x = 2$

and the

absolute minimum = 1 at $x = 1$.

7. $f(x) = x^3 - x^2 - x + 2$; $[-1,0]$

As in Exercise 5, the derivative is 0 at $-\frac{1}{3}$ and 1. But only $-\frac{1}{3}$ is in the interval [-1,0], so there is only one critical point, $-\frac{1}{3}$. The critical point and the endpoints are

$-1, -\frac{1}{3}$, and 0.

$f(-1) = (-1)^3 - (-1)^2 - (-1) + 2$

$= -1 - 1 + 1 + 2 = 1$ (Minimum)

$f\left(-\frac{1}{3}\right) = \left(-\frac{1}{3}\right)^3 - \left(-\frac{1}{3}\right)^2 - \left(-\frac{1}{3}\right) + 2$

$= -\frac{1}{27} - \frac{1}{9} + \frac{1}{3} + 2$

$= -\frac{1}{27} - \frac{3}{27} + \frac{9}{27} + 2$

$= \frac{5}{27} + 2 = \frac{59}{27}$ (Maximum)

$f(0) = 0^3 - 0^2 - 0 + 2 = 2$

The largest of these values, $\frac{59}{27}$, is the maximum. It occurs at $x = -\frac{1}{3}$. The smallest of these values, 1, is the minimum. It occurs at $x = -1$.

Thus, on the interval $[-1,0]$, the

absolute maximum $= \frac{59}{27}$ at $x = -\frac{1}{3}$

and the

absolute minimum = 1 at $x = -1$.

9. $f(x) = 3x - 2;\ [-1,1]$

a) Find $f'(x)$.

$f'(x) = 3$

b) and c)

The derivative exists and is 3 for all real numbers. Note that $f'(x)$ is never 0. Thus, there are no critical points for $f(x) = 3x - 2$, and the maximum and minimum values occur at the endpoints.

d) Find the function values at the endpoints.

$f(-1) = 3(-1) - 2 = -3 - 2 = -5$ (Minimum)

$f(1) = 3\cdot1 - 2 = 3 - 2 = 1$ (Maximum)

Thus, on the interval $[-1,1]$, the

absolute maximum = 1 at $x = 1$

and the

absolute minimum = -5 at $x = -1$.

11. $f(x) = 7 - 4x;\ [-2,5]$

a) Find $f'(x)$.

$f'(x) = -4$

b) and c)

The derivative exists and is -4 for all real numbers. Note that $f'(x)$ is never 0. Thus, there are no critical points for $f(x) = 7 - 4x$, and the maximum and minimum values occur at the endpoints.

d) Find the function values at the endpoints.

$f(-2) = 7 - 4(-2) = 7 + 8 = 15$ (Maximum)

$f(5) = 7 - 4\cdot5 = 7 - 20 = -13$ (Minimum)

Thus, on the interval $[-2,5]$, the

absolute maximum = 15 at $x = -2$

and the

absolute minimum = -13 at $x = 5$.

13. $f(x) = -5;\ [-1,1]$

Note that $f'(x) = 0$ for all real numbers. Thus, all points in $[-1,1]$ are critical points. For all x, $f(x) = -5$.

Thus, absolute maximum = absolute minimum = -5 for all x in $[-1,1]$.

15. $f(x) = x^2 - 6x - 3;\ [-1,5]$

a) $f'(x) = 2x - 6$

b) The derivative exists for all real numbers. We solve $f'(x) = 0$.

$2x - 6 = 0$

$2x = 6$

$x = 3$

c) The critical point and the endpoints are -1, 3, and 5.

d) $f(-1) = (-1)^2 - 6(-1) - 3 = 1 + 6 - 3 = 4$ (Maximum)

$f(3) = 3^2 - 6\cdot3 - 3 = 9 - 18 - 3 = -12$ (Minimum)

$f(5) = 5^2 - 6\cdot5 - 3 = 25 - 30 - 3 = -8$

On the interval $[-1,5]$, the

absolute maximum = 4 at $x = -1$

and the

absolute minimum = -12 at $x = 3$.

17. $f(x) = 3 - 2x - 5x^2;\ [-3,3]$

a) $f'(x) = -2 - 10x$

b) The derivative exists for all real numbers. We solve $f'(x) = 0$.

$-2 - 10x = 0$

$-10x = 2$

$x = -\frac{1}{5}$

c) The critical point and the endpoints are -3, $-\frac{1}{5}$, and 3.

d) $f(-3) = 3 - 2(-3) - 5(-3)^2 = 3 + 6 - 45 = -36$

$f\left(-\frac{1}{5}\right) = 3 - 2\left(-\frac{1}{5}\right) - 5\left(-\frac{1}{5}\right)^2 = 3 + \frac{2}{5} - \frac{1}{5} = 3\frac{1}{5}$ (Maximum)

$f(3) = 3 - 2\cdot3 - 5\cdot3^2 = 3 - 6 - 45 = -48$ (Minimum)

On the interval $[-3,3]$, the

absolute maximum $= 3\frac{1}{5}$ at $x = -\frac{1}{5}$

and the

absolute minimum = -48 at $x = 3$.

19. $f(x) = x^3 - 3x^2;\ [0,5]$

a) $f'(x) = 3x^2 - 6x$

b) The derivative exists for all real numbers. We solve $f'(x) = 0$.

$3x^2 - 6x = 0$

$3x(x - 2) = 0$

$3x = 0$ or $x - 2 = 0$

$x = 0$ or $x = 2$

c) The critical points and endpoints are 0, 2, and 5.

d) $f(0) = 0^3 - 3\cdot 0^2 = 0 - 0 = 0$

$f(2) = 2^3 - 3\cdot 2^2 = 8 - 12 = -4$ (Minimum)

$f(5) = 5^3 - 3\cdot 5^2 = 125 - 75 = 50$ (Maximum)

On the interval [0,5], the

absolute maximum = 50 at x = 5

and the

absolute minimum = -4 at x = 2.

21. $f(x) = x^3 - 3x$; [-5,1]

a) $f'(x) = 3x^2 - 3$

b) The derivative exists for all real numbers. We solve $f'(x) = 0$.

$$3x^2 - 3 = 0$$
$$3(x^2 - 1) = 0$$
$$3(x + 1)(x - 1) = 0$$
$$x + 1 = 0 \quad \text{or} \quad x - 1 = 0$$
$$x = -1 \quad \text{or} \quad x = 1$$

c) The critical points and endpoints are -5, -1, and 1.

d) $f(-5) = (-5)^3 - 3(-5) = -125 + 15 = -110$ (Minimum)

$f(-1) = (-1)^3 - 3(-1) = -1 + 3 = 2$ (Maximum)

$f(1) = 1^3 - 3\cdot 1 = 1 - 3 = -2$

On the interval [-5,1], the

absolute maximum = 2 at x = -1

and the

absolute minimum = -110 at x = -5.

23. $f(x) = 1 - x^3$; [-8,8]

a) $f'(x) = -3x^2$

b) The derivative exists for all real numbers. We solve $f'(x) = 0$.

$$-3x^2 = 0$$
$$x^2 = 0$$
$$x = 0$$

c) The critical point and the endpoints are -8, 0, and 8.

d) $f(-8) = 1 - (-8)^3 = 1 + 512 = 513$ (Maximum)

$f(0) = 1 - 0^3 = 1 - 0 = 1$

$f(8) = 1 - 8^3 = 1 - 512 = -511$ (Minimum)

On the interval [-8,8], the

absolute maximum = 513 at x = -8

and the

absolute minimum = -511 at x = 8.

25. $f(x) = 12 + 9x - 3x^2 - x^3$; [-3,1]

a) $f'(x) = 9 - 6x - 3x^2$

b) The derivative exists for all real numbers. We solve $f'(x) = 0$.

$$9 - 6x - 3x^2 = 0$$
$$3(3 - 2x - x^2) = 0$$
$$3(3 + x)(1 - x) = 0$$
$$3 + x = 0 \quad \text{or} \quad 1 - x = 0$$
$$x = -3 \quad \text{or} \quad 1 = x$$

c) The critical points are the endpoints, -3 and 1.

d) $f(-3) = 12 + 9(-3) - 3(-3)^2 - (-3)^3$

$= 12 - 27 - 27 + 27$

$= -15$ (Minimum)

$f(1) = 12 + 9\cdot 1 - 3\cdot 1^2 - 1^3$

$= 12 + 9 - 3 - 1$

$= 17$ (Maximum)

On the interval [-3,1], the

absolute maximum = 17 at x = 1

and the

absolute minimum = -15 at x = -3.

27. $f(x) = x^4 - 2x^3$; [-2,2]

a) $f'(x) = 4x^3 - 6x^2$

b) The derivative exists for all real numbers. We solve $f'(x) = 0$.

$$4x^3 - 6x^2 = 0$$
$$2x^2(2x - 3) = 0$$
$$2x^2 = 0 \quad \text{or} \quad 2x - 3 = 0$$
$$x = 0 \quad \text{or} \quad x = \frac{3}{2}$$

c) The critical points and endpoints are -2, 0, $\frac{3}{2}$, and 2.

d) $f(-2) = (-2)^4 - 2(-2)^3 = 16 + 16 = 32$ (Maximum)

$f(0) = 0^4 - 2\cdot 0^3 = 0 - 0 = 0$

$f\left(\frac{3}{2}\right) = \left(\frac{3}{2}\right)^4 - 2\left(\frac{3}{2}\right)^3 = \frac{81}{16} - \frac{27}{4} = -\frac{27}{16}$ (Minimum)

$f(2) = 2^4 - 2\cdot 2^3 = 16 - 16 = 0$

On the interval [-2,2], the

absolute maximum = 32 at x = -2

and the

absolute minimum = $-\frac{27}{16}$ at $x = \frac{3}{2}$.

29. $f(x) = x^4 - 2x^2 + 5$; $[-2,2]$

a) $f'(x) = 4x^3 - 4x$

b) The derivative exists for all real numbers. We solve $f'(x) = 0$.

$$4x^3 - 4x = 0$$
$$4x(x^2 - 1) = 0$$
$$4x(x + 1)(x - 1) = 0$$
$$4x = 0 \text{ or } x + 1 = 0 \text{ or } x - 1 = 0$$
$$x = 0 \text{ or } x = -1 \text{ or } x = 1$$

c) The critical points and endpoints are -2, -1, 0, 1, and 2.

d) $f(-2) = (-2)^4 - 2(-2)^2 + 5 = 16 - 8 + 5 = 13$ (Maximum)

$f(-1) = (-1)^4 - 2(-1)^2 + 5 = 1 - 2 + 5 = 4$ (Minimum)

$f(0) = 0^4 - 2 \cdot 0^2 + 5 = 0 - 0 + 5 = 5$

$f(1) = 1^4 - 2 \cdot 1^2 + 5 = 1 - 2 + 5 = 4$ (Minimum)

$f(2) = 2^4 - 2 \cdot 2^2 + 5 = 16 - 8 + 5 = 13$ (Maximum)

On the interval $[-2,2]$, the absolute maximum = 13 at $x = -2$ and $x = 2$ and the absolute minimum = 4 at $x = -1$ and $x = 1$.

31. $f(x) = (x + 3)^{2/3} - 5$; $[-4,5]$

a) $f'(x) = \frac{2}{3}(x + 3)^{-1/3} = \frac{2}{3(x + 3)^{1/3}}$

b) The derivative does not exist for $x = -3$. The equation $f'(x) = 0$ has no solution, so -3 is the only critical point.

c) The critical point and endpoints are -4, -3, and 5.

d) $f(-4) = (-4 + 3)^{2/3} - 5 = (-1)^{2/3} - 5 = 1 - 5 = -4$

$f(-3) = (-3 + 3)^{2/3} - 5 = 0^{2/3} - 5 = 0 - 5 = -5$ (Minimum)

$f(5) = (5 + 3)^{2/3} - 5 = 8^{2/3} - 5 = 4 - 5 = -1$ (Maximum)

On the interval $[-4,5]$, the absolute maximum = -1 at $x = 5$ and the absolute minimum = -5 at $x = -3$.

33. $f(x) = x + \frac{1}{x}$; $[1,20]$

Express the function as $f(x) = x + x^{-1}$.

a) $f'(x) = 1 - x^{-2} = 1 - \frac{1}{x^2}$

b) The derivative exists for all x in $[1,20]$. We solve $f'(x) = 0$.

$$1 - \frac{1}{x^2} = 0$$
$$1 = \frac{1}{x^2}$$
$$x^2 = 1$$
$$x = \pm 1$$

Only 1 is in the interval $[1,20]$.

c) The critical point and endpoints are 1 and 20.

d) $f(1) = 1 + \frac{1}{1} = 1 + 1 = 2$ (Minimum)

$f(20) = 20 + \frac{1}{20} = 20\frac{1}{20}$ (Maximum)

On the interval $[1,20]$, the absolute maximum = $20\frac{1}{20}$ at $x = 20$ and the absolute minimum = 2 at $x = 1$.

35. $f(x) = \frac{x^2}{x^2 + 1}$; $[-2,2]$

a) $f'(x) = \frac{(x^2 + 1)(2x) - 2x(x^2)}{(x^2 + 1)^2}$ (Quotient Rule)

$= \frac{2x^3 + 2x - 2x^3}{(x^2 + 1)^2}$

$= \frac{2x}{(x^2 + 1)^2}$

b) The derivative exists for all real numbers. We solve $f'(x) = 0$.

$$\frac{2x}{(x^2 + 1)^2} = 0$$
$$2x = 0$$
$$x = 0$$

c) The critical point and endpoints are -2, 0, and 2.

d) $f(-2) = \frac{(-2)^2}{(-2)^2 + 1} = \frac{4}{4 + 1} = \frac{4}{5}$ (Maximum)

$f(0) = \frac{0^2}{0^2 + 1} = \frac{0}{1} = 0$ (Minimum)

$f(2) = \frac{2^2}{2^2 + 1} = \frac{4}{4 + 1} = \frac{4}{5}$ (Maximum)

On the interval $[-2,2]$, the absolute maximum = $\frac{4}{5}$ at $x = -2$ and $x = 2$ and the absolute minimum = 0 at $x = 0$.

37. $f(x) = (x + 1)^{1/3}$; $[-2,26]$

a) $f'(x) = \frac{1}{3}(x + 1)^{-2/3} = \frac{1}{3(x + 1)^{2/3}}$

b) The derivative does not exist for $x = -1$. The equation $f'(x) = 0$ has no solution, so -1 is the only critical point.

c) The critical point and endpoints are -2, -1, and 26.

d) $f(-2) = (-2 + 1)^{1/3} = (-1)^{1/3} = -1$ (Minimum)

$f(-1) = (-1 + 1)^{1/3} = 0^{1/3} = 0$

$f(26) = (26 + 1)^{1/3} = 27^{1/3} = 3$ (Maximum)

On the interval [-2,26], the

absolute maximum = 3 at x = 26

and the

absolute minimum = -1 at x = -2.

39. $f(x) = x(70 - x)$

$= 70x - x^2$

When no interval is specified, we use the real line $(-\infty,\infty)$.

a) Find f'(x).

$f'(x) = 70 - 2x$

b) Find the critical points.

The derivative exists for all real numbers. Thus we solve $f'(x) = 0$.

$70 - 2x = 0$

$-2x = -70$

$x = 35$

e) Since there is only one critical point, we can apply Max-Min Principle 2.

Find f''(x).

$f''(x) = -2$

Now the second derivative is constant, so $f''(35) = -2$, and since this is negative, we have a maximum at x = 35.

Find the function value at x = 35.

$f(35) = 35(70 - 35)$

$= 35(35)$

$= 1225$

Thus the

absolute maximum = 1225 at x = 35.

The function has no minimum value.

41. $f(x) = 2x^2 - 40x + 400$

When no interval is specified, we use the real line $(-\infty,\infty)$.

a) Find f'(x).

$f'(x) = 4x - 40$

b) Find the critical points.

The derivative exists for all real numbers. Thus we solve $f'(x) = 0$.

$4x - 40 = 0$

$4x = 40$

$x = 10$

e) Since there is only one critical point, we can apply Max-Min Principle 2.

Find f''(x).

$f''(x) = 4$

Now the second derivative is constant, so $f''(10) = 4$, and since this is positive, we have a minimum at x = 10.

Find the function value at x = 10.

$f(10) = 2\cdot 10^2 - 40\cdot 10 + 400$

$= 200 - 400 + 400$

$= 200$

Thus the

absolute minimum = 200 at x = 10.

The function has no maximum value.

43. $f(x) = x - \frac{4}{3}x^3$; $(0,\infty)$

a) Find f'(x).

$f'(x) = 1 - 4x^2$

b) Find the critical points.

The derivative exists for all real numbers. Thus we solve $f'(x) = 0$.

$1 - 4x^2 = 0$

$-4x^2 = -1$

$x^2 = \frac{1}{4}$

$x = \pm\frac{1}{2}$

e) The interval is not closed. The only critical point in $(0,\infty)$ is $\frac{1}{2}$. Thus, we can apply the second derivative

$f''(x) = -8x$

to determine whether we have a maximum or a minimum. Now f''(x) is <u>negative</u> for all values of x in $(0,\infty)$, thus there is a maximum at $x = \frac{1}{2}$.

$f''\left(\frac{1}{2}\right) = -8\cdot\left(\frac{1}{2}\right) = -4 < 0$

Find the function value at $x = \frac{1}{2}$.

$f\left(\frac{1}{2}\right) = \frac{1}{2} - \frac{4}{3}\left(\frac{1}{2}\right)^3 = \frac{1}{2} - \frac{4}{3}\cdot\frac{1}{8} = \frac{1}{2} - \frac{1}{6} = \frac{1}{3}$

Thus the

absolute maximum = $\frac{1}{3}$ at $x = \frac{1}{2}$.

The function has no minimum value.

45. $f(x) = 17x - x^2$

When no interval is specified, we use the real line $(-\infty,\infty)$.

a) Find f'(x).

$f'(x) = 17 - 2x$

b) Find the critical points.

The derivative exists for all real numbers. Thus we solve $f'(x) = 0$.

$17 - 2x = 0$

$-2x = -17$

$x = \frac{17}{2}$

e) Since there is only one critical point, we can apply Max-Min Principle 2.

Find f''(x).

$f''(x) = -2$

Now the second derivative is constant, so $f''\left(\frac{17}{2}\right) = -2$, and since this is negative, we have a maximum at $x = \frac{17}{2}$.

Find the function value at $x = \frac{17}{2}$.

$f\left(\frac{17}{2}\right) = 17 \cdot \frac{17}{2} - \left(\frac{17}{2}\right)^2$

$= \frac{289}{2} - \frac{289}{4}$

$= \frac{578}{4} - \frac{289}{4}$

$= \frac{289}{4}$

Thus the

absolute maximum $= \frac{289}{4}$ at $x = \frac{17}{2}$.

The function has no minimum value.

47. $f(x) = \frac{1}{3}x^3 - 3x$; $[-2,2]$

a) Find f'(x).

$f'(x) = x^2 - 3$

b) Find the critical points.

The derivative exists for all real numbers. Thus we solve $f'(x) = 0$.

$x^2 - 3 = 0$

$x^2 = 3$

$x = \pm\sqrt{3} \approx \pm 1.732$

Both critical points are in the interval $[-2,2]$.

c) If the interval is closed and there is more than one critical point, then use Max-Min Principle 1.

The critical points and the endpoints are -2, $-\sqrt{3}$, $\sqrt{3}$, and 2.

Next we find the function values at these points.

$f(-2) = \frac{1}{3}(-2)^3 - 3(-2) = -\frac{8}{3} + 6 = \frac{10}{3} = 3.3\overline{3}$

$f(-\sqrt{3}) = \frac{1}{3}(-\sqrt{3})^3 - 3(-\sqrt{3})$

$= \frac{1}{3}(-3\sqrt{3}) + 3\sqrt{3}$

$[(-\sqrt{3})^3 = (-\sqrt{3})(-\sqrt{3})(-\sqrt{3}) = -3\sqrt{3}]$

$= -\sqrt{3} + 3\sqrt{3}$

$= 2\sqrt{3} \approx 2(1.732) = 3.464$ (Maximum)

$f(\sqrt{3}) = \frac{1}{3}(\sqrt{3})^3 - 3\sqrt{3}$

$= \frac{1}{3}(3\sqrt{3}) - 3\sqrt{3}$

$[(\sqrt{3})^3 = \sqrt{3}\cdot\sqrt{3}\cdot\sqrt{3} = 3\sqrt{3}]$

$= \sqrt{3} - 3\sqrt{3}$

$= -2\sqrt{3} \approx -2(1.732) = -3.464$

(Minimum)

$f(2) = \frac{1}{3} \cdot 2^3 - 3\cdot 2 = \frac{8}{3} - 6 = -\frac{10}{3} = -3.3\overline{3}$

The largest of these values, $2\sqrt{3} \approx 3.464$, is the maximum. It occurs at $x = -\sqrt{3}$. The smallest of these values, $-2\sqrt{3} \approx -3.464$, is the minimum. It occurs at $x = \sqrt{3}$.

Thus the

absolute maximum $= 2\sqrt{3}$ at $x = -\sqrt{3}$

and the

absolute minimum $= -2\sqrt{3}$ at $x = \sqrt{3}$.

49. $f(x) = -0.001x^2 + 4.8x - 60$

When no interval is specified, we use the real line $(-\infty,\infty)$.

a) Find f'(x).

$f'(x) = -0.002x + 4.8$

b) Find the critical points.

The derivative exists for all real numbers. Thus we solve $f'(x) = 0$.

$-0.002x + 4.8 = 0$

$-0.002x = -4.8$

$x = 2400$

e) Since there is only one critical point, we can apply Max-Min Principle 2.

Find f''(x).

$f''(x) = -0.002$

Now the second derivative is constant, so $f''(2400) = -0.002$, and since this is negative, we have a maximum at $x = 2400$.

Find the function value at $x = 2400$.

$f(2400) = -0.001(2400)^2 + 4.8(2400) - 60$

$= -5760 + 11{,}520 - 60$

$= 5700$

Thus the

absolute maximum = 5700 at x = 2400.

The function has no minimum value.

51. $f(x) = -\frac{1}{3}x^3 + 6x^2 - 11x - 50$; $(0,3)$

a) Find f'(x).

$f'(x) = -x^2 + 12x - 11$

b) Find the critical points.

The derivative exists for all real numbers. Thus we solve $f'(x) = 0$.

$$-x^2 + 12x - 11 = 0$$
$$x^2 - 12x + 11 = 0$$
$$(x - 11)(x - 1) = 0$$
$$x - 11 = 0 \text{ or } x - 1 = 0$$
$$x = 11 \text{ or } x = 1$$

e) The interval is not closed. The only critical point in (0,3) is 1. Thus, we can apply the second derivative

$$f''(x) = -2x + 12$$

to determine whether we have a maximum or a minimum.

$$f''(1) = -2\cdot 1 + 12 = -2 + 12 = 10 > 0$$

Since the second derivative is positive when $x = 1$, there is a minimum at $x = 1$.

Find the function value at $x = 1$.

$$f(1) = -\frac{1}{3}\cdot 1^3 + 6\cdot 1^2 - 11\cdot 1 - 50$$
$$= -\frac{1}{3} + 6 - 11 - 50$$
$$= -55\,\frac{1}{3}$$

Thus the

absolute minimum $= -55\,\frac{1}{3}$ at $x = 1$.

The function has no maximum value in (0,3).

53. $f(x) = 15x^2 - \frac{1}{2}x^3$; [0,30]

a) Find $f'(x)$.

$$f'(x) = 30x - \frac{3}{2}x^2$$

b) Find the critical points.

The derivative exists for all real numbers. Thus we solve $f'(x) = 0$.

$$30x - \frac{3}{2}x^2 = 0$$
$$60x - 3x^2 = 0$$
$$3x(20 - x) = 0$$
$$3x = 0 \text{ or } 20 - x = 0$$
$$x = 0 \text{ or } x = 20$$

Both critical points are in the interval [0,30].

c) If the interval is closed and there is more than one critical point, then use Max-Min Principle 1.

The critical points and the endpoints are

0, 20, and 30.

Next we find the function values at these points.

$$f(0) = 15\cdot 0^2 - \frac{1}{2}\cdot 0^3$$
$$= 0 \qquad \text{(Minimum)}$$

$$f(20) = 15\cdot 20^2 - \frac{1}{2}\cdot 20^3$$
$$= 15\cdot 400 - \frac{1}{2}\cdot 8000$$
$$= 6000 - 4000$$
$$= 2000 \qquad \text{(Maximum)}$$

$$f(30) = 15\cdot 30^2 - \frac{1}{2}\cdot 30^3$$
$$= 15\cdot 900 - \frac{1}{2}\cdot 27{,}000$$
$$= 13{,}500 - 13{,}500$$
$$= 0 \qquad \text{(Minimum)}$$

The largest of these values, 2000, is the maximum. It occurs at $x = 20$. The smallest of these values is 0. It occurs twice, at $x = 0$ and $x = 30$. Thus, on the interval [0,30], the

absolute maximum = 2000 at $x = 20$

and the

absolute minimum = 0 at $x = 0$ and $x = 30$.

55. $f(x) = 2x + \frac{72}{x}$; $(0,\infty)$

$$= 2x + 72x^{-1}$$

a) Find $f'(x)$.

$$f'(x) = 2 - 72x^{-2}$$
$$= 2 - \frac{72}{x^2}$$

b) Find the critical points.

Now $f'(x)$ exists for all values of x in $(0,\infty)$. Thus the only critical points are those for which $f'(x) = 0$.

$$2 - \frac{72}{x^2} = 0$$
$$2 = \frac{72}{x^2}$$
$$2x^2 = 72 \qquad (\text{Multiplying by } x^2, \text{ since } x \neq 0)$$
$$x^2 = 36$$
$$x = \pm 6$$

e) The only critical point in $(0,\infty)$ is 6. Thus, we can apply the second derivative

$$f''(x) = 144x^{-3} = \frac{144}{x^3}$$

to determine whether we have a maximum or minimum.

$$f''(6) = \frac{144}{6^3} = \frac{144}{216} > 0$$

Since the second derivative is positive when $x = 6$, there is a minimum at $x = 6$.

Find the function value at $x = 6$.

$$f(6) = 2\cdot 6 + \frac{72}{6}$$
$$= 12 + 12$$
$$= 24$$

Thus the

absolute minimum = 24 at $x = 6$.

The function has no maximum value on $(0,\infty)$.

57. $f(x) = x^2 + \frac{432}{x}$; $(0,\infty)$

$= x^2 + 432x^{-1}$

a) Find $f'(x)$.

$f'(x) = 2x - 432x^{-2}$

$= 2x - \frac{432}{x^2}$

b) Find the critical points.

Now $f'(x)$ exists for all values of x in $(0,\infty)$. Thus the only critical points are those for which $f'(x) = 0$.

$2x - \frac{432}{x^2} = 0$

$2x = \frac{432}{x^2}$

$2x^3 = 432$ (Multiplying by x^2, since $x \neq 0$)

$x^3 = 216$

$x = 6$

e) Since there is only one critical point, we can use Max-Min Principle 2 to determine whether we have a maximum or a minimum.

Find $f''(x)$.

$f''(x) = 2 + 864x^{-3}$

$= 2 + \frac{864}{x^3}$

Since

$f''(6) = 2 + \frac{864}{6^3} = 2 + \frac{864}{216} = 2 + 4 = 6$

and $6 > 0$, there is a minimum at $x = 6$.

Find the function value at $x = 6$.

$f(6) = 6^2 + \frac{432}{6}$

$= 36 + 72$

$= 108$

Thus the

absolute minimum = 108 at $x = 6$.

The function has no maximum value.

59. $f(x) = 2x^4 - x$; $[-1,1]$

a) Find $f'(x)$.

$f'(x) = 8x^3 - 1$

b) Find the critical points.

The derivative exists for all real numbers. Thus we solve $f'(x) = 0$.

$8x^3 - 1 = 0$

$8x^3 = 1$

$x^3 = \frac{1}{8}$

$x = \frac{1}{2}$

There is one critical point. It is in the interval $[-1,1]$.

d) On a closed interval Max-Min Principle 1 can always be used.

The critical points and the endpoints are -1, $\frac{1}{2}$, and 1.

We find the function values at these points.

$f(-1) = 2(-1)^4 - (-1) = 2 + 1 = 3$ (Maximum)

$f\left(\frac{1}{2}\right) = 2\left(\frac{1}{2}\right)^4 - \frac{1}{2} = \frac{1}{8} - \frac{1}{2} = -\frac{3}{8}$ (Minimum)

$f(1) = 2(1)^4 - 1 = 2 - 1 = 1$

The largest of these values, 3, is the maximum. It occurs at $x = -1$. The smallest of these values, $-\frac{3}{8}$, is the minimum. It occurs at $x = \frac{1}{2}$.

Thus the

absolute maximum = 3 at $x = -1$

and the

absolute minimum = $-\frac{3}{8}$ at $x = \frac{1}{2}$.

61. $f(x) = \sqrt[3]{x}$; $[0,8]$

$= x^{1/3}$

a) Find $f'(x)$.

$f'(x) = \frac{1}{3}x^{-2/3}$

$= \frac{1}{3\sqrt[3]{x^2}}$

b) Find the critical points.

Point c is a critical point if $f'(c) = 0$ <u>or</u> $f'(c)$ does not exist.

The derivative does not exist for $x = 0$. Thus, 0 is a critical point. Since there are no values of x for which

$\frac{1}{3\sqrt[3]{x^2}} = 0$,

there are no other critical points.

d) On a closed interval, Max-Min Principle 1 can always be used.

The critical points and the endpoints are 0 and 8.

We find the function values at these points.

$f(0) = \sqrt[3]{0} = 0$ (Minimum)

$f(8) = \sqrt[3]{8} = 2$ (Maximum)

Thus the

absolute maximum = 2 at $x = 8$

and the

absolute minimum = 0 at $x = 0$.

63. $f(x) = (x + 1)^3$

When no interval is specified, we use the real line $(-\infty,\infty)$.

a) Find $f'(x)$.

$f'(x) = 3(x + 1)^2 \cdot 1 = 3(x + 1)^2$

b) Find the critical points.

The derivative exists for all real numbers. Thus we solve $f'(x) = 0$.

$$3(x + 1)^2 = 0$$
$$(x + 1)^2 = 0$$
$$x + 1 = 0$$
$$x = -1$$

e) Since there is only one critical point and there are no endpoints, we can try to apply Max-Min Principle 2 using the second derivative:

$f''(x) = 6(x + 1)$.

Now

$f''(-1) = 6(-1 + 1) = 0$,

so Max-Min Principle 2 fails. We cannot use Max-Min Principle 1 because there are no endpoints. We note that $f'(x) = 3(x + 1)^2$ is never negative. Thus, $f(x)$ is increasing everywhere except at $x = -1$, so there is no maximum or minimum. At $x = -1$, the function has a point of inflection.

65. $f(x) = 2x - 3$; $[-1,1]$

a) Find $f'(x)$.

$f'(x) = 2$

b) and c)

The derivative exists and is 2 for <u>all</u> real numbers. Note that $f'(x)$ is never <u>0</u>. Thus, there are no critical points for $f(x) = 2x - 3$, and the maximum and minimum values occur at the endpoints, -1 and 1.

Find the function values at the endpoints.

$f(x) = 2x - 3$

$f(-1) = 2(-1) - 3 = -2 - 3 = -5$ (Minimum)

$f(1) = 2 \cdot 1 - 3 = 2 - 3 = -1$ (Maximum)

Thus, on the interval $[-1,1]$, the

absolute maximum = -1 at $x = 1$

and the

absolute minimum = -5 at $x = -1$.

67. $f(x) = 2x - 3$

Find $f'(x)$.

$f'(x) = 2$

The derivative exists and is 2 for <u>all</u> real numbers. Note that $f'(x)$ is never <u>0</u>. Thus, there are no critical points for $f(x) = 2x - 3$. The interval is $(-\infty,\infty)$. There are no endpoints. We also observe that $f'(x)$ is always positive. Thus $f(x)$ is increasing everywhere, so there is no maximum or minimum.

69. $f(x) = x^{2/3} = \sqrt[3]{x^2}$; $[-1,1]$

Find $f'(x)$.

$f'(x) = \frac{2}{3}x^{-1/3} = \frac{2}{3\sqrt[3]{x}}$

Find the critical points.

[Point c is a critical point if $f'(c) = 0$ <u>or</u> $f'(c)$ does not exist.]

The derivative exists for all real numbers except 0. Thus, 0 is a critical point. Since there are no values of x for which

$$\frac{2}{3\sqrt[3]{x}} = 0$$

there are no other critical points.

On a closed interval Max-Min Principle 1 can always be used. The only critical point is 0. The endpoints are -1 and 1. We find the function value at each of these points.

$f(-1) = \sqrt[3]{(-1)^2} = \sqrt[3]{1} = 1$ (Maximum)

$f(0) = \sqrt[3]{0^2} = \sqrt[3]{0} = 0$ (Minimum)

$f(1) = \sqrt[3]{1^2} = \sqrt[3]{1} = 1$ (Maximum)

Thus the

absolute maximum = 1 at $x = -1$ and $x = 1$

and the

absolute minimum = 0 at $x = 0$.

71. $f(x) = \frac{1}{3}x^3 - x + \frac{2}{3}$; $(-\infty,\infty)$

Find $f'(x)$.

$f'(x) = x^2 - 1$

Find the critical points.

The derivative exists for all real numbers. We solve $f'(x) = 0$.

$$x^2 - 1 = 0$$
$$(x - 1)(x + 1) = 0$$
$$x - 1 = 0 \text{ or } x + 1 = 0$$
$$x = 1 \text{ or } \quad x = -1$$

The case of finding maximum and minimum values when more than one critical point occurs in an interval which is not closed can only be solved with a detailed graph or by techniques beyond the scope of this book. Let us study the graph.

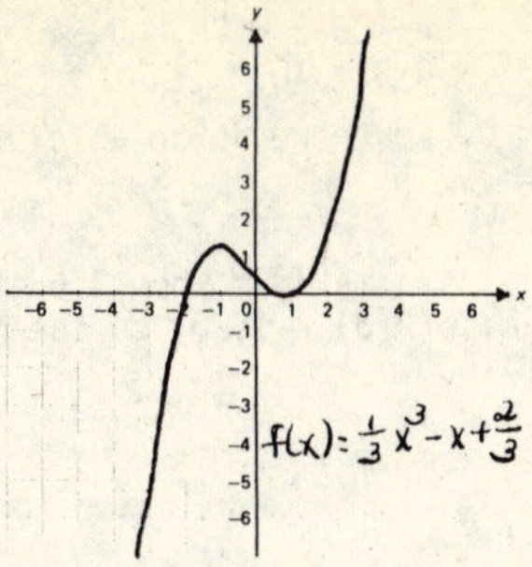

From the graph we observe that on the interval $(-\infty,\infty)$, there is no absolute maximum or absolute minimum.

73. $f(x) = \frac{1}{3}x^3 - 2x^2 + x$; $[0,4]$

Find $f'(x)$.

$f'(x) = x^2 - 4x + 1$

Find the critical points.

The derivative exists for all real numbers. We solve $f'(x) = 0$.

$x^2 - 4x + 1 = 0$

$x = \frac{4 \pm \sqrt{(-4)^2 - 4\cdot1\cdot1}}{2\cdot1}$ (Using the quadratic formula)

$= \frac{4 \pm \sqrt{12}}{2}$

$= \frac{4 \pm 2\sqrt{3}}{2}$

$= 2 \pm \sqrt{3}$ $(2 + \sqrt{3} \approx 3.7,\ 2 - \sqrt{3} \approx 0.3)$

On a closed interval Max-Min Principle 1 can always be used. The endpoints are 0 and 4. The critical points are $2 - \sqrt{3}$ and $2 + \sqrt{3}$. We find the function value at each of these points.

$f(x) = \frac{1}{3}x^3 - 2x^2 + x = x\left[\frac{1}{3}x^2 - 2x + 1\right]$

$f(0) = 0\left[\frac{1}{3}\cdot 0^2 - 2\cdot0 + 1\right] = 0$

$f(2 - \sqrt{3}) = (2 - \sqrt{3})\left[\frac{1}{3}(2-\sqrt{3})^2 - 2(2-\sqrt{3}) + 1\right]$

$= (2 - \sqrt{3})\left[\frac{1}{3}(7-4\sqrt{3}) - 2(2-\sqrt{3}) + 1\right]$

$= (2 - \sqrt{3})\left[\frac{7}{3} - \frac{4}{3}\sqrt{3} - 4 + 2\sqrt{3} + 1\right]$

$= (2 - \sqrt{3})\left[-\frac{2}{3} + \frac{2}{3}\sqrt{3}\right]$

$= -\frac{2}{3}(2 - \sqrt{3})(1 - \sqrt{3})$

$= -\frac{2}{3}(2 - 3\sqrt{3} + 3)$

$= -\frac{2}{3}(5 - 3\sqrt{3})$

$= -\frac{10}{3} + 2\sqrt{3} \approx 0.131$ (Maximum)

$f(2 + \sqrt{3}) = (2 + \sqrt{3})\left[\frac{1}{3}(2+\sqrt{3})^2 - 2(2+\sqrt{3}) + 1\right]$

$= (2 + \sqrt{3})\left[\frac{1}{3}(7+4\sqrt{3}) - 2(2+\sqrt{3}) + 1\right]$

$= (2 + \sqrt{3})\left[\frac{7}{3} + \frac{4}{3}\sqrt{3}) - 4 - 2\sqrt{3} + 1\right]$

$= (2 + \sqrt{3})\left[-\frac{2}{3} - \frac{2}{3}\sqrt{3}\right]$

$= -\frac{2}{3}(2 + \sqrt{3})(1 + \sqrt{3})$

$= -\frac{2}{3}(2 + 3\sqrt{3} + 3)$

$= -\frac{2}{3}(5 + 3\sqrt{3})$

$= -\frac{10}{3} - 2\sqrt{3} \approx -6.797$ (Minimum)

$f(4) = 4\left[\frac{1}{3}\cdot 4^2 - 2\cdot4 + 1\right]$

$= 4\left[\frac{16}{3} - 8 + 1\right]$

$= 4\left[-\frac{5}{3}\right]$

$= -\frac{20}{3} \approx -6.667$

Thus the

absolute maximum $= -\frac{10}{3} + 2\sqrt{3}$ at $x = 2 - \sqrt{3}$

and the

absolute minimum $= -\frac{10}{3} - 2\sqrt{3}$ at $x = 2 + \sqrt{3}$.

75. $f(x) = x^4 - 2x^2$

Find $f'(x)$.

$f'(x) = 4x^3 - 4x$

Find the critical points.

The derivative exists for all real numbers. We solve $f'(x) = 0$.

$4x^3 - 4x = 0$

$4x(x^2 - 1) = 0$

$4x(x - 1)(x + 1) = 0$

$4x = 0$ or $x - 1 = 0$ or $x + 1 = 0$

$x = 0$ or $x = 1$ or $x = -1$

The case of finding maximum and minimum values when more than one critical point occurs in an interval which is not closed can only be solved with a detailed graph or by techniques beyond the scope of this book.

Let us study the graph.

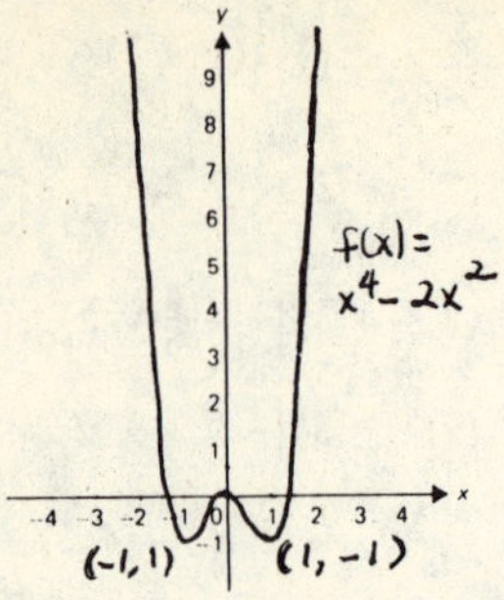

We observe from the graph that the function has no maximum value. The function has an absolute minimum value = -1 when x = 1 and x = -1.

77. $M(t) = -2t^2 + 100t + 180,\ 0 \leqslant t \leqslant 40$

a) $M'(t) = -4t + 100$

b) M'(t) exists for all real numbers. We solve M'(t) = 0.

$$-4t + 100 = 0$$
$$t = 25$$

c) The critical point and endpoints are 0, 25, and 40.

d) $M(0) = -2\cdot 0^2 + 100\cdot 0 + 180 = 180$

$M(25) = -2\cdot 25^2 + 100\cdot 25 + 180 = 1430$ (Maximum)

$M(40) = -2\cdot 40^2 + 100\cdot 40 + 180 = 980$

The maximum productivity for $0 \leqslant t \leqslant 40$ is 1430 units per month at t = 25 years of service.

79. $P(x) = \dfrac{1500}{x^2 - 6x + 10}$

a) $P'(x) = \dfrac{(x^2 - 6x + 10)(0) - (2x - 6)(1500)}{(x^2 - 6x + 10)^2}$ (Quotient Rule)

$= \dfrac{-3000x + 9000}{(x^2 - 6x + 10)^2}$

b) Since $x^2 - 6x + 10 = 0$ has no real-number solutions, P'(x) exists for all real numbers. We solve P'(x) = 0.

$$\frac{-3000x + 9000}{x^2 - 6x + 10} = 0$$
$$-3000x + 9000 = 0$$
$$-3000x = -9000$$
$$x = 3$$

e) We use Max-Min Principle 2.

$f''(x) =$

$\dfrac{(x^2-6x+10)^2(-3000) - 2(x^2-6x+10)(2x-6)(-3000x+9000)}{[(x^2 - 6x + 10)^2]^2}$

$= \dfrac{(x^2-6x+10)[(x^2-6x+10)(-3000)-2(2x-6)(-3000x+9000)]}{(x^2 - 6x + 10)^4}$

$= \dfrac{-3000x^2+18,000x-30,000+12,000x^2-72,000x+108,000}{(x^2 - 6x + 10)^3}$

$= \dfrac{9000x^2 - 54,000x + 78,000}{(x^2 - 6x + 10)^3}$

$f''(3) = \dfrac{9000\cdot 3^2 - 54,000\cdot 3 + 78,000}{(3^2 - 6\cdot 3 + 10)^3}$

$= -3000 < 0$, so there is an absolute maximum at x = 3.

Then total profit is a maximum when 3 units are produced and sold. (P(3) = 1500, so the maximum total profit is $1500.)

81. $y = -6.1x^2 + 752x + 22,620$

a) $\dfrac{dy}{dx} = -12.2x + 752$

b) The derivative exists for all real numbers. We solve $\dfrac{dy}{dx} = 0$.

$$-12.2x + 752 = 0$$
$$-12.2x = -752$$
$$x = \frac{3760}{61}$$

e) We use Max-Min Principle 2.

$$\frac{d^2y}{dx^2} = -12.2$$

Since $\dfrac{d^2y}{dx^2} < 0$ for all values of x, there is an absolute maximum at $x = \dfrac{3760}{61}$. Thus, the most accidents occur at a travel speed of $\dfrac{3760}{61} \approx 61.64$ mph.

83. $g(x) = x\sqrt{x + 3};\ [-3,3]$

$= x(x + 3)^{1/2}$

Find g'(x).

$g'(x) = x \cdot \frac{1}{2}(x + 3)^{-1/2}\cdot 1 + 1\cdot(x + 3)^{1/2}$

$= \dfrac{x}{2(x + 3)^{1/2}} + (x + 3)^{1/2} \cdot \dfrac{2(x + 3)^{1/2}}{2(x + 3)^{1/2}}$

(Multiplying the second term by a form of 1)

$= \dfrac{x}{2(x + 3)^{1/2}} + \dfrac{2(x + 3)}{2(x + 3)^{1/2}}$

$= \dfrac{3x + 6}{2(x + 3)^{1/2}}$, or $\dfrac{3(x + 2)}{2\sqrt{x + 3}}$

Find the critical points.

[Point c is a critical point if f'(c) = 0 <u>or</u> f'(c) does not exist.]

The derivative exists for all real numbers except -3. Thus, -3 is a critical point. To check for other critical points we solve g'(x) = 0, when $x \neq -3$.

$\dfrac{3(x + 2)}{2\sqrt{x + 3}} = 0$ (Assuming $x \neq -3$)

$3(x + 2) = 0$ (Multiplying by $2\sqrt{x + 3}$, since $x \neq -3$)

$$x + 2 = 0$$
$$x = -2$$

On a closed interval Max-Min Principle 1 can always be used. The critical points are -3 and -2. The endpoints are -3 and 3. We find the function value at each of these points.

$g(x) = x\sqrt{x + 3}$

$g(-3) = -3\sqrt{-3 + 3} = -3\sqrt{0} = -3\cdot 0 = 0$

$g(-2) = -2\sqrt{-2 + 3} = -2\sqrt{1} = -2\cdot 1 = -2$ (Min.)

$g(3) = 3\sqrt{3 + 3} = 3\sqrt{6}$ (Max.)

Thus the

absolute maximum is $3\sqrt{6}$ at $x = 3$

and the

absolute minimum is -2 at $x = -2$.

85. $C(x) = (2x + 4) + \left[\frac{2}{x - 6}\right],\ x > 6$

$= 2x + 4 + 2(x - 6)^{-1}$

Find $C'(x)$.

$C'(x) = 2 + 0 - 2(x - 6)^{-2}\cdot 1$

$= 2 - \frac{2}{(x - 6)^2}$

Find the critical points.

The derivative exists for all real numbers in the interval $(6,\infty)$. Thus, we solve $C'(x) = 0$.

$2 - \frac{2}{(x - 6)^2} = 0$

$2(x - 6)^2 - 2 = 0$ [Multiplying by $(x - 6)^2$ since $x \neq 6$]

$(x - 6)^2 - 1 = 0$

$x^2 - 12x + 36 - 1 = 0$

$x^2 - 12x + 35 = 0$

$(x - 5)(x - 7) = 0$

$x - 5 = 0$ or $x - 7 = 0$

$x = 5$ or $x = 7$

The only critical point in $(6,\infty)$ is 7.

We can apply the second derivative

$C''(x) = 4(x - 6)^{-3}$, or $\frac{4}{(x - 6)^3}$

to determine whether we have a maximum or a minimum. Now $C''(x)$ is positive for all values of x in $(6,\infty)$, thus there is a minimum at $x = 7$.

The firm should use 7 "quality units" to minimize its total cost of service.

Exercise Set 3.5

1. Express $Q = xy$ as a function of one variable. First solve $x + y = 50$ for y.

$x + y = 50$

$y = 50 - x$

Then substitute $50 - x$ for y in $Q = xy$.

$Q = xy$

$Q = x(50 - x)$ (Substituting)

$= 50x - x^2$

Find $Q'(x)$, where $Q(x) = 50x - x^2$.

$Q'(x) = 50 - 2x$

This derivative exists for all values of x, thus the only critical points are where

$Q'(x) = 50 - 2x = 0$

$50 = 2x$

$25 = x$

Since there is only one critical point, we can use the second derivative to determine whether we have a maximum. Note that

$Q''(x) = -2$,

which is a constant. Thus $Q''(25)$ is negative, so $Q(25)$ is a maximum.

Now

$Q(x) = x(50 - x)$

$Q(25) = 25(50 - 25)$ (Substituting)

$= 25\cdot 25$

$= 625$ (Maximum)

Thus the maximum product is 625 when $x = 25$. If $x = 25$ then $y = 50 - 25$, or 25. The two numbers are 25 and 25.

3. Since $Q = x(50 - x)$ has no minimum, there is no minimum product. (See Exercise 1.)

5. Let x be one number and y be the other. Since the difference of the numbers must be 4,

$x - y = 4$.

The product, Q, of the two numbers is given by

$Q = xy$.

We must minimize $Q = xy$, where $x - y = 4$.

Express $Q = xy$ as a function of one variable. First solve $x - y = 4$ for y.

$x - y = 4$

$x - 4 = y$

Then substitute $x - 4$ for y in $Q = xy$.

$Q = xy$

$Q = x(x - 4)$ (Substituting)

$= x^2 - 4x$

Find $Q'(x)$, where $Q(x) = x^2 - 4x$.

$Q'(x) = 2x - 4$

This derivative exists for all values of x, thus the only critical points are where

$Q'(x) = 2x - 4 = 0$

$2x = 4$

$x = 2$

Since there is only one critical point, we can use the second derivative to determine whether we have a maximum. Note that

$Q''(x) = 2,$

which is a constant. Thus $Q''(2)$ is positive, so $Q(2)$ is a minimum.

Now

$Q(x) = x(x - 4)$

$Q(2) = 2(2 - 4)$ (Substituting)

$= 2(-2)$

$= -4$ (Minimum)

Thus the minimum product is -4 when $x = 2$. Substitute 2 for x in $x - 4 = y$ to find y.

$x - 4 = y$

$2 - 4 = y$ (Substituting 2 for x)

$-2 = y$

The two numbers which have the minimum product are 2 and -2.

7. Maximize $Q = xy^2$, where x and y are positive numbers, such that $x + y^2 = 1$.

Express Q as a function of one variable. First solve $x + y^2 = 1$ for y^2.

$x + y^2 = 1$

$y^2 = 1 - x$

Then substitute $1 - x$ for y^2 in $Q = xy^2$.

$Q = xy^2$

$Q = x(1 - x)$ (Substituting)

$= x - x^2$

Find $Q'(x)$, where $Q(x) = x - x^2$.

$Q'(x) = 1 - 2x$

This derivative exists for all values of x, thus the only critical points are where

$Q'(x) = 1 - 2x = 0$

$-2x = -1$

$x = \frac{1}{2}$

Since there is only one critical point, we can use the second derivative to determine whether we have a maximum. Note that

$Q''(x) = -2,$

which is a constant. Thus $Q''\left(\frac{1}{2}\right)$ is negative, so $Q\left(\frac{1}{2}\right)$ is a maximum. Now

$Q(x) = x(1 - x)$

$Q\left(\frac{1}{2}\right) = \frac{1}{2}\left(1 - \frac{1}{2}\right)$ (Substituting)

$= \frac{1}{2}\left(\frac{1}{2}\right)$

$= \frac{1}{4}$ (Maximum)

Substitute $\frac{1}{2}$ for x in $x + y^2 = 1$ and solve for y.

$x + y^2 = 1$

$\frac{1}{2} + y^2 = 1$

$y^2 = \frac{1}{2}$

$y = \sqrt{\frac{1}{2}}$ (x and y must be positive)

Thus the maximum value of Q is $\frac{1}{4}$ when $x = \frac{1}{2}$ and $y = \sqrt{\frac{1}{2}}$.

9. Minimize $Q = x^2 + y^2$, where $x + y = 20$.

Express Q as a function of one variable. First solve $x + y = 20$ for y.

$x + y = 20$

$y = 20 - x$

Then substitute $20 - x$ for y in $Q = x^2 + y^2$.

$Q = x^2 + y^2$

$Q = x^2 + (20 - x)^2$ (Substitutingl)

$= x^2 + 400 - 40x + x^2$

$= 2x^2 - 40x + 400$

Find $Q'(x)$, where $Q(x) = 2x^2 - 40x + 400$.

$Q'(x) = 4x - 40$

This derivative exists for all values of x, thus the only critical points are where

$Q'(x) = 4x - 40 = 0$

$4x = 40$

$x = 10$

Since there is only one critical point, we can use the second derivative to determine whether we have a minimum. Note that

$Q''(x) = 4,$

which is a constant. Thus $Q''(10)$ is positive, so $Q(10)$ is a minimum.

Now

$Q(x) = x^2 + (20 - x)^2$

$Q(10) = 10^2 + (20 - 10)^2$ (Substituting)

$= 100 + 100$

$= 200$ (Minimum)

Substitute 10 for x in $y = 20 - x$ to find y.

$y = 20 - x$

$y = 20 - 10$ (Substituting)

$y = 10$

Thus the minimum value of Q is 200 when $x = 10$ and $y = 10$.

11. Maximize $Q = xy$, where x and y are positive numbers, such that $\frac{4}{3}x^2 + y = 16$.

Express Q as a function of one variable. First solve $\frac{4}{3}x^2 + y = 16$ for y.

$\frac{4}{3}x^2 + y = 16$

$y = 16 - \frac{4}{3}x^2$

Then substitute $16 - \frac{4}{3}x^2$ for y in Q = xy.

$Q = xy$

$Q = x\left[16 - \frac{4}{3}x^2\right]$ (Substituting)

$= 16x - \frac{4}{3}x^3$

Find Q'(x), where $Q(x) = 16x - \frac{4}{3}x^3$.

$Q'(x) = 16 - 4x^2$

This derivative exists for all values of x, thus the only critical points are where

$Q'(x) = 16 - 4x^2 = 0$

$-4x^2 = -16$

$x^2 = 4$

$x = \pm 2$

Since there is only one critical point in the domain of Q, 2, we can use the second derivative to determine whether we have a maximum. Note that,

$Q''(x) = -8x$,

and

$Q''(2) = -8 \cdot 2 = -16$.

Since Q''(2) is negative, Q(2) is a maximum.

Now

$Q(x) = x\left[16 - \frac{4}{3}x^2\right]$

$Q(2) = 2\left[16 - \frac{4}{3} \cdot 2^2\right]$ (Substituting)

$= 2\left[16 - \frac{16}{3}\right]$

$= 2\left[\frac{48}{3} - \frac{16}{3}\right]$

$= 2 \cdot \frac{32}{3}$

$= \frac{64}{3}$ (Maximum)

Substitute 2 for x in $y = 16 - \frac{4}{3}x^2$ to find y.

$y = 16 - \frac{4}{3}x^2$

$y = 16 - \frac{4}{3} \cdot 2^2$

$y = 16 - \frac{16}{3}$

$y = \frac{48}{3} - \frac{16}{3}$

$y = \frac{32}{3}$

Thus the maximum value of Q is $\frac{64}{3}$ when x = 2 and $y = \frac{32}{3}$.

13.

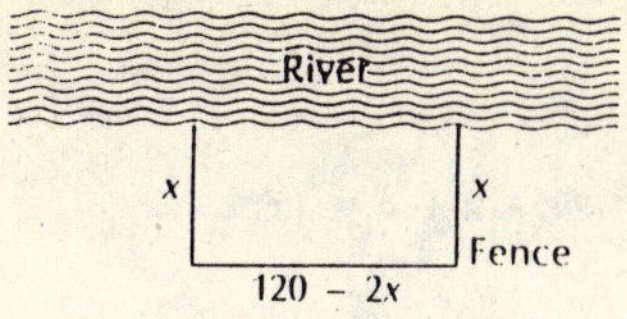

Let x be the width. Then 120 - 2x represents the length.

The area is given by

$A = \ell \cdot w$

$A = (120 - 2x)x$ (Substituting)

$= 120x - 2x^2$

We must maximize $A(x) = 120x - 2x^2$ on the interval (0,60). We consider the interval (0,60) because x is the length of one side and cannot be negative. Since there is only 120 yd of fencing, x cannot be greater than 60. Also, x cannot be 60 because the length of the lot would be 0.

We first find A'(x), where $A(x) = 120x - 2x^2$.

$A'(x) = 120 - 4x$

This derivative exists for all values of x in (0,60). Thus the only critical points are where

$A'(x) = 120 - 4x = 0$

$-4x = -120$

$x = 30$

Since there is only one critical point in the interval, we can use the second derivative to determine whether we have a maximum. Note that

$A''(x) = -4$,

which is a constant. Thus A''(30) is negative, so A(30) is a maximum.

Now

$A(x) = (120 - 2x)x$

$A(30) = (120 - 2 \cdot 30) \cdot 30$ (Substituting)

$= 60 \cdot 30$

$= 1800$ (Maximum)

Note that when x = 30, $120 - 2x = 120 - 2 \cdot 30$, or 60. The maximum area is 1800 yd² when the dimensions are 30 yd by 60 yd.

15. Let x represent the length and y the width. It is helpful to draw a picture.

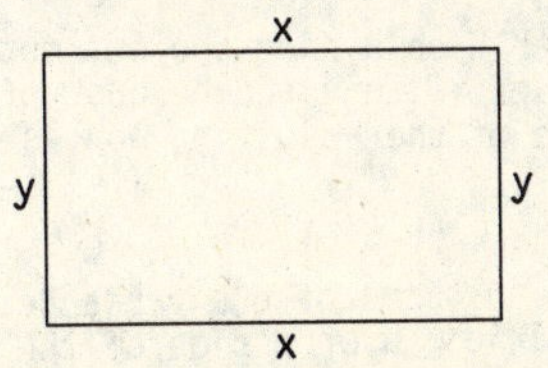

The perimeter is x + x + y + y, or 2x + 2y. Thus

$2x + 2y = 54$

or

$x + y = 27$ $\left[\text{Multiplying by } \frac{1}{2}\right]$

The area is given by A = xy.

First solve $x + y = 27$ for y.

$x + y = 27$

$y = 27 - x$

Then substitute $27 - x$ for y in $A = xy$.

$A = xy$

$A = x(27 - x)$ (Substituting)

$= 27x - x^2$

We must maximize $A(x) = 27x - x^2$ on the interval $(0,27)$. We consider the interval $(0,27)$ because x is the length of one side and cannot be negative. Since the perimeter cannot exceed 54 ft, x cannot be greater than 27. Also, x cannot be 27 because the width of the room would be 0.

Find $A'(x)$, where $A(x) = 27x - x^2$.

$A'(x) = 27 - 2x$

This derivative exists for all values of x in $(0,27)$. Thus the only critical points are where

$A'(x) = 27 - 2x = 0$

$-2x = -27$

$x = \frac{27}{2}$, or 13.5

Since there is only one critical point in the interval, we can use the second derivative to determine whether we have a maximum. Note that

$A''(x) = -2$,

which is a constant. Thus $A''(13.5)$ is negative, so $A(13.5)$ is a maximum.

Now

$A(x) = x(27 - x)$

$A(13.5) = 13.5(27 - 13.5)$ (Substituting)

$= 13.5(13.5)$

$= 182.25$ (Maximum)

Note that when $x = 13.5$, $y = 27 - 13.5$, or 13.5. The maximum area is 182.25 ft² when the dimensions are 13.5 ft by 13.5 ft.

17.

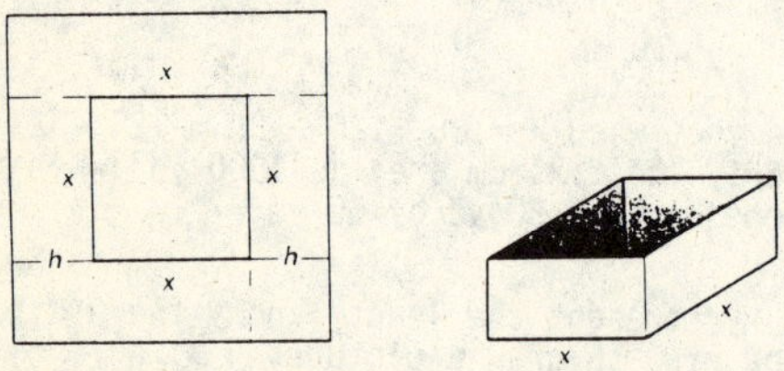

When squares of length h on a side are cut out of the corners, we are left with a square base of length x. The volume of the resulting box is

$V = \ell wh = x \cdot x \cdot h$.

We want to express V in terms of one variable. Note that the overall length of a side of the cardboard is 30 in. We see from the drawing that

$h + x + h = 30$,

or

$x + 2h = 30$.

Solving for h we get

$2h = 30 - x$

$h = \frac{1}{2}(30 - x) = \frac{1}{2} \cdot 30 - \frac{1}{2}x = 15 - \frac{1}{2}x$.

Thus

$V = x \cdot x \cdot \left[15 - \frac{1}{2}x\right] = x^2\left[15 - \frac{1}{2}x\right] = 15x^2 - \frac{1}{2}x^3$.

We must maximize $V(x)$ on the interval $(0,30)$. We first find $V'(x)$.

$V'(x) = 30x - \frac{3}{2}x^2$.

Now $V'(x)$ exists for all x in the interval $(0,30)$, so we set it equal to 0 to find the critical points:

$V'(x) = 30x - \frac{3}{2}x^2 = 0$

$x\left[30 - \frac{3}{2}x\right] = 0$

$x = 0$ or $30 - \frac{3}{2}x = 0$

$x = 0$ or $-\frac{3}{2}x = -30$

$x = 0$ or $x = 20$

The only critical point in $(0,30)$ is 20. Thus we can use the second derivative,

$V''(x) = 30 - 3x$,

to determine whether we have a maximum. Since

$V''(20) = 30 - 3 \cdot 20 = 30 - 60 = -30$,

$V''(20)$ is negative, so $V(20)$ is a maximum, and

$V(20) = 15 \cdot 20^2 - \frac{1}{2} \cdot 20^3$

$= 15 \cdot 400 - \frac{1}{2} \cdot 8000$

$= 6000 - 4000$

$= 2000$

The maximum volume is 2000 in³. The dimensions are

$x = 20$ in. by $x = 20$ in. by $h = 15 - \frac{1}{2} \cdot 20 = 5$ in.

19. We first make a drawing.

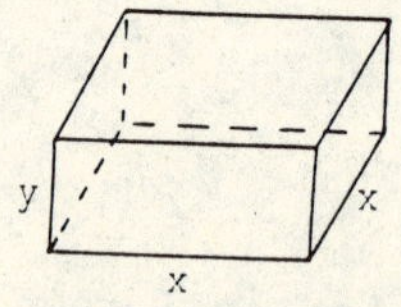

The surface area of the open-top, square-based, rectangular box is

$S = x^2 + 4xy$ (x^2 is area of base; xy is area of one side, $4xy$ is total area of 4 sides)

The volume must be 62.5 cubic inches, and is given by

$V = \ell \cdot w \cdot h = x \cdot x \cdot y = x^2y = 62.5$.

To express S in terms of one variable, we solve $x^2y = 62.5$ for y:

$y = \frac{62.5}{x^2}$

Then

$S(x) = x^2 + 4x\left[\frac{62.5}{x^2}\right]$ (Substituting $\frac{62.5}{x^2}$ for y)

$= x^2 + \frac{250}{x}$, or $x^2 + 250x^{-1}$

Now S is defined only for positive numbers, and the problem dictates that the length x be positive, so we minimize S on the interval $(0,\infty)$.

We first find S'(x).

$S'(x) = 2x - 250x^{-2}$

$= 2x - \frac{250}{x^2}$

Since S'(x) exists for all x in $(0,\infty)$, the only critical points are where $S'(x) = 0$. We solve the following:

$2x - \frac{250}{x^2} = 0$

$x^2\left[2x - \frac{250}{x^2}\right] = x^2 \cdot 0$ (Multiplying by x^2 to clear of fractions)

$2x^3 - 250 = 0$

$2x^3 = 250$

$x^3 = 125$

$x = 5$

Since there is only one critical point, we use the second derivative to determine whether we have a minimum. Note that

$S''(x) = 2 + 500x^{-3}$

$= 2 + \frac{500}{x^3}$

Since S''(x) is positive for all positive values of x, we have a minimum at x = 5.

When x = 5, it follows that y = 2.5:

$y = \frac{62.5}{x^2}$

$= \frac{62.5}{5^2}$ (Substituting 5 for x)

$= \frac{62.5}{25}$

$= 2.5$

Thus the surface area is minimized when x = 5 in. and y = 2.5 in. We find the minimum surface area by substituting 5 for x and 2.5 for y in

$S = x^2 + 4xy$

$S = 5^2 + 4\cdot5\cdot2.5$

$= 25 + 50$

$= 75$

The minimum surface area is 75 in^2 when the dimensions are 5 in. by 5 in. by 2.5 in.

21. $R(x) = 50x - 0.5x^2$

$C(x) = 4x + 10$

$P(x) = R(x) - C(x)$ (Profit = Revenue - Cost)

$= (50x - 0.5x^2) - (4x + 10)$ (Substituting)

$= -0.5x^2 + 46x - 10$

To find the maximum value of P(x), we first find P'(x).

$P'(x) = -x + 46$

This derivative exists for all values of x, but we are actually interested in only the numbers x in $[0,\infty)$ since we cannot produce a negative number of units. Thus we solve $P'(x) = 0$.

$P'(x) = -x + 46 = 0$

$-x = -46$

$x = 46$

Since there is only one critical point, we can use the second derivative to determine whether we have a maximum. Note that

$P''(x) = -1$, a constant.

Thus P''(46) is negative, so P(46) is a maximum.

The maximum profit is given by

$P(46) = -0.5(46)^2 + 46\cdot46 - 10$

$= -1058 + 2116 - 10$

$= 1048$

Thus the maximum profit is \$1048 when 46 units are produced and sold.

23. $R(x) = 2x$

$C(x) = 0.01x^2 + 0.6x + 30$

$P(x) = R(x) - C(x)$

$= 2x - (0.01x^2 + 0.6x + 30)$

$= -0.01x^2 + 1.4x - 30$

To find the maximum value of P(x), we first find P'(x).

$P'(x) = -0.02x + 1.4$

This derivative exists for all values of x, but we are actually interested in only the numbers x in $[0,\infty)$ since we cannot produce a negative number of units. Thus we solve $P'(x) = 0$.

$P'(x) = -0.02x + 1.4 = 0$

$-0.02x = -1.4$

$x = 70$

Since there is only one critical point, we can use the second derivative to determine whether we have a maximum. Note that

$P''(x) = -0.02$, a constant.

Thus P''(70) is negative, so P(70) is a maximum.

The maximum profit is given by

$P(70) = -0.01(70)^2 + 1.4(70) - 30$

$= -49 + 98 - 30$

$= 19$

Thus the maximum profit is \$19 when 70 units are produced and sold.

25. $R(x) = 9x - 2x^2$

$C(x) = x^3 - 3x^2 + 4x + 1$

R(x) and C(x) are in thousands of dollars, and x is in thousands of units.

$P(x) = R(x) - C(x)$

$= (9x - 2x^2) - (x^3 - 3x^2 + 4x + 1)$

$= -x^3 + x^2 + 5x - 1$

To find the maximum value of P(x), we first find P'(x).

$P'(x) = -3x^2 + 2x + 5$

This derivative exists for all values of x, but we are actually interested in only the numbers x in $[0,\infty)$ since we cannot produce a negative number of units. Thus we solve $P'(x) = 0$.

$$P'(x) = -3x^2 + 2x + 5 = 0$$
$$3x^2 - 2x - 5 = 0$$
$$(3x - 5)(x + 1) = 0$$
$$3x - 5 = 0 \text{ or } x + 1 = 0$$
$$x = \frac{5}{3} \text{ or } \quad x = -1$$

There is only one critical point in the interval $[0,\infty)$. We use the second derivative to determine whether we have a maximum. Note that

$P''(x) = -6x + 2$.

Find $P''\left(\frac{5}{3}\right)$.

$$P''\left(\frac{5}{3}\right) = -6 \cdot \frac{5}{3} + 2 = -10 + 2 = -8$$

$P''\left(\frac{5}{3}\right) < 0$, so $P\left(\frac{5}{3}\right)$ is a maximum.

The maximum profit is given by

$$P(x) = -x^3 + x^2 + 5x - 1$$
$$P\left(\frac{5}{3}\right) = -\left(\frac{5}{3}\right)^3 + \left(\frac{5}{3}\right)^2 + 5 \cdot \frac{5}{3} - 1$$
$$= -\frac{125}{27} + \frac{25}{9} + \frac{25}{3} - 1$$
$$= -\frac{125}{27} + \frac{75}{27} + \frac{225}{27} - \frac{27}{27}$$
$$= \frac{148}{27}$$

Note that $\frac{5}{3}$ thousand $\approx$ 1.667 thousand = 1667 and that $\frac{148}{27}$ thousand $\approx$ 5.481 thousand = 5481.

Thus the maximum profit is approximately \$5481 when approximately 1667 units are produced and sold.

27. $p = 150 - 0.5x$ (Price per suit)

$C(x) = 4000 + 0.25x^2$ (Total cost)

a) $\text{Total revenue} = \begin{pmatrix}\text{Number}\\\text{of suits}\end{pmatrix} \cdot \begin{pmatrix}\text{Price}\\\text{per suit}\end{pmatrix}$

$R(x) = x \cdot (150 - 0.5x)$

(Substituting x for number of suits and 150 - 0.5x for price per suit)

$= 150x - 0.5x^2$

b) $\text{Total profit} = \text{Total revenue} - \text{Total cost}$

$$P(x) = R(x) - C(x)$$
$$P(x) = (150x - 0.5x^2) - (4000 + 0.25x^2)$$
$$= 150x - 0.5x^2 - 4000 - 0.25x^2$$
$$= -0.75x^2 + 150x - 4000$$

c) To determine the number of suits required to maximize profit, we first find $P'(x)$.

$P'(x) = -1.5x + 150$

The derivative exists for all real numbers, but we are only interested in numbers x in $[0,\infty)$ because we cannot produce a negative number of suits. Thus we solve

$$P'(x) = -1.5x + 150 = 0$$
$$-1.5x = -150$$
$$x = 100$$

Since there is only one critical point, we use the second derivative to determine whether we have a maximum. Note that

$P''(x) = -1.5$, a constant.

Thus $P''(100)$ is negative, so $P(100)$ is a maximum. The company must produce and sell 100 suits to maximize profit.

d) The maximum profit is given by substituting 100 for x in the profit function.

$$P(x) = -0.75x^2 + 150x - 4000$$
$$P(100) = -0.75 \cdot 100^2 + 150 \cdot 100 - 4000$$
$$= -7500 + 15{,}000 - 4000$$
$$= 3500$$

Thus the maximum profit is \$3500.

e) The price per suit is given by

$p = 150 - 0.5x$

$p = 150 - 0.5(100)$ (Substituting 100 for x)

$= 150 - 50$

$= 100$

The price per suit must be \$100.

29. Let x = the amount by which the price of \$6 should be increased (if x is negative, the price would be decreased). We first express total revenue R as a function of x.

$R(x)$ = (Revenue from tickets) + (Revenue from concessions)

= (Number of people)·(Ticket price) + \$1.50(Number of people)

$= (70{,}000 - 10{,}000x)(6 + x) + 1.50(70{,}000 - 10{,}000x)$

$= 420{,}000 + 70{,}000x - 60{,}000x - 10{,}000x^2 + 105{,}000 - 15{,}000x$

$R(x) = -10{,}000x^2 - 5000x + 525{,}000$

To find x such that $R(x)$ is a maximum, we first find $R'(x)$:

$R'(x) = -20{,}000x - 5000$.

This derivative exists for all real numbers x; thus the only critical points are where $R'(x) = 0$.

$$R'(x) = -20{,}000x - 5000 = 0$$
$$-20{,}000x = 5000$$
$$x = -0.25$$

Since this is the only critical point, we can use the second derivative,

$R''(x) = -20{,}000$,

to determine whether we have a maximum.

Since R''(-0.25) is negative, R(-0.25) is a maximum. Thus to maximize revenue the university should charge

\$6 + (-\$0.25) or \$5.75 per ticket.

That is, this reduced ticket price will get more people into the theater,

70,000 - 10,000(-0.25), or 72,500,

and will result in maximum revenue.

31. Let x = the number of additional trees per acre which should be planted. Then the number of trees planted per acre is represented by (20 + x) and the yield per tree by (30 - x).

$$\begin{matrix}\text{Total}\\ \text{yield}\\ \text{per acre}\end{matrix} = \begin{bmatrix}\text{Yield}\\ \text{per}\\ \text{tree}\end{bmatrix}\cdot\begin{bmatrix}\text{Number of}\\ \text{trees per acre}\end{bmatrix}$$

$$Y(x) = (30 - x)(20 + x)$$
$$= 600 + 10x - x^2$$

To find x such that Y(x) is a maximum we first find Y'(x).

$$Y'(x) = 10 - 2x$$

This derivative exists for all real numbers x; thus the only critical points are where Y'(x) = 0, so we solve that equation.

$$10 - 2x = 0$$
$$-2x = -10$$
$$x = 5$$

Since this is the only critical point, we can use the second derivative,

$$Y''(x) = -2,$$

to determine whether we have a maximum. Since Y''(5) is negative, Y(5) is a maximum. Thus to maximize yield, the apple farm should plant

20 + 5, or 25 trees per acre.

33.

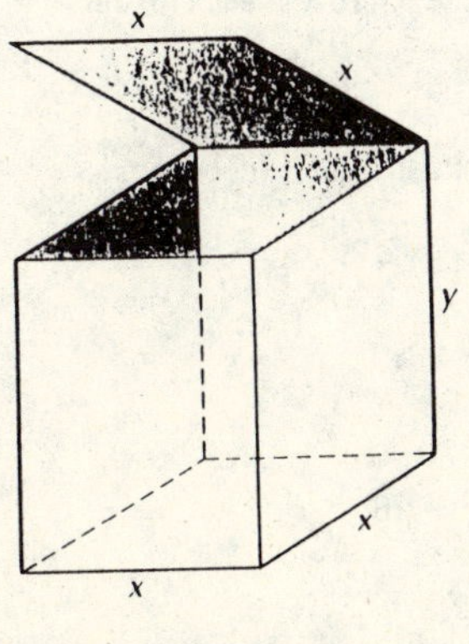

The volume of the box is given by

$$V = x\cdot x\cdot y = x^2y = 320.$$

The area of the base is x^2. The cost of the base is $15x^2$ cents.

The area of the top is x^2. The cost of the top is $10x^2$ cents.

The area of each side is xy; the total area for the four sides is 4xy. The cost of the four sides is 2.5(4xy) cents.

The total cost in cents is given by

$$C = 15x^2 + 10x^2 + 2.5(4xy)$$
$$= 25x^2 + 10xy$$

To express C in terms of one variable, we solve $x^2y = 320$ for y:

$$y = \frac{320}{x^2}$$

Then

$$C(x) = 25x^2 + 10x\left[\frac{320}{x^2}\right] \quad \left(\text{Substituting } \frac{320}{x^2} \text{ for } y\right)$$
$$= 25x^2 + \frac{3200}{x}, \text{ or } 25x^2 + 3200x^{-1}$$

Now C is defined only for positive numbers, and the problem dictates that the length x be positive, so we are minimizing C on the interval $(0,\infty)$.

We first find C'(x).

$$C'(x) = 50x - 3200x^{-2}$$
$$= 50x - \frac{3200}{x^2}$$

Since C'(x) exists for all x in $(0,\infty)$, the only critical points are where C'(x) = 0. Thus we solve the following equation:

$$50x - \frac{3200}{x^2} = 0$$
$$x^2\left[50x - \frac{3200}{x^2}\right] = x^2\cdot 0 \quad (\text{Multiplying by } x^2 \text{ to clear of fractions})$$
$$50x^3 - 3200 = 0$$
$$50x^3 = 3200$$
$$x^3 = 64$$
$$x = 4$$

This is the only critical point, so we can use the second derivative to determine whether we have a minimum.

$$C''(x) = 50 + 6400x^{-3}$$
$$= 50 + \frac{6400}{x^3}$$

Note that this is positive for all positive values of x. Thus we have a minimum at x = 4. When x = 4, it follows that y = 20:

$$y = \frac{320}{x^2}$$
$$y = \frac{320}{4^2} \quad (\text{Substituting 4 for x})$$
$$= \frac{320}{16}$$
$$= 20$$

Thus the cost is minimized when the dimensions are 4 ft by 4 ft by 20 ft.

35. Let A represent the amount deposited in savings accounts and i represent the interest rate paid on the money deposited. If A is directly proportional to i, then there is some positive constant k such that A = ki. The interest earned by the bank is represented by 18%A, or 0.18A. The interest paid by the bank is represented by iA. The profit received by the bank is given by

P = 18%A - iA.

We first express P = 18%A - iA as a function of one variable by substituting ki for A.

$P = 18\%(ki) - i(ki)$

$P = 0.18ki - ki^2$ (k is a constant)

We maximize P on the interval $(0,\infty)$. We first find $P'(i)$.

$P'(i) = 0.18k - 2ki$

Since $P'(i)$ exists for all i in $(0,\infty)$, the only critical points are where $P'(i) = 0$. We solve the following equation.

$$0.18k - 2ki = 0$$
$$-2ki = -0.18k$$
$$i = \frac{-0.18k}{-2k}$$
$$i = 0.09$$

Since there is only one critical point, we can use the second derivative to determine whether we have a maximum. Note that $P''(i) = -2k$, which is a negative constant $(k > 0)$. Thus $P''(0.09)$ is negative, so $P(0.09)$ is a maximum. To maximize profits the bank should pay 9% on its savings accounts.

37. Let x = the lot size. Now the inventory costs are given by

$$C(x) = \begin{matrix}\text{Yearly}\\ \text{carrying}\\ \text{costs}\end{matrix} + \begin{matrix}\text{Yearly}\\ \text{reorder}\\ \text{costs}\end{matrix}$$

We consider each separately:

Yearly carrying costs: The average amount held in stock is x/2, and it costs $20 per pool table for storage. Thus

$$\begin{matrix}\text{Yearly}\\ \text{carrying}\\ \text{costs}\end{matrix} = \begin{bmatrix}\text{Yearly cost}\\ \text{per item}\end{bmatrix}\cdot\begin{bmatrix}\text{Average number}\\ \text{of items}\end{bmatrix}$$
$$= 20 \cdot \frac{x}{2}$$
$$= 10x$$

Yearly reorder costs: Now x = lot size, and suppose there are N reorders each year. Then Nx = 100, and N = 100/x. If there is a fixed cost of $40 for each order plus $16 for each pool table, the cost of each order is 40 + 16x. Thus

$$\begin{matrix}\text{Yearly}\\ \text{reorder}\\ \text{costs}\end{matrix} = \begin{bmatrix}\text{Cost of each}\\ \text{order}\end{bmatrix}\cdot\begin{bmatrix}\text{Number of}\\ \text{reorders}\end{bmatrix}$$
$$= (40 + 16x)\left[\frac{100}{x}\right]$$
$$= \frac{4000}{x} + 1600$$

Hence

$$C(x) = 10x + \frac{4000}{x} + 1600$$
$$= 10x + 4000x^{-1} + 1600$$

We want to find a minimum value of C on the interval [1,100]. We first find $C'(x)$.

$$C'(x) = 10 - 4000x^{-2}$$
$$= 10 - \frac{4000}{x^2}$$

Now $C'(x)$ exists for all x in [1,100], so the only critical points are where $C'(x) = 0$. We solve $C'(x) = 0$.

$$10 - \frac{4000}{x^2} = 0$$
$$10 = \frac{4000}{x^2}$$
$$10x^2 = 4000$$
$$x^2 = 400$$
$$x = \pm 20$$

Now there is only one critical point in the interval [1,100], x = 20, so we can use the second derivative to see if we have a minimum:

$$C''(x) = 8000x^{-3} = \frac{8000}{x^3}$$

Now $C''(x)$ is positive for all x in [1,100], so we do have a minimum at x = 20. Thus to minimize inventory costs, the store should order pool tables 100/20, or 5 times per year. The lot size is 20.

39. Let x = the lot size. Now the inventory costs are given by

$$C(x) = \begin{matrix}\text{Yearly}\\ \text{carrying}\\ \text{costs}\end{matrix} + \begin{matrix}\text{Yearly}\\ \text{reorder}\\ \text{costs}\end{matrix}$$

We consider each separately:

Yearly carrying costs: The average amount held in stock is x/2, and it costs $8 per calculator for storage. Thus

$$\begin{matrix}\text{Yearly}\\ \text{carrying}\\ \text{costs}\end{matrix} = \begin{bmatrix}\text{Yearly cost}\\ \text{per item}\end{bmatrix}\cdot\begin{bmatrix}\text{Aveage number}\\ \text{of items}\end{bmatrix}$$
$$= 8 \cdot \frac{x}{2}$$
$$= 4x$$

Yearly reorder costs: Now x = lot size, and suppose there are N reorders each year. Then Nx = 360, and N = 360/x. If there is a fixed cost of $10 for each order plus $8 for each calculator, the cost of each order is 10 + 8x. Thus

$$\begin{matrix}\text{Yearly}\\ \text{reorder}\\ \text{costs}\end{matrix} = \begin{bmatrix}\text{Cost of each}\\ \text{order}\end{bmatrix}\cdot\begin{bmatrix}\text{Number of}\\ \text{reorders}\end{bmatrix}$$
$$= (10 + 8x)\left[\frac{360}{x}\right]$$
$$= \frac{3600}{x} + 2880$$

Hence

$$C(x) = 4x + \frac{3600}{x} + 2880$$
$$= 4x + 3600x^{-1} + 2880$$

We want to find a minimum value of C on the interval [1,360]. We first find $C'(x)$.

$$C'(x) = 4 - 3600x^{-2}$$
$$= 4 - \frac{3600}{x^2}$$

Now $C'(x)$ exists for all x in [1,360], so the only critical points are where $C'(x) = 0$. We solve $C'(x) = 0$.

$$4 - \frac{3600}{x^2} = 0$$

$$4 = \frac{3600}{x^2}$$

$$4x^2 = 3600$$

$$x^2 = 900$$

$$x = \pm 30$$

Now there is only one critical point in the interval [1,360], x = 30, so we can use the second derivative to see if we have a minimum:

$$C''(x) = 7200x^{-3} = \frac{7200}{x^3}$$

Now C''(x) is positive for all x in [1,360], so we do have a minimum at x = 30. Thus to minimize inventory costs, the store should order calculators 360/30, or 12, times per year. The lot size is 30.

41. Yearly carrying costs $= 9 \cdot \frac{x}{2} = 4.5x$

Yearly reorder costs $= (10 + 8x) \cdot \frac{360}{x} = \frac{3600}{x} + 2880$

Hence

$$C(x) = 4.5x + \frac{3600}{x} + 2880$$

and

$$C'(x) = 4.5 - \frac{3600}{x^2}$$

Now C'(x) exists for all x in [1,360], so the only critical points are where C'(x) = 0. We solve C'(x) = 0.

$$4.5 - \frac{3600}{x^2} = 0$$

$$4.5 = \frac{3600}{x^2}$$

$$4.5x^2 = 3600$$

$$x^2 = 800$$

$$x = \pm\sqrt{800} \approx 28.3$$

Now there is only one critical point in the interval [1,360], $x = \sqrt{800} \approx 28.3$, so we can use the second derivative to see if we have a minimum:

$$C''(x) = 7200x^{-3} = \frac{7200}{x^3}$$

Now C''(x) is positive for all x in [1,360], so we do have a minimum at x ≈ 28.3. Since it does not make sense to reorder 28.3 calculators each time, we consider the two integers closest to 28.3, which are 28 and 29.

C(28) ≈ \$3134.57 and C(29) ≈ \$3134.64

Then the lot size that will minimize cost is 28. The store should reorder 360/28 ≈ 13 times per year.

43.

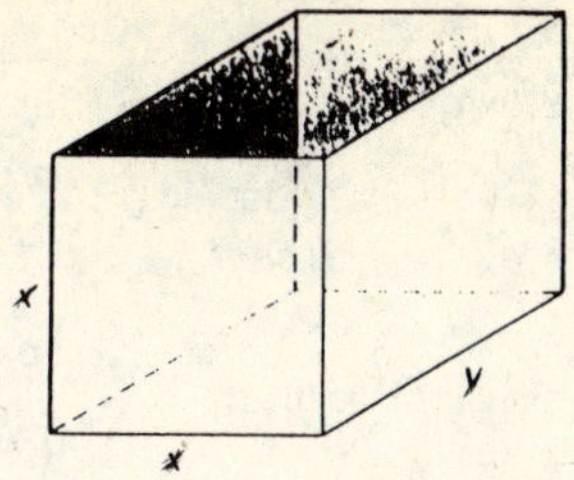

Case I.

If y is the length, the girth is x + x + x + x, or 4x.

Case II.

If x is the length, the girth is x + y + x + y, or 2x + 2y.

Case I.

The combined length and girth is

$y + 4x = 84$.

The volume is

$V = x \cdot x \cdot y$.

We want to express V in terms of one variable. We solve y + 4x = 84 for y.

$$y + 4x = 84$$

$$y = 84 - 4x$$

Thus

$V = x \cdot x \cdot (84 - 4x)$, or $84x^2 - 4x^3$.

To maximize V(x) we first find V'(x).

$$V'(x) = 168x - 12x^2$$

Now V'(x) exists for all x so we set is equal to 0 to find the critical points:

$$V'(x) = 168x - 12x^2 = 0$$

$$12x(14 - x) = 0$$

$$12x = 0 \text{ or } 14 - x = 0$$

$$x = 0 \text{ or } x = 14$$

Since x ≠ 0, we only consider x = 14. We can use the second derivative to determine that V(14) is a maximum. If x = 14, y = 84 - 4·14, or 28. The dimensions are 14 in. by 14 in. by 28 in. The volume is 14 × 14 × 28, or 5488 in^3.

Case II.

The combined length and girth is

$x + 2x + 2y = 84$

or

$3x + 2y = 84$.

The volume is

$V = x \cdot x \cdot y$.

We want to express V in terms of one variable. We solve 3x + 2y = 84 for y.

$$3x + 2y = 84$$

$$2y = 84 - 3x$$

$$y = 42 - 1.5x$$

Thus

$V = x \cdot x \cdot (42 - 1.5x)$, or $42x^2 - 1.5x^3$.

To maximize $V(x)$ we first find $V'(x)$.

$V'(x) = 84x - 4.5x^2$

Now $V'(x)$ exists for all x so we set it equal to 0 to find the critical points:

$$V'(x) = 84x - 4.5x^2 = 0$$
$$3x(28 - 1.5x) = 0$$
$$3x = 0 \text{ or } 28 - 1.5x = 0$$
$$x = 0 \text{ or } \quad -1.5x = -28$$
$$x = 0 \text{ or } \quad x \approx 18.7$$

Since $x \neq 0$, we only consider 18.7. We can use the second derivative to determine that $V(18.7)$ is a maximum. If $x = 18.7$, $y = 42 - 1.5(18.7)$, or 13.95. The dimensions are 18.7 in. by 18.7 in. by 13.95 in. The volume is $18.7 \times 18.7 \times 13.95$, or ≈ 4878 in^3.

Comparing Case I and Case II we see that the maximum volume is 5488 in^3 when the dimensions are 14 in. by 14 in. by 28 in.

45.

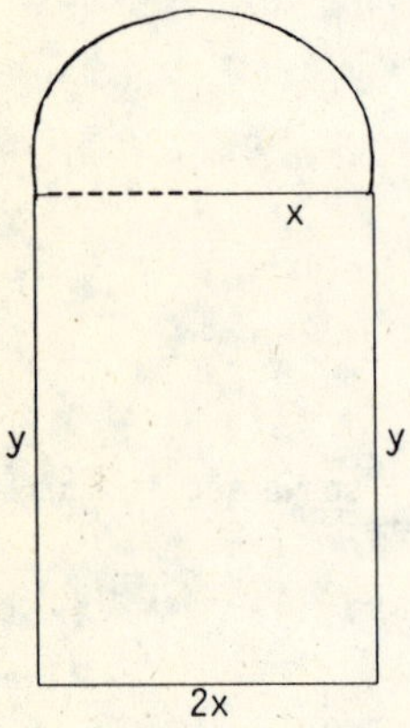

(2x represents the width of the window and the diameter of the circle)

Consider a circle whose radius is x. The circumference of the circle is

$C = \pi(2x)$. $(C = \pi d)$

The perimeter of the semicircle is

$\frac{1}{2} C = \frac{1}{2} \pi(2x) = \pi x$.

The perimeter of the three sides of the rectangle which form the remaining part of the total perimeter of the window is

$2x + y + y$ or $2x + 2y$.

The total perimeter of the window is

$\pi x + 2x + 2y = 24$.

Maximizing the amount of light is the same as maximizing the area of the window. The area of the circle is

$A = \pi x^2$. $(A = \pi r^2)$

The area of the semicircle is

$\frac{1}{2} A = \frac{1}{2} \pi x^2$.

The area of the rectangle is

$2x \cdot y$.

The total area of the Norman window is

$A = \frac{1}{2} \pi x^2 + 2xy$.

To express A in terms of one variable, we solve $\pi x + 2x + 2y = 24$ for y:

$$\pi x + 2x + 2y = 24$$
$$2y = 24 - \pi x - 2x$$
$$y = 12 - \frac{\pi}{2} x - x$$

Then

$$A(x) = \frac{1}{2} \pi x^2 + 2x\left[12 - \frac{\pi}{2} x - x\right] \quad \text{(Substituting)}$$
$$= \frac{1}{2} \pi x^2 + 24x - \pi x^2 - 2x^2$$
$$= \left[-\frac{1}{2}\pi - 2\right]x^2 + 24x$$

We maximize A on the interval $(0,\infty)$. We first find $A'(x)$.

$A'(x) = (-\pi - 4)x + 24$

Since $A'(x)$ exists for all x in $(0,\infty)$, the only critical points are where $A'(x) = 0$. Thus we solve the following equation:

$$(-\pi - 4)x + 24 = 0$$
$$(-\pi - 4)x = -24$$
$$x = \frac{-24}{-\pi - 4}$$
$$x = \frac{24}{\pi + 4} \quad \left[\text{Multiplying by } \frac{-1}{-1}\right]$$

This is the only critical point, so we can use the second derivative to determine whether we have a maximum.

$A''(x) = -\pi - 4$

Note that $-\pi - 4$ is a constant. It is negative for all values of x. Thus $A''\left(\frac{24}{\pi + 4}\right)$ is negative, and we have a maximum at $x = \frac{24}{\pi + 4}$.

When $x = \frac{24}{\pi + 4}$, it follows that $y = \frac{24}{\pi + 4}$.

$$y = 12 - \frac{\pi}{2} x - x$$
$$y = 12 - \frac{\pi}{2}\left[\frac{24}{\pi + 4}\right] - \left[\frac{24}{\pi + 4}\right] \quad \left[\text{Substituting } \frac{24}{\pi + 4} \text{ for } x\right]$$
$$= 12\left[\frac{\pi + 4}{\pi + 4}\right] - \frac{\pi}{2}\left[\frac{24}{\pi + 4}\right] - \left[\frac{24}{\pi + 4}\right] \quad \text{(Multiplying by 1)}$$
$$= \frac{12\pi + 48 - 12\pi - 24}{\pi + 4}$$
$$= \frac{24}{\pi + 4}$$

To maximize the amount of light through the window the dimensions must be

$x = \frac{24}{\pi + 4}$ ft and $y = \frac{24}{\pi + 4}$ ft.

47. Let x be the positive number. Then $\frac{1}{x}$ is the reciprocal of the number, and x^2 is the square of the number.

The sum, S, of the reciprocal and five times the square is given by

$S = \frac{1}{x} + 5x^2$, or $x^{-1} + 5x^2$

We minimize S(x) on the interval $(0,\infty)$. We first find S'(x).

$$S'(x) = -x^{-2} + 10x = -\frac{1}{x^2} + 10x$$

Since S'(x) exists for all x in $(0,\infty)$, the only critical points are where S'(x) = 0. We solve the following equation.

$$-\frac{1}{x^2} + 10x = 0$$

$$x^2\left(-\frac{1}{x^2} + 10x\right) = x^2\cdot 0 \quad \text{(Multiplying by } x^2 \text{ to clear of fractions)}$$

$$-1 + 10x^3 = 0$$

$$10x^3 = 1$$

$$x^3 = \frac{1}{10}$$

$$x = \sqrt[3]{\frac{1}{10}}$$

Since there is only one critical point, we can use the second derivative,

$$S''(x) = 2x^{-3} + 10 = \frac{2}{x^3} + 10$$

to determine whether we have a minimum. Since S''(x) is positive for all positive values of x, the sum is a minimum when $x = \sqrt[3]{\frac{1}{10}}$.

49. Yearly carrying costs $= a\cdot\frac{x}{2} = \frac{ax}{2}$

Yearly reorder costs $= (b + cx)\frac{Q}{x} = \frac{bQ}{x} + cQ$

Hence

$$C(x) = \frac{ax}{2} + \frac{bQ}{x} + cQ = \frac{a}{2}x + bQx^{-1} + cQ$$

We want to find a minimum value of C on the interval [1,Q]. We first find C'(x).

$$C'(x) = \frac{a}{2} - bQx^{-2} = \frac{a}{2} - \frac{bQ}{x^2}$$

Now C'(x) exists for all x in [1,Q], so the only critical points are where C'(x) = 0. We solve C'(x) = 0.

$$\frac{a}{2} - \frac{bQ}{x^2} = 0$$

$$\frac{a}{2} = \frac{bQ}{x^2}$$

$$ax^2 = 2bQ$$

$$x^2 = \frac{2bQ}{a}$$

$$x = \pm\sqrt{\frac{2bQ}{a}}$$

Now there is only one critical point in the interval [1,Q], $x = \sqrt{\frac{2bQ}{a}}$, so we can use the second derivative to see if we have a minimum:

$$C''(x) = 2bQx^{-3} = \frac{2bQ}{x^3}$$

Now C''(x) is positive for all x in [1,Q], so we do have a minimum at $x = \sqrt{\frac{2bQ}{a}}$. Thus to minimize inventory costs, the store should order $\sqrt{\frac{2bQ}{a}}$ units of a product $\frac{Q}{\sqrt{\frac{2bQ}{a}}}$ times per year.

51.

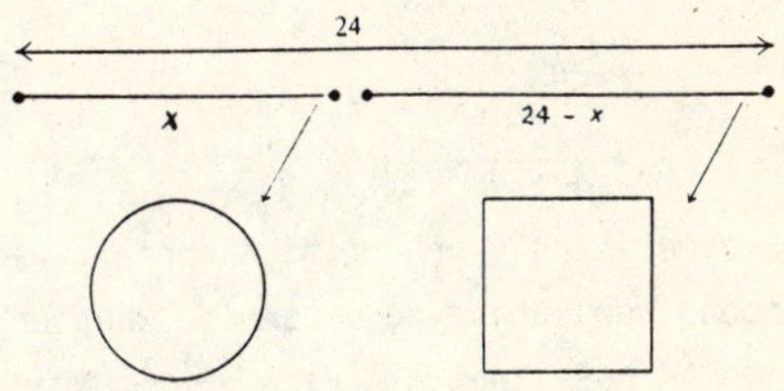

The circumference of the circle is

$x = \pi d$, or $x = 2\pi r$ $\quad (C = \pi d)$

Solving this equation for r, we get $r = \frac{x}{2\pi}$.

The area of the circle is

$$A_C = \pi\left(\frac{x}{2\pi}\right)^2 \qquad (A = \pi r^2)$$

$$= \pi\cdot\frac{x^2}{4\pi^2} = \frac{x^2}{4\pi}$$

The perimeter of the square is 24 - x.

The length of a side of the square is $\frac{24 - x}{4}$.

The area of the square is

$$A_S = \left(\frac{24 - x}{4}\right)^2 \qquad (A = s^2)$$

$$= \frac{576 - 48x + x^2}{16} = \frac{x^2 - 48x + 576}{16}$$

The sum of the areas is

$$A = \frac{x^2}{4\pi} + \frac{x^2 - 48x + 576}{16}$$

$$= \frac{1}{4\pi}x^2 + \frac{1}{16}x^2 - 3x + 36$$

$$= \left(\frac{1}{4\pi} + \frac{1}{16}\right)x^2 - 3x + 36$$

$$= \left(\frac{4}{16\pi} + \frac{\pi}{16\pi}\right)x^2 - 3x + 36$$

$$= \frac{4 + \pi}{16\pi}x^2 - 3x + 36$$

We minimize A on the interval (0,24). We first find A'(x).

$$A'(x) = 2\left(\frac{4 + \pi}{16\pi}\right)x - 3$$

$$= \frac{4 + \pi}{8\pi}x - 3$$

Since A'(x) exists for all x in (0,24), the only critical point is where A'(x) = 0. We solve the following equation.

$$\frac{4 + \pi}{8\pi}x - 3 = 0$$

$$\frac{4 + \pi}{8\pi}x = 3$$

$$(4 + \pi)x = 24\pi$$

$$x = \frac{24\pi}{4 + \pi} \approx 10.55$$

Since there is only one critical point, we can use the second derivative to determine whether we have a minimum. Note that $A''(x) = \frac{4 + \pi}{8\pi}$, which is a positive constant. Thus $A''\left(\frac{24\pi}{4 + \pi}\right)$ is positive, so $A\left(\frac{24\pi}{4 + \pi}\right)$ is a minimum and $x \approx 10.55$ in. and $24 - x \approx 13.45$ in. There is no maximum if the string is to be cut. One would interpret the maximum to be at the endpoint, with the string uncut and used to form a circle.

53.

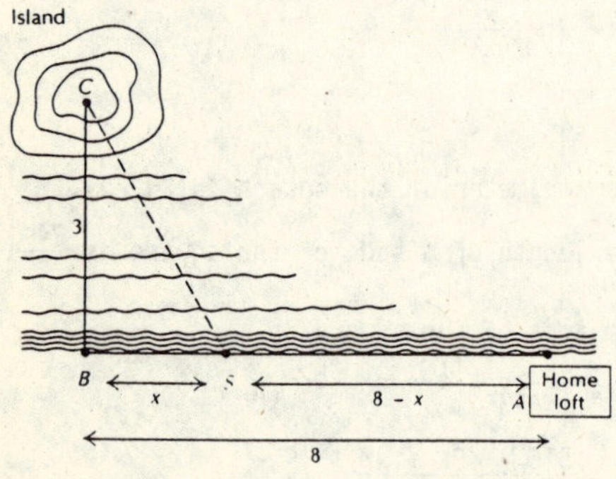

First find the distance over water (from C to S).

$$(CS)^2 = 3^2 + x^2$$

$$CS = \sqrt{9 + x^2}$$

The distance from S to A is 8 - x.

Letting r represent the energy rate over land and 1.28r represent the energy rate over water, we have the following equation for total energy.

Total energy = (Energy rate over water) · (Distance over water) + (Energy rate over land) · (Distance over land)

$$TE = 1.28r \cdot \sqrt{9 + x^2} + r \cdot (8 - x)$$

$$= 1.28r(9 + x^2)^{1/2} + 8r - rx$$

(r is a positive constant)

We minimize the total energy over the interval [0,8]. We find TE'(x).

$$TE'(x) = 1.28r \cdot \frac{1}{2}(9 + x^2)^{-1/2} \cdot 2x - r$$

$$= \frac{1.28rx}{(9 + x^2)^{1/2}} - r$$

Since TE'(x) exists for all x in [0,8], the only critical point is where TE'(x) = 0. We solve the following equation.

$$\frac{1.28rx}{(9 + x^2)^{1/2}} - r = 0$$

$$\frac{1.28rx}{(9 + x^2)^{1/2}} = r$$

$$1.28rx = r(9 + x^2)^{1/2}$$

$$1.28x = (9 + x^2)^{1/2}$$

$$(1.28x)^2 = 9 + x^2$$

$$1.6384x^2 = 9 + x^2$$

$$0.6384x^2 = 9$$

$$x^2 = \frac{9}{0.6384}$$

$$x^2 \approx 14.0977$$

$$x \approx 3.75$$

Since there is only one critical point, we can use the second derivative to determine whether we have a maximum. Note that $TE''(x) = \frac{11.52r}{(9 + x^2)^{3/2}}$ and that $TE''(x) > 0$ for all x in [0,8]. Thus TE(3.75) is a minimum and S should be about 8 - 3.75, or 4.25 miles down shore from A.

55. $C(x) = 8x + 20 + \frac{x^3}{100}$

a) $C'(x) = 8 + 0 + \frac{1}{100} \cdot 3x^2$

$$= 8 + \frac{3}{100}x^2$$

b) $A(x) = \frac{C(x)}{x} = \frac{8x + 20 + \frac{x^3}{100}}{x}$

$$= 8 + \frac{20}{x} + \frac{x^2}{100}$$

$$\text{or } 8 + 20x^{-1} + \frac{1}{100}x^2$$

c) $A'(x) = -20x^{-2} + \frac{1}{100} \cdot 2x$

$$= -\frac{20}{x^2} + \frac{1}{50}x$$

d) Minimize $A(x)$ over $(0,\infty)$. Since $A'(x)$ exists for all x in $(0,\infty)$, we merely solve

$$-\frac{20}{x^2} + \frac{1}{50}x = 0$$

$$\frac{x}{50} = \frac{20}{x^2}$$

$$x^3 = 1000 \qquad (x \neq 0)$$

$$x = 10$$

Since there is only one critical point, we can use the second derivative to determine whether we have a minimum. Note that $A''(x) = \frac{40}{x^3} + \frac{1}{50}$ and that $A''(10) = \frac{3}{50} > 0$. Thus $A(10)$ is a minimum.

$$A(10) = 8 + \frac{20}{10} + \frac{10^2}{100} = 8 + 2 + 1 = 11$$

The minimum average cost is 11 when $x = 10$.

$$C'(10) = 8 + \frac{3}{100} \cdot 10^2 = 8 + 3 = 11$$

The marginal cost at $x = 10$ is 11.

e) Thus $A(10) = C'(10) = 11$.

57. Minimize $Q = x^3 + 2y^3$, where x and y are positive numbers, such that $x + y = 1$.

Express Q as a function of one variable. First solve $x + y = 1$ for y.

$$x + y = 1$$

$$y = 1 - x$$

Then substitute $1 - x$ for y in $Q = x^3 + 2y^3$.

$$Q = x^3 + 2y^3$$

$$Q = x^3 + 2(1 - x)^3 \qquad \text{(Substituting)}$$

$$= x^3 + 2(1 - 3x + 3x^2 - x^3)$$

$$= x^3 + 2 - 6x + 6x^2 - 2x^3$$

$$= -x^3 + 6x^2 - 6x + 2$$

Find $Q'(x)$.

$$Q'(x) = -3x^2 + 12x - 6$$

This derivative exists for all x in the interval $(0,\infty)$, thus the only critical points are where $Q'(x) = 0$.

$$Q'(x) = -3x^2 + 12x - 6 = 0$$

$$x^2 - 4x + 2 = 0$$

We now use the quadratic formula.

$$x = \frac{-(-4) \pm \sqrt{(-4)^2 - 4\cdot1\cdot2}}{2\cdot1}$$

$$= \frac{4 \pm \sqrt{16 - 8}}{2} = \frac{4 \pm \sqrt{8}}{2}$$

$$= \frac{4 \pm 2\sqrt{2}}{2} = 2 \pm \sqrt{2}$$

When $x = 2 + \sqrt{2}$, $y = 1 - (2 + \sqrt{2}) = -1 - \sqrt{2}$.

When $x = 2 - \sqrt{2}$, $y = 1 - (2 - \sqrt{2}) = -1 + \sqrt{2}$.

Since x and y are restricted to positive numbers, we only consider $x = 2 - \sqrt{2}$ and $y = -1 + \sqrt{2}$. The minimum value of Q is found by substituting.

$$Q = x^3 + 2y^3$$

$$= (2 - \sqrt{2})^3 + 2(-1 + \sqrt{2})^3 \qquad \text{(Substituting)}$$

$$= (20 - 14\sqrt{2}) + 2(5\sqrt{2} - 7)$$

$$= 20 - 14\sqrt{2} + 10\sqrt{2} - 14$$

$$= 6 - 4\sqrt{2}$$

Exercise Set 3.6

1. For $y = f(x) = x^2$, $x = 2$, and $\Delta x = 0.01$,

$$\Delta y = f(x + \Delta x) - f(x)$$

$$\Delta y = f(2 + 0.01) - f(2) \qquad \text{(Substituting 2 for x and 0.01 for } \Delta x)$$

$$= f(2.01) - f(2)$$

$$= (2.01)^2 - 2^2$$

$$= 4.0401 - 4$$

$$= 0.0401$$

$$f'(x)\Delta x = 2x\cdot\Delta x \qquad [f(x) = x^2;\ f'(x) = 2x]$$

$$f'(2)\Delta x = 2\cdot2(0.01) \qquad \text{(Substituting 2 for x and 0.01 for } \Delta x)$$

$$= 0.04$$

3. For $y = f(x) = x + x^2$, $x = 3$, and $\Delta x = 0.04$,

$$\Delta y = f(x + \Delta x) - f(x)$$

$$\Delta y = f(3 + 0.04) - f(3) \qquad \text{(Substituting 3 for x and 0.04 for } \Delta x)$$

$$= f(3.04) - f(3)$$

$$= [3.04 + (3.04)^2] - (3 + 3^2)$$

$$= (3.04 + 9.2416) - (3 + 9)$$

$$= 12.2816 - 12$$

$$= 0.2816$$

$$f'(x)\Delta x = (1 + 2x)\Delta x \qquad [f(x) = x + x^2,\ f'(x) = 1 + 2x]$$

$$f'(3)\Delta x = (1 + 2\cdot3)0.04 \qquad \text{(Substituting 3 for x and 0.04 for } \Delta x)$$

$$= 7(0.04)$$

$$= 0.28$$

5. For $y = f(x) = \frac{1}{x^2}$, or x^{-2}, $x = 1$, and $\Delta x = 0.5$,

$$\Delta y = f(x + \Delta x) - f(x)$$

$$\Delta y = f(1 + 0.5) - f(1) \qquad \text{(Substituting 1 for x and 0.5 for } \Delta x)$$

$$= f(1.5) - f(1)$$

$$= \frac{1}{(1.5)^2} - \frac{1}{1^2}$$

$$= \frac{1}{2.25} - 1$$

$$= \frac{4}{9} - \frac{9}{9} \qquad \left[\frac{1}{2.25} = \frac{1}{9/4} = \frac{4}{9};\ 1 = \frac{9}{9}\right]$$

$$= -\frac{5}{9}, \text{ or } -0.556$$

$f'(x)\Delta x = -2x^{-3} \cdot \Delta x$ $[f(x) = x^{-2};\ f'(x) = -2x^{-3}]$

$= -\frac{2}{x^3} \cdot \Delta x$

$f'(1)\Delta x = -\frac{2}{1^3} \cdot 0.5$ (Substituting 1 for x and 0.5 for Δx)

$= -2 \cdot 0.5$

$= -1$

7. For $y = f(x) = 3x - 1$, $x = 4$, and $\Delta x = 2$,

$\Delta y = f(x + \Delta x) - f(x)$

$\Delta y = f(4 + 2) - f(4)$ (Substituting 4 for x and 2 for Δx)

$= f(6) - f(4)$

$= (3 \cdot 6 - 1) - (3 \cdot 4 - 1)$

$= 17 - 11$

$= 6$

$f'(x)\Delta x = 3 \cdot \Delta x$ $[f(x) = 3x - 1;\ f'(x) = 3]$

$f'(4)\Delta x = 3 \cdot 2$ (Substituting 4 for x and 2 for Δx)

$= 6$

9. $C(x) = 0.01x^2 + 0.6x + 30$

$\Delta C = C(x + \Delta x) - C(x)$

$\Delta C = C(70 + 1) - C(70)$ (Substituting 70 for x and 1 for Δx)

$= C(71) - C(70)$

$= [0.01(71^2) + 0.6(71) + 30] - [0.01(70^2) + 0.6(70) + 30]$

$= [50.41 + 42.6 + 30] - [49 + 42 + 30]$

$= 123.01 - 121$

$= \$2.01$

$C'(x) = 0.02x + 0.6$

$C'(70) = 0.02(70) + 0.6$ (Substituting 70 for x)

$= 1.4 + 0.6$

$= \$2.00$

11. $R(x) = 2x$

$\Delta R = R(x + \Delta x) - R(x)$

$\Delta R = R(70 + 1) - R(70)$ (Substituting 70 for x and 1 for Δx)

$= R(71) - R(70)$

$= 2 \cdot 71 - 2 \cdot 70$

$= 142 - 140$

$= \$2$

$R'(x) = 2$ ($R'(x)$ is a constant function)

$R'(70) = \$2$

13. $C(x) = 0.01x^2 + 0.6x + 30$

$R(x) = 2x$

a) $P(x) = R(x) - C(x)$

$= 2x - (0.01x^2 + 0.6x + 30)$

$= 2x - 0.01x^2 - 0.6x - 30$

$= -0.01x^2 + 1.4x - 30$

b) $\Delta P = P(x + \Delta x) - P(x)$

$\Delta P = P(70 + 1) - P(70)$ (Substituting 70 for x and 1 for Δx)

$= P(71) - P(70)$

$= (-0.01 \cdot 71^2 + 1.4 \cdot 71 - 30) - (-0.01 \cdot 70^2 + 1.4 \cdot 70 - 30)$

$= (-50.41 + 99.4 - 30) - (-49 + 98 - 30)$

$= 18.99 - 19$

$= -\$0.01$

$P'(x) = -0.02x + 1.4$

$P'(70) = -0.02 \cdot 70 + 1.4$ (Substituting 70 for x)

$= -1.4 + 1.4$

$= \$0$

15. We first think of the number closest to 19 that is a perfect square. This is 16. We will approximate how y, or $\sqrt{x}$, changes when 16 changes by $\Delta x = 3$. Let

$y = f(x) = \sqrt{x}$, or $x^{1/2}$

Then

$\Delta y = \sqrt{x + \Delta x} - \sqrt{x}$ $[\Delta y = f(x + \Delta x) - f(x)]$

$= \sqrt{x + \Delta x} - y$ (Substituting y for $\sqrt{x}$)

so

$y + \Delta y = \sqrt{x + \Delta x}$ (Adding y)

Now

$\Delta y \approx f'(x)\Delta x$

$\Delta y \approx \frac{1}{2} x^{-1/2}\ \Delta x$ [Substituting $\frac{1}{2} x^{-1/2}$ for $f'(x)$]

$\approx \frac{1}{2\sqrt{x}}\ \Delta x$

Let $x = 16$ and $\Delta x = 3$. Then

$\Delta y \approx \frac{1}{2\sqrt{16}} \cdot 3$ (Substituting 16 for x and 3 for Δx)

$\approx \frac{3}{8}$, or 0.375

So $\sqrt{19} = \sqrt{16 + 3}$

$= \sqrt{x + \Delta x}$

$= y + \Delta y$

$= \sqrt{x} + \Delta y$ (Substituting $\sqrt{x}$ for y)

$\approx \sqrt{16} + 0.375$

$\approx 4 + 0.375$

≈ 4.375

To six decimal places, $\sqrt{19} = 4.358899$. Thus the approximation 4.375 is fairly close.

17. We first think of the number closest to 102 that is a perfect square. This is 100. We will approximate how y, or $\sqrt{x}$, changes when 100 changes by $\Delta x = 2$. Let

$y = f(x) = \sqrt{x}$, or $x^{1/2}$

Then

$\Delta y = \sqrt{x + \Delta x} - \sqrt{x}$ $[\Delta y = f(x + \Delta x) - f(x)]$

$= \sqrt{x + \Delta x} - y$ (Substituting y for $\sqrt{x}$)

so

$y + \Delta y = \sqrt{x + \Delta x}$ (Adding y)

Now

$\Delta y \approx f'(x)\Delta x$

$\Delta y \approx \frac{1}{2} x^{-1/2} \Delta x$ [Substituting $\frac{1}{2} x^{-1/2}$ for f'(x)]

$\approx \frac{1}{2\sqrt{x}} \Delta x$

Let x = 100 and Δx = 2. Then

$\Delta y \approx \frac{1}{2\sqrt{100}} \cdot 2$ (Substituting 100 for x and 2 for Δx)

$\approx \frac{1}{\sqrt{100}}$

$\approx \frac{1}{10}$, or 0.1

So $\sqrt{102} = \sqrt{100 + 2}$

$= \sqrt{x + \Delta x}$

$= y + \Delta y$

$= \sqrt{x} + \Delta y$ (Substituting $\sqrt{x}$ for y)

$\approx \sqrt{100} + 0.1$ (Substituting 100 for x and 0.1 for Δy)

$\approx 10 + 0.1$

≈ 10.1

To six decimal places, $\sqrt{102}$ = 10.099505. Thus the approximation 10.1 is fairly close.

19. We think of the number closest to 10 that is a perfect cube. This is 8. We will approximate how y, or $\sqrt[3]{x}$, changes when 8 changes by Δx = 2. Let

$y = f(x) = \sqrt[3]{x}$, or $x^{1/3}$

Then

$\Delta y = \sqrt[3]{x + \Delta x} - \sqrt[3]{x}$ $[\Delta y = f(x + \Delta x) - f(x)]$

$= \sqrt[3]{x + \Delta x} - y$ (Substituting y for $\sqrt[3]{x}$)

so

$y + \Delta y = \sqrt[3]{x + \Delta x}$ (Adding y)

Now

$\Delta y \approx f'(x)\Delta x$

$\Delta y \approx \frac{1}{3} x^{-2/3} \Delta x$ [Substituting $\frac{1}{3} x^{-2/3}$ for f'(x)]

$\approx \frac{1}{3\sqrt[3]{x^2}} \Delta x$

Let x = 8 and Δx = 2. Then

$\Delta y \approx \frac{1}{3\sqrt[3]{8^2}} \cdot 2$ (Substituting 8 for x and 2 for Δx)

$\approx \frac{1}{3\sqrt[3]{64}} \cdot 2$

$\approx \frac{1}{3\cdot 4} \cdot 2$

$\approx \frac{1}{6}$, or 0.167

So $\sqrt[3]{10} = \sqrt[3]{8 + 2}$

$= \sqrt[3]{x + \Delta x}$

$= y + \Delta y$

$= \sqrt[3]{x} + \Delta y$ (Substituting $\sqrt[3]{x}$ for y)

$\approx \sqrt[3]{8} + 0.167$ (Substituting 8 for x and 0.167 for Δy)

$\approx 2 + 0.167$

≈ 2.167

To six decimal places, $\sqrt[3]{10}$ = 2.154435. Thus the approximation 2.167 is fairly close.

21. $y = (2x^3 + 1)^{3/2}$

$\frac{dy}{dx} = \frac{3}{2}(2x^3 + 1)^{1/2} \cdot 6x^2 = 9x^2(2x^3 + 1)^{1/2}$

$dy = 9x^2(2x^3 + 1)^{1/2}dx$

23. $y = \sqrt[5]{x + 27} = (x + 27)^{1/5}$

$\frac{dy}{dx} = \frac{1}{5}(x + 27)^{-4/5}$

$dy = \frac{1}{5}(x + 27)^{-4/5}dx$

25. $y = x^4 - 2x^3 + 5x^2 + 3x - 4$

$\frac{dy}{dx} = 4x^3 - 6x^2 + 10x + 3$

$dy = (4x^3 - 6x^2 + 10x + 3)dx$

27. $dy = (4x^3 - 6x^2 + 10x + 3)dx$

$dy = (4\cdot 2^3 - 6\cdot 2^2 + 10\cdot 2 + 3)0.1$ (Substituting 2 for x and 0.1 for Δx)

$= (32 - 24 + 20 + 3)0.1$

$= 31(0.1)$

$= 3.1$

29. $A(x) = \frac{13x + 100}{x}$, $x = 100$, $\Delta x = 101 - 100 = 1$

$$A'(x) = \frac{x \cdot 13 - 1 \cdot (13x + 100)}{x^2} = \frac{13x - 13x - 100}{x^2}$$

$$= \frac{-100}{x^2}$$

$$\Delta A \approx A'(x)\Delta x$$

$$\approx -\frac{100}{x^2}\Delta x$$

$$\approx -\frac{100}{100^2} \cdot 1$$

$$\approx -0.01$$

The average cost changes by about -\$0.01. (This is a decrease in cost.)

31. $N(a) = -a^2 + 300a + 6$, $a = 100$, $\Delta a = 101 - 100 = 1$

$$N'(a) = -2a + 300$$

$$\Delta N \approx N'(a)\Delta a$$

$$\approx (-2a + 300)\Delta a$$

$$\approx (-2 \cdot 100 + 300) \cdot 1$$

$$\approx 100$$

About 100 more units will be sold.

33. $V = \frac{4}{3}\pi r^3$, $r = 1$, $\Delta r = 1.2 - 1 = 0.2$

$$V' = \frac{4}{3}\pi \cdot 3r^2 = 4\pi r^2$$

$$\Delta V \approx V'(r)\Delta r$$

$$\Delta V \approx 4\pi r^2 \Delta r$$

$$\approx 4(3.14)(1)^2(0.2)$$

$$\approx 2.512$$

The change in volume is approximately 2.512 cm^3.

35.

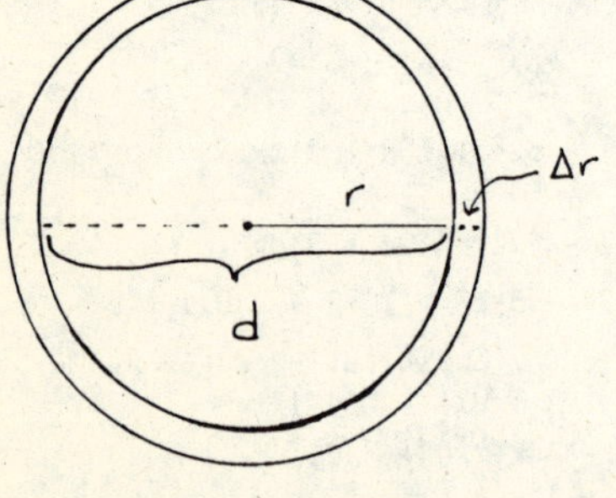

The circumference of the earth, which is the original length of the rope, is given by $C = \pi d$, or $C(r) = 2\pi r$. We need to find the change in the length of the radius, Δr, when the length of the rope is increased 10 ft.

$\Delta C \approx C'(r)\Delta r$ (ΔC represents the change in the length of the rope.)

$10 \approx 2\pi \cdot \Delta r$ [Substituting 10 for ΔC and 2π for $C'(r)$]

$$\frac{10}{2\pi} \approx \Delta r$$

$$\frac{5}{\pi} \approx \Delta r$$

$$1.59 \approx \Delta r$$

Thus the rope is raised approximately 1.59 ft above the earth.

Exercise Set 3.7

1. Differentiate implicitly to find $\frac{dy}{dx}$.

$$xy - x + 2y = 3$$

$$\frac{d}{dx}(xy - x + 2y) = \frac{d}{dx}3$$

(Differentiating both sides with respect to x)

$$\frac{d}{dx}xy - \frac{d}{dx}x + \frac{d}{dx}2y = 0$$

$$x \cdot \frac{dy}{dx} + 1 \cdot y - 1 + 2 \cdot \frac{dy}{dx} = 0$$

$$x \cdot \frac{dy}{dx} + 2 \cdot \frac{dy}{dx} = 1 - y$$

(Getting all terms involving dy/dx alone on one side)

$$(x + 2)\frac{dy}{dx} = 1 - y$$

$$\frac{dy}{dx} = \frac{1 - y}{x + 2}$$

Find the slope of the tangent line to the curve at $\left(-5, \frac{2}{3}\right)$.

$$\frac{dy}{dx} = \frac{1 - y}{x + 2}$$

$\frac{dy}{dx} = \frac{1 - \frac{2}{3}}{-5 + 2}$ $\left[\text{Replacing } x \text{ by } -5 \text{ and } y \text{ by } \frac{2}{3}\right]$

$$= \frac{\frac{1}{3}}{-3} = \frac{1}{3} \cdot \left(-\frac{1}{3}\right) = -\frac{1}{9}$$

3. Differentiate implicitly to find $\frac{dy}{dx}$.

$$x^2 + y^2 = 1$$

$\frac{d}{dx}(x^2 + y^2) = \frac{d}{dx}1$ (Differentiating both sides with respect to x)

$$\frac{d}{dx}x^2 + \frac{d}{dx}y^2 = 0$$

$$2x + 2y\frac{dy}{dx} = 0$$

$$2y\frac{dy}{dx} = -2x$$

$$\frac{dy}{dx} = \frac{-2x}{2y}$$

$$\frac{dy}{dx} = -\frac{x}{y}$$

Find the slope of the tangent line to the curve at $\left(\frac{1}{2}, \frac{\sqrt{3}}{2}\right)$.

$$\frac{dy}{dx} = -\frac{x}{y}$$

$$= -\frac{\frac{1}{2}}{\frac{\sqrt{3}}{2}} \quad \left[\text{Replacing x by } \tfrac{1}{2} \text{ and y by } \tfrac{\sqrt{3}}{2}\right]$$

$$= -\frac{1}{2} \cdot \frac{2}{\sqrt{3}}$$

$$= -\frac{1}{\sqrt{3}}$$

5. Differentiate implicitly to find $\frac{dy}{dx}$.

$$x^2y - 2x^3 - y^3 + 1 = 0$$

$$\frac{d}{dx}(x^2y - 2x^3 - y^3 + 1) = \frac{d}{dx} 0$$

(Differentiating both sides with respect to x)

$$\frac{d}{dx} x^2y - \frac{d}{dx} 2x^3 - \frac{d}{dx} y^3 + \frac{d}{dx} 1 = 0$$

$$x^2 \cdot \frac{dy}{dx} + 2x \cdot y - 6x^2 - 3y^2 \cdot \frac{dy}{dx} + 0 = 0$$

$$x^2 \cdot \frac{dy}{dx} - 3y^2 \cdot \frac{dy}{dx} = 6x^2 - 2xy$$

(Getting all terms involving dy/dx alone on one side)

$$(x^2 - 3y^2)\frac{dy}{dx} = 6x^2 - 2xy$$

$$\frac{dy}{dx} = \frac{6x^2 - 2xy}{x^2 - 3y^2}$$

Find the slope of the tangent line to the curve at (2,-3).

$$\frac{dy}{dx} = \frac{6x^2 - 2xy}{x^2 - 3y^2}$$

$$= \frac{6 \cdot 2^2 - 2 \cdot 2 \cdot (-3)}{2^2 - 3 \cdot (-3)^2} \quad \text{(Replacing x by 2 and y by -3)}$$

$$= \frac{24 + 12}{4 - 27}$$

$$= \frac{36}{-23}$$

$$= -\frac{36}{23}$$

7. Differentiate implicitly to find $\frac{dy}{dx}$.

$$2xy + 3 = 0$$

$$\frac{d}{dx}(2xy + 3) = \frac{d}{dx} 0$$

(Differentiating both sides with respect to x)

$$\frac{d}{dx} 2xy + \frac{d}{dx} 3 = 0$$

$$2\left[x \cdot \frac{dy}{dx} + 1 \cdot y\right] + 0 = 0$$

$$2x \cdot \frac{dy}{dx} + 2y = 0$$

$$2x \cdot \frac{dy}{dx} = -2y$$

$$\frac{dy}{dx} = \frac{-2y}{2x}$$

$$\frac{dy}{dx} = -\frac{y}{x}$$

9. Differentiate implicitly to find $\frac{dy}{dx}$.

$$x^2 - y^2 = 16$$

$$\frac{d}{dx}(x^2 - y^2) = \frac{d}{dx} 16 \quad \text{(Differentiating both sides with respect to x)}$$

$$\frac{d}{dx} x^2 - \frac{d}{dx} y^2 = 0$$

$$2x - 2y \cdot \frac{dy}{dx} = 0$$

$$-2y \cdot \frac{dy}{dx} = -2x$$

$$\frac{dy}{dx} = \frac{-2x}{-2y}$$

$$\frac{dy}{dx} = \frac{x}{y}$$

11. Differentiate implicitly to find $\frac{dy}{dx}$.

$$y^5 = x^3$$

$$\frac{d}{dx} y^5 = \frac{d}{dx} x^3 \quad \text{(Differentiating both sides with respect to x)}$$

$$5y^4 \cdot \frac{dy}{dx} = 3x^2$$

$$\frac{dy}{dx} = \frac{3x^2}{5y^4}$$

13. Differentiate implicitly to find $\frac{dy}{dx}$.

$$x^2y^3 + x^3y^4 = 11$$

$$\frac{d}{dx}(x^2y^3 + x^3y^4) = \frac{d}{dx} 11$$

(Differentiating both sides with respect to x)

$$\frac{d}{dx} x^2y^3 + \frac{d}{dx} x^3y^4 = 0$$

$$x^2 \cdot 3y^2 \cdot \frac{dy}{dx} + 2x \cdot y^3 + x^3 \cdot 4y^3 \cdot \frac{dy}{dx} + 3x^2 \cdot y^4 = 0$$

$$3x^2y^2 \cdot \frac{dy}{dx} + 4x^3y^3 \cdot \frac{dy}{dx} = -2xy^3 - 3x^2y^4$$

$$(3x^2y^2 + 4x^3y^3)\frac{dy}{dx} = -2xy^3 - 3x^2y^4$$

$$\frac{dy}{dx} = \frac{-2xy^3 - 3x^2y^4}{3x^2y^2 + 4x^3y^3}$$

$$= \frac{xy^2(-2y - 3xy^2)}{xy^2(3x + 4x^2y)}$$

$$= \frac{-2y - 3xy^2}{3x + 4x^2y}$$

15. Differentiate implicitly to find $\frac{dp}{dx}$.

$$p^2 + p + 2x = 40$$

$$\frac{d}{dx}(p^2 + p + 2x) = \frac{d}{dx} 40 \quad \text{(Differentiating both sides with respect to x)}$$

$$2p \cdot \frac{dp}{dx} + \frac{dp}{dx} + 2 \cdot 1 = 0$$

$$(2p + 1)\frac{dp}{dx} = -2$$

$$\frac{dp}{dx} = \frac{-2}{2p + 1}$$

17. Differentiate implicitly to find $\frac{dp}{dx}$.

$$(p + 4)(x + 3) = 48$$
$$px + 3p + 4x + 12 = 48$$
$$px + 3p + 4x = 36$$
$$\frac{d}{dx}(px + 3p + 4x) = \frac{d}{dx}36$$

(Differentiating both sides with respect to x)

$$p\cdot 1 + \frac{dp}{dx}\cdot x + 3\cdot\frac{dp}{dx} + 4\cdot 1 = 0$$
$$(x + 3)\frac{dp}{dx} = -p - 4$$
$$\frac{dp}{dx} = \frac{-p - 4}{x + 3}, \text{ or } -\frac{p + 4}{x + 3}$$

19.

$$A^3 + B^3 = 9$$
$$\frac{d}{dt}(A^3 + B^3) = \frac{d}{dt}9$$ (Differentiating with respect to t)
$$3A^2\cdot\frac{dA}{dt} + 3B^2\cdot\frac{dB}{dt} = 0$$
$$3A^2\cdot\frac{dA}{dt} = -3B^2\cdot\frac{dB}{dt}$$
$$\frac{dA}{dt} = \frac{-3B^2}{3A^2}\cdot\frac{dB}{dt}$$
$$\frac{dA}{dt} = -\frac{B^2}{A^2}\cdot\frac{dB}{dt}$$

When A = 2, B = 1:

$$A^3 + B^3 = 9$$
$$2^3 + B^3 = 9$$
$$8 + B^3 = 9$$
$$B^3 = 1$$
$$B = 1$$
$$\frac{dA}{dt} = -\frac{B^2}{A^2}\cdot\frac{dB}{dt}$$
$$= -\frac{1^2}{2^2}\cdot 3$$

(Substituting 2 for A, 1 for B, and 3 for dB/dt)

$$= -\frac{3}{4}$$

21. $R(x) = 50x - 0.5x^2$

$$\frac{dR}{dt} = \frac{d}{dt}(50x - 0.5x^2)$$ (Differentiating with respect to time)
$$= 50\cdot\frac{dx}{dt} - x\cdot\frac{dx}{dt}$$
$$= 50\cdot 20 - 30\cdot 20$$ (Substituting 20 for dx/dt and 30 for x)
$$= 1000 - 600$$
$$= \$400 \text{ per day}$$

$C(x) = 4x + 10$

$$\frac{dC}{dt} = \frac{d}{dt}(4x + 10)$$ (Differentiating with respect to time)
$$= 4\cdot\frac{dx}{dt} + 0$$
$$= 4\cdot\frac{dx}{dt}$$
$$= 4\cdot 20$$ (Substituting 20 for dx/dt)
$$= \$80 \text{ per day}$$

Since $P(x) = R(x) - C(x)$,

$$\frac{dP}{dt} = \frac{dR}{dt} - \frac{dC}{dt}$$
$$= \$400 \text{ per day} - \$80 \text{ per day}$$
$$= \$320 \text{ per day}$$

23. $R(x) = 2x$

$$\frac{dR}{dt} = \frac{d}{dt}(2x)$$ (Differentiating with respect to time)
$$= 2\cdot\frac{dx}{dt}$$
$$= 2\cdot 8$$ (Substituting 8 for dx/dt)
$$= \$16 \text{ per day}$$

$C(x) = 0.01x^2 + 0.6x + 30$

$$\frac{dC}{dt} = \frac{d}{dt}(0.01x^2 + 0.6x + 30)$$ (Differentiating with respect to time)
$$= \frac{d}{dt}0.01x^2 + \frac{d}{dt}0.6x + \frac{d}{dt}30$$
$$= 0.02x\cdot\frac{dx}{dt} + 0.6\cdot\frac{dx}{dt} + 0$$
$$= 0.02\cdot 20\cdot 8 + 0.6\cdot 8$$ (Substituting 20 for x and 8 for dx/dt)
$$= 3.2 + 4.8$$
$$= \$8 \text{ per day}$$

Since $P(x) = R(x) - C(x)$,

$$\frac{dP}{dt} = \frac{dR}{dt} - \frac{dC}{dt}$$
$$= \$16 \text{ per day} - \$8 \text{ per day}$$
$$= \$8 \text{ per day}$$

25. $V = \frac{4}{3}\pi r^3$, $\frac{dr}{dt} = 0.03\ \frac{\text{cm}}{\text{day}}$

We take the derivative of both sides with respect to t.

$$\frac{dV}{dt} = \frac{4}{3}\pi\cdot 3r^2\cdot\frac{dr}{dt}$$
$$= 4\pi r^2\cdot\frac{dr}{dt}$$
$$= 4\pi\cdot(1.2\text{ cm})^2\cdot\left[0.03\ \frac{\text{cm}}{\text{day}}\right]$$
$$= 4\pi\cdot(1.44\text{ cm}^2)\cdot\left[0.03\ \frac{\text{cm}}{\text{day}}\right]$$
$$= 0.1728\pi\ \frac{\text{cm}^3}{\text{day}}$$
$$\approx 0.54\ \frac{\text{cm}^3}{\text{day}}$$

27. $V = \frac{p}{4Lv}(R^2 - r^2)$ (We will assume r, p, L and v are constants)

a) $\frac{dV}{dt} = \frac{d}{dt}\left[\frac{p}{4Lv}(R^2 - r^2)\right]$

$= \frac{p}{4Lv} \cdot \frac{d}{dt}(R^2 - r^2)$

$= \frac{p}{4Lv}\left[\frac{d}{dt}R^2 - \frac{d}{dt}r^2\right]$

$= \frac{p}{4Lv}\left[2R \cdot \frac{dR}{dt} - 0\right]$

$= \frac{2pR}{4Lv} \cdot \frac{dR}{dt}$

$= \frac{2\cdot 100\cdot R}{4\cdot 1\cdot 0.05} \cdot \frac{dR}{dt}$ (Substituting 100 for p, 1 for L, and 0.05 for v)

$= 1000R \cdot \frac{dR}{dt}$

b) $\frac{dV}{dt} = 1000R \cdot \frac{dR}{dt}$

$= 1000\cdot(0.0075)\cdot(-0.0015)$

(Substituting 0.0075 for R and -0.0015 for dR/dt)

$= -0.01125 \ \frac{mm^3}{min}$

29. $D^2 = x^2 + y^2$

After 1 hour,

$D^2 = 25^2 + 60^2$ (Substituting 25 for x and 60 for y)

$D^2 = 625 + 3600$

$D^2 = 4225$

$D = 65$

We also know that $\frac{dx}{dt} = 25 \ \frac{mi}{hr}$ and that $\frac{dy}{dt} = 60 \ \frac{mi}{hr}$.

We differentiate implicitly to find $\frac{dD}{dt}$.

$D^2 = x^2 + y^2$

$\frac{d}{dt}D^2 = \frac{d}{dt}(x^2 + y^2)$

$\frac{d}{dt}D^2 = \frac{d}{dt}x^2 + \frac{d}{dt}y^2$

$2D \cdot \frac{dD}{dt} = 2x \cdot \frac{dx}{dt} + 2y \cdot \frac{dy}{dt}$

$2\cdot 65 \cdot \frac{dD}{dt} = 2\cdot 25\cdot 25 \ \frac{mi}{hr} + 2\cdot 60\cdot 60 \ \frac{mi}{hr}$ (Substituting)

$\frac{dD}{dt} = \frac{1250 + 7200}{130} \ \frac{mi}{hr}$

$\frac{dD}{dt} = 65 \ \frac{mi}{hr}$

31. Differentiate implicitly to find $\frac{dy}{dx}$.

$\sqrt{x} + \sqrt{y} = 1$

$\frac{d}{dx}(x^{1/2} + y^{1/2}) = \frac{d}{dx}1$ (Differentiating both sides with respect to x)

$\frac{1}{2}x^{-1/2} + \frac{1}{2}y^{-1/2}\frac{dy}{dx} = 0$

$\frac{1}{2}y^{-1/2}\frac{dy}{dx} = -\frac{1}{2}x^{-1/2}$

$y^{-1/2}\frac{dy}{dx} = -x^{-1/2}$

$\frac{dy}{dx} = -\frac{x^{-1/2}}{y^{-1/2}}$

$\frac{dy}{dx} = -\frac{y^{1/2}}{x^{1/2}}$, or $-\frac{\sqrt{y}}{\sqrt{x}}$

33. Differentiate implicitly to find $\frac{dy}{dx}$.

$y^3 = \frac{x - 1}{x + 1}$

$\frac{d}{dx}y^3 = \frac{d}{dx}\frac{x - 1}{x + 1}$ (Differentiating both sides with respect to x)

$3y^2 \cdot \frac{dy}{dx} = \frac{(x + 1)\cdot 1 - 1\cdot(x - 1)}{(x + 1)^2}$

$3y^2 \cdot \frac{dy}{dx} = \frac{x + 1 - x + 1}{(x + 1)^2}$

$3y^2 \cdot \frac{dy}{dx} = \frac{2}{(x + 1)^2}$

$\frac{dy}{dx} = \frac{2}{3y^2(x + 1)^2}$

35. Differentiate implicitly to find $\frac{dy}{dx}$.

$x^{3/2} + y^{2/3} = 1$

$\frac{d}{dx}(x^{3/2} + y^{2/3}) = \frac{d}{dx}1$ (Differentiating both sides with respect to x)

$\frac{3}{2}x^{1/2} + \frac{2}{3}y^{-1/3} \cdot \frac{dy}{dx} = 0$

$\frac{2}{3}y^{-1/3} \cdot \frac{dy}{dx} = -\frac{3}{2}x^{1/2}$

$\frac{dy}{dx} = \frac{-\frac{3}{2}x^{1/2}}{\frac{2}{3}y^{-1/3}}$

$\frac{dy}{dx} = -\frac{9}{4}x^{1/2}y^{1/3}$, or

$-\frac{9}{4}\sqrt{x}\sqrt[3]{y}$

37. Differentiate implicitly to find $\frac{dy}{dx}$.

$xy + x - 2y = 4$

$\frac{d}{dx}(xy + x - 2y) = \frac{d}{dx}4$

$x \cdot \frac{dy}{dx} + 1\cdot y + 1 - 2 \cdot \frac{dy}{dx} = 0$

$(x - 2)\frac{dy}{dx} = -1 - y$

$\frac{dy}{dx} = \frac{-1 - y}{x - 2}$, or $\frac{1 + y}{2 - x}$

Differentiate implicitly to find $\frac{d^2y}{dx^2}$.

$$\frac{dy}{dx} = \frac{1 + y}{2 - x}$$

$$\frac{d}{dx}\frac{dy}{dx} = \frac{d}{dx}\frac{1 + y}{2 - x}$$

$$\frac{d^2y}{dx^2} = \frac{(2 - x)\cdot\frac{dy}{dx} - (-1)\cdot(1 + y)}{(2 - x)^2}$$

$$= \frac{(2 - x)\cdot\frac{1 + y}{2 - x} + 1 + y}{(2 - x)^2}$$

(Substituting $\frac{1 + y}{2 - x}$ for $\frac{dy}{dx}$)

$$= \frac{1 + y + 1 + y}{(2 - x)^2}$$

$$= \frac{2 + 2y}{(2 - x)^2}$$

39. Differentiate implicitly to find $\frac{dy}{dx}$.

$$x^2 - y^2 = 5$$

$$\frac{d}{dx}(x^2 - y^2) = \frac{d}{dx}5$$

$$2x - 2y\cdot\frac{dy}{dx} = 0$$

$$-2y\cdot\frac{dy}{dx} = -2x$$

$$\frac{dy}{dx} = \frac{-2x}{-2y}$$

$$\frac{dy}{dx} = \frac{x}{y}$$

Differentiate implicitly to find $\frac{d^2y}{dx^2}$.

$$\frac{dy}{dx} = \frac{x}{y}$$

$$\frac{d}{dx}\frac{dy}{dx} = \frac{d}{dx}\frac{x}{y}$$

$$\frac{d^2y}{dx^2} = \frac{y\cdot 1 - \frac{dy}{dx}\cdot x}{y^2}$$

$$= \frac{y - \frac{x}{y}\cdot x}{y^2} \qquad \left[\text{Substituting } \frac{x}{y} \text{ for } \frac{dy}{dx}\right]$$

$$= \frac{\frac{y^2}{y} - \frac{x^2}{y}}{y^2}$$

$$= \frac{y^2 - x^2}{y}\cdot\frac{1}{y^2}$$

$$= \frac{y^2 - x^2}{y^3}$$

41. See the answer section in the text.

Exercise Set 4.1

1. Graph: $y = 4^x$

First we find some function values.
Note: For

$x = -2,\ y = 4^{-2} = \frac{1}{4^2} = \frac{1}{16} = 0.0625$

$x = -1,\ y = 4^{-1} = \frac{1}{4} = 0.25$

$x = 0,\ y = 4^0 = 1$

$x = 1,\ y = 4^1 = 4$

$x = 2,\ y = 4^2 = 16$

x	y
-2	0.0625
-1	0.25
0	1
1	4
2	16

Plot these points and connect them with a smooth curve.

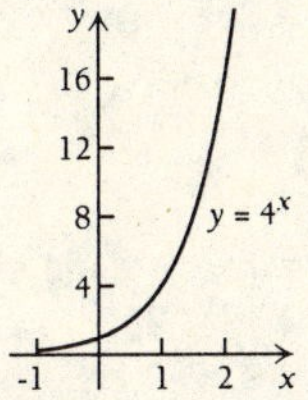

3. Graph: $y = (0.4)^x$

First we find some function values.
Note: For

$x = -2,\ y = (0.4)^{-2} = \frac{1}{(0.4)^2} = 6.25$

$x = -1,\ y = (0.4)^{-1} = \frac{1}{0.4} = 2.5$

$x = 0,\ y = (0.4)^0 = 1$

$x = 1,\ y = (0.4)^1 = 0.4$

$x = 2,\ y = (0.4)^2 = 0.16$

x	y
-2	6.25
-1	2.5
0	1
1	0.4
2	0.16

Plot the points and connect them with a smooth curve.

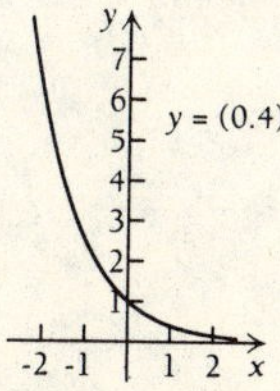

5. Graph: $x = 4^y$

First we find some function values.
Note: For

$y = -2,\ x = 4^{-2} = \frac{1}{4^2} = \frac{1}{16}$

$y = -1,\ x = 4^{-1} = \frac{1}{4}$

$y = 0,\ x = 4^0 = 1$

$y = 1,\ x = 4^1 = 4$

$y = 2,\ x = 4^2 = 16$

x	y
$\frac{1}{16}$	-2
$\frac{1}{4}$	-1
1	0
4	1
16	2

Plot the points and connect them with a smooth curve.

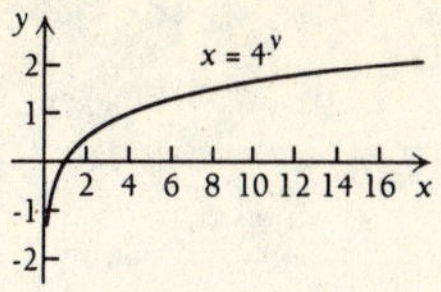

7. $f(x) = e^{3x}$

$f'(x) = 3e^{3x}$ $\left[\frac{d}{dx} e^{f(x)} = f'(x)\, e^{f(x)}\right]$

9. $f(x) = 5e^{-2x}$

$f'(x) = 5\cdot(-2)e^{-2x}$ $\left[\frac{d}{dx} [c\cdot f(x)] = c\, f'(x)\right]$

$\left[\frac{d}{dx} e^{f(x)} = f'(x)\, e^{f(x)}\right]$

$= -10e^{-2x}$

11. $f(x) = 3 - e^{-x}$

$f'(x) = 0 - (-1)e^{-x}$ $\left[\frac{d}{dx} e^{f(x)} = f'(x)\, e^{f(x)}\right]$

$= e^{-x}$

13. $f(x) = -7e^x$

$f'(x) = -7e^x$ $\left[\frac{d}{dx} -7e^x = -7 \cdot \frac{d}{dx} e^x\right]$

$\left[\frac{d}{dx} e^x = e^x\right]$

15. $f(x) = \frac{1}{2} e^{2x}$

$f'(x) = \frac{1}{2} \cdot 2e^{2x}$ $\left[\frac{d}{dx} [c\cdot f(x)] = c\cdot f'(x)\right]$

$\left[\frac{d}{dx} e^{f(x)} = f'(x)\, e^{f(x)}\right]$

$= e^{2x}$

17. $f(x) = x^4 e^x$

$f'(x) = x^4\cdot e^x + 4x^3\cdot e^x$ (Using the Product Rule)

$= x^3 e^x(x + 4)$

19. $f(x) = \frac{e^x}{x^4}$

$f'(x) = \frac{x^4\cdot e^x - 4x^3\cdot e^x}{x^8}$ (Using the Quotient Rule)

$= \frac{x^3 e^x(x - 4)}{x^3\cdot x^5}$ (Factoring both numerator and denominator)

$= \frac{x^3}{x^3} \cdot \frac{e^x(x - 4)}{x^5}$

$= \frac{e^x(x - 4)}{x^5}$ (Simplifying)

21. $f(x) = e^{-x^2+7x}$

$f'(x) = (-2x + 7)e^{-x^2+7x}$ $\left[\frac{d}{dx} e^{f(x)} = f'(x)\, e^{f(x)}\right]$

$[f(x) = -x^2 + 7x,$
$f'(x) = -2x + 7]$

23. $f(x) = e^{-x^2/2}$

$= e^{(-1/2)x^2}$

$f'(x) = \left[-\frac{1}{2} \cdot 2x\right] e^{(-1/2)x^2}$

$= -xe^{-x^2/2}$

25. $y = e^{\sqrt{x-7}}$

$= e^{(x-7)^{1/2}}$

$\frac{dy}{dx} = \frac{1}{2}(x - 7)^{-1/2} \cdot e^{(x-7)^{1/2}}$

$= \frac{e^{(x-7)^{1/2}}}{2(x - 7)^{1/2}}$

$= \frac{e^{\sqrt{x-7}}}{2\sqrt{x - 7}}$

27. $y = \sqrt{e^x - 1}$

$= (e^x - 1)^{1/2}$

$\frac{dy}{dx} = \frac{1}{2}(e^x - 1)^{-1/2} \cdot e^x$ [Extended Power Rule; $\frac{d}{dx}(e^x-1) = e^x-0 = e^x$]

$= \frac{e^x}{2\sqrt{e^x - 1}}$

29. $y = xe^{-2x} + e^{-x} + x^3$

$\frac{dy}{dx} = x\cdot(-2)\cdot e^{-2x} + 1\cdot e^{-2x} + (-1)\cdot e^{-x} + 3x^2$

$= -2xe^{-2x} + e^{-2x} - e^{-x} + 3x^2$

$= (1 - 2x)e^{-2x} - e^{-x} + 3x^2$

31. $y = 1 - e^{-x}$

$\frac{dy}{dx} = 0 - (-1)e^{-x}$

$= e^{-x}$

33. $y = 1 - e^{-kx}$

$\frac{dy}{dx} = 0 - (-k)e^{-kx}$

$= ke^{-kx}$

35. Graph: $f(x) = e^{2x}$

Using a calculator we first find some function values.

Note: For

$x = -2,\ f(-2) = e^{2(-2)} = e^{-4} = 0.0183$

$x = -1,\ f(-1) = e^{2(-1)} = e^{-2} = 0.1353$

$x = 0,\ f(0) = e^{2\cdot 0} = e^0 = 1$

$x = 1,\ f(1) = e^{2\cdot 1} = e^2 = 7.3891$

$x = 2,\ f(2) = e^{2\cdot 2} = e^4 = 54.598$

x	f(x)
-2	0.0183
-1	0.1353
0	1
1	7.3891
2	54.598

Plot the points and connect them with a smooth curve.

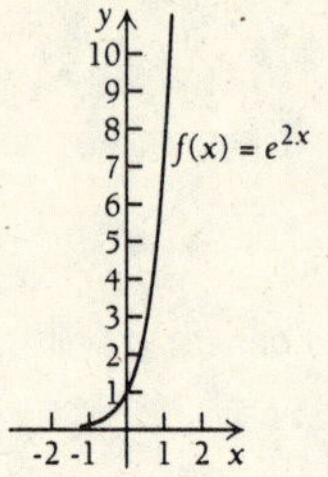

37. Graph: $f(x) = e^{-2x}$

Using a calculator we first find some function values.

Note: For

$x = -2,\ f(-2) = e^{-2(-2)} = e^4 = 54.598$

$x = -1,\ f(-1) = e^{-2(-1)} = e^2 = 7.3891$

$x = 0,\ f(0) = e^{-2\cdot 0} = e^0 = 1$

$x = 1,\ f(1) = e^{-2\cdot 1} = e^{-2} = 0.1353$

$x = 2,\ f(2) = e^{-2\cdot 2} = e^{-4} = 0.0183$

x	f(x)
-2	54.598
-1	7.3891
0	1
1	0.1353
2	0.0183

Plot these points and connect them with a smooth curve.

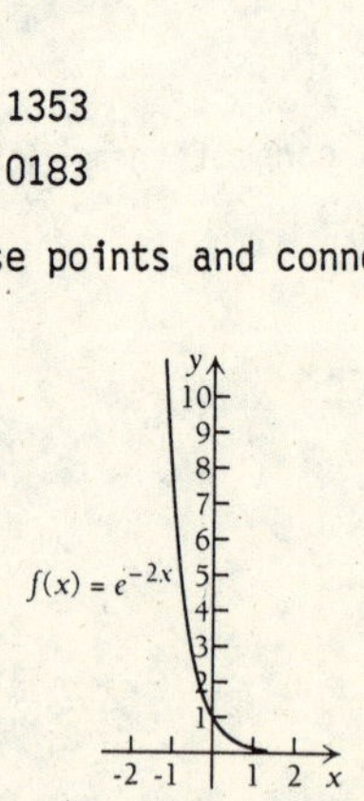

39. Graph: $f(x) = 3 - e^{-x}$, for nonnegative values of x.

Using a calculator we first find some function values.

For

$x = 0,\ f(0) = 3 - e^{-0} = 3 - 1 = 2$

$x = 1,\ f(1) = 3 - e^{-1} = 3 - 0.3679 = 2.6321$

$x = 2,\ f(2) = 3 - e^{-2} = 3 - 0.1353 = 2.8647$

$x = 3,\ f(3) = 3 - e^{-3} = 3 - 0.0498 = 2.9502$

$x = 4$, $f(4) = 3 - e^{-4} = 3 - 0.0183 = 2.9817$

$x = 6$, $f(6) = 3 - e^{-6} = 3 - 0.0025 = 2.9975$

x	f(x)
0	2
1	2.6321
2	2.8647
3	2.9502
4	2.9817
6	2.9975

Plot the points and connect them with a smooth curve.

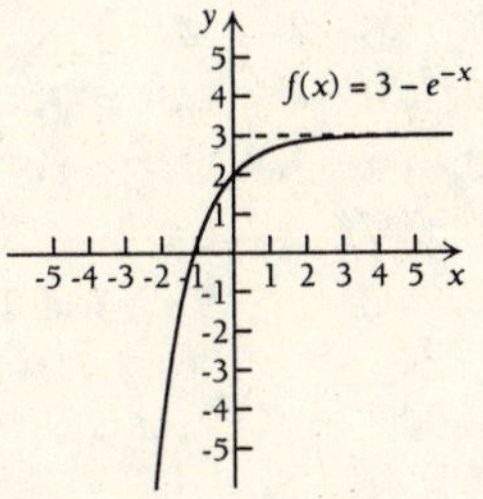

41. We first find the slope of the tangent line at (0,1), f'(0):

$f(x) = e^x$

$f'(x) = e^x$

$f'(0) = e^0 = 1$

Then we find the equation of the line with slope 1 and containing the point (0,1):

$y - y_1 = m(x - x_1)$ (Point-slope equation)

$y - 1 = 1(x - 0)$

$y - 1 = x$

$y = x + 1$

43. $C(t) = 100 - 50e^{-t}$

a) $C'(t) = 0 - 50\cdot(-1)\cdot e^{-t}$
$= 50e^{-t}$

b) $C'(0) = 50e^{-0} = 50e^0 = 50\cdot 1 = \$50$ million

c) $C'(4) = 50e^{-4} = 50(0.018316) \approx \$0.916$ million

45. $x = D(p) = 480e^{-0.003p}$

a) $x = D(120) = 480e^{-0.003(120)}$
$= 480e^{-0.36}$
$\approx 480(0.697676)$
≈ 335

$x = D(180) = 480e^{-0.003(180)}$
$= 480e^{-0.54}$
$\approx 480(0.582748)$
≈ 280

$x = D(340) = 480e^{-0.003(340)}$
$= 480e^{-1.02}$
$\approx 480(0.360595)$
≈ 173

b) We plot the points (120, 335), (180, 280), and (340, 173) and other points as needed. Then we connect the points with a smooth curve.

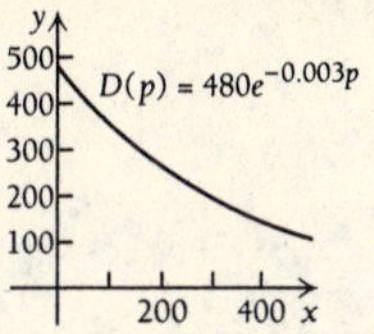

c) $D'(p) = 480(-0.003)e^{-0.003p}$
$= -1.44e^{-0.003p}$

47. $C(t) = 10t^2e^{-t}$

a) $C(0) = 10\cdot 0^2\cdot e^{-0} = 0$ ppm

$C(1) = 10\cdot 1^2\cdot e^{-1}$
$\approx 10(0.367879)$
≈ 3.7 ppm

$C(2) = 10\cdot 2^2\cdot e^{-2}$
$\approx 40(0.135335)$
≈ 5.4 ppm

$C(3) = 10\cdot 3^2\cdot e^{-3}$
$\approx 90(0.049787)$
≈ 4.48 ppm

$C(10) = 10\cdot 10^2\cdot e^{-10}$
$\approx 1000(0.000045)$
≈ 0.05 ppm

b) We plot the points (0,0), (1, 3.7), (2, 5.4), (3, 4.48), and (10, 0.05) and other points as needed. Then we connect the points with a smooth curve.

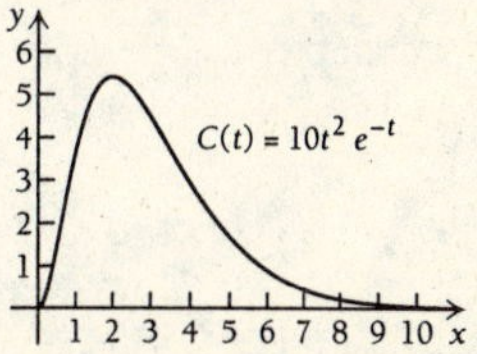

c) $C'(t) = 10t^2(-1)e^{-t} + 20te^{-t}$
$= te^{-t}(-10t + 20)$

d) We find the maximum value of C(t) on $[0,\infty)$.

C'(t) exists for all t in $[0,\infty)$. We solve $C'(t) = 0$.

$te^{-t}(-10t + 20) = 0$

$te^{-t} = 0$ or $-10t + 20 = 0$ $(e^{-t} \neq 0)$

$t = 0$ or $-10t = -20$

$t = 0$ or $x = 2$

Since the function has two critical points and the interval is not closed, we must examine the graph to find the maximum value. We see that the maximum value of the concentration is about 5.4 ppm. It occurs at t = 2 hr.

49. $y = (e^{3x} + 1)^5$

$$\frac{dy}{dx} = 5(e^{3x} + 1)^4 \cdot 3e^{3x}$$
$$= 15e^{3x}(e^{3x} + 1)^4$$

51. $y = \dfrac{e^{3t} - e^{7t}}{e^{4t}}$

$$= \frac{e^{3t}(1 - e^{4t})}{e^{3t} \cdot e^{t}} \quad \text{(Factoring)}$$
$$= \frac{1 - e^{4t}}{e^{t}} \quad \text{(Simplifying)}$$

$$\frac{dy}{dt} = \frac{e^{t}(-4e^{4t}) - e^{t}(1 - e^{4t})}{(e^{t})^2} \quad \text{(Using the Quotient Rule)}$$
$$= \frac{e^{t}(-4e^{4t} - 1 + e^{4t})}{e^{t} \cdot e^{t}}$$
$$= \frac{-3e^{4t} - 1}{e^{t}}$$
$$= -3e^{3t} - e^{-t}$$

53. $y = \dfrac{e^x}{x^2 + 1}$

$$\frac{dy}{dx} = \frac{(x^2 + 1)e^x - 2x \cdot e^x}{(x^2 + 1)^2}$$
$$= \frac{e^x(x^2 - 2x + 1)}{(x^2 + 1)^2}$$
$$= \frac{e^x(x - 1)^2}{(x^2 + 1)^2}$$

55. $f(x) = e^{\sqrt{x}} + \sqrt{e^x}$

$$= e^{x^{1/2}} + e^{x/2}$$

$$f'(x) = \frac{1}{2}\, x^{-1/2}\, e^{x^{1/2}} + \frac{1}{2}\, e^{x/2}$$

$$= \frac{e^{\sqrt{x}}}{2\sqrt{x}} + \frac{\sqrt{e^x}}{2}$$

57. $f(x) = e^{x/2} \cdot \sqrt{x - 1}$

$$= e^{x/2} \cdot (x - 1)^{1/2}$$

$$f'(x) = e^{x/2} \cdot \frac{1}{2}\, (x - 1)^{-1/2} + \frac{1}{2}\, e^{x/2} \cdot (x - 1)^{1/2}$$
$$= \frac{1}{2}\, e^{x/2}\left[(x - 1)^{-1/2} + (x - 1)^{1/2}\right]$$
$$= \frac{1}{2}\, e^{x/2}\left[\frac{1}{\sqrt{x - 1}} + \sqrt{x - 1}\right]$$
$$= \frac{1}{2}\, e^{x/2}\left[\frac{1}{\sqrt{x - 1}} + \sqrt{x - 1} \cdot \frac{\sqrt{x - 1}}{\sqrt{x - 1}}\right]$$
$$= \frac{1}{2}\, e^{x/2}\left[\frac{1}{\sqrt{x - 1}} + \frac{x - 1}{\sqrt{x - 1}}\right]$$
$$= \frac{1}{2}\, e^{x/2}\left[\frac{x}{\sqrt{x - 1}}\right]$$

59. $f(x) = \dfrac{e^x - e^{-x}}{e^x + e^{-x}}$

$$f'(x) = \frac{(e^x + e^{-x})(e^x + e^{-x}) - (e^x - e^{-x})(e^x - e^{-x})}{(e^x + e^{-x})^2}$$
$$= \frac{(e^{2x} + e^0 + e^0 + e^{-2x}) - (e^{2x} - e^0 - e^0 + e^{-2x})}{(e^x + e^{-x})^2}$$
$$= \frac{e^{2x} + 1 + 1 + e^{-2x} - e^{2x} + 1 + 1 - e^{-2x}}{(e^x + e^{-x})^2}$$
$$= \frac{4}{(e^x + e^{-x})^2}$$

61. $e = \lim\limits_{t \to 0} f(t);\ f(t) = (1 + t)^{1/t}$

$$f(1) = (1 + 1)^{1/1} = 2^1 = 2$$
$$f(0.5) = (1 + 0.5)^{1/0.5} = 1.5^2 = 2.25$$
$$f(0.2) = (1 + 0.2)^{1/0.2} = 1.2^5 = 2.48832$$
$$f(0.1) = (1 + 0.1)^{1/0.1} = 1.1^{10} \approx 2.59374$$
$$f(0.001) = (1 + 0.001)^{1/0.001} = (1.001)^{1000} \approx 2.71692$$

63. $f(x) = x^2e^{-x}; \quad [0,4]$

$$f'(x) = x^2 \cdot (-1)e^{-x} + 2x \cdot e^{-x}$$
$$= -x^2e^{-x} + 2xe^{-x}$$

Since $f'(x)$ exists for all x in $[0,4]$, the only critical points are where $f'(x) = 0$.

$$f'(x) = -x^2e^{-x} + 2xe^{-x} = 0$$
$$-e^{-x}(x^2 - 2x) = 0$$
$$x^2 - 2x = 0 \qquad (e^{-x} \neq 0)$$
$$x(x - 2) = 0$$

$$x = 0 \text{ or } x - 2 = 0$$
$$x = 0 \text{ or } \quad x = 2$$

Use Max-Min Principle 1. Find the function values at $x = 0$, $x = 2$, and $x = 4$.

$$f(x) = x^2e^{-x}$$
$$f(0) = 0^2 \cdot e^{-0} = 0 \cdot 1 = 0$$
$$f(2) = 2^2 \cdot e^{-2} \approx 4(0.1353) \approx 0.54$$
$$f(4) = 4^2 \cdot e^{-4} \approx 16(0.0183) \approx 0.29$$

The maximum value of $f(x)$ is $4e^{-2}$, or approximately 0.54, at $x = 2$.

65. Graph: $y = x^2e^{-x}$

From Exercise 63 we know that there is a relative minimum at (0,0) and a relative maximum at $(2, 4e^{-2})$. Also observe that $y \geqslant 0$ for all x.

For

$x = -3$, $y = (-3)^2e^{-(-3)} = 9e^3 \approx 180.77$
$x = -2$, $y = (-2)^2e^{-(-2)} = 4e^2 \approx 29.56$
$x = -1$, $y = (-1)^2e^{-(-1)} = e \approx 2.72$
$x = -0.5$, $y = (-0.5)^2e^{-(-0.5)} = 0.25e^{0.5} \approx 0.41$
$x = 0$, $y = 0^2e^{-0} = 0 \cdot 1 = 0$
$x = 1$, $y = 1^2e^{-1} = e^{-1} \approx 0.37$
$x = 2$, $y = 2^2e^{-2} = 4e^{-2} \approx 0.54$
$x = 3$, $y = 3^2e^{-3} = 9e^{-3} \approx 0.45$

$x = 4$, $y = 4^2e^{-4} = 16e^{-4} \approx 0.29$

$x = 5$, $y = 5^2e^{-5} = 25e^{-5} \approx 0.17$

$x = 10$, $y = 10^2e^{-10} = 100e^{-10} \approx 0.005$

Plot these points and connect them with a smooth curve.

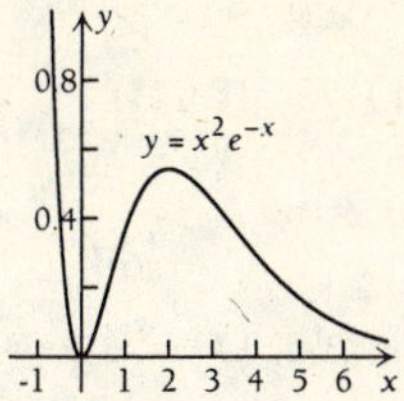

Exercise Set 4.2

1. $\log_2 8 = 3$ (Logarithmic equation)

   $2^3 = 8$ (Exponential equation; 2 is the base, 3 is the exponent)

3. $\log_8 2 = \frac{1}{3}$ (Logarithmic equation)

   $8^{1/3} = 2$ (Exponential equation; 8 is the base, 1/3 is the exponent)

5. $\log_a K = J$ (Logarithmic equation)

   $a^J = K$ (Exponential equation; a is the base, J is the exponent)

7. $\log_b T = v$ (Logarithmic equation)

   $b^v = T$ (Exponential equation; b is the base, v is the exponent)

9. $e^M = b$ (Exponential equation; e is the base, M is the exponent)

   $\log_e b = M$, (Logarithmic equation)

   or $\ln b = M$ (ln b is the abbreviation for $\log_e b$)

11. $10^2 = 100$ (Exponential equation; 10 is the base, 2 is the exponent)

    $\log_{10} 100 = 2$ (Logarithmic equation)

13. $10^{-1} = 0.1$ (Exponential equation; 10 is the base, -1 is the exponent)

    $\log_{10} 0.1 = -1$ (Logarithmic equation)

15. $M^p = V$ (Exponential equation; M is the base, p is the exponent)

    $\log_M V = p$ (Logarithmic equation)

17. $\log_b 15 = \log_b 3\cdot 5$

    $= \log_b 3 + \log_b 5$ (P1)

    $= 1.099 + 1.609$

    $= 2.708$

19. $\log_b \frac{5}{3} = \log_b 5 - \log_b 3$ (P2)

    $= 1.609 - 1.099$

    $= 0.51$

21. $\log_b \frac{1}{5} = \log_b 1 - \log_b 5$ (P2)

    $= 0 - \log_b 5$ (P6)

    $= -\log_b 5$

    $= -1.609$

23. $\log_b \sqrt{b^3} = \log_b b^{3/2}$

    $= \frac{3}{2}$ (P5)

25. $\log_b 5b = \log_b 5 + \log_b b$ (P1)

    $= 1.609 + 1$ (P4)

    $= 2.609$

27. $\log_b 25 = \log_b 5^2$

    $= 2 \log_b 5$ (P3)

    $= 2(1.609)$

    $= 3.218$

29. $\ln 20 = \ln 4\cdot 5$

    $= \ln 4 + \ln 5$ (P1)

    $= 1.3863 + 1.6094$

    $= 2.9957$

31. $\ln \frac{5}{4} = \ln 5 - \ln 4$ (P2)

    $= 1.6094 - 1.3863$

    $= 0.2231$

33. $\ln \frac{1}{4} = \ln 1 - \ln 4$ (P2)

    $= 0 - 1.3863$ (P6)

    $= -1.3863$

35. $\ln 4e = \ln 4 + \ln e$ (P1)

    $= 1.3863 + 1$ (P4)

    $= 2.3863$

37. $\ln \sqrt{e^8} = \ln e^{8/2}$

    $= \ln e^4$

    $= 4$ (P5)

39. $\ln 16 = \ln 4^2$

    $= 2 \ln 4$ (P3)

    $= 2(1.3863)$

    $= 2.7726$

41. $\ln 5894 = 8.681690$ (Using a calculator and rounding to six decimal places)

43. $\ln 0.0182 = -4.006334$

45. $\ln 1.88 = 0.631272$

47. $\ln 0.0188 = -3.973898$

49. $\ln 906 = 6.809039$

51. $\ln 0.011 = -4.509860$

53. $e^t = 100$

$\ln e^t = \ln 100$ (Taking the natural logarithm on both sides)

$t = \ln 100$ (P5)

$t = 4.605170$ (Using a calculator)

$t \approx 4.6$

55. $e^t = 60$

$\ln e^t = \ln 60$ (Taking the natural logarithm on both sides)

$t = \ln 60$ (P5)

$t = 4.094345$ (Using a calculator)

$t \approx 4.1$

57. $e^{-t} = 0.1$

$\ln e^{-t} = \ln 0.1$ (Taking the natural logarithm on both sides)

$-t = \ln 0.1$ (P5)

$t = -\ln 0.1$

$t = -(-2.302585)$ (Using a calculator)

$t = 2.302585$

$t \approx 2.3$

59. $e^{-0.02t} = 0.06$

$\ln e^{-0.02t} = \ln 0.06$ (Taking the natural logarithm on both sides)

$-0.02t = \ln 0.06$ (P5)

$t = \dfrac{\ln 0.06}{-0.02}$

$t = \dfrac{-2.813411}{-0.02}$ (Using a calculator)

$t \approx 141$

61. $y = -6 \ln x$

$\dfrac{dy}{dx} = -6 \cdot \dfrac{1}{x}$ $\left[\dfrac{d}{dx}[c\cdot f(x)] = c\cdot f'(x)\right]$

$= -\dfrac{6}{x}$

63. $y = x^4 \ln x - \dfrac{1}{2}x^2$

$\dfrac{dy}{dx} = x^4 \cdot \dfrac{1}{x} + 4x^3\cdot \ln x - \dfrac{1}{2} \cdot 2x$ (Product Rule on $x^4 \ln x$)

$= x^3 + 4x^3 \ln x - x$

$= x^3(1 + 4 \ln x) - x$

65. $y = \dfrac{\ln x}{x^4}$

$\dfrac{dy}{dx} = \dfrac{x^4 \cdot \frac{1}{x} - 4x^3\cdot \ln x}{x^8}$ (Using the Quotient Rule)

$= \dfrac{x^3 - 4x^3 \ln x}{x^8}$

$= \dfrac{x^3(1 - 4 \ln x)}{x^3\cdot x^5}$ (Factoring both numerator and denominator)

$= \dfrac{x^3}{x^3} \cdot \dfrac{1 - 4 \ln x}{x^5}$

$= \dfrac{1 - 4 \ln x}{x^5}$ (Simplifying)

67. $y = \ln \dfrac{x}{4}$

$y = \ln x - \ln 4$ (P2)

$\dfrac{dy}{dx} = \dfrac{1}{x} - 0$

$= \dfrac{1}{x}$

69. $f(x) = \ln (5x^2 - 7)$

$f'(x) = 10x \cdot \dfrac{1}{5x^2 - 7}$ $\left[\dfrac{d}{dx} \ln g(x) = g'(x) \cdot \dfrac{1}{g(x)}\right]$

$\left[\dfrac{d}{dx} (5x^2 - 7) = 10x\right]$

$= \dfrac{10x}{5x^2 - 7}$

71. $f(x) = \ln (\ln 4x)$

$f'(x) = 4 \cdot \dfrac{1}{4x} \cdot \dfrac{1}{\ln 4x}$ $\left[\dfrac{d}{dx} \ln g(x) = g'(x) \cdot \dfrac{1}{g(x)}\right]$

$\left[\dfrac{d}{dx} \ln 4x = 4 \cdot \dfrac{1}{4x}\right]$

$= \dfrac{1}{x} \cdot \dfrac{1}{\ln 4x}$

$= \dfrac{1}{x \ln 4x}$

73. $f(x) = \ln \left(\dfrac{x^2 - 7}{x}\right)$

$f'(x) = \dfrac{x\cdot 2x - 1\cdot(x^2 - 7)}{x^2} \cdot \dfrac{1}{\frac{x^2 - 7}{x}}$

$\left[\dfrac{d}{dx} \ln g(x) = g'(x) \cdot \dfrac{1}{g(x)}\right]$

$\left[\text{Using Quotient Rule to find } \dfrac{d}{dx}\dfrac{x^2 - 7}{x}\right]$

$= \dfrac{2x^2 - x^2 + 7}{x^2} \cdot \dfrac{x}{x^2 - 7}$ $\left[\dfrac{1}{\frac{x^2 - 7}{x}} = \dfrac{x}{x^2 - 7}\right]$

$= \dfrac{x^2 + 7}{x^2} \cdot \dfrac{x}{x^2 - 7}$

$= \dfrac{x}{x} \cdot \dfrac{x^2 + 7}{x(x^2 - 7)}$

$= \dfrac{x^2 + 7}{x(x^2 - 7)}$

We could also have done this another way using P2:

$$f(x) = \ln \left[\frac{x^2 - 7}{x}\right]$$

$$= \ln (x^2 - 7) - \ln x$$

$$f'(x) = 2x \cdot \frac{1}{x^2 - 7} - \frac{1}{x}$$

$$= \frac{2x}{x^2 - 7} \cdot \frac{x}{x} - \frac{1}{x} \cdot \frac{x^2 - 7}{x^2 - 7} \quad \text{(Multiplying by forms of 1)}$$

$$= \frac{2x^2}{x(x^2 - 7)} - \frac{x^2 - 7}{x(x^2 - 7)}$$

$$= \frac{2x^2 - x^2 + 7}{x(x^2 - 7)}$$

$$= \frac{x^2 + 7}{x(x^2 - 7)}$$

75. $f(x) = e^x \ln x$

$$f'(x) = e^x \cdot \frac{1}{x} + e^x \cdot \ln x \quad \text{(Using the Product Rule)}$$

$$= e^x\left[\frac{1}{x} + \ln x\right]$$

77. $f(x) = \ln (e^x + 1)$

$$f'(x) = e^x \cdot \frac{1}{e^x + 1} \quad \left[\frac{d}{dx} (e^x + 1) = e^x\right]$$

$$= \frac{e^x}{e^x + 1}$$

79. $f(x) = (\ln x)^2$

$$f'(x) = 2(\ln x)^1 \cdot \frac{1}{x} \quad \text{(Extended Power Rule)}$$

$$= \frac{2 \ln x}{x}$$

81. a) $N(a) = 1000 + 200 \ln a,\ a \geqslant 1$

$$N(1) = 1000 + 200 \ln 1 \quad \text{(Substituting 1 for a)}$$

$$= 1000 + 200 \cdot 0$$

$$= 1000 + 0$$

$$= 1000$$

Thus 1000 units were sold after spending \$1000 on advertising.

b) $N(a) = 1000 + 200 \ln a,\ a \geqslant 1$

$$N'(a) = 0 + 200 \cdot \frac{1}{a}$$

$$= \frac{200}{a}$$

$$N'(10) = \frac{200}{10} = 20$$

c) $N'(a) > 0$ for all $a \geqslant 1$. Thus $N(a)$ is an increasing function and has a minimum value of 1000 when $a = 1$. There is no maximum.

83. a) Find R(t).

$$R(t) = \begin{bmatrix}\text{Price}\\ \text{per}\\ \text{unit}\end{bmatrix} \cdot \begin{bmatrix}\text{Target}\\ \text{market}\end{bmatrix} \cdot \begin{bmatrix}\text{Percentage}\\ \text{buying}\end{bmatrix}$$

$$R(t) = 0.5(1{,}000{,}000)(1 - e^{-0.04t})$$

$$= 500{,}000 - 500{,}000e^{-0.04t}$$

b) Find C(t).

$$C(t) = \begin{bmatrix}\text{Advertising costs}\\ \text{per day}\end{bmatrix} \cdot \begin{bmatrix}\text{Number of}\\ \text{days}\end{bmatrix}$$

$$C(t) = 2000t$$

c) Find P(t) and take its derivative.

$$P(t) = R(t) - C(t)$$

$$P(t) = 500{,}000 - 500{,}000e^{-0.04t} - 2000t$$

$$P'(t) = 20{,}000e^{-0.04t} - 2000$$

d) Set the first derivative equal to 0 and solve.

$$20{,}000e^{-0.04t} - 2000 = 0$$

$$20{,}000e^{-0.04t} = 2000$$

$$e^{-0.04t} = \frac{1}{10} = 0.1$$

$$\ln e^{-0.04t} = \ln 0.1$$

$$-0.04t = -2.302585$$

$$t = \frac{-2.302585}{-0.04}$$

$$t \approx 58$$

e) We have only one critical point. So we can use the second derivative to determine whether we have a maximum.

$$P''(t) = -800e^{-0.04t}$$

Now since exponential functions are positive, $e^{-0.04t} > 0$ for all numbers t. Thus, since $-800e^{-0.04t} < 0$ for all numbers t, $P''(t)$ is less than 0 for $t = 58$ and we have a maximum.

The length of the advertising campaign must be 58 days to result in maximum profit.

85. $V(t) = \$58(1 - e^{-1.1t}) + \$20$

(V is the value of the stock after time t, in months.)

a) $V(1) = 58\left[1 - e^{-1.1(1)}\right] + 20$

$$= 58(1 - 0.332871) + 20$$

$$= 58(0.667129) + 20$$

$$= 38.693482 + 20$$

$$= 58.693482$$

$$\approx \$58.69$$

$$V(12) = 58\left[1 - e^{-1.1(12)}\right] + 20$$

$$= 58(1 - e^{-13.2}) + 20$$

$$= 58(1 - 0.000002) + 20$$

$$= 58(0.999998) + 20$$

$$= 57.999884 + 20$$

$$= 77.999884$$

$$\approx \$78.00$$

b) $V'(t) = 58(1.1e^{-1.1t}) + 0$

$$= \$63.80e^{-1.1t}$$

c) $V(t) = 58(1 - e^{-1.1t}) + 20$

$75 = 58(1 - e^{-1.1t}) + 20$ [Substituting 75 for V(t)]

$55 = 58(1 - e^{-1.1t})$

$\frac{55}{58} = 1 - e^{-1.1t}$

$0.948276 = 1 - e^{-1.1t}$

$-0.051724 = -e^{-1.1t}$

$0.051724 = e^{-1.1t}$

$\ln 0.051724 = \ln e^{-1.1t}$ (Taking the natural log on both sides)

$\ln 0.051724 = -1.1t$

$\frac{\ln 0.051724}{-1.1} = t$

$2.7 \text{ months} \approx t$

87. a) $P(t) = 1 - e^{-0.2t}$

$P(1) = 1 - e^{-0.2 \cdot 1}$ (Substituting 1 for t)

$= 1 - e^{-0.2}$

$= 1 - 0.818731$ (Using a calculator)

$= 0.181269$

$\approx 18.1\%$

$P(6) = 1 - e^{-0.2 \cdot 6}$ (Substituting 6 for t)

$= 1 - e^{-1.2}$

$= 1 - 0.301194$ (Using a calculator)

$= 0.698806$

$\approx 69.9\%$

b) $P(t) = 1 - e^{-0.2t}$

$P'(t) = 0 - (-0.2)e^{-0.2t}$

$= 0.2e^{-0.2t}$

c) $P(t) = 1 - e^{-0.2t}$

$0.90 = 1 - e^{-0.2t}$ [Replacing P(t) by 0.90]

$-0.1 = -e^{-0.2t}$ (Adding -1)

$0.1 = e^{-0.2t}$ (Multiplying by -1)

$\ln 0.1 = \ln e^{-0.2t}$ (Taking the natural logarithm on both sides)

$\ln 0.1 = -0.2t$

$\frac{\ln 0.1}{-0.2} = t$

$\frac{-2.302585}{-0.2} = t$ (Using a calculator)

$11.5 \approx t$

Thus it will take approximately 11.5 months for 90% of the doctors to become aware of the new medicine.

89. a) $S(t) = 68 - 20 \ln (t + 1),\ t \geqslant 0$

$S(0) = 68 - 20 \ln (0 + 1)$ (Substituting 0 for t)

$= 68 - 20 \ln 1$

$= 68 - 20 \cdot 0$

$= 68 - 0$

$= 68$

Thus the average score when they initially took the test was 68%.

b) $S(4) = 68 - 20 \ln (4 + 1)$ (Substituting 4 for t)

$= 68 - 20 \ln 5$

$= 68 - 20(1.609438)$ (Using a calculator)

$= 68 - 32.18876$

$\approx 36\%$

c) $S(24) = 68 - 20 \ln (24 + 1)$ (Substituting 24 for t)

$= 68 - 20 \ln 25$

$= 68 - 20(3.218876)$ (Using a calculator)

$= 68 - 64.37752$

$\approx 3.6\%$

d) First we reword the question:

3.6 (the average score after 24 months) is what percent of 68 (the average score when t = 0).

Then we translate and solve:

$3.6 = x \cdot 68$

$\frac{3.6}{68} = x$

$0.052941 = x$

$5\% \approx x$

e) $S(t) = 68 - 20 \ln (t + 1),\ t \geqslant 0$

$S'(t) = 0 - 20 \cdot 1 \cdot \frac{1}{t + 1}$

$= -\frac{20}{t + 1}$

f) $S'(t) < 0$ for all $t \geqslant 0$. Thus S(t) is a decreasing function and has a maximum value of 68% when t = 0.

91. $v(p) = 0.37 \ln p + 0.05$ (p in thousands, v in ft per sec)

a) $v(531) = 0.37 \ln 531 + 0.05$ (Substituting 531 for p)

$= 0.37(6.274762) + 0.05$

$= 2.321662 + 0.05$

$= 2.371662$

≈ 2.37 ft/sec

b) $v(7900) = 0.37 \ln 7900 + 0.05$
$= 0.37(8.974618) + 0.05$
$= 3.320609 + 0.05$
$= 3.370609$
≈ 3.37 ft/sec

c) $v'(p) = 0.37 \cdot \frac{1}{p} + 0$
$= \frac{0.37}{p}$

$v'(p)$ is the acceleration of the walker.

93. $y = (\ln x)^{-4}$

$\frac{dy}{dx} = -4(\ln x)^{-5} \cdot \frac{1}{x}$

$= \frac{-4(\ln x)^{-5}}{x}$

95. $f(t) = \ln (t^3 + 1)^5$

$f'(t) = 5(t^3 + 1)^4 \cdot 3t^2 \cdot \frac{1}{(t^3 + 1)^5}$

$= \frac{15t^2}{(t^3 + 1)}$ (Simplifying)

97. $f(x) = [\ln (x + 5)]^4$

$f'(x) = 4[\ln (x + 5)]^3 \cdot 1 \cdot \frac{1}{x + 5}$

$= \frac{4[\ln (x + 5)]^3}{x + 5}$

99. $f(t) = \ln [(t^3 + 3)(t^2 - 1)]$

$f'(t) = [(t^3+3)(2t) + (3t^2)(t^2-1)] \cdot \frac{1}{(t^3+3)(t^2-1)}$

$= \frac{2t^4 + 6t + 3t^4 - 3t^2}{(t^3 + 3)(t^2 - 1)}$

$= \frac{5t^4 - 3t^2 + 6t}{(t^3 + 3)(t^2 - 1)}$

101. $y = \ln \frac{x^5}{(8x + 5)^2}$

$y = \ln x^5 - \ln (8x + 5)^2$ (Property 2)

$\frac{dy}{dx} = 5x^4 \cdot \frac{1}{x^5} - 2(8x + 5)\cdot 8 \cdot \frac{1}{(8x + 5)^2}$

$= \frac{5}{x} - \frac{16}{8x + 5}$

$= \frac{5}{x} \cdot \frac{8x + 5}{8x + 5} - \frac{16}{8x + 5} \cdot \frac{x}{x}$ (Multiplying by forms of 1)

$= \frac{40x + 25 - 16x}{x(8x + 5)}$

$= \frac{24x + 25}{8x^2 + 5x}$

103. $f(t) = \frac{\ln t^2}{t^2} = t^{-2} \ln t^2$

$f'(t) = t^{-2} \cdot 2t \cdot \frac{1}{t^2} + (-2)t^{-3} \ln t^2$

$= \frac{2t}{t^4} - \frac{2 \ln t^2}{t^3}$

$= \frac{2}{t^3} - \frac{2 \ln t^2}{t^3}$

$= \frac{2 - 2 \ln t^2}{t^3}$, or $\frac{2(1 - \ln t^2)}{t^3}$

105. $y = \frac{x^{n+1}}{n + 1} \left[\ln x - \frac{1}{n + 1}\right]$

$\frac{dy}{dx} = \frac{x^{n+1}}{n + 1} \left[\frac{1}{x} - 0\right] + x^n\left[\ln x - \frac{1}{n + 1}\right]$

$\left[\frac{d}{dx} \frac{x^{n+1}}{n + 1} = \frac{1}{n + 1} \cdot (n + 1)x^n = x^n\right]$

$= \frac{x^n}{n + 1} + x^n \ln x - \frac{x^n}{n + 1}$

$= x^n \ln x$

107. $y = \ln \left[t + \sqrt{1 + t^2}\right] = \ln [t + (1 + t^2)^{1/2}]$

$\frac{dy}{dx} = \left[1 + \frac{1}{2}(1 + t^2)^{-1/2} \cdot 2t\right] \cdot \frac{1}{t + (1 + t^2)^{1/2}}$

$= \left[1 + \frac{t}{(1 + t^2)^{1/2}}\right] \cdot \frac{1}{t + (1 + t^2)^{1/2}}$

$= \left[\frac{(1 + t^2)^{1/2}}{(1 + t^2)^{1/2}} + \frac{t}{(1 + t^2)^{1/2}}\right] \cdot \frac{1}{t+(1 + t^2)^{1/2}}$

$= \frac{t + (1 + t^2)^{1/2}}{(1 + t^2)^{1/2}} \cdot \frac{1}{t + (1 + t^2)^{1/2}}$

$= \frac{1}{(1 + t^2)^{1/2}}$

$= \frac{1}{\sqrt{1 + t^2}}$

109. $f(x) = \ln [\ln x]^3$

$f'(x) = 3[\ln x]^2 \cdot \frac{1}{x} \cdot \frac{1}{[\ln x]^3}$

$= \frac{3}{x \ln x}$

111. $\lim_{h \to 0} \frac{\ln (1 + h)}{h}$

$\ln (1 + h)$ exists only for $1 + h > 0$, or $h > -1$.

The function $\frac{\ln (1 + h)}{h}$ is not continuous at $h = 0$. We use input-output tables.

h	f(h)
-0.9	2.56
-0.7	1.72
-0.5	1.39
-0.3	1.19
-0.1	1.05
-0.01	1.01
-0.001	1.001

(h approaches 0 from the left)

h	f(h)
0.9	0.71
0.7	0.76
0.5	0.81
0.3	0.87
0.1	0.95
0.01	0.995
0.001	0.9995

(h approaches 0 from the right)

From the tables we observe that

$\lim_{h \to 0^-} \frac{\ln(1+h)}{h} = 1$

and

$\lim_{h \to 0^+} \frac{\ln(1+h)}{h} = 1$

Thus, $\lim_{h \to 0} \frac{\ln(1+h)}{h} = 1$.

113. $\sqrt[e]{e} = e^{1/e} \approx 1.444667861$

$e \approx 2.71828183$

For comparison select values of x (x > 0) less than e and greater than e and compute $\sqrt[x]{x}$.

For

$x = 0.5,\quad \sqrt[0.5]{0.5} = 0.25$

$x = 0.8,\quad \sqrt[0.8]{0.8} \approx 0.756593$

$x = 1.5,\quad \sqrt[1.5]{1.5} \approx 1.310371$

$x = 2.0,\quad \sqrt[2]{2} \approx 1.414214$

$x = 2.5,\quad \sqrt[2.5]{2.5} \approx 1.442700$

$x = 2.7,\quad \sqrt[2.7]{2.7} \approx 1.444656$

$x = 2.71,\quad \sqrt[2.71]{2.71} \approx 1.44466538$

$x = 2.72,\quad \sqrt[2.72]{2.72} \approx 1.44466775$

$x = 3.0,\quad \sqrt[3]{3} \approx 1.442250$

$x = 3.5,\quad \sqrt[3.5]{3.5} \approx 1.430369$

For any x > 0 such that $x \neq e$, $\sqrt[e]{e} > \sqrt[x]{x}$.

115. $P = P_0\, e^{kt}$

$\frac{P}{P_0} = e^{kt}$

$\ln \frac{P}{P_0} = \ln e^{kt}$

$\ln P - \ln P_0 = kt$

$\frac{\ln P - \ln P_0}{k} = t$

117. Find $\lim_{x \to \infty} \ln x$.

x	ln x
1	0
10	2.3
100	4.6
1000	6.9
10,000	9.2
100,000	11.5
1,000,000	13.8

(ln x exists only for x > 0)

$\lim_{x \to \infty} \ln x = \infty$

119. Let $a = \ln x$; then $e^a = x$.

$\log x = \log e^a$

$\log x = a \log e$ (P3)

$\frac{\log x}{\log e} = a$ $\left[\text{Multiplying by } \frac{1}{\log e}\right]$

$\frac{\log x}{\log e} = \ln x$ (Substituting ln x for a)

$\frac{\log x}{0.4343} \approx \ln x$

$2.3026 \log x \approx \ln x$ $\left[\frac{1}{0.4343} \approx 2.3026\right]$

Thus, $\ln x = \frac{\log x}{\log e} \approx 2.3026 \log x$.

121. $f(x) = x^2 \ln x,\ x > 0$

$f'(x) = x^2 \cdot \frac{1}{x} + 2x \cdot \ln x$

$= x + 2x \ln x$

Solve $f'(x) = 0$.

$x + 2x \ln x = 0$

$x(1 + 2 \ln x) = 0$

$x = 0$ or $1 + 2 \ln x = 0$

$2 \ln x = -1$

$\ln x = -\frac{1}{2}$

$x = e^{-1/2}$

Only $e^{-1/2}$ is in the interval $(0,\infty)$.

We use Max-Min Principle 2.

$f''(x) = 1 + 2x \cdot \frac{1}{x} + 2 \ln x$

$= 1 + 2 + 2 \ln x$

$= 3 + 2 \ln x$

Since $f''(x) > 0$ for all x in $(0,\infty)$, there is a minimum value at $x = e^{-1/2}$.

Find the function value at $x = e^{-1/2}$.

$f(x) = x^2 \ln x$

$f(e^{-1/2}) = (e^{-1/2})^2 \ln e^{-1/2}$

$= e^{-1} \cdot \left(-\frac{1}{2}\right)$

$= -\frac{1}{2e}$

Exercise Set 4.3

1. The solution of $\frac{dQ}{dt} = kQ$ is $Q(t) = ce^{kt}$, where t is the time. At t = 0, we have some "initial" population Q(0) that we will represent by Q_0. Thus $Q_0 = Q(0) = ce^{k\cdot 0} = ce^0 = c\cdot 1 = c$.

Thus, $Q_0 = c$, so we can express Q(t) as $Q(t) = Q_0\, e^{kt}$.

3. The balance P grows at the rate given by

$\frac{dP}{dt} = 0.09P.$

a) $P(t) = P_0\, e^{0.09t}$ (Substituting 0.09 for k)

b) $P(1) = 1000\, e^{0.09 \cdot 1}$ (Substituting 1000 for P_0 and 1 for t)

$= 1000\, e^{0.09}$

$= 1000(1.094174)$ (Using a calculator)

≈ 1094.17

The balance after 1 year is $1094.17.

$P(2) = 1000\, e^{0.09 \cdot 2}$ (Substituting 1000 for P_0 and 2 for t)

$= 1000\, e^{0.18}$

$= 1000(1.197217)$ (Using a calculator)

≈ 1197.22

The balance after 2 years is $1197.22.

c) $T = \frac{\ell n\ 2}{k}$

$= \frac{0.693147}{0.09}$ (Substituting 0.693147 for ℓn 2 and 0.09 for k)

≈ 7.7

The investment of $1000 will double itself in 7.7 years.

5. $k = \frac{\ell n\ 2}{T}$

$= \frac{0.693147}{10}$ (Substituting 0.693147 for ℓn 2 and 10 for T)

$= 0.0693147$

$\approx 6.9\%$

The annual interest rate is 6.9%.

7. It is estimated that the number of franchises N will increase at a rate of 10% per year. That is

$\frac{dN}{dt} = 0.10N.$

a) $N(t) = N_0\, e^{kt}$

$N(t) = 50\, e^{0.10t}$ (Substituting 50 for N_0 and 0.10 for k)

b) $N(20) = 50\, e^{0.10 \cdot 20}$ (Substituting 20 for t)

$= 50\, e^2$

$= 50(7.389056)$ (Using a calculator)

≈ 369

Thus in 20 years there will be 369 franchises.

c) $T = \frac{\ell n\ 2}{k}$

$= \frac{0.693147}{0.10}$ (Substituting 0.693147 for ℓn 2 and 0.10 for k)

≈ 6.9

The initial number of franchises will double in 6.9 years.

9. $T = \frac{\ell n\ 2}{k}$

$= \frac{0.693147}{0.10}$ (Substituting 0.693147 for ℓn 2 and 0.10 for k)

≈ 6.9

The demand for oil in the U.S. will be double that of 1990 6.9 years after 1990.

11. a) The exponential growth function is $V(t) = V_0\, e^{kt}$. We will express V(t) in thousands of dollars and t as the number of years after 1947. Since $V_0 = 84$ thousand, we have

$V(t) = 84\, e^{kt}.$

In 1987 (t = 40), we know that V(t) = 53,900 thousand. We substitute and solve for k.

$53{,}900 = 84\, e^{k(40)}$

$\frac{1925}{3} = e^{40k}$

$\ell n\ \frac{1925}{3} = \ell n\ e^{40k}$

$\ell n\ \frac{1925}{3} = 40k$

$\frac{\ell n\ \frac{1925}{3}}{40} = k$

$\frac{6.4641}{40} \approx k$

$0.161602 \approx k$

The exponential growth rate is about 0.161602, or 16.1602%. The exponential growth function is $V(t) = 84\, e^{0.161602t}$, where V is in thousands of dollars and t is the number of years after 1947.

b) In 1997, t = 50. We find V(50).

$V(50) = 84\, e^{0.161602(50)} = 84\, e^{8.0801}$

84(3229.5562) ≈ $271,282 thousand, or $271,282,000

(Answers may vary slightly due to rounding.)

c) We will use the expression relating growth rate k and doubling time T. (We could also set V(t) = 168 and solve for t.)

$T = \frac{\ell n\ 2}{k} \approx \frac{0.6931}{0.161602} \approx 4.3$ years

d) $1 billion = $1,000,000 thousand. We set V(t) = 1,000,000 and solve for t.

$1{,}000{,}000 = 84\, e^{0.161602t}$

$\frac{250{,}000}{21} = e^{0.161602t}$

$\ell n\ \frac{250{,}000}{21} = \ell n\ e^{0.161602t}$

$\ell n\ \frac{250{,}000}{21} = 0.161602t$

$\frac{\ell n\ \frac{250{,}000}{21}}{0.161602} = t$

$\frac{9.3847}{0.161602} \approx t$

$58 \approx t$

The value of the painting will be \$1 billion about 58 years after 1947.

13. a) $P(t) = P_0 e^{kt}$

$184.50 = 100 e^{k\cdot 10}$ (Substituting 100 for P_0, 184.50 for $P(t)$ and 10 for t, 1977 - 1967 = 10)

$\frac{184.50}{100} = e^{10k}$

$1.845 = e^{10k}$

$\ln 1.845 = \ln e^{10k}$

$\ln 1.845 = 10k$

$\frac{\ln 1.845}{10} = k$

$\frac{0.612479}{10} = k$ (Using a calculator)

$0.0612479 = k$

$0.061248 \approx k$

Thus, $P(t) = \$100\, e^{0.061248t}$.

b) $P(t) = 100e^{0.061248t}$

$P(30) = 100e^{0.061248(30)}$ (Substituting 30 for t, 1997 - 1967 = 30)

$= 100\, e^{1.83744}$

$= 100(6.280440)$ (Using a calculator)

≈ 628.04

Thus, in 1997 the same goods and services will cost \$628.04.

c) $T = \frac{\ln 2}{k}$

$= \frac{0.693147}{0.061248}$ (Substituting 0.693147 for $\ln 2$ and 0.061248 for k)

≈ 11.3

Thus, in 11.3 years the same goods and services will cost double that of 1967.

15. $V(t) = V_0 e^{kt}$

$V(t) = 24\, e^{0.08(372)}$ (Substituting 24 for P_0, 0.08 for k, and 372 for t, 1998 - 1626 = 372)

$= 24\, e^{29.76}$

$= 24(8.406279 \times 10^{12})$ (Using a calculator)

$= 2.017507 \times 10^{14}$

$= 202{,}000{,}000{,}000{,}000$

Manhattan Island will be worth \$202,000,000,000,000 in 1998.

17. Find k, the growth rate.

$S(t) = S_0 e^{kt}$

$363{,}000 = 29{,}303\, e^{k\cdot 15}$ (Substituting 29,303 for S_0, 363,000 for S(t), and 15 for t, 1985 - 1970 = 15)

$\frac{363{,}000}{29{,}303} = e^{15k}$

$\ln \frac{363{,}000}{29{,}303} = \ln e^{15k}$

$2.516713 = 15k$

$\frac{2.516713}{15} = k$

$0.16778 \approx k$

The growth rate was 16.778%.

Find the average salary in 1997.

$S(t) = 29{,}303\, e^{0.16778t}$

$S(27) = 29{,}303\, e^{0.16778(27)}$ (Substituting 27 for t, 1997 - 1970 = 27)

$= 29{,}303\, e^{4.53006}$

$= 29{,}303(92.764127)$ (Using a calculator)

$\approx 2{,}718{,}267$

The average salary will be \$2,718,267 in 1997.

Find the average salary in 2000.

$S(t) = 29{,}303\, e^{0.16778t}$

$S(30) = 29{,}303\, e^{0.16778(30)}$ (Substituting 30 for t, 2000 - 1970 = 30)

$= 29{,}303\, e^{5.0334}$

$\approx 29{,}303(153.453870)$

$\approx 4{,}496{,}659$

The average salary will be \$4,496,659 in 2000.

19. $P(t) = \frac{100\%}{1 + 49\, e^{-0.13t}}$

a) $P(0) = \frac{100\%}{1 + 49\, e^{-0.13(0)}}$

$= \frac{100\%}{1 + 49\, e^{0}}$

$= \frac{100\%}{1 + 49\cdot 1}$

$= \frac{100\%}{50}$

$= 2\%$

b) $P(5) = \frac{100\%}{1 + 49\, e^{-0.13(5)}}$

$= \frac{100\%}{1 + 49\, e^{-0.65}}$

$\approx 3.8\%$

$$P(10) = \frac{100\%}{1 + 49\ e^{-0.13(10)}}$$
$$= \frac{100\%}{1 + 49\ e^{-1.3}}$$
$$\approx 7.0\%$$

$$P(20) = \frac{100\%}{1 + 49\ e^{-0.13(20)}}$$
$$= \frac{100\%}{1 + 49\ e^{-2.6}}$$
$$\approx 21.6\%$$

$$P(30) = \frac{100\%}{1 + 49\ e^{-0.13(30)}}$$
$$= \frac{100\%}{1 + 49\ e^{-3.9}}$$
$$\approx 50.2\%$$

$$P(50) = \frac{100\%}{1 + 49\ e^{-0.13(50)}}$$
$$= \frac{100\%}{1 + 49\ e^{-6.5}}$$
$$\approx 93.1\%$$

$$P(60) = \frac{100\%}{1 + 49\ e^{-0.13(60)}}$$
$$= \frac{100\%}{1 + 49\ e^{-7.8}}$$
$$\approx 98.0\%$$

c) $P'(t) =$

$$\frac{(1 + 49\ e^{-0.13t})(0) - 49(-0.13)\ e^{-0.13t}\cdot 100\%}{(1 + 49\ e^{-0.13t})^2}$$
$$= \frac{637\%\cdot e^{-0.13t}}{(1 + 49\ e^{-0.13t})^2}, \text{ or } \frac{6.37\ e^{-0.13t}}{(1 + 49\ e^{-0.13t})^2}$$

d) The derivative $P'(t)$ exists for all real numbers. The equation $P'(t) = 0$ has no solution. Thus, the function has no critical points and hence no relative extrema. $P'(t) > 0$ for all real numbers, so $P(t)$ is increasing on $[0,\infty)$. The second derivative can be used to show that the graph has an inflection point at (29.9, 50%), or (29.9, 0.5). The function is concave up on (0, 29.9) and concave down on (29,9, ∞).

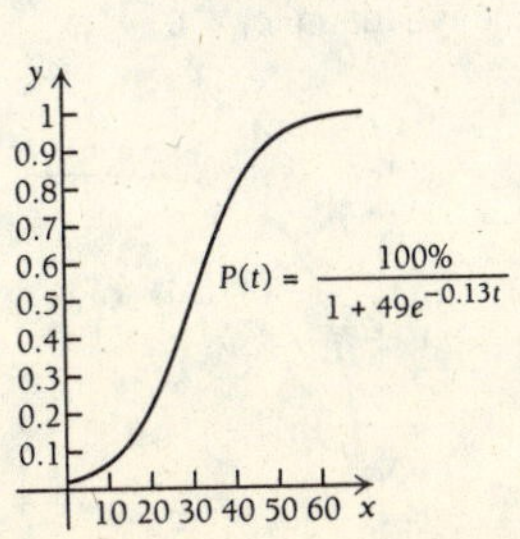

21. $T = \frac{\ell n\ 2}{k}$

$= \frac{0.693147}{0.035}$ (Substituting 0.693147 for ℓn 2 and 0.035 for k, 3.5% = 0.035)

≈ 19.8

The population of Central America will double in 19.8 years.

23. It was estimated that the population P was growing at a rate of 1% per year. That is

$$\frac{dP}{dt} = 0.01P.$$

a) $P(t) = P_0\ e^{kt}$

$P(t) = 209\ e^{0.01t}$ (Substituting 209 for P_0 and 0.01 for k)

b) $P(40) = 209\ e^{0.01\ 40}$ (Substituting 40 for t, 1999 - 1959 = 40)

$= 209\ e^{0.4}$

$= 209(1.491825)$ (Using a calculator)

≈ 312

Thus, in 1999 the population of Russia will be approximately 312 million.

c) $T = \frac{\ell n\ 2}{k}$

$= \frac{0.693147}{0.01}$ (Substituting 0.693147 for ℓn 2 and 0.01 for k)

≈ 69.3

The population will double that of 1959 in approximately 69.3 years.

25. $R(b) = e^{21.4b}$ (See Example 7)

$80 = e^{21.4b}$ [Substituting 80 for R(b)]

$\ell n\ 80 = \ln e^{21.4b}$

$\ell n\ 80 = 21.4b$

$\frac{\ell n\ 80}{21.4} = b$

$\frac{4.382027}{21.4} = b$ (Using a calculator)

$0.20 \approx b$ (Rounding to the nearest hundredth)

Thus when the blood alcohol level is 0.20%, the risk of having an accident is 80%.

27. $P(t) = P_0\ e^{kt}$

$118 = 107\ e^{k\cdot 4}$ (Substituting 107 for P_0, 118 for P(t), and 4 for t, 1984 - 1980 = 4)

$\frac{118}{107} = e^{4k}$

$\ell n\ \frac{118}{107} = \ell n\ e^{4k}$

$0.097856 = 4k$ (Using a calculator)

$\frac{0.097856}{4} = k$

$0.024464 \approx k$

Thus, $P(t) = 107\ e^{0.024464t}$

b) $P(t) = 107\ e^{0.024464t}$

$P(6) = 107\ e^{0.024464(16)}$ (Substituting 16 for t, 1996 - 1980 = 16)

$= 107\ e^{0.391424}$

$= 107(1.479086)$ (Using a calculator)

≈ 158

Thus in 1996 the population of Tempe, Arizona, will be 158 thousand.

29. $P(t) = \dfrac{2500}{1 + 5.25\ e^{-0.32t}}$

a) $P(0) = \dfrac{2500}{1 + 5.25\ e^{-0.32(0)}}$

$= \dfrac{2500}{1 + 5.25\ e^{0}}$

$= \dfrac{2500}{1 + 5.25(1)}$

$= 400$

$P(1) = \dfrac{2500}{1 + 5.25\ e^{-0.32(1)}}$

$= \dfrac{2500}{1 + 5.25\ e^{-0.32}}$

≈ 520

$P(5) = \dfrac{2500}{1 + 5.25\ e^{-0.32(5)}}$

$= \dfrac{2500}{1 + 5.25\ e^{-1.6}}$

≈ 1214

$P(10) = \dfrac{2500}{1 + 5.25\ e^{-0.32(10)}}$

$= \dfrac{2500}{1 + 5.25\ e^{-3.2}}$

≈ 2059

$P(15) = \dfrac{2500}{1 + 5.25\ e^{-0.32(15)}}$

$= \dfrac{2500}{1 + 5.25\ e^{-4.8}}$

≈ 2396

$P(20) = \dfrac{2500}{1 + 5.25\ e^{-0.32(20)}}$

$= \dfrac{2500}{1 + 5.25\ e^{-6.4}}$

≈ 2478

b) $P'(t) =$

$\dfrac{(1+5.25\ e^{-0.32t})(0)-5.25(-0.32)\ e^{-0.32t}(2500)}{(1 + 5.25\ e^{-0.32t})^2}$

(Quotient Rule)

$= \dfrac{4200\ e^{-0.32t}}{(1 + 5.25\ e^{-0.32t})^2}$

c) The derivative $P'(t)$ exists for all real numbers. The equation $P'(t) = 0$ has no solution. Thus, the function has no critical points and hence no relative extrema. $P'(t) > 0$ for all real numbers, so $P(t)$ is increasing on $[0,\infty)$. The second derivative can be used to show that the graph has an inflection point at (5.18, 1250). The function is concave up on (0, 5.18) and concave down on (5.18, ∞).

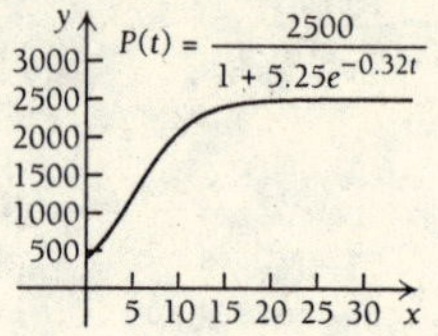

31. $P(t) = 100\%(1 - e^{-0.4t})$

a) $P(0) = 100\%(1 - e^{-0.4(0)}) = 100\%(1 - e^{0}) = 100\%(1 - 1) = 100\%(0) = 0\%$

$P(1) = 100\%(1 - e^{-0.4(1)}) = 100\%(1 - e^{-0.4}) \approx 33.0\%$

$P(2) = 100\%(1 - e^{-0.4(2)}) = 100\%(1 - e^{-0.8}) \approx 55.1\%$

$P(3) = 100\%(1 - e^{-0.4(3)}) = 100\%(1 - e^{-1.2}) \approx 69.9\%$

$P(5) = 100\%(1 - e^{-0.4(5)}) = 100\%(1 - e^{-2}) \approx 86.5\%$

$P(12) = 100\%(1 - e^{-0.4(12)}) = 100\%(1 - e^{-4.8}) \approx 99.2\%$

$P(16) = 100\%(1 - e^{-0.4(16)}) = 100\%(1 - e^{-6.4}) \approx 99.8\%$

b) $P'(t) = 100\%[-(-0.4)\ e^{-0.4t}]$

$= 100\%(0.4)\ e^{-0.4t}$

$= 0.4\ e^{-0.4t}$ (100% = 1)

c) The derivative $P'(t)$ exists for all real numbers. The equation $P'(t) = 0$ has no solution. Thus, the function has no critical points and hence no relative extrema. $P'(t) > 0$ for all real numbers, so $P(t)$ is increasing on $[0,\infty)$. $P''(t) = -0.16\ e^{-0.4t}$, so $P''(t) < 0$ for all real numbers and hence is concave down on $[0,\infty)$.

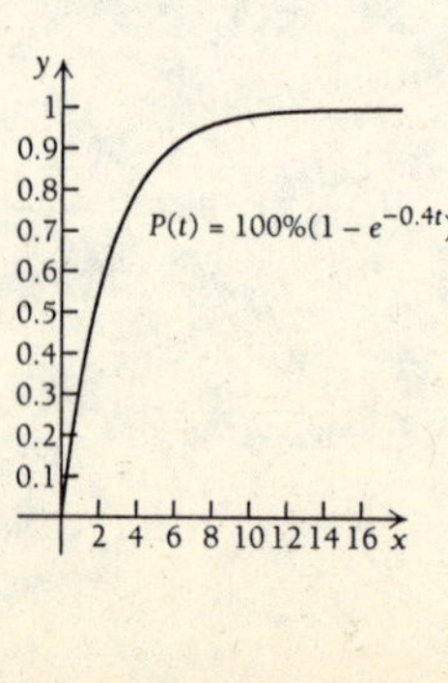

33. Effective annual yield $= i = e^k - 1$

$= e^{0.14} - 1$ (Substituting 0.14 for k)

$= 1.150274 - 1$

$= 0.150274$

$\approx 15.03\%$

35. Effective annual yield $= i = e^k - 1$

$9.42\% = e^k - 1$ (Substituting 9.42% for i)

$0.0942 = e^k - 1$

$1.0942 = e^k$

$\ell n\ 1.0942 = \ell n\ e^k$

$0.090024 = k$

$9.0\% \approx k$

The investment was invested at 9%.

37. We can find a general expression relating the growth rate k and the tripling time T_3 by solving the following equation.

$3\,P_0 = P_0\,e^{kT_3}$

$3 = e^{kT_3}$

$\ell n\ 3 = \ell n\ e^{kT_3}$

$\ell n\ 3 = kT_3$

$\frac{\ell n\ 3}{k} = T_3$

39. Answers depend on particular data.

41. If an amount grows at a rate of 100% per day, this means it doubles each day. The doubling time T is 1 day, or 24 hours.

The growth rate k and the doubling time T are related by

$k = \frac{\ln 2}{T}$

$= \frac{\ell n\ 2}{24}$ (Substituting for T)

≈ 0.028881

A growth rate of 100% per day corresponds to a growth rate of approximately 2.9% per hour.

Exercise Set 4.4

1. $P_0 = Pe^{-kt}$

$P_0 = \$5000\ e^{-0.09(20)}$ (Substituting)

$= \$5000\ e^{-1.8}$

$\approx \$5000(0.165299)$

$\approx \$826.49$

3. $P_0 = Pe^{-kt}$

$P_0 = \$60{,}000\ e^{-0.12(8)}$ (Substituting)

$= \$60{,}000\ e^{-0.96}$

$\approx \$60{,}000(0.382893)$

$\approx \$22{,}973.57$

5. $V(t) = \$40{,}000\ e^{-t}$

a) Let t = 0.

$V(0) = 40{,}000\ e^{-0}$

$= 40{,}000 \cdot 1$

$= \$40{,}000$

The initial cost was $40,000.

b) Let t = 2.

$V(2) = 40{,}000\ e^{-2}$

$= 40{,}000(0.135335)$ (Using a calculator)

$\approx \$5413$

The salvage value after 2 years is $5413.

7. $P_0 = Pe^{-kt}$

$= \$80{,}000\ e^{-0.083(13)}$

$= \$80{,}000\ e^{-1.079}$

$\approx \$80{,}000(0.339935)$

$\approx \$27{,}194.82$

9. $T = \frac{\ell n\ 2}{k}$

$= \frac{0.693147}{0.096}$ (Substituting 0.693147 for ℓn 2 and 0.096 for k, 9.6% = 0.096)

≈ 7.2

The half-life of iodine-131 is 7.2 days.

11. $k = \frac{\ell n\ 2}{T}$

$= \frac{0.693147}{3}$ (Substituting 0.693147 for ℓn 2 and 3 for T)

≈ 0.23, or 23%

The decay rate of polonium is 23% per minute.

13. $P(t) = P_0e^{-kt}$

$P(20) = 1000\ e^{-0.23(20)}$ (Substituting 1000 for P_0, 0.23 for k and 20 for t)

$= 1000e^{-4.6}$

$= 1000(0.010052)$ (Using a calculator)

≈ 10.1

Thus 10.1 grams of polonium will remain after 20 minutes.

15. $N(t) = N_0 e^{-0.0001205t}$ [See Example 3 (b), Sec. 4.4]

If a piece of wood has lost 90% of its carbon-14 from an initial amount P_0, then 10% P_0 is the amount present. To find the age of the wood, we solve the following equation for t:

$$10\% P_0 = P_0 e^{-0.0001205t}$$ [Substituting 10% P_0 for P(t)]

$$0.1 = e^{-0.0001205t}$$

$$\ell n\ 0.1 = \ell n\ e^{-0.0001205t}$$

$$\ell n\ 0.1 = -0.0001205t$$

$$-2.302585 = -0.0001205t$$ (Using a calculator)

$$\frac{-2.302585}{-0.0001205} = t$$

$$19{,}109 \approx t$$

Thus, the piece of wood is about 19,109 years old.

17. If an aritfact has lost 60% of its carbon-14 from an initial amount P_0, then 40% P_0 is the amount present. To find the age t we solve the following equation for t.

$$40\% P_0 = P_0 e^{-0.0001205t}$$ [Substituting 40% P_0 for P(t)] [See Example 3 (b), Sec. 4.4]

$$0.4 = e^{-0.0001205t}$$

$$\ln 0.4 = \ln e^{-0.0001205t}$$

$$\ln 0.4 = -0.0001205t$$

$$-0.916291 = -0.0001205t$$ (Using a calculator)

$$\frac{-0.916291}{-0.0001205} = t$$

$$7604 \approx t$$

The artifact is about 7604 years old.

19. a) When A decomposes at a rate proportional to the amount of A present, we know that

$$\frac{dA}{dt} = -kA.$$

The solution of this equation is

$$A = A_0 e^{-kt}.$$

b) We first find k. The half-life of A is 3 hr.

$$k = \frac{\ln 2}{T}$$

$$k = \frac{0.693147}{3}$$ (Substituting 0.693147 for ln 2 and 3 for T)

$$\approx 0.23, \text{ or } 23\%$$

We now substitute 8 for A_0, 1 for A, and 0.23 for k and solve for t.

$$A = A_0 e^{-kt}$$

$$1 = 8 e^{-0.23t}$$ (Substituting)

$$\frac{1}{8} = e^{-0.23t}$$

$$0.125 = e^{-0.23t}$$

$$\ln 0.125 = \ln e^{-0.23t}$$

$$-2.079442 = -0.23t$$ (Using a calculator)

$$\frac{-2.079442}{-0.23} = t$$

$$9 \approx t$$

After 9 hr there will be 1 gram left.

21. a) $W = W_0 e^{-0.008t}$

$k = 0.008$, or 0.8%

The starving animal loses 0.8% of its weight each day.

b) $W = W_0 e^{-0.008t}$

$W = W_0 e^{-0.008(30)}$ (Substituting 30 for t)

$W = W_0 e^{-0.24}$

$W = 0.786628\ W_0$ (Using a calculator)

$W \approx 78.7\%\ W_0$

Thus, after 30 days, 78.7% of the initial weight remains.

23. a) $I = I_0 e^{-\mu x}$

$I = I_0 e^{-1.4(1)}$ (Substituting 1.4 for μ and 1 for x)

$I = I_0 e^{-1.4}$

$I = I_0(0.246597)$ (Using a calculator)

$I \approx 25\%\ I_0$

$I = I_0 e^{-1.4(2)}$ (Substituting 1.4 for μ and 2 for x)

$I = I_0 e^{-2.8}$

$I = I_0(0.060810)$ (Using a calculator)

$I \approx 6.1\%\ I_0$

$I = I_0 e^{-1.4(3)}$ (Substituting 1.4 for μ and 3 for x)

$I = I_0 e^{-4.2}$

$I = I_0(0.014996)$ (Using a calculator)

$I \approx 1.5\%\ I_0$

$I = I_0 e^{-\mu x}$

$I = I_0 e^{-1.4(10)}$ (Substituting 1.4 for μ and 10 for x)

$I = I_0 e^{-14}$

$I = I_0(0.00000083)$ (Using a calculator)

$I \approx 0.00008\%\ I_0$

25. a) $T(t) = ae^{-kt} + C$ (Newton's Law of Cooling)

At $t = 0$, $T = 100^0$. We solve the following equation for a.

$100 = ae^{-k \cdot 0} + 75$ (Substituting 100 for T, 0 for t, and 75 for C)

$25 = ae^0$

$25 = a$ $(e^0 = 1)$

Thus, $T(t) = 25\, e^{-kt} + 75$.

b) Now we find k using the fact that at $t = 10$, $T = 90^0$.

$T(t) = 25\, e^{-kt} + 75$

$90 = 25\, e^{-k \cdot 10} + 75$

(Substituting 90 for T, and 10 for t)

$15 = 25\, e^{-10k}$

$\frac{15}{25} = e^{-10k}$

$0.6 = e^{-10k}$

$\ln 0.6 = \ln e^{-10k}$

$\ln 0.6 = -10k$

$\frac{\ln 0.6}{-10} = k$

$\frac{-0.510826}{-10} = k$

$0.05 \approx k$

Thus, $T(t) = 25\, e^{-0.05t} + 75$.

c) $T(t) = 25\, e^{-0.05t} + 75$

$T(20) = 25\, e^{-0.05(20)} + 75$ (Substituting 20 for t)

$= 25\, e^{-1} + 75$

$= 25(0.367879) + 75$ (Using a calculator)

$= 9.196975 + 75$

≈ 84.2

The temperature after 20 minutes is 84.2^0.

d) $T(t) = 25\, e^{-0.05t} + 75$

$80 = 25\, e^{-0.05t} + 75$ (Substituting 80 for T)

$5 = 25\, e^{-0.05t}$

$\frac{5}{25} = e^{-0.05t}$

$0.2 = e^{-0.05t}$

$\ln 0.2 = \ln e^{-0.05t}$

$\ln 0.2 = -0.05t$

$\frac{\ln 0.2}{-0.05} = t$

$\frac{-1.609438}{-0.05} = t$

$32 \approx t$

It takes 32 minutes for the liquid to cool to 80^0.

27. We first find a in the equation $T(t) = ae^{-kt} + C$.

Assuming the temperature of the body was normal when the murder occurred, we have $T = 98.6^0$ at $t = 0$. Thus

$98.6^0 = ae^{-k \cdot 0} + 60^0$ (Room temperature is 60^0)

so

$a = 38.6^0$.

Thus T is given by $T(t) = 38.6\, e^{-kt} + 60$.

We want to find the number of hours N since the murder was committed. To find N we must first determine k. From the two temperature readings, we have

$85.9 = 38.6\, e^{-kN} + 60$, or $25.9 = 38.6\, e^{-kN}$

$83.4 = 38.6\, e^{-k(N+1)} + 60$, or $23.4 = 38.6\, e^{-k(N+1)}$

Dividing the first equation by the second, we get

$\frac{25.9}{23.4} = \frac{38.6\, e^{-kN}}{38.6e^{-k(N+1)}} = e^{-kN+k(N+1)} = e^k$.

We solve this equation for k:

$\ln \frac{25.9}{23.4} = \ln e^k$

$\ln 1.106838 = k$

$0.10 \approx k$

Now we substitute 0.10 for k in the equation $25.9 = 38.6\, e^{-kN}$ and solve for N.

$25.9 = 38.6\, e^{-0.10N}$

$\frac{25.9}{38.6} = e^{-0.10N}$

$\ln \frac{25.9}{38.6} = \ln e^{-0.10N}$

$\ln 0.670984 = -0.10N$

$\frac{-0.399009}{-0.10} = N$

$4 \text{ hr} \approx N$

The coroner arrived at 11 P.M., so the murder was committed about 7 P.M.

29. a) $P(t) = P_0\, e^{-kt}$

$385{,}000 = 453{,}000\, e^{-k \cdot 10}$

[Substituting 385,000 for P, 453,000 for P_0, and 10 for t (1980 - 1970 = 10)]

$\frac{385{,}000}{453{,}000} = e^{-10k}$

$\frac{385}{453} = e^{-10k}$

$\ln \frac{385}{453} = \ln e^{-10k}$

$\ln 0.849890 = -10k$

$\frac{\ln 0.849890}{-10} = k$

$\frac{-0.162649}{-10} = k$

$0.0162649 \approx k$

Thus, $P(t) = 453{,}000\, e^{-0.0162649t}$, where t = years since 1970.

b) $P(t) = 453{,}000\ e^{-0.0162649t}$

$P(20) = 453{,}000\ e^{-0.0162649(30)}$ (Substituting 30 for t)

$= 453{,}000\ e^{-0.487947}$

$= 453{,}000(0.613885)$ (Using a calculator)

$\approx 278{,}090$

Thus, the population of Cincinnati in 2000 will be approximately 278,090.

c) $P(t) = 453{,}000\ e^{-0.0162649t}$

$1 = 453{,}000\ e^{-0.0162649t}$ (Substituting 1 for P(t))

$\frac{1}{453{,}000} = e^{-0.0162649t}$

$\ln \frac{1}{453{,}000} = \ln e^{-0.0162649t}$

$\ln 1 - \ln 453{,}000 = -0.0162649t$

$-\ln 453{,}000 = -0.0162649t$

$\frac{\ln 453{,}000}{0.0162649} = t$

$\frac{13.023647}{0.0162649} = t$

$801 \approx t$

Thus, the population of Cincinnati will be only 1 person in 801 years.

31. $P(t) = P_0\ e^{kt}$ (Exponential growth equation)

$P(t) = 258\ e^{0.01t}$, where P(t) is in millions and t is the number of years after 1980.

In 1970, $t = 1970 - 1980 = -10$:

$P(-10) = 258\ e^{0.01(-10)}$

$= 258\ e^{-0.1}$

$\approx 258(0.904837)$

≈ 233

In 1970 the population was about 233 million.

In 1940, $t = 1940 - 1980 = -40$:

$P(-40) = 258\ e^{0.01(-40)}$

$= 258\ e^{-0.4}$

$\approx 258(0.670320)$

≈ 173

In 1940 the population was about 173 million.

33. a) $P(t) = 50\ e^{-0.004t}$

$P(375) = 50\ e^{-0.004(375)}$ (Substituting 375 for t)

$= 50\ e^{-1.5}$

$= 50(0.223130)$

≈ 11

After 375 days, 11 watts will be available.

b) $T = \frac{\ln 2}{k}$

$= \frac{0.693147}{0.004}$ (Substituting 0.693147 for ln 2 and 0.004 for k)

≈ 173

The half-life of the power supply is 173 days.

c) $P(t) = 50\ e^{-0.004t}$

$10 = 50\ e^{-0.004t}$ [Substituting 10 for P(t)]

$\frac{10}{50} = e^{-0.004t}$

$0.2 = e^{-0.004t}$

$\ln 0.2 = \ln e^{-0.004t}$

$\ln 0.2 = -0.004t$

$\frac{\ln 0.2}{-0.004} = t$

$\frac{-1.609438}{-0.004} = t$ (Using a calculator)

$402 \approx t$

The satellite can stay in operation 402 days.

d) When $t = 0$,

$P = 50\ e^{-0.004(0)}$ (Substituting 0 for t)

$= 50\ e^0$

$= 50 \cdot 1$

$= 50$

At the beginning, the power output was 50 watts.

35. We solve $D(p) = S(p)$.

$480\ e^{-0.003p} = 150\ e^{0.004p}$

$\frac{480}{150} = \frac{e^{0.004p}}{e^{-0.003p}}$

$3.2 = e^{0.007p}$

$\ell n\ 3.2 = \ell n\ e^{0.007p}$

$\ell n\ 3.2 = 0.007p$

$\frac{\ell n\ 3.2}{0.007} = p$

$166.16 \approx p$

Now find D(p) or S(p) for p = 166.16. We will find S(p).

$S(166.16) = 150\ e^{0.004(166.16)}$

$= 150\ e^{0.66464}$

$\approx 150(1.943791)$

≈ 292

The equilibrium point is ($166.16, 292).

Exercise Set 4.5

1. $5^4 = e^{4 \cdot \ell n\ 5}$ (Theorem 13: $a^x = e^{x \cdot \ell n\ a}$)

$\approx e^{4(1.609438)}$ (Using a calculator)

$\approx e^{6.4378}$

3. $(3.4)^{10} = e^{10\cdot \ell n\, 3.4}$ (Theorem 13: $a^x = e^{x\cdot \ell n\, a}$)

$\approx e^{10(1.223775)}$ (Using a calculator)

$\approx e^{12.238}$

5. $4^k = e^{k\cdot \ell n\, 4}$ (Theorem 13: $a^x = e^{x\cdot \ell n\, a}$)

$\approx e^{k(1.386294)}$ (Using a calculator)

$\approx e^{1.3863k}$

7. $8^{kT} = e^{kT\cdot \ell n\, 8}$ (Theorem 13: $a^x = e^{x\cdot \ell n\, a}$)

$\approx e^{kT(2.079442)}$ (Using a calculator)

$\approx e^{2.0794kT}$

9. $y = 6^x$

$\frac{dy}{dx} = (\ell n\, 6)6^x$ $\left[\text{Theorem 14: } \frac{dy}{dx} a^x = (\ell n\, a)a^x\right]$

11. $f(x) = 10^x$

$f'(x) = (\ell n\, 10)10^x$ (Theorem 14)

13. $f(x) = x(6.2)^x$

$f'(x) = x\left[\frac{d}{dx}(6.2)^x\right] + \left[\frac{d}{dx}x\right](6.2)^x$ (Product Rule)

$= x\cdot(\ell n\, 6.2)(6.2)^x + 1\cdot(6.2)^x$ — $(\ell n\, 6.2)(6.2)^x$: (Theorem 14)

$= (6.2)^x[x\, \ell n\, 6.2 + 1]$

15. $y = x^3 10^x$

$\frac{dy}{dx} = x^3\cdot(\ell n\, 10)10^x + 3x^2\cdot 10^x$ (Product Rule) — $(\ell n\, 10)10^x$: (Theorem 14)

$= 10^x x^2(x\, \ell n\, 10 + 3)$

17. $y = \log_4 x$

$\frac{dy}{dx} = \frac{1}{\ell n\, 4}\cdot\frac{1}{x}$ $\left[\text{Theorem 16: } \frac{d}{dx}\log_a x = \frac{1}{\ell n\, a}\cdot\frac{1}{x}\right]$

19. $f(x) = 2\log x$

$f'(x) = 2\cdot\frac{d}{dx}\log x$

$= 2\cdot\frac{1}{\ell n\, 10}\cdot\frac{1}{x}$ ($\log x = \log_{10}x$) (Theorem 16)

$= \frac{2}{\ell n\, 10}\cdot\frac{1}{x}$

21. $f(x) = \log\frac{x}{3}$

$f(x) = \log x - \log 3$ (P2)

$f'(x) = \frac{1}{\ell n\, 10}\cdot\frac{1}{x} - 0$ ($\log x = \log_{10}x$) (Theorem 16)

$= \frac{1}{\ell n\, 10}\cdot\frac{1}{x}$

23. $y = x^3\log_8 x$

$\frac{dy}{dx} = x^3\left[\frac{1}{\ell n\, 8}\cdot\frac{1}{x}\right] + 3x^2\cdot\log_8 x$ (Product Rule) — $\left[\frac{1}{\ell n\, 8}\cdot\frac{1}{x}\right]$: (Theorem 16)

$= x^2\left[\frac{1}{\ell n\, 8} + 3\log_8 x\right]$

25. $N(t) = 250{,}000\left(\frac{1}{4}\right)^t$

$N'(t) = 250{,}000\,\frac{d}{dx}\left(\frac{1}{4}\right)^t$

$= 250{,}000\cdot\left[\ell n\,\frac{1}{4}\right]\left(\frac{1}{4}\right)^t$ (Theorem 14)

$= 250{,}000(\ln 1 - \ell n\, 4)\left(\frac{1}{4}\right)^t$ (P2)

$= -250{,}000(\ln 4)\left(\frac{1}{4}\right)^t$ ($\ell n\, 1 = 0$)

27. $R = \log\frac{I}{I_0}$

$R = \log\frac{10^5\cdot I_0}{I_0}$ (Substituting $10^5\cdot I_0$ for I)

$= \log 10^5$

$= 5$ (P5)

The magnitude on the Richter scale is 5.

29. a) $I = I_0\, 10^R$

$I = I_0\, 10^7$ (Substituting 7 for R)

$= 10^7\, I_0$

b) $I = I_0\, 10^R$

$I = I_0\, 10^8$ (Substituting 8 for R)

$= 10^8\, I_0$

c) The intensity in (b), $10^8\, I_0$, is 10 times that in (a).

$10^8\, I_0 = 10\cdot 10^7\, I_0$

d) $I = I_0\, 10^R$

$\frac{dI}{dR} = I_0\cdot\frac{d}{dR}10^R$ (I_0 is a constant)

$= I_0\cdot(\ell n\, 10)10^R$ (Theorem 14)

$= (I_0\, \ell n\, 10)10^R$

31. $R = \log\frac{I}{I_0}$

$R = \log I - \log I_0$ (P2)

$\frac{dR}{dI} = \frac{1}{\ell n\, 10}\cdot\frac{1}{I} - 0$ (Theorem 16) (I_0 is a constant)

$= \frac{1}{\ell n\, 10}\cdot\frac{1}{I}$

33. $y = m\log x + b$

$\frac{dy}{dx} = m\cdot\frac{d}{dx}\log x + 0$ (m and b are constants)

$= m\left[\frac{1}{\ell n\, 10}\cdot\frac{1}{x}\right]$ (Theorem 16)

$= \frac{m}{\ell n\, 10}\cdot\frac{1}{x}$

35. $f(x) = 3^{2x} = (3^x)^2$

$f'(x) = 2(3^x)\cdot(\ln 3)3^x$

$= 2(\ln 3)(3^x)^2$

$= 2(\ln 3)(3^{2x})$

37. $y = x^x, \quad x > 0$

$y = e^{x \ln x}$ (Theorem 13: $a^x = e^{x \ln a}$)

$\frac{dy}{dx} = \left[x \cdot \frac{1}{x} + 1 \cdot \ln x\right]e^{x \ln x}$

$= (1 + \ln x)x^x$ (Substituting x^x for $e^{x \ln x}$)

39. $f(x) = x^{e^x}, \quad x > 0$

$f(x) = e^{e^x \ln x}$ (Theorem 13: $a^x = e^{x \ln a}$)

$f'(x) = \left[e^x \cdot \frac{1}{x} + e^x \ln x\right]e^{e^x \ln x}$

$= e^x\left[\frac{1}{x} + \ln x\right]x^{e^x}$ (Substituting x^{e^x} for $e^{e^x \ln x}$)

$= e^x x^{e^x}\left[\ln x + \frac{1}{x}\right]$

41. $y = \log_a f(x), \quad f(x) > 0$

$a^y = f(x)$ (Exponential equation)

$e^{y \ln a} = f(x)$ (Theorem 13)

Differentiate implicitly to find dy/dx.

$\frac{d}{dx} e^{y \ln a} = \frac{d}{dx} f(x)$

$\ln a \cdot \frac{dy}{dx} \cdot e^{y \ln a} = f'(x)$

$\ln a \cdot \frac{dy}{dx} \cdot f(x) = f'(x)$ (Substituting $f(x)$ for $e^{y \ln a}$)

$\frac{dy}{dx} = \frac{1}{\ln a} \cdot \frac{f'(x)}{f(x)}$

Exercise Set 4.6

1. $x = D(p) = 400 - p; \ p = \$125$

a) To find the elasticity, we first find D'(p):

$D'(p) = -1$

Then we substitute -1 for D'(p) and 400 - p for D(p) in the expression for elasticity:

$E(p) = -\frac{pD'(p)}{D(p)} = -\frac{p\cdot(-1)}{400 - p} = \frac{p}{400 - p}$

b) $E(125) = \frac{125}{400 - 125} = \frac{125}{275} = \frac{5}{11}$

Since $E(125) < 1$, the demand is inelastic.

c) Total revenue is a maximum at the value(s) of p for which E(p) = 1. We solve E(p) = 1.

$\frac{p}{400 - p} = 1$

$p = 400 - p$

$2p = 400$

$p = \$200$

3. $x = D(p) = 200 - 4p; \ p = \$46$

a) $D'(p) = -4$

$E(p) = -\frac{pD'(p)}{D(p)} = -\frac{p\cdot(-4)}{200 - 4p} = \frac{4p}{200 - 4p} = \frac{p}{50 - p}$

b) $E(46) = \frac{46}{50 - 46} = \frac{46}{4} = \frac{23}{2}$

Since $E(46) > 1$, the demand is elastic.

c) Solve E(p) = 1.

$\frac{p}{50 - p} = 1$

$p = 50 - p$

$2p = 50$

$p = \$25$

5. $x = D(p) = \frac{400}{p} = 400p^{-1}; \ p = \$50$

a) $D'(p) = -400p^{-2} = -\frac{400}{p^2}$

$E(p) = -\frac{pD'(p)}{D(p)} = -\frac{p\left[-\frac{400}{p^2}\right]}{\frac{400}{p}} = \frac{\frac{400}{p}}{\frac{400}{p}} = 1$

b) $E(50) = 1$

Since E(50) = 1, demand has unit elasticity.

c) E(p) = 1 for all p > 0, so total revenue is a maximum for all p > 0.

7. $x = D(p) = \sqrt{500 - p} = (500 - p)^{1/2}; \ p = \$400$

a) $D'(p) = \frac{1}{2}(500 - p)^{-1/2}(-1) = -\frac{1}{2\sqrt{500 - p}}$

$E(p) = -\frac{pD'(p)}{D(p)} = -\frac{p\left[-\frac{1}{2\sqrt{500 - p}}\right]}{\sqrt{500 - p}} = \frac{\frac{p}{2\sqrt{500 - p}}}{\sqrt{500 - p}} = \frac{p}{2(500 - p)} = \frac{p}{1000 - 2p}$

b) $E(400) = \frac{400}{1000 - 2\cdot400} = \frac{400}{200} = 2$

Since $E(400) > 1$, the demand is elastic.

c) Solve E(p) = 1.

$\frac{p}{1000 - 2p} = 1$

$p = 1000 - 2p$

$3p = 1000$

$p = \frac{\$1000}{3} \approx \$333.33$

9. $x = D(p) = 100\ e^{-0.25p}; \ p = \$10$

a) $D'(p) = 100(-0.25)\ e^{-0.25p} = -25\ e^{-0.25p}$

$E(p) = -\frac{pD'(p)}{D(p)} = -\frac{p(-25\ e^{-0.25p})}{100\ e^{-0.25p}} = \frac{25p}{100} = \frac{p}{4}$

b) $E(10) = \frac{10}{4} = \frac{5}{2}$

Since $E(10) > 1$, the demand is elastic.

c) Solve $E(p) = 1$.

$$\frac{p}{4} = 1$$

$$p = \$4$$

11. $x = D(p) = \frac{100}{(p + 3)^2} = 100(p + 3)^{-2}$; $p = \$1$

a) $D'(p) = 100(-2)(p + 3)^{-3} = -\frac{200}{(p + 3)^3}$

$$E(p) = -\frac{pD'(p)}{D(p)} = -\frac{p\left[-\frac{200}{(p + 3)^3}\right]}{\frac{100}{(p + 3)^2}} =$$

$$\frac{\frac{200p}{(p + 3)^3}}{\frac{100}{(p + 3)^2}} = \frac{2p}{p + 3}$$

b) $E(1) = \frac{2 \cdot 1}{1 + 3} = \frac{2}{4} = \frac{1}{2}$

Since $E(1) < 1$, the demand is inelastic.

c) Solve $E(p) = 1$.

$$\frac{2p}{p + 3} = 1$$

$$2p = p + 3$$

$$p = \$3$$

13. $x = D(p) = 967 - 25p$

a) $D'(p) = -25$

$$E(p) = -\frac{pD'(p)}{D(p)} = -\frac{p \cdot (-25)}{967 - 25p} = \frac{25p}{967 - 25p}$$

b) We set $E(p) = 1$ and solve for p.

$$\frac{25p}{967 - 25p} = 1$$

$$25p = 967 - 25p$$

$$50p = 967$$

$$p = 19.34¢$$

c) Demand is elastic when $E(p) > 1$. Test a value on either side of 19.34.

$$E(19) = \frac{25 \cdot 19}{967 - 25 \cdot 19} \approx 0.97 < 1$$

$$E(20) = \frac{25 \cdot 20}{967 - 25 \cdot 20} \approx 1.07 > 1$$

Thus, demand is elastic for $p > 19.34¢$.

d) Demand is inelastic when $E(p) < 1$. Using the calculations in part c), we see that demand is inelastic for $p < 19.34¢$.

e) Since $E(19.34¢) = 1$, total revenue is a maximum when $p = 19.34¢$.

f) At a price of 20¢ per cookie the demand is elastic. (See part c).) Thus, a small increase in price will cause the total revenue to decrease.

15. $x = D(p) = \sqrt{200 - p^3}$, or $(200 - p^3)^{1/2}$

a) $D'(p) = \frac{1}{2}(200 - p^3)^{-1/2}(-3p^2)$

$$= -\frac{3p^2}{2\sqrt{200 - p^3}}$$

$$E(p) = -\frac{pD'(p)}{D(p)} = -\frac{p\left[-\frac{3p^2}{2\sqrt{200 - p^3}}\right]}{\sqrt{200 - p^3}} =$$

$$\frac{3p^3}{2(200 - p^3)}$$

b) $E(3) = \frac{3(3)^3}{2(200 - 3^3)} = \frac{81}{2 \cdot 173} = \frac{81}{346} \approx 0.234$

c) $E(3) < 1$, so the demand is inelastic at $p = \$3$. Therefore, a small price increase will cause total revenue to increase.

17. a) $x = D(p) = \frac{k}{p^n} = kp^{-n}$

$$D'(p) = -nkp^{-n-1} = -\frac{nk}{p^{n+1}}$$

$$E(p) = -\frac{pD'(p)}{D(p)} = -\frac{p\left[-\frac{nk}{p^{n+1}}\right]}{\frac{k}{p^n}} = \frac{\frac{nk}{p^n}}{\frac{k}{p^n}} = n$$

b) No, $E(p)$ is a constant, n.

c) Solve $E(p) = 1$.

$$n = 1$$

Thus, total revenue is a maximum for $p = \$1$.

19. We will use implicit differentiation.

$$L(p) = \ln D(p)$$

$$L'(p) = D'(p) \cdot \frac{1}{D(p)}$$

$$L'(p)D(p) = D'(p)$$

$$E(p) = -\frac{pD'(p)}{D(p)} = -\frac{pL'(p)D(p)}{D(p)} = -pL'(p)$$

Exercise Set 5.1

1. $\int x^6\,dx$

$= \frac{x^{6+1}}{6+1} + C \qquad \left[\int x^r\,dx = \frac{x^{r+1}}{r+1} + C\right]$

$= \frac{x^7}{7} + C$

3. $\int 2\,dx$

$= 2x + C \qquad \left[\int k\,dx,\ (k \text{ a constant}) = kx + C\right]$

5. $\int x^{1/4}\,dx$

$= \frac{x^{1/4+1}}{\frac{1}{4}+1} + C \qquad \left[\int x^r\,dx = \frac{x^{r+1}}{r+1} + C\right]$

$= \frac{x^{5/4}}{\frac{5}{4}} + C$

$= \frac{4}{5}\,x^{5/4} + C$

7. $\int (x^2 + x - 1)\,dx$

$= \int x^2\,dx + \int x\,dx - \int 1\,dx$

(The integral of a sum is the sum of the integrals.)

$= \frac{x^3}{3} + \frac{x^2}{2} - x + C \longleftarrow$ DON'T FORGET THE C!

$\left[\int x^r\,dx = \frac{x^{r+1}}{r+1} + C\right]$

$\left[\int k\,dx,\ (k \text{ a constant}) = dx + C\right]$

9. $\int (t^2 - 2t + 3)\,dt$

$= \int t^2\,dt - \int 2t\,dt + \int 3\,dt$

(The integral of a sum is the sum of the integrals.)

$= \frac{t^3}{3} - 2 \cdot \frac{t^2}{2} + 3t + C$

$\left[\int x^r\,dx = \frac{x^{r+1}}{r+1} + C\right]$

$\left[\int k\,dx,\ (k \text{ a constant}) = kx + C\right]$

$= \frac{t^3}{3} - t^2 + 3t + C$

11. $\int 5\,e^{8x}\,dx$

$= \frac{5}{8}\,e^{8x} + C \qquad \left[\int be^{ax}\,dx = \frac{b}{a}\,e^{ax} + C\right]$

13. $\int (x^3 - x^{8/7})\,dx$

$= \int x^3\,dx - \int x^{8/7}\,dx$

(The integral of a sum is the sum of the integrals.)

$= \frac{x^4}{4} - \frac{x^{8/7\,+\,1}}{\frac{8}{7}+1} + C \qquad \left[\int x^r\,dx = \frac{x^{r+1}}{r+1} + C\right]$

$= \frac{x^4}{4} - \frac{x^{15/7}}{\frac{15}{7}} + C$

$= \frac{x^4}{4} - \frac{7}{15}\,x^{15/7} + C$

15. $\int \frac{1000}{x}\,dx$

$= 1000 \int \frac{1}{x}\,dx$ (The integral of a constant times a function is the constant times the integral.)

$= 1000 \ln x + C \qquad \left[\int \frac{1}{x}\,dx = \ln x + C,\ x > 0;\right.$ we generally consider $\left. x > 0\right]$

17. $\int \frac{dx}{x^2} = \int \frac{1}{x^2}\,dx = \int x^{-2}\,dx$

$= \frac{x^{-2+1}}{-2+1} + C \qquad \left[\int x^r\,dx = \frac{x^{r+1}}{r+1} + C\right]$

$= \frac{x^{-1}}{-1} + C$

$= -x^{-1} + C$, or $-\frac{1}{x} + C$

19. $\int \sqrt{x}\,dx = \int x^{1/2}\,dx$

$= \frac{x^{1/2+1}}{\frac{1}{2}+1} + C$

$= \frac{x^{3/2}}{\frac{3}{2}} + C$

$= \frac{2}{3}\,x^{3/2} + C$

21. $\int \frac{-6}{\sqrt[3]{x^2}}\,dx = \int \frac{-6}{x^{2/3}}\,dx = \int -6x^{-2/3}\,dx$

$= -6 \int x^{-2/3}\,dx$

$= -6 \cdot \frac{x^{-2/3+1}}{-\frac{2}{3}+1} + C$

$= -6 \cdot \frac{x^{1/3}}{\frac{1}{3}} + C$

$= -6 \cdot 3x^{1/3} + C$

$= -18x^{1/3} + C$

23. $\int 8\,e^{-2x}\,dx = \frac{8}{-2}\,e^{-2x} + C$

$= -4\,e^{-2x} + C$

25. $\int\left[x^2 - \frac{3}{2}\sqrt{x} + x^{-4/3}\right] dx$

$= \int x^2\, dx - \frac{3}{2}\int x^{1/2}\, dx + \int x^{-4/3}\, dx$

$= \frac{x^{2+1}}{2+1} - \frac{3}{2}\cdot\frac{x^{1/2+1}}{\frac{1}{2}+1} + \frac{x^{-4/3+1}}{-\frac{4}{3}+1} + C$

$= \frac{x^3}{3} - \frac{3}{2}\cdot\frac{x^{3/2}}{\frac{3}{2}} + \frac{x^{-1/3}}{-\frac{1}{3}} + C$

$= \frac{x^3}{3} - x^{3/2} - 3x^{-1/3} + C$

27. Find the function f such that $f'(x) = x - 3$ and $f(2) = 9$.

We first find f(x) by integrating.

$f(x) = \int (x - 3)\, dx$

$= \int x\, dx - \int 3\, dx$

$= \frac{x^2}{2} - 3x + C$

The condition $f(2) = 9$ allows us to find C.

$f(x) = \frac{x^2}{2} - 3x + C$

$f(2) = \frac{2^2}{2} - 3\cdot 2 + C = 9$ [Substituting 2 for x and 9 for f(2)]

$2 - 6 + C = 9$

$C = 13$

Thus, $f(x) = \frac{x^2}{2} - 3x + 13$.

29. Find the function f such that $f'(x) = x^2 - 4$ and $f(0) = 7$.

We first find f(x) by integrating.

$f(x) = \int (x^2 - 4)\, dx$

$= \int x^2\, dx - \int 4\, dx$

$= \frac{x^3}{3} - 4x + C$

The condition $f(0) = 7$ allows us to find C.

$f(x) = \frac{x^3}{3} - 4x + C$

$f(0) = \frac{0^3}{3} - 4\cdot 0 + C = 7$ [Substituting 0 for x and 7 for f(0)]

Solving for C we get $C = 7$.

Thus, $f(x) = \frac{x^3}{3} - 4x + 7$.

31. $C'(x) = x^3 - 2x$

We integrate to find C(x) using K for the integration constant to avoid confusion with the cost function C.

$C(x) = \int C'(x)\, dx$

$= \int (x^3 - 2x)\, dx$

$= \frac{x^4}{4} - x^2 + K$

Fixed costs are \$100. This means $C(0) = 100$. This allows us to determine the value of K.

$C(0) = \frac{0^4}{4} - 0^2 + K = 100$

[Substituting 0 for x and 100 for C(0)]

Solving for K we get $K = 100$.

Thus, $C(x) = \frac{x^4}{4} - x^2 + 100$.

33. $R'(x) = x^2 - 3$

a) We integrate to find R(x).

$R(x) = \int R'(x)\, dx$

$= \int (x^2 - 3)\, dx$

$= \frac{x^3}{3} - 3x + C$

The condition $R(0) = 0$ allows us to find C.

$R(0) = \frac{0^3}{3} - 3\cdot 0 + C = 0$

[Substituting 0 for x and 0 for R(0)]

Solving for C we get $C = 0$.

Thus, $R(x) = \frac{x^3}{3} - 3x$.

b) If you sell no products, you make no money.

35. $D'(p) = -\frac{4000}{p^2} = -4000p^{-2}$

We integrate to find D(p).

$D(p) = \int -4000p^{-2}\, dp$

$= -4000\int p^{-2}\, dp$

$= -4000\cdot\frac{p^{-1}}{-1} + C$

$= \frac{4000}{p} + C$

When p = \$4 per unit, D(p) = 1003 units, or D(4) = 1003. We use this to find C.

$D(4) = \frac{4000}{4} + C = 1003$

$1000 + C = 1003$

$C = 3$

$D(p) = \frac{4000}{p} + 3$

37. $\frac{dE}{dt} = 30 - 10t$, where t = the number of hours the operator has been at work.

a) We find E(t) by integrating E'(t).

$E(t) = \int E'(t)\,dt$

$= \int (30 - 10t)\,dt$

$= 30t - 5t^2 + C$

The condition E(2) = 72 allows us to find C.

$E(2) = 30\cdot 2 - 5\cdot 2^2 + C = 72$

[Substituting 2 for t and 72 for E(2)]

Solving for C we get:

$60 - 20 + C = 72$

$40 + C = 72$

$C = 32$

Thus, $E(t) = 30t - 5t^2 + 32$.

b) $E(t) = 30t - 5t^2 + 32$

$E(3) = 30\cdot 3 - 5\cdot 3^2 + 32$ (Substituting 3 for t)

$= 90 - 45 + 32$

$= 77$

The operator's efficiency after 3 hours is 77%.

$E(5) = 30\cdot 5 - 5\cdot 5^2 + 32$ (Substituting 5 for t)

$= 150 - 125 + 32$

$= 57$

The operator's efficiency after 5 hours is 57%.

39. $v(t) = 3t^2$, $s(0) = 4$

We find s(t) by integrating v(t).

$s(t) = \int v(t)\,dt$

$= \int 3t^2\,dt$

$= 3 \cdot \frac{t^3}{3} + C$

$= t^3 + C$

The condition s(0) = 4 allows us to find C.

$s(0) = 0^3 + C = 4$ [Substituting 0 for t and 4 for s(0)]

Solving for C, we get C = 4.

Thus, $s(t) = t^3 + 4$.

41. $a(t) = 4t$, $v(0) = 20$

We find v(t) by integrating a(t).

$v(t) = \int a(t)\,dt$

$= \int 4t\,dt$

$= 4 \cdot \frac{t^2}{2} + C$

$= 2t^2 + C$

The condition v(0) = 20 allows us to find C.

$v(0) = 2\cdot 0^2 + C = 20$ [Substituting 0 for t and 20 for v(0)]

Solving for C, we get C = 20.

Thus, $v(t) = 2t^2 + 20$.

43. $a(t) = -2t + 6$, $v(0) = 6$, and $s(0) = 10$

We find v(t) by integrating a(t).

$v(t) = \int a(t)\,dt$

$= \int (-2t + 6)\,dt$

$= -t^2 + 6t + C_1$

The condition v(0) = 6 allows us to find C_1.

$v(0) = -(0)^2 + 6\cdot 0 + C_1 = 6$

[Substituting 0 for t and 6 for v(0)]

Solving for C_1, we get $C_1 = 6$.

Thus, $v(t) = -t^2 + 6t + 6$.

We find s(t) by integrating v(t).

$s(t) = \int (v(t)\,dt$

$= \int (-t^2 + 6t + 6)\,dt$

$= -\frac{t^3}{3} + 3t^2 + 6t + C_2$

The condition s(0) = 10 allows us to find C_2.

$s(0) = -\frac{0^3}{3} + 3\cdot 0^2 + 6\cdot 0 + C_2 = 10$

[Substituting 0 for t and 10 for s(0)]

Solving for C_2, we get $C_2 = 10$.

Thus, $s(t) = -\frac{1}{3}t^3 + 3t^2 + 6t + 10$.

45. $a(t) = -32$ ft/sec²

$v(0)$ = initial velocity = v_0

$s(0)$ = initial height = s_0

We find v(t) by integrating a(t).

$v(t) = \int a(t)\,dt$

$= \int (-32)\,dt$

$= -32t + C_1$

The condition $v(0) = v_0$ allows us to find C_1.

$v(0) = -32\cdot 0 + C_1 = v_0$ [Substituting 0 for t and v_0 for v(0)]

$C_1 = v_0$

Thus, $v(t) = -32t + v_0$.

We find s(t) by integrating v(t).

$s(t) = \int v(t)\,dt$

$= \int (-32t + v_0)\,dt$

$= -16t^2 + v_0t + C_2$ (v_0 is a constant)

The condition $s(0) = s_0$ allows us to find C_2.

$s(0) = -16\cdot0^2 + v_0\cdot0 + C_2 = s_0$ [Substituting 0 for t and s_0 for s(0)]

$C_2 = s_0$

Thus, $s(t) = -16t^2 + v_0t + s_0$.

47. $a(t) = k$ (Constant acceleration)

$v(t) = \int a(t)\,dt = \int k\,dt = kt$ [v(0) = 0, thus C = 0]

$s(t) = \int v(t)\,dt = \int kt\,dt = k\cdot\frac{t^2}{2} = \frac{1}{2}kt^2$

[s(0) = 0, thus C = 0]

We know that

$a(t) = k = \dfrac{60\text{ mph}}{\frac{1}{2}\text{ min}}$

and that

$t = \frac{1}{2}$ min.

Thus

$s(t) = \frac{1}{2}kt^2$

$s\left(\frac{1}{2}\text{ min}\right) = \frac{1}{2}\cdot\dfrac{60\text{ mph}}{\frac{1}{2}\text{ min}}\cdot\left(\frac{1}{2}\text{ min}\right)^2$

$= \frac{1}{2}\cdot\frac{60\text{ mi}}{\text{hr}}\cdot\frac{1}{2}\text{ min}$

$= \frac{1}{2}\cdot\frac{60\text{ mi}}{\text{hr}}\cdot\frac{1}{120}\text{ hr}$

$= \frac{60}{240}\text{ mi}$

$= \frac{1}{4}\text{ mi}$

The car travels $\frac{1}{4}$ mi during that time.

49. $M'(t) = 0.2t - 0.003t^2$

a) We integrate to find M(t).

$M(t) = \int (0.2t - 0.003t^2)\,dt$

$= 0.1t^2 - 0.001t^3 + C$

We use M(0) = 0 to find C.

$M(0) = 0.1(0)^2 - 0.001(0)^3 + C = 0$

$C = 0$

$M(t) = 0.1t^2 - 0.001t^3$

b) $M(8) = 0.1(8)^2 - 0.001(8)^3$

$= 6.4 - 0.512$

$= 5.888$

≈ 6 words

51. Find the function f such that

$f'(t) = t^{\sqrt{3}}$ and $f(0) = 8$.

We first find f'(x) by integrating.

$f(t) = \int f'(t)\,dt$

$= \int t^{\sqrt{3}}\,dt$

$= \dfrac{t^{\sqrt{3}+1}}{\sqrt{3}+1} + C$

The condition f(0) = 8 allows us to find C.

$f(t) = \dfrac{t^{\sqrt{3}+1}}{\sqrt{3}+1} + C$

$f(0) = \dfrac{0^{\sqrt{3}+1}}{\sqrt{3}+1} + C = 8$ [Substituting 0 for t and 8 for f(t)]

Solving for C we get C = 8.

Thus, $f(t) = \dfrac{t^{\sqrt{3}+1}}{\sqrt{3}+1} + 8$.

53. $\int (x - 1)^2 x^3\,dx$

$= \int (x^2 - 2x + 1)x^3\,dx$

$= \int (x^5 - 2x^4 + x^3)\,dx$

$= \int x^5\,dx - \int 2x^4\,dx + \int x^3\,dx$

$= \frac{x^6}{6} - 2\cdot\frac{x^5}{5} + \frac{x^4}{4} + C$

$= \frac{1}{6}x^6 - \frac{2}{5}x^5 + \frac{1}{4}x^4 + C$

55. $\int \dfrac{(t + 3)^2}{\sqrt{t}}\,dt$

$= \int \dfrac{t^2 + 6t + 9}{t^{1/2}}\,dt$

$= \int (t^{3/2} + 6t^{1/2} + 9t^{-1/2})\,dt$

$= \int t^{3/2}\,dt + \int 6t^{1/2}\,dt + \int 9t^{-1/2}\,dt$

$= \dfrac{t^{5/2}}{5/2} + 6\cdot\dfrac{t^{3/2}}{3/2} + 9\cdot\dfrac{t^{1/2}}{1/2} + C$

$= \frac{2}{5}t^{5/2} + 4t^{3/2} + 18t^{1/2} + C$

57. $\int (t + 1)^3\,dt$

$= \int (t^3 + 3t^2 + 3t + 1)\,dt$

$= \int t^3\,dt + \int 3t^2\,dt + \int 3t\,dt + \int dt$

$= \frac{t^4}{4} + 3\cdot\frac{t^3}{3} + 3\cdot\frac{t^2}{2} + t + C$

$= \frac{1}{4}t^4 + t^3 + \frac{3}{2}t^2 + t + C$

59. $\int be^{ax}\,dx = \frac{b}{a}e^{ax} + C$ (See Theorem 2.)

61. $\int \sqrt[3]{64x^4}\,dx$

$= \int 4\sqrt[3]{x^4}\,dx$

$= 4\int x^{4/3}\,dx$

$= 4\cdot\frac{x^{7/3}}{7/3} + C$

$= \frac{12}{7}x^{7/3} + C$

63. $\int \frac{t^3+8}{t+2}\,dt$

$= \int \frac{(t+2)(t^2-2t+4)}{(t+2)}\,dt$

$= \int (t^2 - 2t + 4)\,dt$

$= \int t^2\,dt - \int 2t\,dt + \int 4\,dt$

$= \frac{t^3}{3} - 2\cdot\frac{t^2}{2} + 4t + C$

$= \frac{1}{3}t^3 - t^2 + 4t + C$

Exercise Set 5.2

1. Find the area under the curve $y = 4$ on the interval $[1,3]$.

$A(x) = \int 4\,dx$

$= 4x + C$

Since we know that $A(1) = 0$ (there is no area above the number 1), we can substitute for x and A(x) to determine C.

$A(1) = 4\cdot 1 + C = 0$ [Substituting 1 for x and 0 for A(1)]

Solving for C we get:

$4 + C = 0$

$C = -4$

Thus, $A(x) = 4x - 4$.

Then the area on the interval $[1,3]$ is A(3).

$A(3) = 4\cdot 3 - 4$ (Substituting 3 for x)

$= 12 - 4$

$= 8$

3. Find the area under the curve $y = 2x$ on the interval $[1,3]$.

$A(x) = \int 2x\,dx$

$= x^2 + C$

Since we know that $A(1) = 0$ (there is no area above the number 1), we can substitute for x and A(x) to determine C.

$A(1) = 1^2 + C = 0$ [Substituting 1 for x and 0 for A(1)]

Solving for C we get:

$1 + C = 0$

$C = -1$

Thus, $A(x) = x^2 - 1$.

Then the area on the interval $[1,3]$ is A(3).

$A(3) = 3^2 - 1$ (Substituting 3 for x)

$= 9 - 1$

$= 8$

5. Find the area under the curve $y = x^2$ on the interval $[0,5]$.

$A(x) = \int x^2\,dx$

$= \frac{x^3}{3} + C$

Since we know that $A(0) = 0$ (there is no area above the number 0), we can substitute for x and A(x) to determine C.

$A(0) = \frac{0^3}{3} + C = 0$ [Substituting 0 for x and 0 for A(0)]

Solving for C, we get $C = 0$.

Thus, $A(x) = \frac{x^3}{3}$.

The area on the interval $[0,5]$ is A(5).

$A(5) = \frac{5^3}{3}$ (Substituting 5 for x)

$= \frac{125}{3}$, or $41\frac{2}{3}$

7. Find the area under the curve $y = x^3$ on the interval $[0,1]$.

$A(x) = \int x^3\,dx$

$= \frac{x^4}{4} + C$

Since we know that $A(0) = 0$, we can substitute for x and A(x) to determine C.

$A(0) = \frac{0^4}{4} + C = 0$ [Substituting 0 for x and 0 for A(0)]

Solving for C, we get $C = 0$.

Thus, $A(x) = \frac{x^4}{4}$.

The area on the interval $[0,1]$ is A(1).

$A(1) = \frac{1^4}{4}$ (Substituting 1 for x)

$= \frac{1}{4}$

9. Find the area under the curve $y = 4 - x^2$ on the interval $[-2,2]$.

$A(x) = \int (4 - x^2)\,dx$

$= 4x - \frac{x^3}{3} + C$

Since we know that $A(-2) = 0$ (there is no area above the number -2), we can substitute for x and A(x) to determine C.

$$A(-2) = 4(-2) - \frac{(-2)^3}{3} + C = 0$$

[Substituting -2 for x and 0 for A(-2)]

Solving for C we get:

$$-8 + \frac{8}{3} + C = 0$$

$$-\frac{24}{3} + \frac{8}{3} + C = 0$$

$$-\frac{16}{3} + C = 0$$

$$C = \frac{16}{3}$$

Thus, $A(x) = 4x - \frac{x^3}{3} + \frac{16}{3}$.

The area on the interval [-2,2] is A(2).

$$A(2) = 4 \cdot 2 - \frac{2^3}{3} + \frac{16}{3} \quad \text{(Substituting 2 for x)}$$

$$= 8 - \frac{8}{3} + \frac{16}{3}$$

$$= \frac{24}{3} - \frac{8}{3} + \frac{16}{3}$$

$$= \frac{32}{3}, \text{ or } 10\frac{2}{3}$$

11. Find the area under the curve $y = e^x$ on the interval [0,3].

$$A(x) = \int e^x\, dx$$

$$= e^x + C$$

Since we know that A(0) = 0 (there is no area above the number 0), we can substitute for x and A(x) to determine C.

$$A(0) = e^0 + C = 0 \quad \text{[Substituting 0 for x and 0 for A(0)]}$$

$$1 + C = 0 \quad (e^0 = 1)$$

$$C = -1$$

Thus, $A(x) = e^x - 1$.

The area on the interval [0,3] is A(3).

$$A(3) = e^3 - 1$$

$$= 20.085537 - 1 \quad \text{(Using a calculator)}$$

$$\approx 19.086$$

13. Find the area under the curve $y = \frac{1}{x}$ on the interval [1,3].

$$A(x) = \int \frac{1}{x}\, dx$$

$$= \ln x + C$$

Since we know that A(1) = 0 (there is no area above the number 1), we can substitute for x and A(x) to determine C.

$$A(1) = \ln 1 + C = 0 \quad \text{[Substituting 1 for x and 0 for A(1)]}$$

$$0 + C = 0 \quad (\ln 1 = 0)$$

$$C = 0$$

Thus, $A(x) = \ln x$.

The area on the interval [1,3] is A(3).

$$A(3) = \ln 3 \quad \text{(Substituting 3 for x)}$$

$$\approx 1.098612 \quad \text{(Using a calculator)}$$

15. Find the area under the curve $y = x^2 - 4x$ on the interval [-4,-1].

$$A(x) = \int (x^2 - 4x)\, dx$$

$$= \frac{x^3}{3} - 4 \cdot \frac{x^2}{2} + C$$

$$= \frac{x^3}{3} - 2x^2 + C$$

Since we know that A(-4) = 0, we can substitute for x and A(x) to determine C.

$$A(-4) = \frac{(-4)^3}{3} - 2(-4)^2 + C = 0$$

[Substituting -4 for x and 0 for A(-4)]

$$-\frac{64}{3} - 32 + C = 0$$

$$C = \frac{96}{3} + \frac{64}{3}$$

$$C = \frac{160}{3}$$

Thus, $A(x) = \frac{x^3}{3} - 2x^2 + \frac{160}{3}$.

The area on the interval [-4,-1] is A(-1).

$$A(-1) = \frac{(-1)^3}{3} - 2(-1)^2 + \frac{160}{3}$$

(Substituting -1 for x)

$$= -\frac{1}{3} - 2 + \frac{160}{3}$$

$$= -\frac{1}{3} - \frac{6}{3} + \frac{160}{3}$$

$$= \frac{153}{3}$$

$$= 51$$

17. The shaded region represents an antiderivative. It also represents velocity, the antiderivative of acceleration.

19. The shaded region represents an antiderivative. It also represents total energy used in time t.

21. The shaded region represents an antiderivative. It also represents total revenue.

23. The shaded region represents an antiderivative. It also represents the amount of drug in blood.

25. a) We find C(x) by integrating C'(x).

$$C'(x) = 100 - 0.2x$$

$$C(x) = 100x - 0.1x^2 + K$$

(Using K for the integration constant)

Since we know that C(0) = 0, we can substitute for x and C(x) to determine K.

$C(0) = 100 \cdot 0 - 0.1 \cdot 0^2 + K = 0$

[Substituting 0 for x and 0 for C(0)]

Solving for K we get K = 0.

Thus, $C(x) = 100x - 0.1x^2$.

b) We find R(x) by integrating R'(x).

$R'(x) = 100 + 0.2x$

$R(x) = 100x + 0.1x^2 + C$

Since we know that R(0) = 0, we can substitute for x and R(x) to determine C.

$R(0) = 100 \cdot 0 + 0.1 \cdot 0^2 + C = 0$

[Substituting 0 for x and 0 for R(0)]

Solving for C we get C = 0.

Thus, $R(x) = 100x + 0.1x^2$.

c) $P(x) = R(x) - C(x)$

$P(x) = (100x + 0.1x^2) - (100x - 0.1x^2)$ (Substituting)

$= 100x + 0.1x^2 - 100x + 0.1x^2$

$= 0.2x^2$

d) $P(x) = 0.2x^2$

$P(1000) = 0.2(1000)^2$ (Substituting 1000 for x)

$= 0.2(1{,}000{,}000)$

$= 200{,}000$

The total profit from the production and sale of 1000 stereos is $200,000.

27. a) We find s(t) by integrating.

$s(t) = \int v(t)\, dt$

$= \int (3t^2 + 2t)\, dt$

$= t^3 + t^2 + C$

Since the particle starts out from the origin, we know that s(0) = 0 (there is no area above the number 0). We can substitute for t and s(t) to determine C.

$s(0) = 0^3 + 0^2 + C = 0$ [Substituting 0 for t and 0 for s(0)]

Solving for C we get C = 0.

Thus, $s(t) = t^3 + t^2$.

b) $s(t) = t^3 + t^2$

$s(5) = 5^3 + 5^2$ (Substituting 5 for t)

$= 125 + 25$

$= 150$

29. Find the area under the curve $y = \frac{x^2 - 1}{x - 1}$ on the interval [2,3].

$A(x) = \int \frac{x^2 - 1}{x - 1}\, dx$

$= \int \frac{(x + 1)(x - 1)}{x - 1}\, dx$

$= \int (x + 1)\, dx$

$= \frac{x^2}{2} + x + C$

Since we know that A(2) = 0, we can substitute for x and A(x) to determine C.

$A(2) = \frac{2^2}{2} + 2 + C = 0$ [Substituting 2 for x and 0 for A(2)]

Solving for C we get:

$2 + 2 + C = 0$

$C = -4$

Thus, $A(x) = \frac{x^2}{2} + x - 4$.

The area on the interval [2,3] is A(3).

$A(3) = \frac{3^2}{2} + 3 - 4$ (Substituting 3 for x)

$= \frac{9}{2} + \frac{6}{2} - \frac{8}{2}$

$= \frac{7}{2}$, or $3\frac{1}{2}$

31. Find the area under the curve $y = (x - 1)\sqrt{x}$ on the interval [4,16].

$A(x) = \int (x - 1)\sqrt{x}\, dx$

$= \int (x - 1)x^{1/2}\, dx$

$= \int (x^{3/2} - x^{1/2})\, dx$

$= \frac{x^{5/2}}{5/2} - \frac{x^{3/2}}{3/2} + C$

$= \frac{2}{5} x^{5/2} - \frac{2}{3} x^{3/2} + C$

Since we know that A(4) = 0, we can substitute for x and A(x) to determine C.

$A(4) = \frac{2}{5} \cdot 4^{5/2} - \frac{2}{3} \cdot 4^{3/2} + C = 0$ (Substituting)

$\frac{2}{5} \cdot 32 - \frac{2}{3} \cdot 8 + C = 0$

$[4^{5/2} = (2^2)^{5/2} = 2^5 = 32]$

$[4^{3/2} = (2^2)^{3/2} = 2^3 = 8]$

$\frac{64}{5} - \frac{16}{3} + C = 0$

$\frac{192}{15} - \frac{80}{15} + C = 0$

$\frac{112}{15} + C = 0$

$C = -\frac{112}{15}$

Thus, $A(x) = \frac{2}{5} x^{5/2} - \frac{2}{3} x^{3/2} - \frac{112}{15}$.

The area on the interval [4,16] is A(16).

$A(16) = \frac{2}{5} \cdot 16^{5/2} - \frac{2}{3} \cdot 16^{3/2} - \frac{112}{15}$

$= \frac{2}{5} \cdot 2^{10} - \frac{2}{3} \cdot 2^{6} - \frac{112}{15}$

$= \frac{2^{11}}{5} - \frac{2^7}{3} - \frac{112}{15}$

$= \frac{3 \cdot 2^{11}}{15} - \frac{5 \cdot 2^7}{15} - \frac{112}{15}$

$= \frac{6144}{15} - \frac{640}{15} - \frac{112}{15}$

$= \frac{5392}{15}$

$= 359\frac{7}{15}$

33. Find the area under the curve $y = \frac{\sqrt[3]{x^2} - 1}{\sqrt[3]{x}}$ on the interval [1,8].

$A(x) = \int \frac{\sqrt[3]{x^2} - 1}{\sqrt[3]{x}}\, dx$

$= \int \frac{x^{2/3} - 1}{x^{1/3}}\, dx$

$= \int (x^{1/3} - x^{-1/3})\, dx$

$= \int x^{1/3}\, dx - \int x^{-1/3}\, dx$

$= \frac{x^{4/3}}{4/3} - \frac{x^{2/3}}{2/3} + C$

$= \frac{3}{4} x^{4/3} - \frac{3}{2} x^{2/3} + C$

Since we know that A(1) = 0, we can substitute for x and A(x) to determine C.

$A(1) = \frac{3}{4} \cdot 1^{4/3} - \frac{3}{2} \cdot 1^{2/3} + C = 0$ (Substituting)

$\frac{3}{4} - \frac{3}{2} + C = 0$

$C = -\frac{3}{4} + \frac{3}{2}$

$C = -\frac{3}{4} + \frac{6}{4}$

$C = \frac{3}{4}$

Thus, $A(x) = \frac{3}{4} x^{4/3} - \frac{3}{2} x^{2/3} + \frac{3}{4}$

The area on the interval [1,8] is A(8).

$A(8) = \frac{3}{4} \cdot 8^{4/3} - \frac{3}{2} \cdot 8^{2/3} + \frac{3}{4}$

$= \frac{3}{4} \cdot 2^4 - \frac{3}{2} \cdot 2^2 + \frac{3}{4}$

$= 12 - 6 + \frac{3}{4}$

$= 6\frac{3}{4}$

Exercise Set 5.3

1. $\int_0^1 (x - x^2)\, dx$

$= \left[\frac{x^2}{2} - \frac{x^3}{3}\right]_0^1$

$= \left(\frac{1^2}{2} - \frac{1^3}{3}\right) - \left(\frac{0^2}{2} - \frac{0^3}{3}\right)$

(Substituting 0 for x)

(Substituting 1 for x)

$= \left(\frac{1}{2} - \frac{1}{3}\right) - (0 - 0)$

$= \frac{3}{6} - \frac{2}{6}$

$= \frac{1}{6}$

3. $\int_{-1}^1 (x^2 - x^4)\, dx$

$= \left[\frac{x^3}{3} - \frac{x^5}{5}\right]_{-1}^1$

$= \left[\frac{1^3}{3} - \frac{1^5}{5}\right] - \left[\frac{(-1)^3}{3} - \frac{(-1)^5}{5}\right]$

$= \left(\frac{1}{3} - \frac{1}{5}\right) - \left(-\frac{1}{3} + \frac{1}{5}\right)$

$= \frac{1}{3} - \frac{1}{5} + \frac{1}{3} - \frac{1}{5}$

$= \frac{2}{3} - \frac{2}{5}$

$= \frac{10}{15} - \frac{6}{15}$

$= \frac{4}{15}$

5. $\int_b^a e^t\, dt$

$= [e^t]_a^b$

$= e^b - e^a$

7. $\int_a^b 3t^2\, dt$

$= \left[3 \cdot \frac{t^3}{3}\right]_a^b$

$= [t^3]_a^b$

$= b^3 - a^3$

9. $\int_1^e \left(x + \frac{1}{x}\right) dx$

$= \left[\frac{x^2}{2} + \ln x\right]_1^e$

$= \left(\frac{e^2}{2} + \ln e\right) - \left(\frac{1^2}{2} + \ln 1\right)$

$= \frac{e^2}{2} + 1 - \frac{1}{2}$ (ln e = 1, ln 1 = 0)

$= \frac{e^2}{2} + \frac{1}{2}$

11. $\int_0^1 \sqrt{x}\, dx$

$= \int_0^1 x^{1/2}\, dx$

$= \left[\frac{x^{3/2}}{3/2}\right]_0^1$

$= \left[\frac{2}{3} x^{3/2}\right]_0^1$

$= \frac{2}{3} \cdot 1^{3/2} - \frac{2}{3} \cdot 0^{3/2}$

$= \frac{2}{3} \cdot 1 - \frac{2}{3} \cdot 0$

$= \frac{2}{3} - 0$

$= \frac{2}{3}$

13. $\int_0^1 \frac{10}{17} t^3\, dt$

$= \frac{10}{17} \int_0^1 t^3\, dt$ (Property 1)

$= \frac{10}{17} \left[\frac{t^4}{4}\right]_0^1$

$= \frac{10}{17}\left(\frac{1^4}{4} - \frac{0^4}{4}\right)$

$= \frac{10}{17}\left(\frac{1}{4} - 0\right)$

$= \frac{10}{17} \cdot \frac{1}{4}$

$= \frac{10}{68}$

$= \frac{5}{34}$

15. Find the area under $y = x^3$ on $[0,2]$.

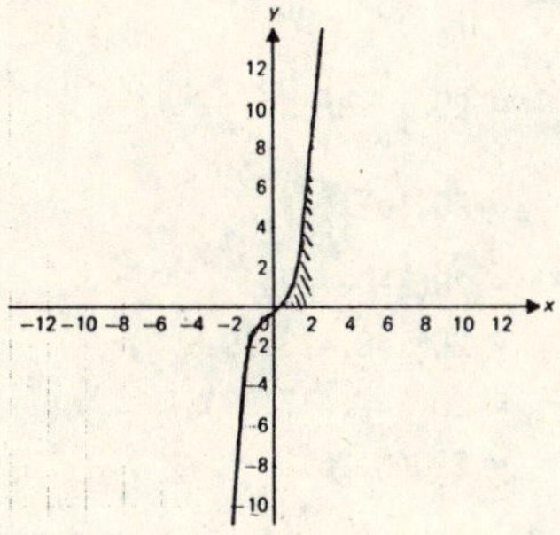

$\int_0^2 x^3\, dx$

$= \left[\frac{x^4}{4}\right]_0^2$

$= \frac{2^4}{4} - \frac{0^4}{4}$

$= 4 - 0$

$= 4$

17. Find the area under $y = x^2 + x + 1$ on $[2,3]$.

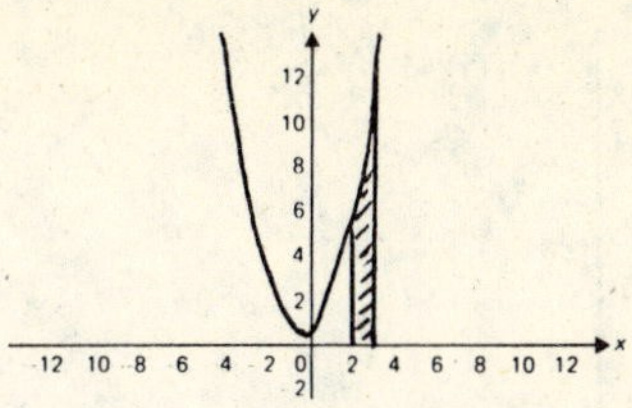

$\int_2^3 (x^2 + x + 1)\, dx$

$= \left[\frac{x^3}{3} + \frac{x^2}{2} + x\right]_2^3$

$= \left(\frac{3^3}{3} + \frac{3^2}{2} + 3\right) - \left(\frac{2^3}{3} + \frac{2^2}{2} + 2\right)$

$= \left(9 + \frac{9}{2} + 3\right) - \left(\frac{8}{3} + 2 + 2\right)$

$= 12 + \frac{9}{2} - \frac{8}{3} - 4$

$= 8 + \frac{27}{6} - \frac{16}{6}$

$= 8 + \frac{11}{6}$

$= 8 + 1\frac{5}{6}$

$= 9\frac{5}{6}$

19. Find the area under $y = 5 - x^2$ on $[-1,2]$.

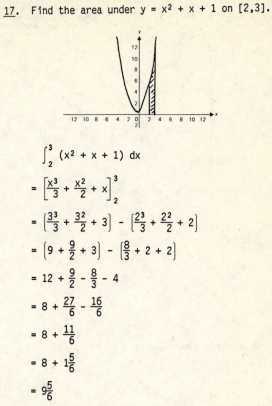

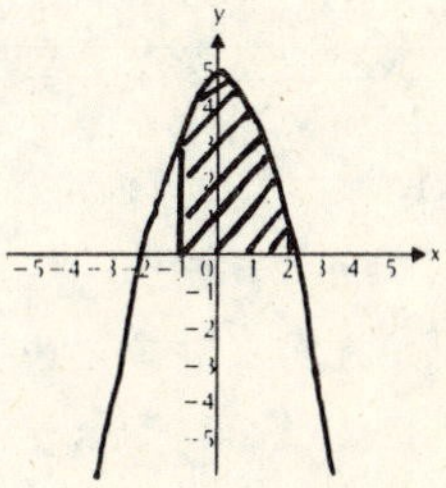

$\int_{-1}^2 (5 - x^2)\, dx$

$= \left[5x - \frac{x^3}{3}\right]_{-1}^2$

$= \left[5 \cdot 2 - \frac{2^3}{3}\right] - \left[5(-1) - \frac{(-1)^3}{3}\right]$

$= \left(10 - \frac{8}{3}\right) - \left(-5 + \frac{1}{3}\right)$

$= \left(\frac{30}{3} - \frac{8}{3}\right) - \left(-\frac{15}{3} + \frac{1}{3}\right)$

$= \frac{22}{3} - \left(-\frac{14}{3}\right)$

$= \frac{22}{3} + \frac{14}{3}$

$= \frac{36}{3}$

$= 12$

21. Find the area under $y = e^x$ on $[-1,5]$.

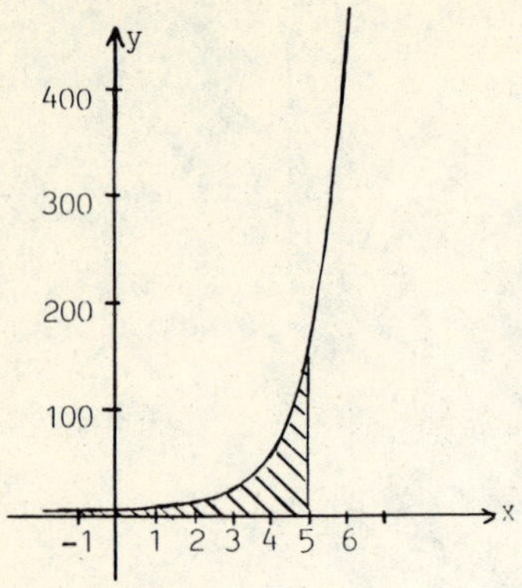

$\int_{-1}^{5} e^x \, dx$

$= \left[e^x\right]_{-1}^{5}$

$= e^5 - e^{-1}$

$= e^5 - \frac{1}{e}$

23. Find the area under $y = 2x - \frac{1}{x^2}$ on $[1,3]$.

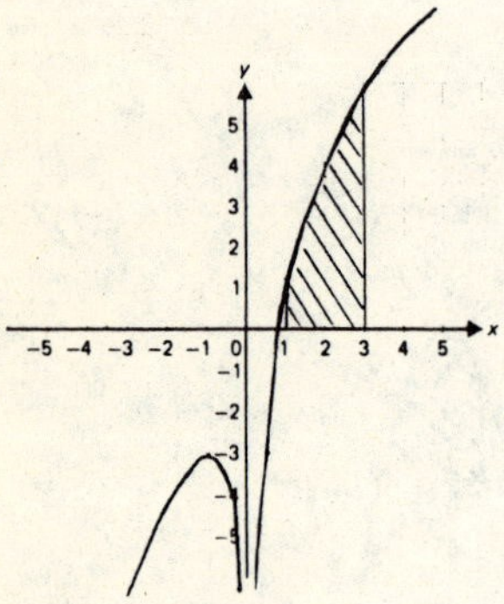

$\int_{1}^{3} \left(2x - \frac{1}{x^2}\right) dx$

$= \int_{1}^{3} (2x - x^{-2}) \, dx$

$= \left[x^2 - \frac{x^{-1}}{-1}\right]_{1}^{3}$

$= \left[x^2 + \frac{1}{x}\right]_{1}^{3}$

$= \left(3^2 + \frac{1}{3}\right) - \left(1^2 + \frac{1}{1}\right)$

$= \left(9 + \frac{1}{3}\right) - (1 + 1)$

$= 9\frac{1}{3} - 2$

$= 7\frac{1}{3}$

25. Find the area under

$f(x) = \begin{cases} 4 - x^2, & \text{if } x < 0 \\ 4, & \text{if } x \geq 0 \end{cases}$ on $[-2,3]$.

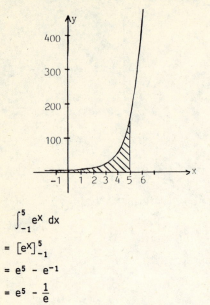

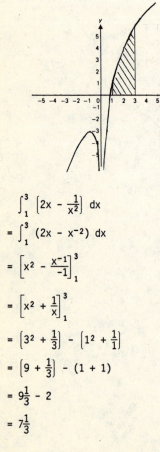

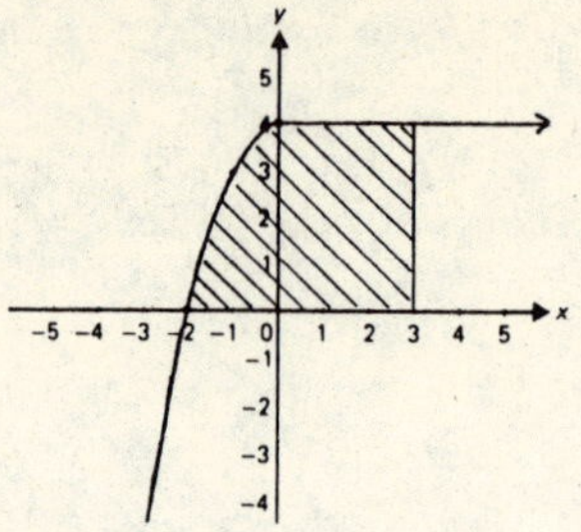

$\int_{-2}^{3} f(x) \, dx$

$= \int_{-2}^{0} f(x) \, dx + \int_{0}^{3} f(x) \, dx$

$= \int_{-2}^{0} (4 - x^2) \, dx + \int_{0}^{3} 4 \, dx$

$= \left[4x - \frac{x^3}{3}\right]_{-2}^{0} + \left[4x\right]_{0}^{3}$

$= \left\{\left[4\cdot0 - \frac{0^3}{3}\right] - \left[4(-2) - \frac{(-2)^3}{3}\right]\right\} + \left\{[4\cdot3] - [4\cdot0]\right\}$

$= (0 - 0) - \left(-8 + \frac{8}{3}\right) + 12 - 0$

$= -\left(-\frac{24}{3} + \frac{8}{3}\right) + 12$

$= -\left(-\frac{16}{3}\right) + 12$

$= \frac{16}{3} + \frac{36}{3}$

$= \frac{52}{3}$, or $17\frac{1}{3}$

27. a) Accumulated sales through day 5 are

$\int_{0}^{5} S'(t) \, dt = \int_{0}^{5} 20e^t \, dt$

$= 20 \int_{0}^{5} e^t \, dt$

$= 20 \left[e^t\right]_{0}^{5}$

$= 20(e^5 - e^0)$

$= 20(148.413159 - 1)$

$= 20(147.413159)$

$\approx \$2948.26$

b) Accumulated sales from the 2nd through the 5th day are

$\int_{1}^{5} S'(t) \, dt = \int_{1}^{5} 20e^t \, dt$

$= 20 \int_{1}^{5} e^t \, dt$

$= 20 \left[e^t\right]_{1}^{5}$

$= 20(e^5 - e^1)$

$= 20(148.413159 - 2.718282)$

$= 20(145.694877)$

$\approx \$2913.90$

c) Accumulated sales through day k are

$$\int_0^k S'(t)\,dt = \int_0^k 20e^t\,dt$$
$$= 20\int_0^k e^t\,dt$$
$$= 20\left[e^t\right]_0^k$$
$$= 20(e^k - e^0)$$
$$= 20(e^k - 1)$$

We set this equal to \$20,000 and solve for k.

$$20(e^k - 1) = 20{,}000$$
$$e^k - 1 = 1000$$
$$e^k = 1001$$
$$\ln e^k = \ln 1001$$
$$k = \ln 1001$$
$$k = 6.908755$$

Thus, accumulated sales will exceed \$20,000 on the 7th day.

29. $C'(x) = 0.0003x^2 - 0.2x + 50$

$$C(x) = \int_{200}^{400} (0.0003x^2 - 0.2x + 50)\,dx$$
$$= \left[0.0003\cdot\frac{x^3}{3} - 0.2\cdot\frac{x^2}{2} + 50x\right]_{200}^{400}$$
$$= \left[0.0001x^3 - 0.1x^2 + 50x\right]_{200}^{400}$$
$$= [0.0001(400^3) - 0.1(400^2) + 50(400)] - [0.0001(200^3) - 0.1(200^2) + 50(200)]$$
$$= (6400 - 16{,}000 + 20{,}000) - (800 - 4000 + 10{,}000)$$
$$= 10{,}400 - 6800$$
$$= \$3600$$

31. $R'(x) = 200x - 1080$

$$R(x) = \int_{1000}^{1300} (200x - 1080)dx$$
$$= \left[100x^2 - 1080x\right]_{1000}^{1300}$$
$$= [100(1300)^2 - 1080\cdot 1300] - [100(1000)^2 - 1080\cdot 1000]$$
$$= (169{,}000{,}000 - 1{,}404{,}000) - (100{,}000{,}000 - 1{,}080{,}000)$$
$$= 167{,}596{,}000 - 98{,}920{,}000$$
$$= \$68{,}676{,}000$$

33. $s(t) = \int v(t)\,dt$

To find the distance traveled (s) from the first through the third hour we integrate. We find the area under $v(t) = 4t^3 + 2t$ on [0,3].

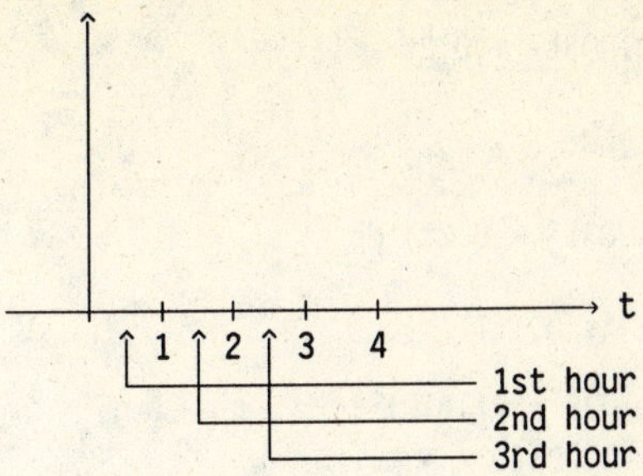

$$\int_0^3 v(t)\,dt$$
$$= \int_0^3 (4t^3 + 2t)\,dt$$
$$= \left[t^4 + t^2\right]_0^3$$
$$= (3^4 + 3^2) - (0^4 + 0^2)$$
$$= 81 + 9$$
$$= 90$$

35. a) $P'(t) = 1200e^{0.32t}$

$$P(t) = \int_0^{20} 1200e^{0.32t}\,dt$$
$$= \left[\frac{1200}{0.32}e^{0.32t}\right]_0^{20}$$
$$= \left[3750e^{0.32t}\right]_0^{20}$$
$$= 3750e^{0.32(20)} - 3750e^{0.32(0)}$$
$$= 3750e^{6.4} - 3750$$
$$\approx 2{,}253{,}169$$

b) The population through day k is

$$\int_0^k 1200e^{0.32t}\,dt$$
$$= \left[\frac{1200}{0.32}e^{0.32t}\right]_0^k$$
$$= \left[3750e^{0.32t}\right]_0^k$$
$$= 3750e^{0.32k} - 3750e^{0.32(0)}$$
$$= 3750e^{0.32k} - 3750$$

We set this equal to 4,000,000 and solve for k.

$$3750e^{0.32k} - 3750 = 4{,}000{,}000$$
$$3750e^{0.32k} = 4{,}003{,}750$$
$$e^{0.32k} = \frac{4{,}003{,}750}{3750}$$
$$\ln e^{0.32k} = \ln\frac{4{,}003{,}750}{3750}$$
$$0.32k = \ln\frac{4{,}003{,}750}{3750}$$
$$k = \frac{\ln\frac{4{,}003{,}750}{3750}}{0.32}$$
$$k \approx 21.8 \text{ days}$$

37. $M'(t) = -0.003t^2 + 0.2t$

$\int_0^{10} M'(t)\,dt$

$= \int_0^{10} (-0.003t^2 + 0.2t)\,dt$

$= \left[-0.001t^3 + 0.1t^2\right]_0^{10}$

$= [-0.001(10)^3 + 0.1(10)^2] - [-0.001(0)^3 + 0.1(0)^2]$

$= [-1 + 10] - [0 + 0]$

$= 9$

Thus, 9 words are memorized in the first 10 minutes.

39. $\int_1^2 (4x + 3)(5x - 2)\,dx$

$= \int_1^2 (20x^2 + 7x - 6)\,dx$

$= \left[20 \cdot \frac{x^3}{3} + 7 \cdot \frac{x^2}{2} - 6x\right]_1^2$

$= \left(20\cdot \frac{2^3}{3} + 7\cdot \frac{2^2}{2} - 6\cdot 2\right) - \left(20\cdot \frac{1^3}{3} + 7\cdot \frac{1^2}{2} - 6\cdot 1\right)$

$= \left(\frac{160}{3} + 14 - 12\right) - \left(\frac{20}{3} + \frac{7}{2} - 6\right)$

$= \left(53\frac{1}{3} + 14 - 12\right) - \left(\frac{40}{6} + \frac{21}{6} - \frac{36}{6}\right)$

$= 55\frac{1}{3} - 4\frac{1}{6}$

$= 51\frac{1}{6}$, or $\frac{307}{6}$

41. $\int_0^1 (t + 1)^3\,dt$

$= \int_0^1 (t^3 + 3t^2 + 3t + 1)\,dt$

$= \left[\frac{t^4}{4} + t^3 + \frac{3}{2}t^2 + t\right]_0^1$

$= \left(\frac{1^4}{4} + 1^3 + \frac{3}{2} \cdot 1^2 + 1\right) - \left(\frac{0^4}{4} + 0^3 + \frac{3}{2} \cdot 0^2 + 0\right)$

$= \left(\frac{1}{4} + 1 + \frac{3}{2} + 1\right) - 0$

$= 3\frac{3}{4}$, or $\frac{15}{4}$

43. $\int_1^3 \frac{t^5 - t}{t^3}\,dt$

$= \int_1^3 (t^2 - t^{-2})\,dt$

$= \left[\frac{t^3}{3} - \frac{t^{-1}}{-1}\right]_1^3$

$= \left[\frac{t^3}{3} + \frac{1}{t}\right]_1^3$

$= \left(\frac{3^3}{3} + \frac{1}{3}\right) - \left(\frac{1^3}{3} + \frac{1}{1}\right)$

$= 9\frac{1}{3} - 1\frac{1}{3}$

$= 8$

45. $\int_3^5 \frac{x^2 - 4}{x - 2}\,dx$

$= \int_3^5 \frac{(x - 2)(x + 2)}{x - 2}\,dx$

$= \int_3^5 (x + 2)\,dx$

$= \left[\frac{x^2}{2} + 2x\right]_3^5$

$= \left(\frac{5^2}{2} + 2\cdot 5\right) - \left(\frac{3^2}{2} + 2\cdot 3\right)$

$= \left(\frac{25}{2} + 10\right) - \left(\frac{9}{2} + 6\right)$

$= \frac{45}{2} - \frac{21}{2}$

$= \frac{24}{2}$

$= 12$

47. See the answer section in the text.

49. See the answer section in the text.

Exercise Set 5.4

1. First graph the system of equations and shade the region bounded by the graphs.

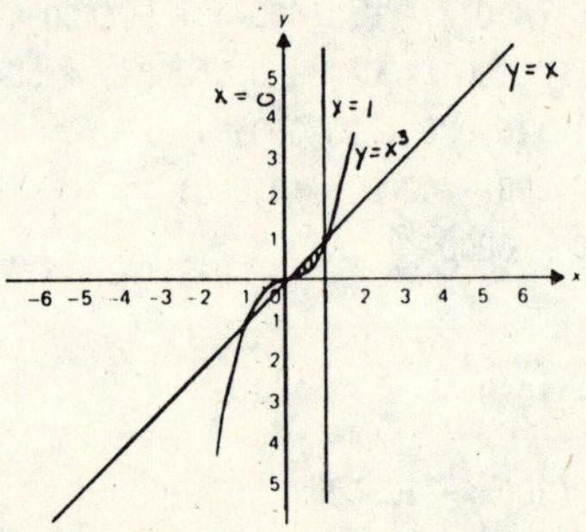

$y = x$

$y = x^3$

$x = 0$

$x = 1$

Here the boundaries are easily determined by looking at the graph. Note which is the upper graph. Here it is $x \geqslant x^3$ over the interval [0,1].

$\int_0^1 (x - x^3)\,dx$

$= \left[\frac{x^2}{2} - \frac{x^4}{4}\right]_0^1$

$= \left(\frac{1^2}{2} - \frac{1^4}{4}\right) - \left(\frac{0^2}{2} - \frac{0^4}{4}\right)$

$= \frac{1}{2} - \frac{1}{4}$

$= \frac{1}{4}$

3. First graph the system of equations and shade the region bounded by the graphs.

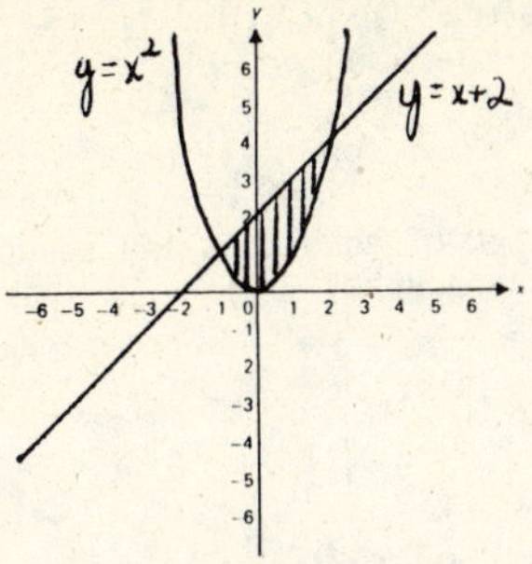

$y = x + 2$

$y = x^2$

Here the boundaries are easily determined by the graph. Note which is the upper graph. Here it is $(x + 2) \geqslant x^2$ over the interval $[-1,2]$.

Compute the area as follows:

$$\int_{-1}^{2} [(x + 2) - x^2]\,dx$$

$$= \int_{-1}^{2} (-x^2 + x + 2)\,dx$$

$$= \left[-\frac{x^3}{3} + \frac{x^2}{2} + 2x\right]_{-1}^{2}$$

$$= \left[-\frac{2^3}{3} + \frac{2^2}{2} + 2\cdot2\right] - \left[-\frac{(-1)^3}{3} + \frac{(-1)^2}{2} + 2(-1)\right]$$

$$= \left(-\frac{8}{3} + 2 + 4\right) - \left(\frac{1}{3} + \frac{1}{2} - 2\right)$$

$$= -\frac{8}{3} + 2 + 4 - \frac{1}{3} - \frac{1}{2} + 2$$

$$= -\frac{9}{3} - \frac{1}{2} + 2 + 4 + 2$$

$$= 4\frac{1}{2}, \text{ or } \frac{9}{2}$$

5. First graph the system of equations and shade the region bounded by the graphs.

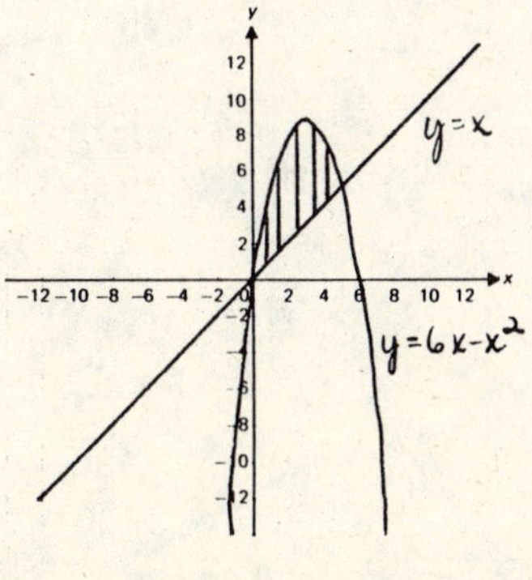

$y = 6x - x^2$

$y = x$

Here the boundaries are easily determined by the graph. Note which is the upper graph. Here it is $(6x - x^2) \geqslant x$ over the interval $[0,5]$.

Compute the area as follows:

$$\int_{0}^{5} [(6x - x^2) - x]\,dx$$

$$= \int_{0}^{5} (-x^2 + 5x)\,dx$$

$$= \left[-\frac{x^3}{3} + 5\cdot\frac{x^2}{2}\right]_{0}^{5}$$

$$= \left(-\frac{5^3}{3} + 5\cdot\frac{5^2}{2}\right) - \left(-\frac{0^3}{3} + 5\cdot\frac{0^2}{2}\right)$$

$$= \left(-\frac{125}{3} + \frac{125}{2}\right) - 0$$

$$= -\frac{250}{6} + \frac{375}{6}$$

$$= \frac{125}{6}$$

7. First graph the system of equations and shade the region bounded by the graphs.

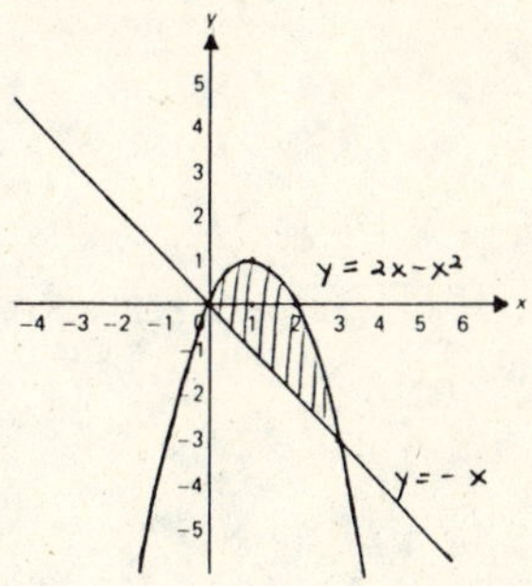

$y = 2x - x^2$

$y = -x$

The boundaries are easily determined by looking at the graph. Note which is the upper graph over the shaded region. Here it is $(2x - x^2) \geqslant -x$ over the interval $[0,3]$.

Compute the area as follows:

$$\int_{0}^{3} [(2x - x^2) - (-x)]\,dx$$

$$= \int_{0}^{3} (3x - x^2)\,dx$$

$$= \left[\frac{3}{2}x^2 - \frac{x^3}{3}\right]_{0}^{3}$$

$$= \left(\frac{3}{2}\cdot 3^2 - \frac{3^3}{3}\right) - \left(\frac{3}{2}\cdot 0^2 - \frac{0^3}{3}\right)$$

$$= \frac{27}{2} - \frac{27}{3}$$

$$= \frac{81}{6} - \frac{54}{6}$$

$$= \frac{27}{6}$$

$$= \frac{9}{2}$$

9. First graph the system of equations and shade the region bounded by the graphs.

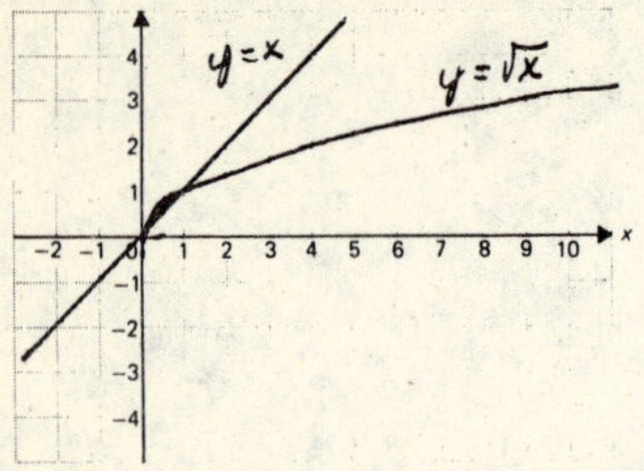

$y = x$
$y = \sqrt{x}$

The boundaries are easily determined by looking at the graph. Note which is the upper graph over the shaded region. Here it is $\sqrt{x} \geqslant x$ over the interval [0,1].

Compute the area as follows:

$$\int_0^1 (\sqrt{x} - x)\,dx$$

$$= \int_0^1 (x^{1/2} - x)\,dx$$

$$= \left[\frac{2}{3}x^{3/2} - \frac{1}{2}x^2\right]_0^1$$

$$= \left(\frac{2}{3}\cdot 1^{3/2} - \frac{1}{2}\cdot 1^2\right) - \left(\frac{2}{3}\cdot 0^{3/2} - \frac{1}{2}\cdot 0^2\right)$$

$$= \frac{2}{3} - \frac{1}{2} - 0$$

$$= \frac{4}{6} - \frac{3}{6}$$

$$= \frac{1}{6}$$

11. Graph the system of equations and shade the region bounded by the graphs.

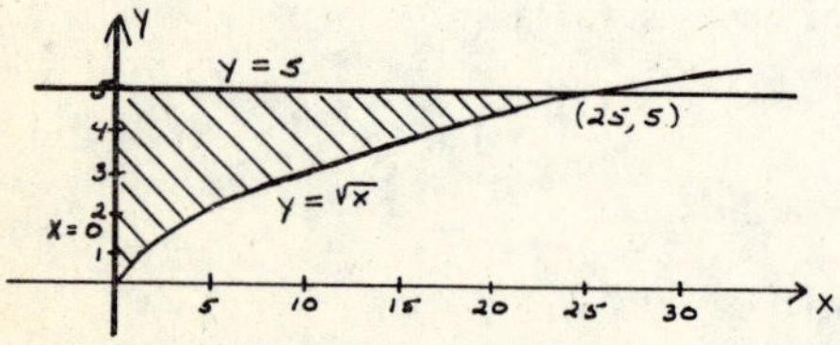

$y = 5$
$y = \sqrt{x}$
$x = 0$

The boundaries are easily determined by looking at the graphs. Here $5 \geqslant \sqrt{x}$ over the interval [0,25].

Compute the area as follows:

$$\int_0^{25} (5 - \sqrt{x})\,dx$$

$$= \int_0^{25} (5 - x^{1/2})\,dx$$

$$= \left[5x - \frac{x^{3/2}}{\frac{3}{2}}\right]_0^{25}$$

$$= \left[5x - \frac{2}{3}x^{3/2}\right]_0^{25}$$

$$= \left(5\cdot 25 - \frac{2}{3}\cdot 25^{3/2}\right) - \left(5\cdot 0 - \frac{2}{3}\cdot 0^{3/2}\right)$$

$$= 125 - \frac{250}{3} - 0 \qquad [25^{3/2} = (5^2)^{3/2} = 5^3 = 125]$$

$$= \frac{375}{3} - \frac{250}{3}$$

$$= \frac{125}{3}, \text{ or } 41\frac{2}{3}$$

13. First graph the system of equations and shade the region bounded by the graphs.

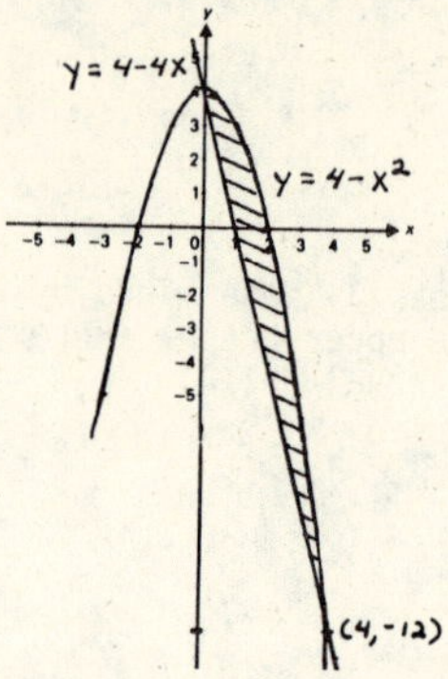

$y = 4 - x^2$
$y = 4 - 4x$

Then determine the first coordinates of possible points of intersection by solving the system of equations as follows. At the points of intersection, $y = 4 - x^2$ and $y = 4 - 4x$, so

$$4 - x^2 = 4 - 4x$$
$$0 = x^2 - 4x$$
$$0 = x(x - 4)$$

$x = 0$ or $x = 4$

Thus the interval with which we are concerned is [0,4]. Note that $4 - x^2 \geqslant 4 - 4x$ over the interval [0,4].

Compute the area as follows:

$$\int_0^4 [(4 - x^2) - (4 - 4x)]\,dx$$

$$= \int_0^4 (-x^2 + 4x)\,dx$$

$$= \left[-\frac{x^3}{3} + 2x^2\right]_0^4$$

$$= \left(-\frac{4^3}{3} + 2\cdot 4^2\right) - \left(-\frac{0^3}{3} + 2\cdot 0^2\right)$$

$$= -\frac{64}{3} + 32$$

$$= -\frac{64}{3} + \frac{96}{3}$$

$$= \frac{32}{3}$$

15. First graph the system of equations and shade the region bounded by the graphs.

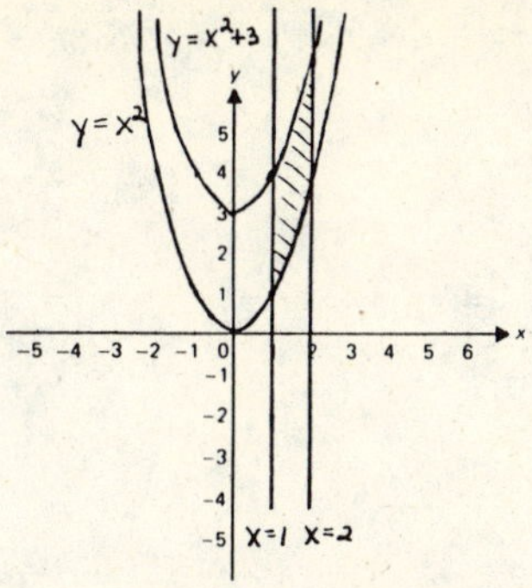

$y = x^2 + 3$

$y = x^2$

$x = 1$

$x = 2$

From the graph we can easily determine the interval with which we are concerned. Here $x^2 + 3 \geqslant x^2$ over the interval [1,2].

Compute the area as follows:

$$\int_1^2 [(x^2 + 3) - x^2]\, dx$$

$$= \int_1^2 3\, dx$$

$$= [3x]_1^2$$

$$= 3\cdot 2 - 3\cdot 1$$

$$= 6 - 3$$

$$= 3$$

17. $f(x) \geqslant g(x)$ on [-5,-1], and $g(x) \geqslant f(x)$ on [-1,3]. We use two integrals to find the total area.

$$\int_{-5}^{-1} [(x^3 + 3x^2 - 9x - 12) - (4x + 3)]\, dx +$$

$$\int_{-1}^{3} [(4x + 3) - (x^3 + 3x^2 - 9x - 12)]\, dx$$

$$= \int_{-5}^{-1} (x^3 + 3x^2 - 13x - 15)\, dx +$$

$$\int_{-1}^{3} (-x^3 - 3x^2 + 13x + 15)\, dx$$

$$= \left[\frac{x^4}{4} + x^3 - \frac{13x^2}{2} - 15x\right]_{-5}^{-1} +$$

$$\left[-\frac{x^4}{4} - x^3 + \frac{13x^2}{2} + 15x\right]_{-1}^{3}$$

$$= \left[\frac{(-1)^4}{4} + (-1)^3 - \frac{13(-1)^2}{2} - 15(-1)\right] -$$

$$\left[\frac{(-5)^4}{4} + (-5)^3 - \frac{13(-5)^2}{2} - 15(-5)\right] +$$

$$\left[-\frac{3^4}{4} - 3^3 + \frac{13(3)^2}{2} + 15\cdot 3\right] -$$

$$\left[-\frac{(-1)^4}{4} - (-1)^3 + \frac{13(-1)^2}{2} + 15(-1)\right]$$

$$= \left[\frac{1}{4} - 1 - \frac{13}{2} + 15\right] - \left[\frac{625}{4} - 125 - \frac{325}{2} + 75\right] +$$

$$\left[-\frac{81}{4} - 27 + \frac{117}{2} + 45\right] - \left[-\frac{1}{4} + 1 + \frac{13}{2} - 15\right]$$

$$= \frac{31}{4} + \frac{225}{4} + \frac{225}{4} + \frac{31}{4}$$

$$= 128$$

19. $f(x) \geqslant g(x)$ on [1,4]. We find the area.

$$\int_1^4 [(4x - x^2) - (x^2 - 6x + 8)]\, dx$$

$$= \int_1^4 (-2x^2 + 10x - 8)\, dx$$

$$= \left[-\frac{2x^3}{3} + 5x^2 - 8x\right]_1^4$$

$$= \left[-\frac{2\cdot 4^3}{3} + 5\cdot 4^2 - 8\cdot 4\right] - \left[-\frac{2\cdot 1^3}{3} + 5\cdot 1^2 - 8\cdot 1\right]$$

$$= \left[-\frac{128}{3} + 80 - 32\right] - \left[-\frac{2}{3} + 5 - 8\right]$$

$$= \frac{16}{3} + \frac{11}{3} = \frac{27}{3}$$

$$= 9$$

21. a) $-0.003t^2 \geqslant -0.009t^2$

Thus, $M'(t) \geqslant m'(t)$, so subject B has the higher rate of memorization.

b) $$\int_0^{10} [(-0.003t^2 + 0.2t)-(-0.009t^2 + 0.2t)]\, dt$$

$$= \int_0^{10} 0.006t^2\, dt$$

$$= [0.002t^3]_0^{10}$$

$$= 0.002\cdot 10^3 - 0.002\cdot 0^3$$

$$= 2 - 0$$

$$= 2$$

Subject B memorizes 2 more words than subject A during the first 10 minutes.

23. First graph the system of equations and shade the region bounded by the graph.

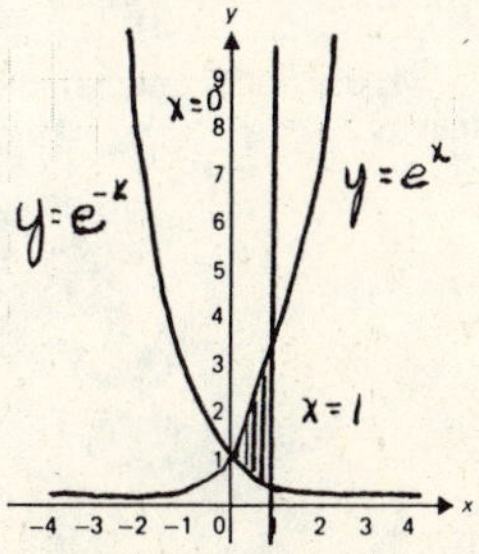

$y = e^x$

$y = e^{-x}$

$x = 0$

$x = 1$

From the graph we can easily determine the interval with which we are concerned. Here $e^x \geqslant e^{-x}$ over the interval [0,1].

Compute the area as follows:

$$\int_0^1 (e^x - e^{-x})\, dx$$

$$= [e^x + e^{-x}]_0^1$$

$$= (e^1 + e^{-1}) - (e^0 + e^{-0})$$

$$= \left[e + \frac{1}{e}\right] - (1 + 1)$$

$$= e + \frac{1}{e} - 2$$

$$= \frac{e^2 - 2e + 1}{e}$$

$$= \frac{(e - 1)^2}{e}$$

25. First graph the system of equations and shade the region bounded by the graph.

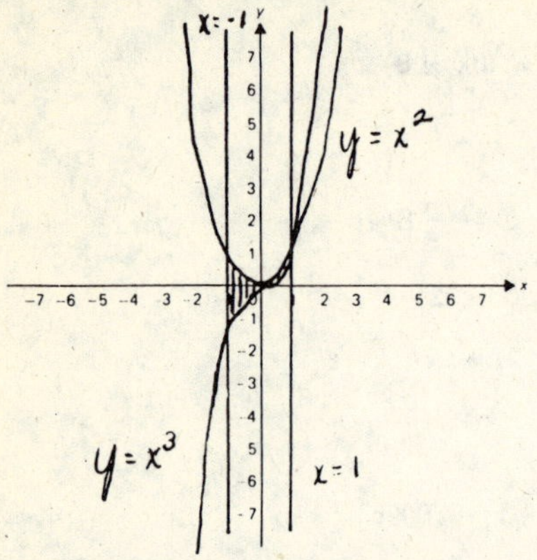

$y = x^2$
$y = x^3$
$x = -1$
$x = 1$

From the graph we can easily determine the interval with which we are concerned. Here $x^2 \geqslant x^3$ over the interval $[-1,1]$.

Compute the area as follows:

$$\int_{-1}^{1} (x^2 - x^3)\, dx$$

$$= \left[\frac{x^3}{3} - \frac{x^4}{4}\right]_{-1}^{1}$$

$$= \left(\frac{1^3}{3} - \frac{1^4}{4}\right) - \left(\frac{(-1)^3}{3} - \frac{(-1)^4}{4}\right)$$

$$= \left(\frac{1}{3} - \frac{1}{4}\right) - \left(-\frac{1}{3} - \frac{1}{4}\right)$$

$$= \frac{1}{3} - \frac{1}{4} + \frac{1}{3} + \frac{1}{4}$$

$$= \frac{2}{3}$$

27. First graph the system of equations and shade the region bounded by the graph.

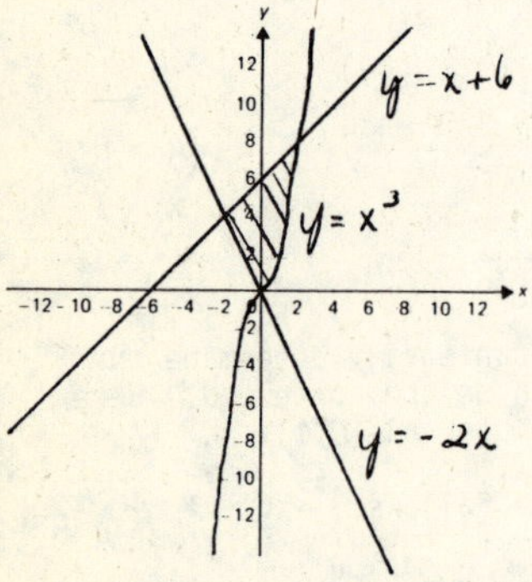

$y = x + 6$
$y = -2x$
$y = x^3$

From the graph we can determine the two intervals with which we are concerned. Here $(x + 6) \geqslant -2x$ over the interval $[-2,0]$ and $(x + 6) \geqslant x^3$ over the interval $[0,2]$.

We compute each area and then add.

$$\int_{-2}^{0} [(x + 6) - (-2x)]\, dx$$

$$= \int_{-2}^{0} (3x + 6)\, dx$$

$$= \left[\frac{3}{2}x^2 + 6x\right]_{-2}^{0}$$

$$= \left(\frac{3}{2}\cdot 0^2 + 6\cdot 0\right) - \left[\frac{3}{2}(-2)^2 + 6(-2)\right]$$

$$= 0 - (6 - 12)$$

$$= -(-6)$$

$$= 6$$

$$\int_{0}^{2} [(x + 6) - x^3]\, dx$$

$$= \int_{0}^{2} (-x^3 + x + 6)\, dx$$

$$= \left[-\frac{x^4}{4} + \frac{x^2}{2} + 6x\right]_{0}^{2}$$

$$= \left(-\frac{2^4}{4} + \frac{2^2}{2} + 6\cdot 2\right) - \left(-\frac{0^4}{4} + \frac{0^2}{2} + 6\cdot 0\right)$$

$$= -4 + 2 + 12 - 0$$

$$= 10$$

The total area is 6 + 10, or 16.

29. First we find the first coordinates of the relative extrema.

$$y = 3x^5 - 20x^3$$

$$\frac{dy}{dx} = 15x^4 - 60x^2$$

The derivative exists for all real numbers. We solve $\frac{dy}{dx} = 0$.

$$15x^4 - 60x^2 = 0$$

$$15x^2(x^2 - 4) = 0$$

$$15x^2(x + 2)(x - 2) = 0$$

$$x^2 = 0 \quad \text{or} \quad x + 2 = 0 \quad \text{or} \quad x - 2 = 0$$

$$x = 0 \quad \text{or} \quad x = -2 \quad \text{or} \quad x = 2$$

We use the Second Derivative Test.

$$\frac{d^2y}{dx^2} = 60x^3 - 120x$$

When $x = -2$, $\frac{d^2y}{dx^2} = 60(-2)^3 - 120(-2) = -240 < 0$, so there is a relative maximum at $x = -2$. When $x = 2$, $\frac{d^2y}{dx^2} = 60\cdot 2^3 - 120\cdot 2 = 240 > 0$, so there is a relative minimum at $x = 2$.

When $x = 0$, $\frac{d^2y}{dx^2} = 60\cdot 0^3 - 120\cdot 0 = 0$, so the test fails. We use the first derivative and test a value in $(-2,0)$ and a value in $(0,2)$.

Test -1, $15\cdot 1^4 - 60\cdot 1^2 = -45 < 0$

Test 1, $15\cdot 1^4 - 60\cdot 1^2 = -45 < 0$

Thus, the function is decreasing on both intervals, so there is no relative extremum at $x = 0$.

We graph the region. We use different scales on the x- and y-axes. Note that the equation of the x-axis is y = 0.

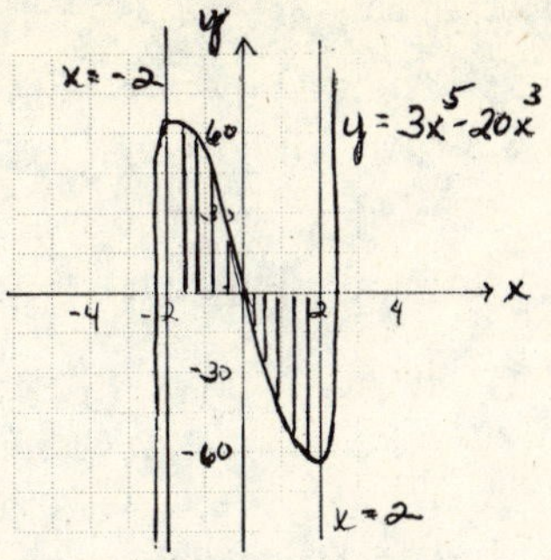

$y = 3x^5 - 20x^3$

$y = 0$

$x = -2$

$x = 2$

On the interval [-2,0], $3x^5 - 20x^3 \geqslant 0$, and on the interval [0,2], $0 \geqslant 3x^5 - 20x^3$. We evaluate two integrals to find the total area.

$$\int_{-2}^{0} [(3x^5 - 20x^3) - 0]\, dx + \int_{0}^{2} [0 - (3x^5 - 20x^3)\, dx$$

$$= \int_{-2}^{0} (3x^5 - 20x^3)\, dx + \int_{0}^{2} (-3x^5 + 20x^3)\, dx$$

$$= \left[\frac{x^6}{2} - 5x^4\right]_{-2}^{0} + \left[-\frac{x^6}{2} + 5x^4\right]_{0}^{2}$$

$$= \left[\frac{0^6}{2} - 5\cdot 0^4\right] - \left[\frac{(-2)^6}{2} - 5(-2)^4\right] + \left[-\frac{2^6}{2} + 5\cdot 2^4\right] - \left[-\frac{0^6}{2} + 5\cdot 0^4\right]$$

$= (0 - 0) - (32 - 80) + (-32 + 80) - (-0 + 0)$

$= 0 + 48 + 48 - 0$

$= 96$

31. See the answer section in the text.

Exercise Set 5.5

1. $\int \frac{3x^2\, dx}{7 + x^3}$

Let $u = 7 + x^3$, then $du = 3x^2\, dx$.

$= \int \frac{du}{u}$ (Substituting u for $7 + x^3$ and du for $3x^2\, dx$)

$= \int \frac{1}{u}\, du$

$= \ell n\, u + C$ (Using Formula C)

$= \ell n\, (7 + x^3) + C$

3. $\int e^{4x}\, dx$

Let $u = 4x$, then $du = 4\, dx$.

We do not have 4 dx. We only have dx and need to supply a 4. We do this by multiplying by $\frac{1}{4} \cdot 4$ as follows.

$\frac{1}{4} \cdot 4 \int e^{4x}\, dx$ (Multiplying by 1)

$= \frac{1}{4} \int 4e^{4x}\, dx$

$= \frac{1}{4} \int e^{4x}(4\, dx)$

$= \frac{1}{4} \int e^u\, du$ (Substituting u for 4x and du for 4 dx)

$= \frac{1}{4} e^u + C$ (Using Formula B)

$= \frac{1}{4} e^{4x} + C$

5. $\int e^{x/2}\, dx = \int e^{(1/2)x}\, dx$

Let $u = \frac{1}{2}x$, then $du = \frac{1}{2}\, dx$.

We do not have $\frac{1}{2}$ dx. We only have dx and need to supply a $\frac{1}{2}$ by multiplying by $2 \cdot \frac{1}{2}$ as follows.

$2 \cdot \frac{1}{2} \int e^{x/2}\, dx$ (Multiplying by 1)

$= 2 \int \frac{1}{2} e^{x/2}\, dx$

$= 2 \int e^{x/2} \left[\frac{1}{2}\, dx\right]$

$= 2 \int e^u\, du$ [Substituting u for $\frac{x}{2}$ and du for $\frac{1}{2}$ dx]

$= 2\, e^u + C$ (Using Formula B)

$= 2\, e^{x/2} + C$

7. $\int x^3 e^{x^4}\, dx$

Let $u = x^4$, then $du = 4x^3\, dx$.

We do not have $4x^3$ dx. We only have x^3 dx and need to supply a 4. We do this by multiplying by $\frac{1}{4} \cdot 4$ as follows.

$\frac{1}{4} \cdot 4 \int x^3 e^{x^4}\, dx$ (Multiplying by 1)

$= \frac{1}{4} \int 4x^3 e^{x^4}\, dx$

$= \frac{1}{4} \int e^{x^4}(4x^3\, dx)$

$= \frac{1}{4} \int e^u\, du$ (Substituting u for x^4 and du for $4x^3$ dx)

$= \frac{1}{4} \cdot e^u + C$ (Using Formula B)

$= \frac{1}{4} e^{x^4} + C$

9. $\int t^2 e^{-t^3}\,dt$

Let $u = -t^3$, then $du = -3t^2\,dt$.

We do not have $-3t^2\,dt$. We only have $t^2\,dt$. We need to supply a -3 by multiplying by $-\frac{1}{3}\cdot(-3)$ as follows.

$-\frac{1}{3}\cdot(-3)\int t^2 e^{-t^3}\,dt$ (Multiplying by 1)

$= -\frac{1}{3}\int -3t^2 e^{-t^3}\,dt$

$= -\frac{1}{3}\int e^{-t^3}(-3t^2\,dt)$

$= -\frac{1}{3}\int e^u\,du$ (Substituting u for $-t^3$ and du for $-3t^2\,dt$)

$= -\frac{1}{3}e^u + C$ (Using Formula B)

$= -\frac{1}{3}e^{-t^3} + C$

11. $\int \frac{\ln 4x\,dx}{x}$

Let $u = \ln 4x$, then $du = \frac{1}{x}\,dx$.

$= \int \ln 4x\left[\frac{1}{x}\,dx\right]$

$= \int u\,du$ $\left[\text{Substituting } u \text{ for } \ln 4x \text{ and } du \text{ for } \frac{1}{x}\,dx\right]$

$= \frac{u^2}{2} + C$ (Using Formula A)

$= \frac{(\ln 4x)^2}{2} + C$

13. $\int \frac{dx}{1+x}$

Let $u = 1 + x$, then $du = dx$.

$= \int \frac{du}{du}$ (Substituting u for $1 + x$ and du for dx)

$= \int \frac{1}{u}\,du$

$= \ln u + C$ (Using Formula C)

$= \ln(1 + x) + C$

15. $\int \frac{dx}{4-x}$

Let $u = 4 - x$, then $du = -dx$.

We do not have $-dx$. We only have dx and need to supply a -1 by multiplying by $-1\cdot(-1)$ as follows.

$-1\cdot(-1)\int \frac{dx}{4-x}$ (Multiplying by 1)

$= -1\int -1\cdot\frac{dx}{4-x}$

$= -\int \frac{1}{4-x}(-dx)$

$= -\int \frac{1}{u}\,du$ (Substituting u for $4 - x$ and du for $-dx$)

$= -\ln u + C$ (Using Formula C)

$= -\ln(4 - x) + C$

17. $\int t^2(t^3 - 1)^7\,dt$

Let $u = t^3 - 1$, then $du = 3t^2\,dt$.

We do not have $3t^2\,dt$. We only have $t^2\,dt$. We need to supply a 3 by multiplying by $\frac{1}{3}\cdot 3$ as follows.

$\frac{1}{3}\cdot 3\int t^2(t^3 - 1)^7\,dt$ (Multiplying by 1)

$= \frac{1}{3}\int 3t^2(t^3 - 1)^7\,dt$

$= \frac{1}{3}\int (t^3 - 1)^7\,3t^2\,dt$

$= \frac{1}{3}\int u^7\,du$ (Substituting u for $t^3 - 1$ and du for $3t^2\,dt$)

$= \frac{1}{3}\cdot\frac{u^8}{8} + C$ (Using Formula A)

$= \frac{1}{24}(t^3 - 1)^8 + C$

19. $\int (x^4 + x^3 + x^2)^7(4x^3 + 3x^2 + 2x)\,dx$

Let $u = x^4 + x^3 + x^2$, then $du = (4x^3 + 3x^2 + 2x)\,dx$.

$= \int u^7\,du$ [Substituting u for $x^4 + x^3 + x^2$ and du for $(4x^3 + 3x^2 + 2x)\,dx$]

$= \frac{u^8}{8} + C$ (Using Formula A)

$= \frac{1}{8}(x^4 + x^3 + x^2)^8 + C$

21. $\int \frac{e^x\,dx}{4 + e^x}$

Let $u = 4 + e^x$, then $du = e^x\,dx$.

$= \int \frac{du}{u}$ (Substituting u for $4 + e^x$ and du for $e^x\,dx$)

$= \int \frac{1}{u}\,du$

$= \ln u + C$ (Using Formula C)

$= \ln(4 + e^x) + C$

23. $\int \frac{\ln x^2}{x}\,dx$

Let $u = \ln x^2$, then $du = \left[2x\cdot\frac{1}{x^2}\right]dx = \frac{2}{x}\,dx$.

We do not have $\frac{2}{x}\,dx$. We only have $\frac{1}{x}\,dx$ and need to supply a 2 by multiplying by $\frac{1}{2}\cdot 2$ as follows.

$\frac{1}{2}\cdot 2\int \frac{\ln x^2}{x}\,dx$ (Multiplying by 1)

$= \frac{1}{2}\int 2\cdot\frac{\ln x^2}{x}\,dx$

$= \frac{1}{2}\int \ln x^2\cdot\frac{2}{x}\,dx$

$= \frac{1}{2}\int u\,du$ $\left[\text{Substituting } u \text{ for } \ln x^2 \text{ and } du \text{ for } \frac{2}{x}\,dx\right]$

$= \frac{1}{2}\cdot\frac{u^2}{2} + C$ (Using Formula A)

$= \frac{u^2}{4} + C$

$= \frac{1}{4} (\ln x^2)^2 + C$

or $\frac{1}{4} (2 \ln x)^2 + C = \frac{1}{4} \cdot 4(\ln x)^2 + C$

$= (\ln x)^2 + C$

25. $\int \frac{dx}{x \ln x}$

Let $u = \ln x$, then $du = \frac{1}{x}\, dx$.

$= \int \frac{1}{\ln x} \left[\frac{1}{x}\, dx\right]$

$= \int \frac{1}{u}\, du$ [Substituting u for $\ln x$ and du for $\frac{1}{x}\, dx$]

$= \ln u + C$ (Using Formula C)

$= \ln (\ln x) + C$

27. $\int \sqrt{ax + b}\, dx$, or $\int (ax + b)^{1/2}\, dx$

Let $u = ax + b$, then $du = a\, dx$.

We do not have a dx. We only have dx and need to supply an a by multiplying by $\frac{1}{a} \cdot a$ as follows.

$\frac{1}{a} \cdot a \int \sqrt{ax + b}\, dx$ (Multiplying by 1)

$= \frac{1}{a} \int a\sqrt{ax + b}\, dx$

$= \frac{1}{a} \int \sqrt{ax + b}\, (a\, dx)$

$= \frac{1}{a} \int \sqrt{u}\, du$ (Substituting u for ax + b and du for adx)

$= \frac{1}{a} \int u^{1/2}\, du$

$= \frac{1}{a} \cdot \frac{u^{3/2}}{\frac{3}{2}} + C$ (Using Formula A)

$= \frac{2}{3a} \cdot u^{3/2} + C$

$= \frac{2}{3a} (ax + b)^{3/2} + C$

29. $\int b\, e^{ax}\, dx$

$= b \int e^{ax}\, dx$

Let $u = ax$, then $du = a\, dx$.

We do not have a dx. We only have dx and need to supply an a by multiplying by $\frac{1}{a} \cdot a$ as follows.

$= b \cdot \frac{1}{a} \cdot a \int e^{ax}\, dx$

$= \frac{b}{a} \int a\, e^{ax}\, dx$

$= \frac{b}{a} \int e^{ax}\, (a\, dx)$

$= \frac{b}{a} \int e^{u}\, du$ (Substituting u for ax and du for adx)

$= \frac{b}{a} e^{u} + C$ (Using Formula B)

$= \frac{b}{a} e^{ax} + C$

31. $\int \frac{3x^2\, dx}{(1 + x^3)^5}$

Let $u = 1 + x^3$, then $du = 3x^2\, dx$.

$= \int \frac{1}{(1 + x^3)^5} \cdot 3x^2\, dx$

$= \int \frac{1}{u^5}\, du$ (Substituting u for $1 + x^3$ and du for $3x^2\, dx$)

$= \int u^{-5}\, du$

$= \frac{u^{-4}}{-4} + C$ (Using Formula A)

$= -\frac{1}{4u^4} + C$

$= -\frac{1}{4(1 + x^3)^4} + C$

33. $\int 7x \sqrt[3]{4 - x^2}\, dx$

$= 7 \int x \sqrt[3]{4 - x^2}\, dx$

Let $u = 4 - x^2$, then $du = -2x\, dx$.

We supply a -2 by multiplying by $-\frac{1}{2}(-2)$.

$= 7\left[-\frac{1}{2}\right](-2) \int x \sqrt[3]{4 - x^2}\, dx$

$= 7\left[-\frac{1}{2}\right] \int -2x \sqrt[3]{4 - x^2}\, dx$

$= -\frac{7}{2} \int \sqrt[3]{4 - x^2}\, (-2x\, dx)$

$= -\frac{7}{2} \int \sqrt[3]{u}\, du$ (Substituting u for $4 - x^2$ and du for $-2x\, dx$)

$= -\frac{7}{2} \int u^{1/3}\, du$

$= -\frac{7}{2} \cdot \frac{u^{4/3}}{\frac{4}{3}} + C$ (Using Formula A)

$= -\frac{21}{8} (4 - x^2)^{4/3} + C$

35. $\int_0^1 2x\, e^{x^2}\, dx$

First find the indefinite integral.

$\int 2x\, e^{x^2}\, dx$

Let $u = x^2$, then $du = 2x\, dx$.

$= \int e^{x^2}\, (2x\, dx)$

$= \int e^{u}\, du$ (Substituting u for x^2 and du for $2x\, dx$)

$= e^{u} + C$ (Using Formula B)

$= e^{x^2} + C$

Then evaluate the definite integral on [0,1].

$\int_0^1 2x\, e^{x^2}\, dx$

$= \left[e^{x^2}\right]_0^1$

$= e^{1^2} - e^{0^2}$

$= e - 1$

37. $\int_0^1 x(x^2 + 1)^5\, dx$

First find the indefinite integral.

$\int x(x^2 + 1)^5\, dx$

Let $u = x^2 + 1$, then $du = 2x\, dx$.

We only have x dx and need to supply a 2 by multiplying by $\frac{1}{2} \cdot 2$.

$\frac{1}{2} \cdot 2 \int x(x^2 + 1)^5\, dx$ (Multiplying by 1)

$= \frac{1}{2} \int 2x(x^2 + 1)^5\, dx$

$= \frac{1}{2} \int (x^2 + 1)^5 \cdot 2x\, dx$

$= \frac{1}{2} \int u^5\, du$ (Substituting u for $x^2 + 1$ and du for 2x dx)

$= \frac{1}{2} \cdot \frac{u^6}{6} + C$ (Using Formula A)

$= \frac{(x^2 + 1)^6}{12} + C$

Then evaluate the definite integral on [0,1].

$\int_0^1 x(x^2 + 1)^5\, dx$

$= \left[\frac{(x^2 + 1)^6}{12}\right]_0^1$

$= \frac{(1^2 + 1)^6}{12} - \frac{(0^2 + 1)^6}{12}$

$= \frac{64}{12} - \frac{1}{12}$

$= \frac{63}{12}$

$= \frac{21}{4}$

39. $\int_1^3 \frac{dt}{1 + t}$

First find the indefinite integral.

$\int \frac{dt}{1 + t}$

Let $u = 1 + t$, then $du = dt$.

$= \int \frac{du}{u}$ (Substituting u for 1 + t and du for dt)

$= \int \frac{1}{u}\, du$

$= \ln u + C$ (Using Formula C)

$= \ln (1 + t) + C$

Then evaluate the definite integral on [1,3].

$\int_1^3 \frac{dt}{1 + t}$

$= \left[\ln (1 + t)\right]_1^3$

$= \ln (1 + 3) - \ln (1 + 1)$

$= \ln 4 - \ln 2$

$= \ln \frac{4}{2}$

$= \ln 2$

41. $\int_1^4 \frac{2x + 1}{x^2 + x - 1}\, dx$

First find the indefinite integral.

$\int \frac{2x + 1}{x^2 + x - 1}\, dx$

Let $u = x^2 + x - 1$, then $du = (2x + 1)\, dx$.

$= \int \frac{1}{x^2 + x - 1} (2x + 1)\, dx$

$= \int \frac{1}{u}\, du$ [Substituting u for $x^2 + x - 1$ and du for $(2x + 1)$ dx]

$= \ln u + C$

$= \ln (x^2 + x - 1) + C$

Then evaluate the definite integral on [1,4].

$\int_1^4 \frac{2x + 1}{x^2 + x - 1}\, dx$

$= \left[\ln (x^2 + x - 1)\right]_1^4$

$= \ln (4^2 + 4 - 1) - \ln (1^2 + 1 - 1)$

$= \ln 19 - \ln 1$

$= \ln 19$ $(\ln 1 = 0)$

43. $\int_0^b e^{-x}\, dx$

First find the indefinite integral.

$\int e^{-x}\, dx$

Let $u = -x$, then $du = -dx$.

We only have dx and need to supply a -1 by multiplying by $-1 \cdot (-1)$.

$-1 \cdot (-1) \int e^{-x}\, dx$

$= - \int -e^{-x}\, dx$

$= - \int e^{-x}\, (-dx)$

$= - \int e^u\, du$ (Substituting u for -x and du for -dx)

$= -e^u + C$ (Using Formula B)

$= -e^{-x} + C$

Then evaluate the definite integral on [0,b].

$\int_0^b e^{-x}\, dx$

$= \left[-e^{-x}\right]_0^b$

$= (-e^{-b}) - (-e^{-0})$

$= -e^{-b} + e^0$

$= -e^{-b} + 1$

$= 1 - \frac{1}{e^b}$

45. $\int_0^b m\, e^{-mx}\, dx$

First find the indefinite integral.

$\int m\, e^{-mx}\, dx$ (m is a constant)

Let $u = -mx$, then $du = -m\, dx$.

We only have m dx and need to supply a -1 by multiplying by $-1\cdot(-1)$.

$-1\cdot(-1) \int m\, e^{-mx}\, dx$

$= - \int -m\, e^{-mx}\, dx$

$= - \int e^{-mx}\, (-m\, dx)$

$= - \int e^u\, du$ (Substituting u for -mx and du for -m dx)

$= -e^u + C$ (Using Formula B)

$= -e^{-mx} + C$

Then evaluate the definite integral on [0,b].

$\int_0^b m\, e^{-mx}\, dx$

$= \left[-e^{-mx}\right]_0^b$

$= (-e^{-mb}) - (-e^{-m\cdot 0})$

$= -e^{-mb} + 1$ $(e^{-m\cdot 0} = e^0 = 1)$

$= 1 - e^{-mb}$

$= 1 - \frac{1}{e^{mb}}$

47. $\int_0^4 (x - 6)^2\, dx$

First find the indefinite integral.

$\int (x - 6)^2\, dx$

Let $u = x - 6$, then $du = dx$.

$= \int u^2\, du$ (Substituting u for x - 6 and du for dx)

$= \frac{u^3}{3} + C$

$= \frac{(x - 6)^3}{3} + C$

Then evaluate the definite integral on [0,4].

$\int_0^4 (x - 6)^2\, dx$

$= \left[\frac{(x - 6)^3}{3}\right]_0^4$

$= \frac{(4 - 6)^3}{3} - \frac{(0 - 6)^3}{3}$

$= -\frac{8}{3} - \left[-\frac{216}{3}\right]$

$= -\frac{8}{3} + \frac{216}{3}$

$= \frac{208}{3}$

49. $\int_0^2 \frac{3x^2\, dx}{(1 + x^3)^5}$

From Exercise 31 we know that the indefinite integral is

$\int \frac{3x^2\, dx}{(1 + x^3)^5} = -\frac{1}{4(1 + x^3)^4} + C.$

Now we evaluate the definite integral on [0,2].

$\int_0^2 \frac{3x^2\, dx}{(1 + x^3)^5}$

$= \left[-\frac{1}{4(1 + x^3)^4}\right]_0^2$

$= \left[-\frac{1}{4(1 + 2^3)^4}\right] - \left[-\frac{1}{4(1 + 0^3)^4}\right]$

$= -\frac{1}{26{,}244} + \frac{1}{4}$

$= \frac{6560}{26{,}244}$

$= \frac{1640}{6561}$

51. $\int_0^{\sqrt{7}} 7x \sqrt[3]{1 + x^2}\, dx = 7 \int_0^{\sqrt{7}} x \sqrt[3]{1 + x^2}\, dx$

First find the indefinite integral.

$7 \int x \sqrt[3]{1 + x^2}\, dx$

Let $u = 1 + x^2$, then $du = 2x\, dx$.

$= 7 \cdot \frac{1}{2} \cdot 2 \int x \sqrt[3]{1 + x^2}\, dx$

$= 7 \cdot \frac{1}{2} \int 2x \sqrt[3]{1 + x^2}\, dx$

$= \frac{7}{2} \int \sqrt[3]{1 + x^2}\, (2x\, dx)$

$= \frac{7}{2} \int \sqrt[3]{u}\, du$

$= \frac{7}{2} \int u^{1/3}\, du$

$= \frac{7}{2} \cdot \frac{u^{4/3}}{\frac{4}{3}} + C$

$= \frac{21}{8} u^{4/3} + C$

$= \frac{21}{8} (1 + x^2)^{4/3} + C$

Then evaluate the definite integral on $[0, \sqrt{7}]$.

$7 \int_0^{\sqrt{7}} x \sqrt[3]{1 + x^2}\, dx$

$= \left[\frac{21}{8} (1 + x^2)^{4/3}\right]_0^{\sqrt{7}}$

$= \frac{21}{8} (1 + (\sqrt{7})^2)^{4/3} - \frac{21}{8} (1 + 0^2)^{4/3}$

$= \frac{21}{8} \cdot 8^{4/3} - \frac{21}{8} \cdot 1^{4/3}$

$= \frac{21}{8} \cdot 16 - \frac{21}{8} \cdot 1$

$= 42 - \frac{21}{8}$

$= \frac{315}{8}$

53. $D(p) = \int \frac{-2000p}{\sqrt{25 - p^2}}\,dp$

Let $u = 25 - p^2$, then $du = -2p\,dp$.
Note that $-2000 = -2 \cdot 1000$.

$= \int \frac{1000}{\sqrt{25 - p^2}} \cdot (-2p)\,dp$

$= \int \frac{1000}{\sqrt{u}}\,du$ (Substituting)

$= 1000 \int u^{-1/2}\,du$

$= 1000 \frac{u^{1/2}}{\frac{1}{2}} + C$

$= 2000\,u^{1/2} + C$

$= 2000\sqrt{25 - p^2} + C$

Use $D(3) = 13{,}000$ to find C.

$D(3) = 2000\sqrt{25 - 3^2} + C = 13{,}000$

$2000\sqrt{16} + C = 13{,}000$

$2000 \cdot 4 + C = 13{,}000$

$8000 + C = 13{,}000$

$C = 5000$

Then $D(p) = 2000\sqrt{25 - p^2} + 5000$.

55. a) We will evaluate each integral separately and then do the operations to find P(T).

Find $\int_0^T R'(t)\,dt$: First find the indefinite integral.

$\int 4000t\,dt = \frac{4000t^2}{2} + C = 2000t^2 + C$

Then evaluate the definite integral on [0,T].

$[2000t^2]_0^T = 2000T^2 - 2000 \cdot 0^2$

$= 2000T^2$

Find $\int_0^T V'(t)\,dt$: First find the indefinite integral.

$\int 25{,}000\,e^{-0.1t}\,dt$

Let $u = -0.1t$, then $du = -0.1\,dt$.

$= \int 25{,}000(-0.1)\left[\frac{1}{-0.1}\right] e^{-0.1t}\,dt$

$= \int 25{,}000\left[\frac{1}{-0.1}\right] e^{-0.1t}\,(-0.1)\,dt$

$= 25{,}000\left[\frac{1}{-0.1}\right] \int e^{-0.1t}\,(-0.1)\,dt$

$= 25{,}000(-10) \int e^u\,du$ $\left[\text{Substituting; } \frac{1}{-0.1} = -10\right]$

$= -250{,}000\,e^u + C$

$= -250{,}000\,e^{-0.1t} + C$

Then evaluate the definite integral on [0,T].

$-250{,}000\left[e^{-0.1t}\right]_0^T$

$= -250{,}000(e^{-0.1T} - e^{-0.1(0)})$

$= -250{,}000(e^{-0.1T} - 1)$ $(e^0 = 1)$

$P(T) = 2000T^2 + [-250{,}000(e^{-0.1T} - 1)] - 250{,}000$

$= 2000T^2 - 250{,}000\,e^{-0.1T} + 250{,}000 - 250{,}000$

$= 2000T^2 - 250{,}000\,e^{-0.1T}$

b) $P(10) = 2000(10)^2 - 250{,}000\,e^{-0.1(10)}$

$= 2000 \cdot 100 - 250{,}000\,e^{-1}$

$= 200{,}000 - 250{,}000\,e^{-1}$

$\approx \$108{,}030$

57. The area of the shaded region is the area between the curves $y = 0$ (the x-axis) and $y = x\sqrt{16 - x^2}$. On [-4,0], $x(16 - x^2) \leqslant 0$, and on [0,4], $x(16 - x^2 \geqslant 0$. We will calculate each portion of the area separately and then add them. On [-4,0]: First find the indefinite integral.

$\int (0 - x\sqrt{16 - x^2})\,dx$

$= \int -x\sqrt{16 - x^2}\,dx$

Let $u = 16 - x^2$, then $du = -2x\,dx$.

$= \frac{1}{2}(2) \int -x\sqrt{16 - x^2}\,dx$

$= \frac{1}{2} \int \sqrt{16 - x^2}\,(-2x)\,dx$

$= \frac{1}{2} \int \sqrt{u}\,du$

$= \frac{1}{2} \int u^{1/2}\,du$

$= \frac{1}{2} \cdot \frac{u^{3/2}}{\frac{3}{2}} + C$

$= \frac{1}{3} u^{3/2} + C$

$= \frac{1}{3}(16 - x^2)^{3/2} + C$

Then find the definite integral on [-4,0].

$\frac{1}{3}\left[(16 - x^2)^{3/2}\right]_{-4}^{0}$

$= \frac{1}{3}\left[(16 - 0^2)^{3/2} - (16 - (-4)^2)^{3/2}\right]$

$= \frac{1}{3}(16^{3/2} - 0^{3/2})$

$= \frac{1}{3} \cdot 64$

$= \frac{64}{3}$

On [0,4]: First find the indefinite integral.

$\int (x\sqrt{16 - x^2} - 0)\,dx$

$= \int x\sqrt{16 - x^2}\,dx$

Let $u = 16 - x^2$, then $du = -2x\,dx$.

$= -\frac{1}{2}(-2)\int x\sqrt{16 - x^2}\,dx$

$= -\frac{1}{2}\int \sqrt{16 - x^2}\,(-2x)\,dx$

$= -\frac{1}{2}\int \sqrt{u}\,du$

$= -\frac{1}{2}\int u^{1/2}\,du$

$= -\frac{1}{2}\frac{u^{3/2}}{\frac{3}{2}} + C$

$= -\frac{1}{3}u^{3/2} + C$

$= -\frac{1}{3}(16 - x^2)^{3/2} + C$

Then find the definite integral on [0,4].

$-\frac{1}{3}\Big[(16 - x^2)^{3/2}\Big]_0^4$

$= -\frac{1}{3}\Big[(16 - 4^2)^{3/2} - (16 - 0^2)^{3/2}\Big]$

$= -\frac{1}{3}(0^{3/2} - 16^{3/2})$

$= -\frac{1}{3}(-64)$

$= \frac{64}{3}$

The total area is $\frac{64}{3} + \frac{64}{3} = \frac{128}{3}$.

59. $\int 5x\sqrt{1 - 4x^2}\,dx$, or $5\int x(1 - 4x^2)^{1/2}\,dx$

Let $u = 1 - 4x^2$, then $du = -8x\,dx$.

We do not have $-8x\,dx$. We only have $x\,dx$ and need to supply a -8 by multiplying by $-\frac{1}{8}\cdot(-8)$ as follows.

$-\frac{1}{8}\cdot(-8)\cdot 5\int x(1 - 4x^2)^{1/2}\,dx$ (Multiplying by 1)

$= -\frac{5}{8}\int (1 - 4x^2)^{1/2}(-8x\,dx)$

$= -\frac{5}{8}\int u^{1/2}\,du$ (Substituting u for $1 - 4x^2$ and du for $-8x\,dx$)

$= -\frac{5}{8}\cdot\frac{u^{3/2}}{3/2} + C$ (Using Formula A)

$= -\frac{10}{24}u^{3/2} + C$

$= -\frac{5}{12}u^{3/2} + C$

$= -\frac{5}{12}(1 - 4x^2)^{3/2} + C$

61. $\int \frac{x^2}{e^{x^3}}\,dx$, or $\int x^2 e^{-x^3}\,dx$

Let $u = -x^3$, then $du = -3x^2\,dx$.

We do not have $-3x^2\,dx$. We only have $x^2\,dx$ and need to supply a -3 by multiplying by $-\frac{1}{3}\cdot(-3)$ as follows.

$-\frac{1}{3}\cdot(-3)\cdot\int x^2 e^{-x^3}\,dx$ (Multiplying by 1)

$= -\frac{1}{3}\int e^{-x^3}(-3x^2\,dx)$

$= -\frac{1}{3}\int e^u\,du$ (Substituting u for $-x^3$ and du for $-3x^2\,dx$)

$= -\frac{1}{3}\cdot e^u + C$ (Using Formula B)

$= -\frac{1}{3}e^{-x^3} + C$

63. $\int \frac{e^{1/t}}{t^2}\,dt$, or $\int e^{t^{-1}}\,t^{-2}\,dt$

Let $u = t^{-1}$, then $du = -t^{-2}\,dt$.

We do not have $-t^{-2}\,dt$. We only have $t^{-2}\,dt$ and need to supply a -1 by multiplying by $-1\cdot(-1)$ as follows.

$-1\cdot(-1)\cdot\int e^{t^{-1}}\,t^{-2}\,dt$ (Multiplying by 1)

$= -\int e^{t^{-1}}(-t^{-2}\,dt)$

$= -\int e^u\,du$ (Substituting u for t^{-1} and du for $-t^{-2}\,dt$)

$= -e^u + C$

$= -e^{t^{-1}} + C$

$= -e^{1/t} + C$

65. $\int \frac{dx}{x(\ell n\,x)^4}$, or $\int \frac{1}{x}(\ell n\,x)^{-4}\,dx$

Let $u = \ell n\,x$, then $du = \frac{1}{x}\,dx$.

$= \int u^{-4}\,du$ [Substituting u for $\ell n\,x$ and du for $\frac{1}{x}\,dx$]

$= \frac{u^{-3}}{-3} + C$ (Using Formula A)

$= -\frac{1}{3}u^{-3} + C$

$= -\frac{1}{3}(\ell n\,x)^{-3} + C$

67. $\int x^2 \sqrt{x^3 + 1}\, dx$, or $\int x^2(x^3 + 1)^{1/2}\, dx$

Let $u = x^3 + 1$, then $du = 3x^2\, dx$.

We do not have $3x^2\, dx$. We only have $x^2\, dx$ and need to supply a 3 by multiplying by $\frac{1}{3} \cdot 3$ as follows.

$\frac{1}{3} \cdot 3 \int x^2(x^3 + 1)^{1/2}\, dx$ (Multiplying by 1)

$= \frac{1}{3} \int (x^3 + 1)^{1/2}\, 3x^2\, dx$

$= \frac{1}{3} \int u^{1/2}\, du$ (Substituting u for $x^3 + 1$ and du for $3x^2\, dx$)

$= \frac{1}{3} \cdot \frac{u^{3/2}}{3/2} + C$ (Using Formula A)

$= \frac{2}{9} u^{3/2} + C$

$= \frac{2}{9} (x^3 + 1)^{3/2} + C$

69. $\int \frac{x - 3}{(x^2 - 6x)^{1/3}}\, dx$, or $\int (x - 3)(x^2 - 6x)^{-1/3}\, dx$

Let $u = x^2 - 6x$, then $du = (2x - 6)\, dx$, or $2(x - 3)\, dx$.

We do not have $2(x - 3)\, dx$. We only have $(x - 3)\, dx$ and need to supply a 2 by multiplying by $\frac{1}{2} \cdot 2$.

$\frac{1}{2} \cdot 2 \int (x - 3)(x^2 - 6x)^{-1/3}\, dx$ (Multiplying by 1)

$= \frac{1}{2} \int (x^2 - 6x)^{-1/3}\, 2(x - 3)\, dx$

$= \frac{1}{2} \int u^{-1/3}\, du$ [Substituting u for $x^2 - 6x$ and du for $2(x - 3)\, dx$]

$= \frac{1}{2} \cdot \frac{u^{2/3}}{2/3} + C$

$= \frac{3}{4} u^{2/3} + C$

$= \frac{3}{4} (x^2 - 6x)^{2/3} + C$

71. $\int \frac{t^3 \ln (t^4 + 8)}{t^4 + 8}\, dt$

Let $u = \ln (t^4 + 8)$, then $du = \frac{4t^3}{t^4 + 8}\, dt$.

We only have $\frac{t^3}{t^4 + 8}\, dt$ and need to supply a 4 by multiplying by $\frac{1}{4} \cdot 4$.

$\frac{1}{4} \cdot 4 \int \frac{t^3 \ln (t^4 + 8)}{t^4 + 8}\, dt$ (Multiplying by 1)

$= \frac{1}{4} \int \ln (t^4 + 8)\, \frac{4t^3}{t^4 + 8}\, dt$

$= \frac{1}{4} \int u\, du$ [Substituting u for $\ln (t^4 + 8)$ and du for $\frac{4t^3}{t^4 + 8}\, dt$]

$= \frac{1}{4} \cdot \frac{u^2}{2} + C$ (Using Formula A)

$= \frac{1}{8} u^2 + C$

$= \frac{1}{8} [\ln (t^4 + 8)]^2 + C$

73. $\int \frac{x^2 + 6x}{(x + 3)^2}\, dx$

$= \int \frac{x^2 + 6x + 9 - 9}{x^2 + 6x + 9}\, dx$ (Adding 9 - 9 in the numerator)

$= \int \left[\frac{x^2 + 6x + 9}{x^2 + 6x + 9} - \frac{9}{x^2 + 6x + 9}\right] dx$

$= \int \left[1 - \frac{9}{(x + 3)^2}\right] dx$

$= \int dx - 9 \int (x + 3)^{-2}\, dx$

Let $u = x + 3$, then $du = dx$.

$= \int du - 9 \int u^{-2}\, du$ (Substituting)

$= u - 9 \cdot \frac{u^{-1}}{-1} + C$ (Using Formula A)

$= u + \frac{9}{u} + C$

$= (x + 3) + \frac{9}{x + 3} + K$

$= x + \frac{9}{x + 3} + C$ $(3 + K = C$, a constant)

75. $\int \frac{t - 5}{t - 4}\, dt$

We first divide.

$$\begin{array}{r} 1 \\ t - 4 \enclose{longdiv}{t - 5} \\ \underline{t - 4} \\ -1 \end{array}$$

Thus, $\frac{t - 5}{t - 4} = 1 - \frac{1}{t - 4}$.

$= \int \left[1 - \frac{1}{t - 4}\right] dt$

$= \int dt - \int (t - 4)^{-1}\, dt$

Let $u = t - 4$, then $du = dt$.

$= \int du - \int u^{-1}\, du$ (Substituting)

$= u - \ln u + C$ (Using Formulas A and C)

$= (t - 4) - \ln (t - 4) + K$

$= t - \ln (t - 4) + C$ $(-4 + K = C$, a constant)

77. $\int \frac{dx}{e^x + 1}$, or $\int \frac{1}{e^x + 1}\, dx$

Note: $\frac{1}{e^x + 1} \cdot \frac{e^{-x}}{e^{-x}} = \frac{e^{-x}}{e^0 + e^{-x}} = \frac{e^{-x}}{1 + e^{-x}}$

$= \int \frac{e^{-x}}{1 + e^{-x}}\, dx$

Let $u = 1 + e^{-x}$, then $du = -e^{-x}\, dx$.

We only have $e^{-x}\, dx$ and need to supply a -1 by multiplying by $-1 \cdot (-1)$ as follows.

$= -1\cdot(-1) \int \frac{e^{-x}}{1 + e^{-x}}\,dx$

$= - \int \frac{1}{1 + e^{-x}} \cdot (-e^{-x})\,dx$

$= - \int \frac{1}{u}\,du$ (Substituting)

$= - \ell n\, u + C$ (Using Formula C)

$= - \ell n\,(1 + e^{-x}) + C$

79. $\int \frac{(\ell n\, x)^n}{x}\,dx$, or $\int \frac{1}{x} (\ell n\, x)^n\,dx$

Let $u = \ell n\, x$, then $du = \frac{1}{x}\,dx$.

$= \int u^n\,du$ (Substituting)

$= \frac{u^{n+1}}{n + 1} + C$ (Using Formula A)

$= \frac{(\ell n\, x)^{n+1}}{n + 1} + C$

81. $\int \frac{dx}{x\, \ell n\, x[\ell n\,(\ell n\, x)]}$

Let $u = \ell n\,(\ell n\, x)$, then $du = \frac{1}{x\, \ell n\, x}\,dx$.

$= \int \frac{1}{u}\,du$ (Substituting)

$= \ell n\, u + C$

$= \ell n\,[\ell n\,(\ell n\, x)] + C$

83. $\int 9x(7x^2 + 9)^n\,dx$

$= 9 \int x(7x^2 + 9)^n\,dx$

Let $u = 7x^2 + 9$, then $du = 14x\,dx$.

We only have x dx and need to supply a 14 by multiplying by $\frac{1}{14} \cdot 14$ as follows.

$\frac{1}{14} \cdot 14 \cdot 9 \int x(7x^2 + 9)^n\,dx$

$= \frac{9}{14} \int (7x^2 + 9)^n\, 14x\,dx$

$= \frac{9}{14} \int u^n\,du$ (Substituting)

$= \frac{9}{14} \cdot \frac{u^{n+1}}{n + 1} + C$ (Using Formula A)

$= \frac{9(7x^2 + 9)^{n+1}}{14(n + 1)} + C$

Exercise Set 5.6

1. $\int 5x\, e^{5x}\,dx = \int x(5e^{5x}\,dx)$

Let

$u = x$ and $dv = 5e^{5x}\,dx$.

Then

$du = dx$ and $v = e^{5x}$.

$$\int \underset{u}{x}\, \underset{dv}{(5e^{5x}\,dx)} = \underset{u}{x}\cdot\underset{v}{e^{5x}} - \int \underset{v}{e^{5x}} \cdot \underset{du}{dx}$$

[Using Theorem 7: $\int u\,dv = uv - \int v\,du$]

$= xe^{5x} - \frac{1}{5} e^{5x} + C$

3. $\int x^3(3x^2\,dx)$

Let

$u = x^3$ and $dv = 3x^2\,dx$.

Then

$du = 3x^2\,dx$ and $v = x^3$.

$$\int \underset{u}{x^3}\, \underset{dv}{(3x^2\,dx)} = \underset{u}{x^3}\cdot\underset{v}{x^3} - \int \underset{v}{x^3}\cdot\underset{du}{3x^2\,dx}$$

(Integration by Parts)

$= x^6 - \int 3x^5\,dx$

$= x^6 - 3 \int x^5\,dx$

$= x^6 - 3 \cdot \frac{x^6}{6} + C$

$= x^6 - \frac{1}{2} x^6 + C$

$= \frac{1}{2} x^6 + C$

This problem can also be worked with substitution or with Formula A.

$\int x^3(3x^2\,dx)$

Let $u = x^3$, then $du = 3x^2\,dx$.

$= \int u\,du$ (Substituting u for x^3 and du for $3x^2\,dx$)

$= \frac{u^2}{2} + C$

$= \frac{(x^3)^2}{2} + C$

$= \frac{1}{2} x^6 + C$

$\int x^3(3x^2\,dx)$

$= \int 3x^5\,dx$

$= 3 \int x^5\,dx$

$= 3 \cdot \frac{x^6}{6} + C$ (Using Formula A)

$= \frac{1}{2} x^6 + C$

5. $\int xe^{2x}\,dx$

Let

$u = x$ and $dv = e^{2x}\,dx$.

Then

$du = dx$ and $v = \frac{1}{2}e^{2x}$. $\left[\int be^{ax}\,dx = \frac{b}{a}e^{ax} + C\right]$

u dv u v v du

$$\int x\,e^{2x}\,dx = x \cdot \frac{1}{2}e^{2x} - \int \frac{1}{2}e^{2x}\,dx$$

(Integration by Parts)

$$= \frac{1}{2}xe^{2x} - \frac{\frac{1}{2}}{2}e^{2x} + C$$

$\left[\int be^{ax}\,dx = \frac{b}{a}e^{ax} + C\right]$

$$= \frac{1}{2}xe^{2x} - \frac{1}{4}e^{2x} + C$$

7. $\int xe^{-2x}\,dx$

Let

$u = x$ and $dv = e^{-2x}\,dx$.

Then

$du = dx$ and $v = -\frac{1}{2}e^{-2x}$. $\left[\int be^{ax}\,dx = \frac{b}{a}e^{ax} + C\right]$

u dv u v v du

$$\int x\,e^{-2x}\,dx = x\cdot\left(-\frac{1}{2}e^{-2x}\right) - \int\left(-\frac{1}{2}e^{-2x}\right)dx$$

(Integration by Parts)

$$= -\frac{1}{2}xe^{-2x} - \frac{-\frac{1}{2}}{-2}e^{-2x} + C$$

$\left[\int be^{ax}\,dx = \frac{b}{a}e^{ax} + C\right]$

$$= -\frac{1}{2}xe^{-2x} - \frac{1}{4}e^{-2x} + C$$

9. $\int x^2 \ln x\,dx = \int (\ln x)\,x^2\,dx$

Let

$u = \ln x$ and $dv = x^2\,dx$.

Then

$du = \frac{1}{x}\,dx$ and $v = \frac{x^3}{3}$.

u dv u v v du

$$\int (\ln x)\,x^2\,dx = \ln x \cdot \frac{x^3}{3} - \int \frac{x^3}{3}\cdot\frac{1}{x}\,dx$$

(Integration by Parts)

$$= \frac{x^3}{3}\ln x - \frac{1}{3}\int x^2\,dx$$

$$= \frac{x^3}{3}\ln x - \frac{1}{3}\cdot\frac{x^3}{3} + C$$

$$= \frac{x^3}{3}\ln x - \frac{x^3}{9} + C$$

11. $\int x \ln x^2\,dx = \int (\ln x^2)\,x\,dx$

Let

$u = \ln x^2$ and $dv = x\,dx$.

Then

$du = 2x \cdot \frac{1}{x^2}\,dx$ and $v = \frac{x^2}{2}$.

$= \frac{2}{x}\,dx$

u dv u v v du

$$\int (\ln x^2)\,x\,dx = (\ln x^2)\cdot\frac{x^2}{2} - \int \frac{x^2}{2}\cdot\frac{2}{x}\,dx$$

(Integration by Parts)

$$= \frac{x^2}{2}\ln x^2 - \int x\,dx$$

$$= \frac{x^2}{2}\ln x^2 - \frac{x^2}{2} + C$$

13. $\int \ln(x + 3)\,dx$

Let

$u = \ln(x + 3)$ and $dv = dx$.

Then

$du = \frac{1}{x + 3}\,dx$ and $v = x + 3$.

(Choosing $x + 3$ as an antiderivative of dv)

u dv u v

$$\int \ln(x + 3)\,dx = [\ln(x + 3)]\cdot[x + 3] -$$

v du

$$\int (x + 3)\cdot\frac{1}{x + 3}\,dx$$

(Integration by Parts)

$$= (x + 3)\ln(x + 3) - \int dx$$

$$= (x + 3)\ln(x + 3) - x + C$$

15. $\int (x + 2)\ln x\,dx = \int (\ln x)(x + 2)\,dx$

Let

$u = \ln x$ and $dv = (x + 2)\,dx$.

Then

$du = \frac{1}{x}\,dx$ and $v = \frac{(x + 2)^2}{2}$.

u dv

$$\int (\ln x)(x + 2)\,dx$$

u v v du

$$= (\ln x)\cdot\frac{(x + 2)^2}{2} - \int \frac{(x + 2)^2}{2}\cdot\frac{1}{x}\,dx$$

(Integration by Parts)

$$= \frac{x^2 + 4x + 4}{2}\ln x - \int \frac{x^2 + 4x + 4}{2x}\,dx$$

$$= \frac{x^2 + 4x + 4}{2}\ln x - \int\left(\frac{x}{2} + 2 + \frac{2}{x}\right)dx$$

$$= \frac{x^2 + 4x + 4}{2}\ln x - \frac{1}{2}\int x\,dx - 2\int dx - 2\int \frac{1}{x}\,dx$$

$= \frac{x^2 + 4x + 4}{2} \ln x - \frac{1}{2} \cdot \frac{x^2}{2} - 2 \cdot x - 2 \cdot \ln x + C$

$= \left[\frac{x^2 + 4x + 4}{2} - 2\right] \ln x - \frac{x^2}{4} - 2x + C$

$= \left[\frac{x^2}{2} + 2x\right] \ln x - \frac{x^2}{4} - 2x + C$

17. $\int (x - 1) \ln x \, dx = \int (\ln x)(x - 1) \, dx$

Let

$u = \ln x$ and $dv = (x - 1)\, dx$.

Then

$du = \frac{1}{x}\, dx$ and $v = \frac{(x - 1)^2}{2}$.

u dv

$\int (\ln x)(x - 1)\, dx$

u v v du

$= (\ln x) \cdot \frac{(x - 1)^2}{2} - \int \frac{(x - 1)^2}{2} \cdot \frac{1}{x}\, dx$

(Integration by Parts)

$= \frac{x^2 - 2x + 1}{2} \ln x - \int \frac{x^2 - 2x + 1}{2x}\, dx$

$= \frac{x^2 - 2x + 1}{2} \ln x - \int \left[\frac{x}{2} - 1 + \frac{1}{2x}\right] dx$

$= \frac{x^2 - 2x + 1}{2} \ln x - \frac{1}{2} \int x\, dx + \int dx - \frac{1}{2} \int \frac{1}{x}\, dx$

$= \frac{x^2 - 2x + 1}{2} \ln x - \frac{1}{2} \cdot \frac{x^2}{2} + x - \frac{1}{2} \ln x + C$

$= \left[\frac{x^2 - 2x + 1}{2} - \frac{1}{2}\right] \ln x - \frac{x^2}{4} + x + C$

$= \left[\frac{x^2}{2} - x\right] \ln x - \frac{x^2}{4} + x + C$

19. $\int x \sqrt{x + 2}\, dx$

Let

$u = x$ and $dv = \sqrt{x + 2}\, dx$
$= (x + 2)^{1/2}\, dx$.

Then

$du = dx$ and $v = \frac{(x + 2)^{3/2}}{3/2}$
$= \frac{2}{3} (x + 2)^{3/2}$.

u dv

$\int x \sqrt{x + 2}\, dx$

u v v du

$= x \cdot \frac{2}{3} (x + 2)^{3/2} - \int \frac{2}{3} (x + 2)^{3/2}\, dx$

(Integration by Parts)

$= \frac{2}{3} x(x + 2)^{3/2} - \frac{2}{3} \int (x + 2)^{3/2}\, dx$

$= \frac{2}{3} x(x + 2)^{3/2} - \frac{2}{3} \cdot \frac{(x + 2)^{5/2}}{5/2} + C$

$= \frac{2}{3} x(x + 2)^{3/2} - \frac{4}{15} (x + 2)^{5/2} + C$

21. $\int x^3 \ln 2x\, dx = \int (\ln 2x)(x^3\, dx)$

Let

$u = \ln 2x$ and $dv = x^3\, dx$.

Then

$du = \frac{1}{x}\, dx$ and $v = \frac{x^4}{4}$.

u dv u v v du

$\int (\ln 2x)(x^3\, dx) = (\ln 2x) \cdot \frac{x^4}{4} - \int \frac{x^4}{4} \cdot \frac{1}{x}\, dx$

(Integration by Parts)

$= \frac{x^4}{4} \ln 2x - \frac{1}{4} \int x^3\, dx$

$= \frac{x^4}{4} \ln 2x - \frac{1}{4} \cdot \frac{x^4}{4} + C$

$= \frac{x^4}{4} \ln 2x - \frac{x^4}{16} + C$

23. $\int x^2 e^x\, dx$

Let

$u = x^2$ and $dv = e^x\, dx$.

Then

$du = 2x\, dx$ and $v = e^x$.

u dv u v v du

$\int x^2 e^x\, dx = x^2 e^x - \int e^x \cdot 2x\, dx$

(Integration by Parts)

$= x^2 e^x - \int 2xe^x\, dx$

We evaluate $\int 2xe^x\, dx$ using the Integration by Parts formula.

$\int 2xe^x\, dx$

Let

$u = 2x$ and $dv = e^x\, dx$.

Then

$du = 2\, dx$ and $v = e^x$.

u dv u v v du

$\int 2x e^x\, dx = 2x \cdot e^x - \int 2e^x\, dx$

$= 2xe^x - 2e^x + K$

Thus,

$\int x^2 e^x\, dx = x^2e^x - (2xe^x - 2e^x + K)$
$= x^2 e^x - 2xe^x + 2e^x + C \quad (C = -K)$

Since we have an integral $\int f(x)g(x)\, dx$ where $f(x)$, or x^2, can be differentiated repeatedly to a derivative that is eventually 0 and $g(x)$, or e^x, can be integrated repeatedly easily, we can use tabular integration.

f(x) and repeated derivatives		g(x) and repeated integrals
x^2	+	e^x
$2x$	-	e^x
2	+	e^x
0		e^x

We add the products along the arrows, making the alternating sign changes.

$$\int x^2 e^x\,dx = x^2e^x - 2xe^x + 2e^x + C$$

25. $\int x^2 e^{2x}\,dx$

Let

$u = x^2$ and $dv = e^{2x}\,dx$.

Then

$du = 2x\,dx$ and $v = \frac{1}{2}e^{2x}$.

$$\overset{u}{\int x^2}\ \overset{dv}{e^{2x}\,dx} = \overset{u}{x^2}\cdot\overset{v}{\frac{1}{2}e^{2x}} - \int \overset{v}{\frac{1}{2}e^{2x}}\cdot\overset{du}{2x\,dx}$$

(Integration by Parts)

$$= \frac{1}{2}x^2e^{2x} - \int x e^{2x}\,dx$$

We integrate $xe^{2x}\,dx$ using the Integration by Parts formula.

$$\int xe^{2x}\,dx$$

Let

$u = x$ and $dv = e^{2x}\,dx$.

Then

$du = dx$ and $v = \frac{1}{2}e^{2x}$.

$$\int \overset{u}{x}\ \overset{dv}{e^{2x}\,dx} = \overset{u}{x}\cdot\overset{v}{\frac{1}{2}e^{2x}} - \int \overset{v}{\frac{1}{2}e^{2x}}\cdot\overset{du}{dx}$$

$$= \frac{1}{2}xe^{2x} - \frac{1}{2}\cdot\frac{1}{2}e^{2x} + K$$

$$= \frac{1}{2}xe^{2x} - \frac{1}{4}e^{2x} + K$$

Thus,

$$\int x^2e^{2x}\,dx = \frac{1}{2}x^2e^{2x} - \frac{1}{2}xe^{2x} + \frac{1}{4}e^{2x} + C$$

$(C = 2K)$

We could also use tabular integration.

f(x) and repeated derivatives		g(x) and repeated integrals
x^2	+	e^{2x}
$2x$	-	$\frac{1}{2}e^{2x}$
2	+	$\frac{1}{4}e^{2x}$
0		$\frac{1}{8}e^{2x}$

$$\int x^2e^{2x}\,dx$$

$$= x^2\cdot\frac{1}{2}e^{2x} - 2x\cdot\frac{1}{4}e^{2x} + 2\cdot\frac{1}{8}e^{2x} + C$$

$$= \frac{1}{2}x^2e^{2x} - \frac{1}{2}xe^{2x} + \frac{1}{4}e^{2x} + C$$

27. $\int x^3e^{-2x}\,dx$

We will use tabular integration.

f(x) and repeated derivatives		g(x) and repeated integrals
x^3	+	e^{-2x}
$3x^2$	-	$-\frac{1}{2}e^{-2x}$
$6x$	+	$\frac{1}{4}e^{-2x}$
6	-	$-\frac{1}{8}e^{-2x}$
0		$\frac{1}{16}e^{-2x}$

$$\int x^3e^{-2x}\,dx$$

$$= x^3\left[-\frac{1}{2}e^{-2x}\right] - 3x^2\left[\frac{1}{4}e^{-2x}\right] + 6x\left[-\frac{1}{8}e^{-2x}\right] - 6\left[\frac{1}{16}e^{-2x}\right] + C$$

$$= -\frac{1}{2}x^3e^{-2x} - \frac{3}{4}x^2e^{-2x} - \frac{3}{4}xe^{-2x} - \frac{3}{8}e^{-2x} + C$$

$$= e^{-2x}\left[-\frac{1}{2}x^3 - \frac{3}{4}x^2 - \frac{3}{4}x - \frac{3}{8}\right] + C$$

29. $\int (x^4 + 1)e^{3x}\,dx$

We will use tabular integration.

f(x) and repeated derivatives		g(x) and repeated integrals
$x^4 + 1$	+	e^{3x}
$4x^3$	-	$\frac{1}{3}e^{3x}$
$12x^2$	+	$\frac{1}{9}e^{3x}$
$24x$	-	$\frac{1}{27}e^{3x}$
24	+	$\frac{1}{81}e^{3x}$
0		$\frac{1}{243}e^{3x}$

$$\int (x^4+1)e^{3x}\,dx = (x^4+1)\cdot\frac{1}{3}e^{3x} - 4x^3\cdot\frac{1}{9}e^{3x} + 12x^2\cdot\frac{1}{27}e^{3x} - 24x\cdot\frac{1}{81}e^{3x} + 24\cdot\frac{1}{243}e^{3x} + C$$

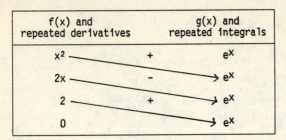

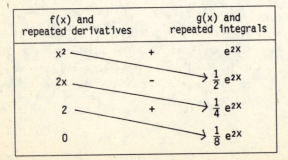

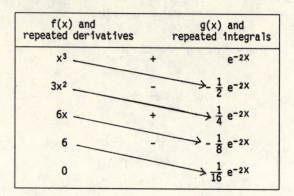

$= e^{3x}\left[\frac{1}{3}(x^4 + 1) - \frac{4}{9}x^3 + \frac{4}{9}x^2 - \frac{8}{27}x + \frac{8}{81}\right] + C,$

or $e^x\left[\frac{1}{3}x^4 - \frac{4}{9}x^3 + \frac{4}{9}x^2 - \frac{8}{27}x + \frac{35}{81}\right] + C$

$\left[\text{Adding } \frac{1}{3} \text{ and } \frac{8}{81}\right]$

31. $\int_1^2 x^2 \ln x\, dx$

In Exercise 9 above we found the indefinite integral.

$\int x^2 \ln x\, dx = \frac{x^3}{3}\ln x - \frac{x^3}{9} + C$

Evaluate the definite integral.

$\int_1^2 x^2 \ln x\, dx = \left[\frac{x^3}{3}\ln x - \frac{x^3}{9}\right]_1^2$

$= \left(\frac{8}{3}\ln 2 - \frac{8}{9}\right) - \left(\frac{1}{3}\ln 1 - \frac{1}{9}\right)$

$= \frac{8}{3}\ln 2 - \frac{8}{9} + \frac{1}{9}$

$= \frac{8}{3}\ln 2 - \frac{7}{9}$

33. $\int_2^6 \ln(x + 3)\, dx$

In Exercise 13 above we found the indefinite integral.

$\int \ln(x + 3)\, dx = (x + 3)\ln(x + 3) - x + C$

Evaluate the definite integral.

$\int_2^6 \ln(x + 3)\, dx = [(x + 3)\ln(x + 3) - x]_2^6$

$= (9 \ln 9 - 6) - (5 \ln 5 - 2)$

$= 9 \ln 9 - 6 - 5 \ln 5 + 2$

$= 9 \ln 9 - 5 \ln 5 - 4$

35. a) We first find the indefinite integral.

$\int xe^x\, dx$

Let

$u = x$ and $dv = e^x\, dx$.

Then

$du = dx$ and $v = e^x$.

$\int xe^x\, dx = xe^x - \int e^x\, dx$

$= xe^x - e^x + C$

b) Evaluate the definite integral.

$\int_0^1 xe^x\, dx = [xe^x - e^x]_0^1$

$= (1 \cdot e^1 - e^1) - (0 \cdot e^0 - e^0)$

$= 0 - (-1)$

$= 1$

37. $C(x) = \int 4x\sqrt{x + 3}\, dx$

Let

$u = 4x$ and $dv = \sqrt{x + 3}\, dx$.

Then

$du = 4\, dx$ and $v = \frac{2}{3}(x + 3)^{3/2}$.

$C(x) = 4x \cdot \frac{2}{3}(x + 3)^{3/2} - \int \frac{2}{3}(x + 3)^{3/2} \cdot 4\, dx$

$= \frac{8}{3}x(x + 3)^{3/2} - \frac{8}{3}\int (x + 3)^{3/2}\, dx$

$= \frac{8}{3}x(x + 3)^{3/2} - \frac{8}{3} \cdot \frac{2}{5}(x + 3)^{5/2} + C$

$= \frac{8}{3}x(x + 3)^{3/2} - \frac{16}{15}(x + 3)^{5/2} + C$

Use $C(13) = \$1126.40$ to find C.

$C(13) = \frac{8}{3} \cdot 13(13 + 3)^{3/2} - \frac{16}{15}(13 + 3)^{5/2} + C =$ 1126.40

$\frac{104}{3}(16)^{3/2} - \frac{16}{15}(16)^{5/2} + C = 1126.40$

$\frac{104}{3} \cdot 64 - \frac{16}{15} \cdot 1024 + C = 1126.40$

$\frac{6656}{3} - \frac{16{,}384}{15} + C = 1126.40$

$\frac{16{,}896}{15} + C = 1126.40$

$1126.40 + C = 1126.40$

$C = 0$

$C(x) = \frac{8}{3}x(x + 3)^{3/2} - \frac{16}{15}(x + 3)^{5/2}$

39. a) We first find the indefinite integral.

$\int 10te^{-t}\, dt = 10\int te^{-t}\, dt$

Let

$u = t$ and $dv = e^{-t}\, dt$.

Then

$du = dt$ and $v = -e^{-t}$.

$10\int te^{-t}\, dt = 10\left[t(-e^{-t}) - \int -e^{-t}\, dt\right]$

$= 10\left[-te^{-t} + \int e^{-t}\, dt\right]$

$= 10(-te^{-t} - e^{-t} + K)$

$= -10te^{-t} - 10e^{-t} + C \quad (C = 10K)$

Then evaluate the definite integral.

$\int_0^T te^{-t}\, dt = [-10te^{-t} - 10e^{-t}]_0^T$

$= (-10Te^{-T} - 10e^{-T}) -$

$(-10 \cdot 0 \cdot e^{-0} - 10e^{-0})$

$= (-10Te^{-T} - 10e^{-T}) - (0 - 10)$

$= -10Te^{-T} - 10e^{-T} + 10$

$= -10[e^{-T}(T + 1) - 1]$

b) Substitute 4 for T.

$$\int_0^4 te^{-t}\,dt = -10[e^{-4}(4+1) - 1]$$
$$= -50e^{-4} + 10$$
$$\approx -50(0.018316) + 10$$
$$\approx -0.915800 + 10$$
$$\approx 9.084$$

41. $\int \sqrt{x}\,\ell n\,x\,dx = \int (\ell n\,x)\,x^{1/2}\,dx$

Let

$u = \ell n\,x$ and $dv = x^{1/2}\,dx$.

Then

$du = \frac{1}{x}\,dx$ and $v = \frac{2}{3}x^{3/2}$.

$$\int (\ell n\,x)\,x^{1/2}\,dx = (\ell n\,x)\left[\frac{2}{3}x^{3/2}\right] - \int \frac{2}{3}x^{3/2}\cdot\frac{1}{x}\,dx$$
$$= \frac{2}{3}x^{3/2}\,\ell n\,x - \frac{2}{3}\int x^{1/2}\,dx$$
$$= \frac{2}{3}x^{3/2}\,\ell n\,x - \frac{2}{3}\cdot\frac{x^{3/2}}{3/2} + C$$
$$= \frac{2}{3}x^{3/2}\,\ell n\,x - \frac{4}{9}x^{3/2} + C, \text{ or}$$
$$\frac{2}{9}x^{3/2}(3\,\ell n\,x - 2) + C$$

(Factoring)

43. $\int \frac{te^t}{(t+1)^2}\,dt = \int te^t(t+1)^{-2}\,dt$

Let

$u = te^t$ and $dv = (t+1)^{-2}\,dt$.

Then

$du = (te^t + e^t)\,dt$ and $v = \frac{(t+1)^{-1}}{-1}$

$= e^t(t+1)\,dt$ $\quad = -\frac{1}{t+1}$.

$$\int te^t(t+1)^{-2}\,dt$$
$$= te^t\left[-\frac{1}{t+1}\right] - \int\left[-\frac{1}{t+1}\right]e^t(t+1)\,dt$$
$$= -\frac{te^t}{t+1} + \int e^t\,dt$$
$$= -\frac{te^t}{t+1} + e^t + C$$
$$= e^t\left[1 - \frac{t}{t+1}\right] + C$$
$$= \frac{e^t}{t+1} + C$$

45. $\int \frac{\ell n\,x}{\sqrt{x}}\,dx = \int (\ell n\,x)x^{-1/2}\,dx$

Let

$u = \ell n\,x$ and $dv = x^{-1/2}\,dx$.

Then

$du = \frac{1}{x}\,dx$ and $v = 2x^{1/2}$.

$$\int (\ell n\,x)\,x^{-1/2}\,dx = (\ell n\,x)\,2x^{1/2} - \int 2x^{1/2}\cdot\frac{1}{x}\,dx$$
$$= 2x^{1/2}\,\ell n\,x - 2\int x^{-1/2}\,dx$$
$$= 2x^{1/2}\,\ell n\,x - 2\cdot\frac{x^{1/2}}{1/2} + C$$
$$= 2\sqrt{x}\,\ell n\,x - 4\sqrt{x} + C$$

47. $\int \frac{13t^2 - 48}{\sqrt[5]{4t+7}}\,dt = \int (13t^2 - 48)(4t+7)^{-1/5}\,dt$

Use tabular integration.

f(x) and repeated derivatives		g(x) and repeated integrals
$13t^2 - 48$	+	$(4t+7)^{-1/5}$
$26t$	-	$\frac{5}{16}(4t+7)^{4/5}$
26	+	$\frac{25}{576}(4t+7)^{9/5}$
0		$\frac{125}{32{,}256}(4t+7)^{14/5}$

$$\int \frac{13t^2 - 48}{\sqrt[5]{4t+7}}\,dt = (13t^2 - 48)\left[\frac{5}{16}(4t+7)^{4/5}\right] -$$
$$26t\left[\frac{25}{576}(4t+7)^{9/5}\right] + 26\left[\frac{125}{32{,}256}(4t+7)^{14/5}\right] + C$$
$$= (13t^2 - 48)\left[\frac{5}{16}(4t+7)^{4/5}\right] -$$
$$\frac{325}{288}\,t(4t+7)^{9/5} + \frac{1625}{16{,}128}(4t+7)^{14/5} + C$$

49. $\int x^n e^x\,dx$

Let

$u = x^n$ and $dv = e^x\,dx$.

Then

$du = nx^{n-1}\,dx$ and $v = e^x$.

$$\int x^n e^x\,dx = x^n e^x - \int e^x(nx^{n-1})\,dx$$
$$= x^n e^x - n\int x^{n-1} e^x\,dx$$

51. See the answer section in the text.

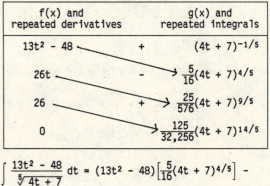

Exercise Set 5.7

1. $\int xe^{-3x}\,dx$

This integral fits Formula 6 in Table 1.

$$\int xe^{ax}\,dx = \frac{1}{a^2}\cdot e^{ax}(ax - 1) + C$$

In our integral a = -3, so we have, by the formula,

$$\int xe^{-3x}\,dx = \frac{1}{(-3)^2}\cdot e^{-3x}(-3x - 1) + C$$
$$= \frac{1}{9}e^{-3x}(-3x - 1) + C$$
$$\text{or } -\frac{1}{9}e^{-3x}(3x + 1) + C$$

3. $\int 5^x\,dx$

This integral fits Formula 11 in Table 1.

$$\int a^x\,dx = \frac{a^x}{\ln a} + C,\ a > 0,\ a \neq 1$$

In our integral a = 5, so we have, by the formula,

$$\int 5^x\,dx = \frac{5^x}{\ln 5} + C$$

5. $\int \frac{1}{16 - x^2}\,dx$

This integral fits Formula 15 in Table 1.

$$\int \frac{1}{a^2 - x^2}\,dx = \frac{1}{2a}\ln\left[\frac{a + x}{a - x}\right] + C$$

In our integral a = 4, so we have, by the formula,

$$\int \frac{1}{16 - x^2}\,dx = \int \frac{1}{4^2 - x^2}\,dx$$
$$= \frac{1}{2\cdot 4}\ln\frac{4 + x}{4 - x} + C$$
$$= \frac{1}{8}\ln\frac{4 + x}{4 - x} + C$$

7. $\int \frac{x}{5 - x}\,dx$

This integral fits Formula 18 in Table 1.

$$\int \frac{x}{ax + b}\,dx = \frac{b}{a^2} + \frac{x}{a} - \frac{b}{a^2}\ln(ax + b) + C$$

In our integral a = -1 and b = 5, so we have, by the formula,

$$\int \frac{x}{5 - x}\,dx = \frac{5}{(-1)^2} + \frac{x}{(-1)} - \frac{5}{(-1)^2}\ln(-1\cdot x+5) + C$$
$$= 5 - x - 5\ln(5 - x) + C$$

9. $\int \frac{1}{x(5 - x)^2}\,dx$

This integral fits Formula 21 in Table 1.

$$\int \frac{1}{x(ax + b)^2}\,dx = \frac{1}{b(ax + b)} + \frac{1}{b^2}\ln\left[\frac{x}{ax + b}\right] + C$$

In our integral a = -1 and b = 5, so we have, by the formula,

$$\int \frac{1}{x(5 - x)^2}\,dx = \int \frac{1}{x(-x + 5)^2}\,dx$$
$$= \frac{1}{5(-x + 5)} + \frac{1}{5^2}\ln\left[\frac{x}{-x + 5}\right] + C$$
$$= \frac{1}{5(5 - x)} + \frac{1}{25}\ln\left[\frac{x}{5 - x}\right] + C$$

11. $\int \ln 3x\,dx$

$$= \int (\ln 3 + \ln x)\,dx$$
$$= \int \ln 3\,dx + \int \ln x\,dx$$
$$= (\ln 3)\,x + \int \ln x\,dx$$

The integral in the second term fits Formula 8 in Table 1.

$$\int \ln x\,dx = x\ln x - x + C$$

$$\int \ln 3x\,dx = (\ln 3)\,x + \int \ln x\,dx$$
$$= (\ln 3)\,x + x\ln x - x + C$$

13. $\int x^4 e^{5x}\,dx$

This integral fits Formula 7 in Table 1.

$$\int x^n e^{ax}\,dx = \frac{x^n e^{ax}}{a} - \frac{n}{a}\int x^{n-1} e^{ax}\,dx$$

In our integral n = 4 and a = 5, so we have, by the formula,

$$\int x^4 e^{5x}\,dx$$
$$= \frac{x^4 e^{5x}}{5} - \frac{4}{5}\int x^3 e^{5x}\,dx$$

In the integral in the second term where n = 3 and a = 5, we again apply Formula 7.

$$= \frac{x^4 e^{5x}}{5} - \frac{4}{5}\left[\frac{x^3 e^{5x}}{5} - \frac{3}{5}\int x^2 e^{5x}\,dx\right]$$

We continue to apply Formula 7.

$$= \frac{x^4 e^{5x}}{5} - \frac{4}{25}x^3 e^{5x} + \frac{12}{25}\left[\frac{x^2 e^{5x}}{5} - \frac{2}{5}\int x\,e^{5x}\,dx\right]$$

$$= \frac{x^4 e^{5x}}{5} - \frac{4}{25}x^3 e^{5x} + \frac{12}{125}x^2 e^{5x} - \frac{24}{125}\left[\frac{x\,e^{5x}}{5} - \frac{1}{5}\int x^0 e^{5x}\,dx\right]$$

$$= \frac{x^4 e^{5x}}{5} - \frac{4}{25}x^3 e^{5x} + \frac{12}{125}x^2 e^{5x} - \frac{24}{625}x\,e^{5x} + \frac{24}{625}\int e^{5x}\,dx$$

We now apply Formula 5, $\int e^{ax}\,dx = \frac{1}{a}\cdot e^{ax} + C$.

$$= \frac{x^4 e^{5x}}{5} - \frac{4}{25}x^3 e^{5x} + \frac{12}{125}x^2 e^{5x} - \frac{24}{625}x\,e^{5x} + \frac{24}{3125}e^{5x} + C$$

15. $\int x^3 \ln x \, dx$

This integral fits Formula 10 in Table 1.

$$\int x^n \ln x \, dx = x^{n+1}\left[\frac{\ln x}{n+1} - \frac{1}{(n+1)^2}\right] + C, \; n \neq 1$$

In our integral n = 3, so we have, by the formula,

$$\int x^3 \ln x \, dx = x^{3+1}\left[\frac{\ln x}{3+1} - \frac{1}{(3+1)^2}\right] + C$$

$$= x^4\left[\frac{\ln x}{4} - \frac{1}{16}\right] + C$$

17. $\int \frac{dx}{\sqrt{x^2 + 7}}$

This integral fits Formula 12 in Table 1.

$$\int \frac{1}{\sqrt{x^2 + a^2}} dx = \ln (x + \sqrt{x^2 + a^2}) + C$$

In our integral $a^2 = 7$, so we have, by the formula,

$$\int \frac{dx}{\sqrt{x^2 + 7}} = \ln (x + \sqrt{x^2 + 7}) + C$$

19. $\int \frac{10 \, dx}{x(5 - 7x)^2} = 10 \int \frac{1}{x(-7x + 5)^2} dx$

This integral fits Formula 21 in Table 1.

$$\int \frac{1}{x(ax + b)^2} dx = \frac{1}{b(ax + b)} + \frac{1}{b^2} \ln \left|\frac{x}{ax + b}\right| + C$$

In our integral a = -7 and b = 5, so we have, by the formula,

$$10 \int \frac{1}{x(-7x + 5)^2} dx$$

$$= 10 \left[\frac{1}{5(-7x + 5)} + \frac{1}{5^2} \ln \left|\frac{x}{-7x + 5}\right| + C\right]$$

$$= \frac{2}{5 - 7x} + \frac{2}{5} \ln \left|\frac{x}{5 - 7x}\right| + C$$

21. $\int \frac{-5}{4x^2 - 1} dx = -5 \int \frac{1}{4x^2 - 1} dx$

This integral almost fits Formula 14 in Table 1.

$$\int \frac{1}{x^2 - a^2} dx = \frac{1}{2a} \ln \left|\frac{x - a}{x + a}\right| + C$$

But the x^2 coefficient needs to be 1. We factor out 4 as follows. Then we apply Formula 14.

$$-5 \int \frac{1}{4x^2 - 1} dx = -5 \int \frac{1}{4\left(x^2 - \frac{1}{4}\right)} dx$$

$$= -\frac{5}{4} \int \frac{1}{x^2 - \frac{1}{4}} dx \qquad \left(a^2 = \frac{1}{4}, \; a = \frac{1}{2}\right)$$

$$= -\frac{5}{4}\left[\frac{1}{2 \cdot \frac{1}{2}} \ln \left|\frac{x - \frac{1}{2}}{x + \frac{1}{2}}\right|\right] + C$$

$$= -\frac{5}{4} \ln \left|\frac{x - 1/2}{x + 1/2}\right| + C$$

23. $\int \sqrt{4m^2 + 16} \, dm$

This integral almost fits Formula 22 in Table 1.

$$\int \sqrt{x^2 + a^2} \, dx$$

$$= \frac{1}{2}\left[x\sqrt{x^2 + a^2} + a^2 \ln \left|x + \sqrt{x^2 + a^2}\right|\right] + C$$

But the x^2 coefficient needs to be 1. We factor out 4 as follows. Then we apply Formula 22.

$$\int \sqrt{4m^2 + 16} \, dm$$

$$= \int \sqrt{4(m^2 + 4)} \, dm$$

$$= 2 \int \sqrt{m^2 + 4} \, dm$$

$$= 2 \cdot \frac{1}{2}\left[m\sqrt{m^2 + 4} + 4 \ln \left|m + \sqrt{m^2 + 4}\right|\right] + C$$

$$= m\sqrt{m^2 + 4} + 4 \ln \left|m + \sqrt{m^2 + 4}\right| + C$$

25. $\int \frac{-5 \ln x}{x^3} dx = -5 \int x^{-3} \ln x \, dx$

This integral fits Formula 10 in Table 1.

$$\int x^n \ln x \, dx = x^{n+1}\left[\frac{\ln x}{n+1} - \frac{1}{(n+1)^2}\right] + C, \; n \neq -1$$

In our integral n = -3, so we have, by the formula,

$$-5 \int x^{-3} \ln x \, dx$$

$$= -5\left[x^{-3+1}\left(\frac{\ln x}{-3+1} - \frac{1}{(-3+1)^2}\right)\right] + C$$

$$= -5\left[x^{-2}\left(\frac{\ln x}{-2} - \frac{1}{4}\right)\right] + C$$

$$= \frac{5 \ln x}{2x^2} + \frac{5}{4x^2} + C$$

27. $\int \frac{e^x}{x^{-3}} dx = \int x^3 e^x \, dx$

This integral fits Formula 7 in Table 1.

$$\int x^n e^{ax} dx = \frac{x^n e^{ax}}{a} - \frac{n}{a} \int x^{n-1} e^{ax} dx$$

In our integral n = 3 and a = 1, so we have, by the formula,

$$\int x^3 e^x \, dx$$

$$= x^3 e^x - 3 \int x^2 e^x \, dx$$

We continue to apply Formula 7.

$$= x^3 e^x - 3\left[x^2 e^x - 2 \int x e^x \, dx\right] \qquad (n = 2, \; a = 1)$$

$$= x^3 e^x - 3x^2 e^x + 6 \int x e^x \, dx$$

$$= x^3 e^x - 3x^2 e^x + 6\left[x e^x - \int x^0 e^x \, dx\right] \qquad (n = 1, \; a = 1)$$

$$= x^3 e^x - 3x^2 e^x + 6x e^x - 6 \int e^x \, dx$$

$$= x^3 e^x - 3x^2 e^x + 6x e^x - 6 e^x + C$$

29. $S(p) = \int \frac{100p}{(20 - p)^2}\,dp,\ 0 \leqslant p \leqslant 19$

$= 100 \int \frac{p}{(20 - p)^2}\,dp$

This integral fits Formula 19 in Table 1.

$\int \frac{x}{(ax + b)^2}\,dx = \frac{b}{a^2(ax + b)} + \frac{1}{a^2}\,\ell n\,(ax + b) + C$

In our integral $a = -1$ and $b = 20$.

$100 \int \frac{p}{(20 - p)^2}$

$= 100\left[\frac{20}{(-1)^2(-1\cdot p + 20)} + \frac{1}{(-1)^2}\,\ell n\,(-1\cdot p + 20)\right] + C$

$= 100\left[\frac{20}{20 - p} + \ell n\,(20 - p)\right] + C$

Use $S(19) = 2000$ to find C.

$S(19) = 100\left[\frac{20}{20 - 19} + \ell n\,(20 - 19)\right] + C = 2000$

$100\left[\frac{20}{1} + \ell n\,1\right] + C = 2000$

$100(20 + 0) + C = 2000$

$2000 + C = 2000$

$C = 0$

$S(p) = 100\left[\frac{20}{20 - p} + \ell n\,(20 - p)\right]$

31. $\int \frac{8}{3x^2 - 2x}\,dx = 8 \int \frac{1}{x(3x - 2)}\,dx$

This integral fits Formula 20 in Table 1.

$\int \frac{1}{x(ax + b)}\,dx = \frac{1}{a}\,\ell n\,\left|\frac{x}{ax + b}\right| + C$

In our integral $a = 3$ and $b = -2$, so we have, by the formula,

$8 \int \frac{1}{x(3x - 2)}\,dx = 8\left[\frac{1}{-2}\,\ell n\,\left|\frac{x}{3x - 2}\right|\right] + C$

$= -4\,\ell n\,\left|\frac{x}{3x - 2}\right| + C$

33. $\int \frac{dx}{x^3 - 4x^2 + 4x} = \int \frac{1}{x(x^2 - 4x + 4)}\,dx$

$= \int \frac{1}{x(x - 2)^2}\,dx$

This integral fits Formula 21 in Table 1.

$\int \frac{1}{x(ax + b)^2}\,dx = \frac{1}{b(ax + b)} + \frac{1}{b^2}\,\ell n\,\left|\frac{x}{ax + b}\right| + C$

In our integral $a = 1$ and $b = -2$, so we have, by the formula,

$\int \frac{1}{x(x - 2)^2}\,dx = \frac{1}{-2(x - 2)} + \frac{1}{(-2)^2}\,\ell n\,\left|\frac{x}{x - 2}\right| + C$

$= -\frac{1}{2(x - 2)} + \frac{1}{4}\,\ell n\,\left|\frac{x}{x - 2}\right| + C$

35. $\int \frac{-e^{-2x}\,dx}{9 - 6e^{-x} + e^{-2x}} = \int \frac{e^{-x}(-e^{-x})\,dx}{(e^{-x} - 3)^2}$

This integral fits Formula 19 in Table 1.

$\int \frac{x}{(ax + b)^2}\,dx = \frac{b}{a^2(ax + b)} + \frac{1}{a^2}\,\ell n\,(ax + b) + C$

In our integral x is represented by e^{-x}, dx is represented by $-e^{-x}\,dx$, $a = 1$, and $b = -3$.

$\int \frac{e^{-x}(-e^{-x})\,dx}{(e^{-x} - 3)^2}$

$= \frac{-3}{1^2(1\cdot e^{-x} - 3)} + \frac{1}{1^2}\,\ell n\,(1\cdot e^{-x} - 3) + C$

$= \frac{-3}{e^{-x} - 3} + \ell n\,(e^{-x} - 3) + C$

Exercise Set 5.8

1. a) $f(x) = \frac{1}{x^2}$

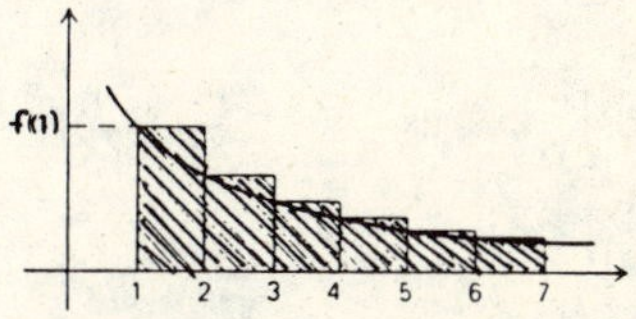

In the drawing [1,7] has been divided into 6 subintervals, each having width 1 $\left[\Delta x = \frac{7 - 1}{6} = 1\right]$.

The heights of the rectangles shown are

$f(1) = \frac{1}{1^2} = 1$

$f(2) = \frac{1}{2^2} = \frac{1}{4} = 0.2500$

$f(3) = \frac{1}{3^2} = \frac{1}{9} \approx 0.1111$

$f(4) = \frac{1}{4^2} = \frac{1}{16} = 0.0625$

$f(5) = \frac{1}{5^2} = \frac{1}{25} = 0.0400$

$f(6) = \frac{1}{6^2} = \frac{1}{36} \approx 0.0278$

The area of the region under the curve over [1,7] is approximately the sum of the areas of the 6 rectangles.

Area of each rectangle:

1st rectangle = $1\cdot 1 = 1$

$[f(1) = 1 \text{ and } \Delta x = 1]$

2nd rectangle = $0.2500\cdot 1 = 0.2500$

$[f(2) = 0.2500 \text{ and } \Delta x = 1]$

3rd rectangle = $0.1111\cdot 1 = 0.1111$

4th rectangle = $0.0625\cdot 1 = 0.0625$

5th rectangle = $0.0400\cdot 1 = 0.0400$

6th rectangle = $0.0278\cdot 1 = 0.0278$

The total area is $1 + 0.2500 + 0.1111 + 0.0625 + 0.0400 \pm 0.0278$, or 1.4914.

b) $\int_1^7 (dx/x^2)$

$= \int_1^7 \frac{1}{x^2}\,dx$

$= \int_1^7 x^{-2}\,dx$

$= \left[\frac{x^{-1}}{-1}\right]_1^7$

$= \left[-\frac{1}{x}\right]_1^7$

$= -\frac{1}{7} - \left(-\frac{1}{1}\right)$

$= -\frac{1}{7} + 1$

$= \frac{6}{7}$

≈ 0.8571

3. $y = 2x^3;\ [-1,1]$

$y_{av} = \frac{1}{b - a}\int_a^b f(x)\,dx$

$y_{av} = \frac{1}{1 - (-1)}\int_{-1}^1 2x^3\,dx$ [a = -1, b = 1, and f(x) = $2x^3$]

$= \frac{1}{2}\cdot 2\int_{-1}^1 x^3\,dx$

$= \int_{-1}^1 x^3\,dx$

$= \left[\frac{x^4}{4}\right]_{-1}^1$

$= \frac{1^4}{4} - \frac{(-1)^4}{4}$

$= \frac{1}{4} - \frac{1}{4}$

$= 0$

5. $y = e^x;\ [0,1]$

$y_{av} = \frac{1}{b - a}\int_a^b f(x)\,dx$

$y_{av} = \frac{1}{1 - 0}\int_0^1 e^x\,dx$ [a = 0, b = 1, and f(x) = e^x]

$= \int_0^1 e^x\,dx$

$= [e^x]_0^1$

$= e^1 - e^0$

$= e - 1$

7. $y = x^2 - x + 1;\ [0,2]$

$y_{av} = \frac{1}{b - a}\int_a^b f(x)\,dx$

$y_{av} = \frac{1}{2 - 0}\int_0^2 (x^2 - x + 1)\,dx$ [a = 0, b = 2, and f(x) = $x^2 - x + 1$]

$= \frac{1}{2}\left[\frac{x^3}{3} - \frac{x^2}{2} + x\right]_0^2$

$= \frac{1}{2}\left[\left(\frac{2^3}{3} - \frac{2^2}{2} + 2\right) - \left(\frac{0^3}{3} - \frac{0^2}{2} + 0\right)\right]$

$= \frac{1}{2}\left(\frac{8}{3} - 2 + 2\right)$

$= \frac{1}{2}\cdot\frac{8}{3}$

$= \frac{4}{3}$

9. $y = 3x + 1;\ [2,6]$

$y_{av} = \frac{1}{b - a}\int_a^b f(x)\,dx$

$y_{av} = \frac{1}{6 - 2}\int_2^6 (3x + 1)\,dx$ [a = 2, b = 6, and f(x) = $3x + 1$]

$= \frac{1}{4}\left[\frac{3}{2}x^2 + x\right]_2^6$

$= \frac{1}{4}\left[\left(\frac{3}{2}\cdot 6^2 + 6\right) - \left(\frac{3}{2}\cdot 2^2 + 2\right)\right]$

$= \frac{1}{4}[(54 + 6) - (6 + 2)]$

$= \frac{1}{4}(60 - 8)$

$= \frac{1}{4}\cdot 52$

$= 13$

11. $y = x^n;\ [0,1]$

$y_{av} = \frac{1}{b - a}\int_a^b f(x)\,dx$

$y_{av} = \frac{1}{1 - 0}\int_0^1 x^n\,dx$ [a = 0, b = 1, and f(x) = x^n]

$= \int_0^1 x^n\,dx$

$= \left[\frac{x^{n+1}}{n + 1}\right]_0^1$

$= \frac{1^{n+1}}{n + 1} - \frac{0^{n+1}}{n + 1}$

$= \frac{1}{n + 1} - 0$

$= \frac{1}{n + 1}$

13. $W(t) = -6t^2 + 12t + 90$, t in [0,4]

a) $W(0) = -6\cdot 0^2 + 12\cdot 0 + 90$

$= 90$ words per minute

b) We first find $W'(t)$.

$W'(t) = -12t + 12$

Then we find the critical points. Since $W'(t)$ exists for all values of t in $[0,4]$, we solve $W'(t) = 0$.

$$-12t + 12 = 0$$
$$-12t = -12$$
$$t = 1 \quad \text{(Critical point)}$$

Find the function values at 0, 1, and 4.

$W(0) = 90$

$W(1) = -6\cdot 1^2 + 12\cdot 1 + 90$
$= 96$ (Maximum)

$W(4) = -6\cdot 4^2 + 12\cdot 4 + 90$
$= 42$

The maximum speed is 96 words per minute at $t = 1$ minute.

c) $W_{av} = \frac{1}{b - a}\int_a^b W(t)\,dt$

$= \frac{1}{4 - 0}\int_0^4 (-6t^2 + 12t + 90)\,dt$

$= \frac{1}{4}\Big[-2t^3 + 6t^2 + 90t\Big]_0^4$

$= \frac{1}{4}\Big[(-2\cdot 4^3 + 6\cdot 4^2 + 90\cdot 4) - (-2\cdot 0^3 + 6\cdot 0^3 + 90\cdot 0)\Big]$

$= \frac{1}{4}\Big[(-128 + 96 + 360) - (-0 + 0 + 0)\Big]$

$= \frac{1}{4}\cdot 328$

$= 82$ words per minute

15. $P(t) = 241e^{0.009t}$

$P_{av} = \frac{1}{b - a}\int_a^b P(t)\,dt$

$P_{av} = \frac{1}{10 - 0}\int_0^{10} 241e^{0.009t}\,dt$ (a = 0, b = 10; 1996-1986 = 10)

$= \frac{1}{10}\cdot 241\int_0^{10} e^{0.009t}\,dt$

$= 24.1\left[\frac{e^{0.009t}}{0.009}\right]_0^{10}$ (Using Formula 5 in Table 1)

$= \frac{24.1}{0.009}\Big[e^{0.009t}\Big]_0^{10}$

$= \frac{24.1}{0.009}\Big[e^{0.009(10)} - e^{0.009(0)}\Big]$

$= \frac{24.1}{0.009}(e^{0.09} - 1)$

$= 2677.8(1.094174 - 1)$ (Using a calculator)

$= 2677.8(0.094174)$

≈ 252

The average value of the population from 1986 to 1996 is approximately 252 million.

17. a) $S(t) = t^2$, t in $[0,10]$

We first find $S'(t)$.

$S'(t) = 2t$

Then we find the critical points.

Since $S'(t)$ exists for all values of t in $[0,10]$, we solve $S'(t) = 0$.

$2t = 0$

$t = 0$ (Critical point)

Find the function values at 0 and 10.

$S(t) = t^2$

$S(0) = 0^2 = 0$ Minimum

$S(10) = 10^2 = 100$ Maximum

Thus, the maximum score is 100 after the student studies 10 hours.

b) $S(t)_{av} = \frac{1}{b - a}\int_a^b S(t)\,dt$

$S_{av} = \frac{1}{10 - 0}\int_0^{10} t^2\,dt$

$= \frac{1}{10}\left[\frac{t^3}{3}\right]_0^{10}$

$= \frac{1}{10}\left[\frac{10^3}{3} - \frac{0^3}{3}\right]$

$= \frac{1}{10}\cdot\frac{10^3}{3}$

$= \frac{10^2}{3}$

$= \frac{100}{3}$, or $33\frac{1}{3}$

19. $\int_a^b f(x)\,dx$

$\approx \Delta x\left[\frac{f(a)}{2} + f(x_2) + f(x_3) + \cdots + f(x_n) + \frac{f(b)}{2}\right]$

$\int_0^5 (x^2 + 1)\,dx$ $(a = 0,\ b = 5,\ \Delta x = 1,\ f(x) = x^2 + 1)$

$\approx 1\cdot\left[\frac{f(0)}{2} + f(1) + f(2) + f(3) + f(4) + \frac{f(5)}{2}\right]$

$(x_1 = 0,\ x_2 = 1,\ x_3 = 2,$ etc.$)$

$\approx \frac{1}{2} + 2 + 5 + 10 + 17 + \frac{26}{2}$

$\left[\begin{array}{l} f(0) = 0^2 + 1 = 1 \\ f(1) = 1^2 + 1 = 2 \\ f(2) = 2^2 + 1 = 5 \\ f(3) = 3^2 + 1 = 10 \\ f(4) = 4^2 + 1 = 17 \\ f(5) = 5^2 + 1 = 26 \end{array}\right]$

$\approx 47\frac{1}{2}$

Exercise Set 6.1

1. $D(x) = -\frac{5}{6}x + 10$

$S(x) = \frac{1}{2}x + 2$

a) To find the equilibrium point we set $D(x) = S(x)$ and solve.

$-\frac{5}{6}x + 10 = \frac{1}{2}x + 2$

$10 - 2 = \frac{1}{2}x + \frac{5}{6}x$

$8 = \frac{4}{3}x$ $\quad \left[\frac{1}{2} + \frac{5}{6} = \frac{8}{6} = \frac{4}{3}\right]$

$\frac{3}{4}\cdot 8 = \frac{3}{4}\cdot\frac{4}{3}x$ $\quad$ [Multiplying by $\frac{3}{4}$]

$6 = x$

Thus $x_E = 6$ units. To find p_E we substitute x_E into $D(x)$ or $S(x)$. Here we use $D(x)$.

$p_E = D(x_E) = D(6) = -\frac{5}{6}\cdot 6 + 10$

$= -5 + 10$

$= \$5$ per unit

Thus the equilibrium point is (6,\$5).

b) The consumer's surplus is

$\int_0^{x_E} D(x)\,dx - x_E\,p_E$,

or

$\int_0^6 \left(-\frac{5}{6}x + 10\right)dx - 6\cdot 5$

[Substituting $\left(-\frac{5}{6}x + 10\right)$ for $D(x)$, 6 for x_E, and 5 for p_E]

$= \left[-\frac{5x^2}{12} + 10x\right]_0^6 - 30$

$= \left[\left(-\frac{5\cdot 6^2}{12} + 10\cdot 6\right) - \left(-\frac{5\cdot 0^2}{12} + 10\cdot 0\right)\right] - 30$

$= (-15 + 60) - 30$

$= \$15$

c) The producer's surplus is

$x_E\,p_E - \int_0^{x_E} S(x)\,dx$

or

$6\cdot 5 - \int_0^6 \left(\frac{1}{2}x + 2\right)dx$

[Substituting $\left(\frac{1}{2}x + 2\right)$ for $S(x)$, 6 for x_E, and 5 for p_E]

$= 30 - \left[\frac{x^2}{4} + 2x\right]_0^6$

$= 30 - \left[\left(\frac{6^2}{4} + 2\cdot 6\right) - \left(\frac{0^2}{4} + 2\cdot 0\right)\right]$

$= 30 - (9 + 12)$

$= 30 - 21$

$= \$9$

3. $D(x) = (x - 4)^2$

$S(x) = x^2 + 2x + 6$

a) To find the equilibrium point we set $D(x) = S(x)$ and solve.

$(x - 4)^2 = x^2 + 2x + 6$

$x^2 - 8x + 16 = x^2 + 2x + 6$

$-8x + 16 = 2x + 6$

$10 = 10x$

$1 = x$

Thus $x_E = 1$ unit. To find p_E we substitute x_E into $D(x)$ or $S(x)$. Here we use $D(x)$.

$p_E = D(x_E) = D(1) = (1 - 4)^2 = (-3)^2$

$= \$9$ per unit

Thus the equilibrium point is (1,\$9).

b) The consumer's surplus is

$\int_0^{x_E} D(x)\,dx - x_E\,p_E$,

or

$\int_0^1 (x - 4)^2\,dx - 1\cdot 9$ $\quad$ [Substituting $(x - 4)^2$ for $D(x)$, 1 for x_E and 9 for p_E]

$= \int_0^1 (x^2 - 8x + 16)\,dx - 9$

$= \left[\frac{x^3}{3} - 4x^2 + 16x\right]_0^1 - 9$

$= \left[\left(\frac{1}{3} - 4 + 16\right) - (0 - 0 + 0)\right] - 9$

$= 12\frac{1}{3} - 9$

$= 3\frac{1}{3}$

$= \$3.33$

c) The producer's surpus is

$x_E\,p_E - \int_0^{x_E} S(x)\,dx$

or

$1\cdot 9 - \int_0^1 (x^2 + 2x + 6)\,dx$

[Substituting $x^2 + 2x + 6$ for $S(x)$, 1 for x_E and 9 for p_E]

$= 9 - \left[\frac{x^3}{3} + x^2 + 6x\right]_0^1$

$= 9 - \left[\left(\frac{1}{3} + 1 + 6\right) - (0 + 0 + 0)\right]$

$= 9 - 7\frac{1}{3} = 1\frac{2}{3} = \$1.67$

5. $D(x) = (x - 6)^2$
$S(x) = x^2$

a) To find the equilibrium point we set $D(x) = S(x)$ and solve.

$$(x - 6)^2 = x^2$$
$$x^2 - 12x + 36 = x^2$$
$$-12x + 36 = 0$$
$$36 = 12x$$
$$3 = x$$

Thus $x_E = 3$ units. To find p_E we substitute x_E into $D(x)$ or $S(x)$. Here we use $S(x)$.

$p_E = S(x_E) = S(3) = 3^2 = \$9$ per unit

Thus the equilibrium point is (3,\$9).

b) The consumer's surplus is

$\int_0^{x_E} D(x)\,dx - x_E\,p_E$,

or

$\int_0^3 (x - 6)^2\,dx - 3\cdot 9$ [Substituting $(x - 6)^2$ for $D(x)$, 3 for x_E, and 9 for p_E]

$= \int_0^3 (x^2 - 12x + 36)\,dx - 27$

$= \left[\frac{x^3}{3} - 6x^2 + 36x\right]_0^3 - 27$

$= \left[\left(\frac{3^3}{3} - 6\cdot 3^2 + 36\cdot 3\right) - \left(\frac{0^3}{3} - 6\cdot 0^2 + 36\cdot 0\right)\right] - 27$

$= 9 - 54 + 108 - 27$
$= \$36$

c) The producer's surplus is

$x_E\,p_E - \int_0^{x_E} S(x)\,dx$

or

$3\cdot 9 - \int_0^3 x^2\,dx$ [Substituting x^2 for $S(x)$, 3 for x_E, and 9 for p_E]

$= 27 - \left[\frac{x^3}{3}\right]_0^3$

$= 27 - \left(\frac{3^3}{3} - \frac{0^3}{3}\right)$

$= 27 - 9$
$= \$18$

7. $D(x) = 1000 - 10x$
$S(x) = 250 + 5x$

a) To find the equilibrium point we set $D(x) = S(x)$ and solve.

$$1000 - 10x = 250 + 5x$$
$$750 = 15x$$
$$50 = x$$

Thus $x_E = 50$ units. To find p_E we substitute x_E into $D(x)$ or $S(x)$. Here we use $D(x)$.

$p_E = D(x_E) = 1000 - 10\cdot 50$
$= 1000 - 500$
$= \$500$ per unit

Thus the equilibrium point is (50, \$500).

b) The consumer's surplus is

$\int_0^{x_E} D(x)\,dx - x_E\,p_E$

or

$\int_0^{50} (1000 - 10x)\,dx - 50\cdot 500$

$= \left[1000x - 5x^2\right]_0^{50} - 25{,}000$

$= \left[(1000\cdot 50 - 5(50)^2) - (1000\cdot 0 - 5(0)^2)\right] - 25{,}000$

$= 50{,}000 - 12{,}500 - 0 - 25{,}000$
$= \$12{,}500$

c) The producer's surplus is

$x_E\,p_E - \int_0^{x_E} S(x)\,dx$

or

$50\cdot 500 - \int_0^{50} (250 + 5x)\,dx$

$= 25{,}000 - \left[250x + \frac{5x^2}{2}\right]_0^{50}$

$= 25{,}000 - \left[\left(250\cdot 50 + \frac{5(50)^2}{2}\right) - \left(250\cdot 0 + \frac{5(0)^2}{2}\right)\right]$

$= 25{,}000 - 12{,}500 - 6250 + 0$
$= \$6250$

9. $D(x) = 5 - x,\ 0 \leqslant x \leqslant 5$
$S(x) = \sqrt{x + 7}$

a) To find the equilibrium point we set $D(x) = S(x)$ and solve.

$$5 - x = \sqrt{x + 7}$$
$$(5 - x)^2 = (\sqrt{x + 7})^2$$
$$25 - 10x + x^2 = x + 7$$
$$x^2 - 11x + 18 = 0$$
$$(x - 2)(x - 9) = 0$$
$$x - 2 = 0 \quad \text{or} \quad x - 9 = 0$$
$$x = 2 \quad \text{or} \quad x = 9$$

Only 2 is in the domain of $D(x)$, so $x_E = 2$ units. To find p_E we substitute x_E into $D(x)$ or $S(x)$. Here we use $D(x)$.

$p_E = D(x_E) = 5 - 2 = \$3$

Thus the equilibrium point is (2,\$3).

b) The consumer's surplus is

$$\int_0^{x_E} D(x)\,dx - x_E\,p_E$$

or

$$\int_0^2 (5 - x)\,dx - 2\cdot 3$$

$$= \left[5x - \frac{x^2}{2}\right]_0^2 - 6$$

$$= \left[\left[5\cdot 2 - \frac{2^2}{2}\right] - \left[5\cdot 0 - \frac{0^2}{2}\right]\right] - 6$$

$$= 8 - 0 - 6$$

$$= \$2$$

c) The producer's surplus is

$$x_E\,p_E - \int_0^{x_E} S(x)\,dx$$

or

$$2\cdot 3 - \int_0^2 \sqrt{x + 7}\,dx$$

$$= 6 - \left[\frac{2}{3}(x + 7)^{3/2}\right]_0^2$$

$$= 6 - \left[\frac{2}{3}(2 + 7)^{3/2} - \frac{2}{3}(0 + 7)^{3/2}\right]$$

$$= 6 - \frac{2}{3}\cdot 27 + \frac{2}{3}\cdot 7^{3/2}$$

$$= 6 - 18 + \frac{2}{3}\cdot 7^{3/2}$$

$\approx \$0.35$ (Using a calculator)

11. $D(x) = e^{-x+4.5}$

$S(x) = e^{x-5.5}$

a) To find the equilibrium point we set $D(x) = S(x)$ and solve.

$$e^{-x+4.5} = e^{x-5.5}$$

$$\ln e^{-x+4.5} = \ln e^{x-5.5}$$

$$-x + 4.5 = x - 5.5$$

$$10 = 2x$$

$$5 = x$$

Thus x_E = 5 units. To find p_E we substitute x_E into D(x) or S(x). Here we use S(x).

$p_E = S(x_E) = S(5) = e^{5-5.5} = e^{-0.5} \approx \$0.61$

Thus the equilibrium point is (5,\$0.61).

b) The consumer's surplus is

$$\int_0^{x_E} D(x)\,dx - x_E\,p_E,$$

or

$$\int_0^5 (e^{-x+4.5})\,dx - 5(0.61)$$

$$= \left[-e^{-x+4.5}\right]_0^5 - 3.05$$

$$= \left[(-e^{-5+4.5}) - (-e^{-0+4.5})\right] - 3.05$$

$$= (-e^{-0.5} + e^{4.5}) - 3.05$$

$$\approx -0.606531 + 90.017131 - 3.05$$

$$\approx \$86.36$$

c) The producer's surplus is

$$x_E\,p_E - \int_0^{x_E} S(x)\,dx$$

or

$$5(0.61) - \int_0^5 e^{x-5.5}\,dx$$

$$= 3.05 - \left[e^{x-5.5}\right]_0^5$$

$$= 3.05 - (e^{5-5.5} - e^{0-5.5})$$

$$= 3.05 - (e^{-0.5} - e^{-5.5})$$

$$\approx 3.05 - (0.606531 - 0.004087)$$

$$\approx 3.05 - 0.602444$$

$$\approx \$2.45$$

13. See the answer section in the text.

Exercise Set 6.2

---

1. $P(t) = P_0\,e^{kt}$

$P(3) = 100\,e^{0.09(3)}$ (Substituting 3 for t and 100 for P_0)

$$= 100\,e^{0.27}$$

$$\approx 100(1.309964)$$

$$\approx \$131.00$$

3. $$\int_0^T P_0\,e^{kt}\,dt = \frac{P_0}{k}(e^{kT} - 1)$$

$$\int_0^{20} 100\,e^{0.09t}\,dt = \frac{100}{0.09}(e^{0.09(20)} - 1)$$

(Substituting 100 for P_0, 20 for T, and 0.09 for k)

$$= 1111.11\,(e^{1.8} - 1)$$

$$\approx 1111.11(6.049647 - 1)$$

$$\approx 1111.11(5.049647)$$

$$\approx \$5610.71$$

5. $$\int_0^T P_0\,e^{kt}\,dt = \frac{P_0}{k}(e^{kT} - 1)$$

$$\int_0^{40} 1000\,e^{0.085t}\,dt = \frac{1000}{0.085}(e^{0.085(40)} - 1)$$

(Substituting 1000 for P_0, 40 for T, and 0.085 for k)

$$= 11{,}764.71\,(e^{3.4} - 1)$$

$$\approx 11{,}764.71(29.964100 - 1)$$

$$\approx 11{,}764.71(28.964100)$$

$$\approx \$340{,}754.12$$

7. $50{,}000 = \int_0^{20} P_0\, e^{0.085t}\, dt$

$50{,}000 = \frac{P_0}{0.085}(e^{0.085(20)} - 1)$

$4250 = P_0(e^{1.7} - 1)$

$4250 \approx P_0(5.473547 - 1)$

$4250 \approx P_0(4.473947)$

$\frac{4250}{4.473947} \approx P_0$

$\$949.94 \approx P_0$

9. $40{,}000 = \int_0^{30} P_0\, e^{0.09t}\, dt$

$40{,}000 = \frac{P_0}{0.09}(e^{0.09(30)} - 1)$

$3600 = P_0(e^{2.7} - 1)$

$3600 \approx P_0(14.879732 - 1)$

$3600 \approx P_0(13.879732)$

$\frac{3600}{13.879732} \approx P_0$

$\$259.37 \approx P_0$

11. $P_0 = Pe^{-kt}$

$P_0 = 50{,}000\, e^{-0.09(20)}$ (Substituting 50,000 for P, 0.09 for k and 20 for t)

$P_0 = 50{,}000\, e^{-1.8}$

$P_0 \approx 50{,}000\,(0.165299)$

$P_0 \approx \$8264.94$

13. $P_0 = Pe^{-kt}$

$P_0 = 60{,}000\, e^{-0.088(8)}$ (Substituting 60,000 for P, 0.088 for k, and 8 for t)

$P_0 = 60{,}000\, e^{-0.704}$

$P_0 \approx \$29{,}676.18$

15. Accumulated present value $= \int_0^T P\, e^{-kt}\, dt = \frac{P}{k}(1 - e^{-kT})$

$\int_0^{10} 2700\, e^{-0.09t}\, dt = \frac{2700}{0.09}\left[1 - e^{-0.09(10)}\right]$

$= 30{,}000(1 - e^{-0.9})$

$\approx 30{,}000(1 - 0.406570)$

(Using a calculator)

$\approx 30{,}000(0.593430)$

$\approx \$17{,}802.90$

Thus the accumulated present value is $17,802.90. Note that if we had waited to round until after we had multiplied by 30,000, the answer would have been $17,802.91.

17. $\int_0^T P\, e^{-kt}\, dt = \frac{P}{k}(1 - e^{-kT})$

$\int_0^{30} 45{,}000\, e^{-0.08t}\, dt = \frac{45{,}000}{0.08}\left[1 - e^{-0.08(30)}\right]$

$= 562{,}500(1 - e^{-2.4})$

$\approx 562{,}500(1 - 0.090718)$

(Using a calculator)

$\approx 562{,}500(0.909282)$

$\approx \$511{,}471.13$

If we had not rounded until after we had multiplied by 562,500, the answer would have been $511,471.15.

19. $\int_0^T P_0\, e^{kt}\, dt = \frac{P_0}{k}(e^{kT} - 1)$

$\int_0^{14} 101{,}000{,}000\, e^{0.12t}\, dt = \frac{101{,}000{,}000}{0.12}(e^{0.12(14)} - 1)$

$= 841{,}666{,}666.7(e^{1.68} - 1)$

$= 841{,}666{,}666.7(5.365556)$

$= 841{,}666{,}666.7(4.365556)$

$\approx 3{,}674{,}343{,}000$

Thus from 1990 to 2004 the world will use approximately 3,674,343,000 tons of aluminum ore.

21. $75{,}000{,}000{,}000 = \frac{101{,}000{,}000}{0.12}(e^{0.12T} - 1)$

$75{,}000{,}000{,}000 \approx 841{,}666{,}666.7\,(e^{0.12T} - 1)$

$\frac{75{,}000{,}000{,}000}{841{,}666{,}666.7} \approx e^{0.12T} - 1$

$89.108911 \approx e^{0.12T} - 1$

$89.108911 \approx e^{0.12T}$

$\ell n\, 89.108911 \approx \ln e^{0.12T}$

$\ell n\, 89.108911 \approx 0.12T$

$\frac{\ell n\, 89.108911}{0.12} \approx T$

$\frac{4.489859}{0.12} \approx T$

$37.4 \approx T$

$38 \approx T$

Thus in 38 years from 1990 the world reserves of aluminum ore will be exhausted.

23. $\int_0^T P\, e^{-kt}\, dt = \frac{P}{k}(1 - e^{-kT})$

$\int_0^{20} 1 \cdot e^{-0.00003t}\, dt = \frac{1}{0.00003}\left[1 - e^{-0.00003(20)}\right]$

$= 33{,}333.33(1 - e^{-0.0006})$

$\approx 33{,}333.33(1 - 0.999400)$

$\approx 33{,}333.33(0.00059982)$

≈ 19.994

The total amount of radioactive buildup is about 19.994 lb.

25. $\int_0^T R(t)\, e^{k(T-t)}\, dt$

$= \int_0^{30} (2000t + 7)\, e^{0.08(30-t)}\, dt$

$= \int_0^{30} 2000t\, e^{0.08(30-t)}\, dt + \int_0^{30} 7\, e^{0.08(30-t)}\, dt$

$= 2000 \int_0^{30} t\, e^{0.08(30-t)}\, dt + 7 \int_0^{30} e^{0.08(30-t)}\, dt$

We will use integration by parts to evaluate the first integral.

$\int t\, e^{0.08(30-t)}\, dt$

Let

$u = t$ and $dv = e^{0.08(30-t)}$

Then

$du = dt$ and $v = -\frac{1}{0.08} e^{0.08(30-t)}$

$\int t\, e^{0.08(30-t)}\, dt$

$= -\frac{1}{0.08} t\, e^{0.08(30-t)} - \int -\frac{1}{0.08} e^{0.08(30-t)}\, dt$

$= -\frac{1}{0.08} t\, e^{0.08(30-t)} - \frac{1}{0.08} \cdot \frac{1}{0.08} e^{0.08(30-t)} +$ C

$= -12.5\, t\, e^{0.08(30-t)} - 156.25\, e^{0.08(30-t)} + C$

Find the definite integral.

$\int_0^{30} t\, e^{0.08(30-t)}\, dt$

$= \left[-12.5t\, e^{0.08(30-t)} - 156.25\, e^{0.08(30-t)}\right]_0^{30}$

$= \left[(-12.5(30)\, e^{0.08(30-30)} - 156.25\, e^{0.08(30-30)} - (-12.5(0)\, e^{0.08(30-0)} - 156.25\, e^{0.08(30-0)}\right]$

$= (-375\, e^0 - 156.25\, e^0) - (0 - 156.25\, e^{2.4})$

$\approx -375 - 156.25 - 0 + 1722.37$

≈ 1191.12

From the process of integration by parts we know

$\int e^{0.08(30-t)}\, dt = -\frac{1}{0.08} e^{0.08(30-t)} + C$

$= -12.5\, e^{0.08(30-t)} + C$

Find the definite integral.

$\int_0^{30} e^{0.08(30-t)}\, dt$

$= \left[-12.5\, e^{0.08(30-t)}\right]_0^{30}$

$= \left[-12.5\, e^{0.08(30-30)} - (-12.5\, e^{0.08(30-0)}\right]$

$= -12.5\, e^0 + 12.5\, e^{2.4}$

$\approx -12.5 + 137.79$

≈ 125.29

Then

$2000 \int_0^{30} t\, e^{0.08(30-t)}\, dt + 7 \int_0^{30} e^{0.08(30-t)}\, dt$

$\approx 2000(1191.12) + 7(125.29)$

$\approx \$2{,}383{,}117$

If we had done all the calculations before rounding, the result would have been $2,383,120.

27. $\int_0^T R(t)\, e^{-k(T-t)}\, dt$

$= \int_0^{20} t\, e^{-0.08(20-t)}\, dt$

Use integration by parts.

Let

$u = t$ and $dv = e^{-0.08(20-t)}\, dt$

Then

$du = dt$ and $v = \frac{1}{0.08} e^{-0.08(20-t)}$

$= \frac{1}{0.08} t\, e^{-0.08(20-t)} - \int \frac{1}{0.08} e^{-0.08(20-t)}\, dt$

$= \frac{1}{0.08} t\, e^{-0.08(20-t)} - \frac{1}{0.08} \cdot \frac{1}{0.08} e^{-0.08(20-t)} + C$

$= 12.5t\, e^{-0.08(20-t)} - 156.25\, e^{-0.08(20-t)} + C$

Find the definite integral.

$\int_0^{20} t\, e^{-0.08(20-t)}\, dt$

$= \left[12.5t\, e^{-0.08(20-t)} - 156.25\, e^{-0.08(20-t)}\right]_0^{20}$

$= \left[(12.5(20)\, e^{-0.08(20-20)} - 156.25\, e^{-0.08(20-20)}) - (12.5(0)\, e^{-0.08(20-0)} - 156.25\, e^{-0.08(20-0)})\right]$

$= (250\, e^0 - 156.25\, e^0) - (0 - 156.25\, e^{-1.6})$

$\approx 250 - 156.25 - 0 + 31.55$

$\approx \$125.30$

29. See the answer section in the text.

Exercise Set 6.3

1. $\int_3^{\infty} \frac{dx}{x^2}$

$= \lim_{b\to\infty} \int_3^b x^{-2}\, dx$

$= \lim_{b\to\infty} \left[\frac{x^{-1}}{-1}\right]_3^b$

$= \lim_{b\to\infty} \left[-\frac{1}{x}\right]_3^b$

$= \lim_{b\to\infty} \left(-\frac{1}{b} - \left(-\frac{1}{3}\right)\right)$

$= \frac{1}{3}$ $\left[\text{As } b \to \infty,\ -\frac{1}{b} \to 0 \text{ and } -\frac{1}{b} + \frac{1}{3} \to \frac{1}{3}\right]$

The limit does exist. Thus the improper integral in convergent.

3. $\int_3^\infty \frac{dx}{x} = \lim_{b\to\infty} \int_3^b \frac{1}{x}\,dx$

$= \lim_{b\to\infty} [\ln x]_3^b$

$= \lim_{b\to\infty} (\ln b - \ln 3)$

Note that ln b increases indefinitely as b increases. Therefore, the limit does not exist. If the limit does not exist, we say the improper integral is divergent.

5. $\int_0^\infty 3\,e^{-3x}\,dx$

$= \lim_{b\to\infty} \int_0^b 3\,e^{-3x}\,dx$

$= \lim_{b\to\infty} \left[\frac{3}{-3} e^{-3x}\right]_0^b$

$= \lim_{b\to\infty} [-e^{-3x}]_0^b$

$= \lim_{b\to\infty} [-e^{-3b} - (-e^{-3\cdot 0})]$

$= \lim_{b\to\infty} (-e^{-3b} + 1)$

$= \lim_{b\to\infty} \left(1 - \frac{1}{e^{3b}}\right)$

$= 1$ $\left[\text{As } b \to \infty,\ e^{3b} \to \infty, \text{ so } \frac{1}{e^{3b}} \to 0 \text{ and } \left(1 - \frac{1}{e^{3b}}\right) \to 1\right]$

The limit does exist. Thus the improper integral is convergent.

7. $\int_1^\infty \frac{dx}{x^3}$

$= \lim_{b\to\infty} \int_1^b x^{-3}\,dx$

$= \lim_{b\to\infty} \left[\frac{x^{-2}}{-2}\right]_1^b$

$= \lim_{b\to\infty} \left[-\frac{1}{2x^2}\right]_1^b$

$= \lim_{b\to\infty} \left[-\frac{1}{2b^2} - \left(-\frac{1}{2\cdot 1^2}\right)\right]$

$= \lim_{b\to\infty} \left(-\frac{1}{2b^2} + \frac{1}{2}\right)$

$= \frac{1}{2}$ $\left(\text{As } b \to \infty,\ 2b^2 \to \infty, \text{ so } -\frac{1}{2b^2} \to 0\right)$

The limit does exist. Thus the improper integral is convergent.

9. $\int_0^\infty \frac{dx}{1 + x} = \lim_{b\to\infty} \int_0^b \frac{1}{1 + x}\,dx$

$= \lim_{b\to\infty} [\ln (1 + x)]_0^b$

$= \lim_{b\to\infty} [\ln (1 + b) - \ln (1 + 0)]$

$= \lim_{b\to\infty} [\ln (1 + b) - \ln 1]$

$= \lim_{b\to\infty} \ln (1 + b)$

Note that ln (1 + b) increases indefinitely as b increases. Therefore, the limit does not exist. Thus, the improper integral is divergent.

11. $\int_1^\infty 5x^{-2}\,dx$

$= \lim_{b\to\infty} \int_1^b 5x^{-2}\,dx$

$= \lim_{b\to\infty} \left[5 \cdot \frac{x^{-1}}{-1}\right]_1^b$

$= \lim_{b\to\infty} \left[-\frac{5}{x}\right]_1^b$

$= \lim_{b\to\infty} \left[-\frac{5}{b} - \left(-\frac{5}{1}\right)\right]$

$= \lim_{b\to\infty} \left(-\frac{5}{b} + 5\right)$

$= 5$ $\left[\text{As } b \to \infty,\ -\frac{5}{b} \to 0, \text{ so } \left(-\frac{5}{b} + 5\right) \to 5\right]$

The limit does exist. Thus the improper integral is convergent.

13. $\int_0^\infty e^x\,dx = \lim_{b\to\infty} \int_0^b e^x\,dx$

$= \lim_{b\to\infty} [e^x]_0^b$

$= \lim_{b\to\infty} (e^b - e^0)$

$= \lim_{b\to\infty} (e^b - 1)$

As $b \to \infty$, $e^b \to \infty$. Thus the limit does not exist. The improper integral is divergent.

15. $\int_3^\infty x^2\,dx = \lim_{b\to\infty} \int_3^b x^2\,dx$

$= \lim_{b\to\infty} \left[\frac{x^3}{3}\right]_3^b$

$= \lim_{b\to\infty} \left(\frac{b^3}{3} - \frac{3^3}{3}\right)$

$= \lim_{b\to\infty} \frac{b^3}{3} - 9$

Since b^3 increases indefinitely as b increases, the limit does not exist. The improper integral is divergent.

17. $\int_0^\infty x\, e^x\, dx = \lim_{b\to\infty} \int_0^b x\, e^x\, dx$

$= \lim_{b\to\infty} \left[e^x(x - 1)\right]_0^b$

(Using Integration by Parts or Formula 6)

$= \lim_{b\to\infty} \left[e^b(b - 1) - e^0(0 - 1)\right]$

$= \lim_{b\to\infty} \left[e^b(b - 1) + 1\right]$

Since $e^b(b - 1)$ increases indefinitely as b increases, the limit does not exist. The improper integral is divergent.

19. $\int_0^\infty m\, e^{-mx}\, dx,\ m > 0$

$= \lim_{b\to\infty} \int_0^b m\, e^{-mx}\, dx$

$= \lim_{b\to\infty} \left[\frac{m}{-m} e^{-mx}\right]_0^b$

$= \lim_{b\to\infty} \left[-e^{-mx}\right]_0^b$

$= \lim_{b\to\infty} \left[-e^{-mb} - (-e^{-m\cdot 0})\right]$

$= \lim_{b\to\infty} \left(1 - \frac{1}{e^{mb}}\right)$

$= 1$ $\left[\text{As } b \to \infty,\ e^{mb} \to \infty.\ \text{Thus } \frac{1}{e^{mb}} \to 0 \text{ and } \left(1 - \frac{1}{e^{mb}}\right) \to 1\right]$

The limit does exist. Thus the improper integral is convergent.

21. The area is given by

$\int_0^\infty 2x\, e^{-x^2}\, dx$

$= \lim_{b\to\infty} \int_0^b 2x\, e^{-x^2}\, dx$

(We use the substitution $u = -x^2$ to integrate.)

$= \lim_{b\to\infty} \left[-e^{-x^2}\right]_0^b$

$= \lim_{b\to\infty} \left[-e^{-b^2} - (-e^{-0^2})\right]$

$= \lim_{b\to\infty} \left(-\frac{1}{e^{b^2}} + 1\right)$

$= 1$ $\left[\text{As } b \to \infty,\ -\frac{1}{e^{b^2}} \to 0 \text{ and } -\frac{1}{e^{b^2}} + 1 \to 1\right]$

The area of the region is 1.

23. $\frac{P}{k} = \frac{3600}{0.08}$ (Substituting 3600 for P and 0.08 for k)

$= 45{,}000$

The accumulated present value is \$45,000.

25. $P(x) = \int_0^\infty 200\, e^{-0.032x}\, dx$

$= \lim_{b\to\infty} \int_0^b 200\, e^{-0.032x}\, dx$

$= \lim_{b\to\infty} \left[\frac{200}{-0.032} e^{-0.032x}\right]_0^b$

$= \lim_{b\to\infty} \left[-6250\, e^{-0.032b} - (-6250\, e^{-0.032(0)})\right]$

$= \lim_{b\to\infty} \left(-\frac{6250}{e^{0.032b}} + 6250\right)$

$= 6250$ $\left[\text{As } b \to \infty,\ -\frac{6250}{e^{0.032b}} \to 0\right]$

The total profit would be \$6250.

27. $C(x) = \int_1^\infty 3600x^{-1.8}\, dx$

$= \lim_{b\to\infty} \int_1^b 3600x^{-1.8}\, dx$

$= \lim_{b\to\infty} \left[\frac{3600}{-0.8} x^{-0.8}\right]_1^b$

$= \lim_{b\to\infty} \left[-4500b^{-0.8} - (-4500)\, 1^{-0.8}\right]$

$= \lim_{b\to\infty} \left(-\frac{4500}{b^{0.8}} + 4500\right)$

$= 4500$ $\left[\text{As } b \to \infty,\ -\frac{4500}{b^{0.8}} \to 0\right]$

The total cost would be \$4500.

29. $\int_0^T P\, e^{-kt}\, dt = \frac{P}{k}(1 - e^{-kT})$

$= \frac{P}{k}\left(1 - \frac{1}{e^{kT}}\right)$

As $T \to \infty$, the buildup of radioactive material approaches a limiting value P/k.

$\frac{P}{k} = \frac{1}{0.00003}$ (Substituting 1 for P and 0.00003 for k)

(0.003% = 0.00003)

$\approx 33{,}333$ lb

31. $\int_0^\infty \frac{dx}{x^{2/3}} = \lim_{b\to\infty} \int_0^b x^{-2/3}\, dx$

$= \lim_{b\to\infty} \left[\frac{x^{1/3}}{1/3}\right]_0^b$

$= \lim_{b\to\infty} \left[3x^{1/3}\right]_0^b$

$= \lim_{b\to\infty} (3b^{1/3} - 3\cdot 0^{1/3})$

$= \lim_{b\to\infty} 3\sqrt[3]{b}$

Now, as $b \to \infty$, we know that $\sqrt[3]{b} \to \infty$. Thus, the limit does not exist. The improper integral is divergent.

33. $\int_0^\infty \frac{dx}{(x+1)^{3/2}}$

$= \lim_{b\to\infty} \int_0^b (x+1)^{-3/2}\, dx$

$= \lim_{b\to\infty} \left[\frac{(x+1)^{-1/2}}{-1/2}\right]_0^b$

$= \lim_{b\to\infty} \left[-\frac{2}{\sqrt{x+1}}\right]_0^b$

$= \lim_{b\to\infty} \left[-\frac{2}{\sqrt{b+1}} - \left(-\frac{2}{\sqrt{0+1}}\right)\right]$

$= \lim_{b\to\infty} \left[-\frac{2}{\sqrt{b+1}} + 2\right]$

$= 2$ $\quad$ [As $b \to \infty$, $\sqrt{b+1} \to \infty$, so $-\frac{2}{\sqrt{b+1}} \to 0$]

The limit does exist. The improper integral is convergent.

35. $\int_0^\infty x\, e^{-x^2}\, dx$

$= \int_0^\infty -\frac{1}{2} \cdot (-2)x\, e^{-x^2}\, dx$ (Multiplying by 1)

$= -\frac{1}{2}\int_0^\infty e^{-x^2}(-2x)\, dx$

$= -\frac{1}{2} \lim_{b\to 0} \left[e^{-x^2}\right]_0^b$ (Using substitution where $u = -x^2$ and $du = -2x\, dx$)

$= -\frac{1}{2} \lim_{b\to\infty} \left(e^{-b^2} - e^{-0^2}\right)$

$= -\frac{1}{2} \lim_{b\to\infty} \left(\frac{1}{e^{b^2}} - 1\right)$

$= -\frac{1}{2}(-1)$ $\quad$ [As $b \to \infty$, $e^{b^2} \to \infty$, so $\frac{1}{e^{b^2}} \to 0$ and $\left(\frac{1}{e^{b^2}} - 1\right) \to -1$]

$= \frac{1}{2}$

The limit does exist. The improper integral is convergent.

37. $\int_0^\infty E(t)\, dt$

$= \int_0^\infty t\, e^{-kt}\, dt$

$= \lim_{b\to\infty} \int_0^b t\, e^{-kt}\, dt$

$= \lim_{b\to\infty} \left[\frac{1}{(-k)^2} \cdot e^{-kt}(-kt - 1)\right]_0^b$

$= \lim_{b\to\infty} \left[-\frac{kt+1}{k^2 e^{kt}}\right]_0^b$

$= \lim_{b\to\infty} \left[-\frac{kb+1}{k^2 e^{kb}} - \left(-\frac{k\cdot 0+1}{k^2 e^{k\cdot 0}}\right)\right]$

$= \lim_{b\to\infty} \left[-\frac{kb+1}{k^2 e^{kb}} + \frac{1}{k^2}\right]$

$= \frac{1}{k^2}$ $\quad$ [Input-output tables show that as $b \to \infty$, $-\frac{kb+1}{k^2 e^{kb}} \to 0$.]

The integral represents the total dose of the drug.

39. See the answer section in the text.

Exercise Set 6.4

1. $f(x) = 2x, \quad [0,1]$

$P([0,1]) = \int_0^1 f(x)\, dx$

$= \int_0^1 2x\, dx$

$= \left[x^2\right]_0^1$

$= 1^2 - 0^2$

$= 1$

3. $f(x) = \frac{1}{3}, \quad [4,7]$

$P([4,7]) = \int_4^7 f(x)\, dx$

$= \int_4^7 \frac{1}{3}\, dx$

$= \left[\frac{1}{3}x\right]_4^7$

$= \frac{1}{3}\cdot 7 - \frac{1}{3}\cdot 4$

$= \frac{7}{3} - \frac{4}{3}$

$= \frac{3}{3}$

$= 1$

5. $f(x) = \frac{3}{26} x^2$, [1,3]

$$P([1,3]) = \int_1^3 f(x)\,dx$$
$$= \int_1^3 \frac{3}{26} x^2\,dx$$
$$= \left[\frac{3}{26} \cdot \frac{x^3}{3}\right]_1^3$$
$$= \left[\frac{x^3}{26}\right]_1^3$$
$$= \frac{3^3}{26} - \frac{1^3}{26}$$
$$= \frac{27}{26} - \frac{1}{26}$$
$$= \frac{26}{26}$$
$$= 1$$

7. $f(x) = \frac{1}{x}$, [1,e]

$$P([1,e]) = \int_1^e f(x)\,dx$$
$$= \int_1^e \frac{1}{x}\,dx$$
$$= [\ln x]_1^e$$
$$= \ln e - \ln 1$$
$$= 1 - 0$$
$$= 1$$

9. $f(x) = \frac{3}{2} x^2$, [-1,1]

$$P([-1,1]) = \int_{-1}^1 f(x)\,dx$$
$$= \int_{-1}^1 \frac{3}{2} x^2\,dx$$
$$= \left[\frac{3}{2} \cdot \frac{x^3}{3}\right]_{-1}^1$$
$$= \left[\frac{x^3}{2}\right]_{-1}^1$$
$$= \frac{1^3}{2} - \frac{(-1)^3}{2}$$
$$= \frac{1}{2} + \frac{1}{2}$$
$$= 1$$

11. $f(x) = 3e^{-3x}$, $[0,\infty)$

$$P\big[[0,\infty)\big] = \int_0^\infty f(x)\,dx$$
$$= \int_0^\infty 3e^{-3x}\,dx$$
$$= \lim_{b\to\infty} \int_0^b 3e^{-3x}\,dx$$
$$= \lim_{b\to\infty} \left[\frac{3}{-3} e^{-3x}\right]_0^b$$
$$= \lim_{b\to\infty} [-e^{-3x}]_0^b$$
$$= \lim_{b\to\infty} [-e^{-3b} - (-e^{-3\cdot 0})]$$
$$= \lim_{b\to\infty} [-e^{-3b} + 1]$$
$$= \lim_{b\to\infty} \left[-\frac{1}{e^{3b}} + 1\right]$$
$$= 1$$

[As $b \to \infty$, $e^{3b} \to \infty$, so $-\frac{1}{e^{3b}} \to 0$ and $\left(-\frac{1}{e^{3b}} + 1\right) \to 1$]

13. $f(x) = kx$, [1,3]

Find k such that $\int_1^3 kx\,dx = 1$.

$$\int_1^3 x\,dx = \left[\frac{x^2}{2}\right]_1^3$$
$$= \frac{3^2}{2} - \frac{1^2}{2} = \frac{9}{2} - \frac{1}{2} = \frac{8}{2} = 4$$

Thus $k = \frac{1}{4}$ and $f(x) = \frac{1}{4} x$.

15. $f(x) = kx^2$, [-1,1]

Find k such that $\int_{-1}^1 kx^2\,dx = 1$.

$$\int_{-1}^1 x^2\,dx = \left[\frac{x^3}{3}\right]_{-1}^1$$
$$= \frac{1^3}{3} - \frac{(-1)^3}{3} = \frac{1}{3} + \frac{1}{3} = \frac{2}{3}$$

Thus $k = \frac{1}{2/3} = \frac{3}{2}$ and $f(x) = \frac{3}{2} x^2$.

17. $f(x) = k$, [2,7]

Find k such that $\int_2^7 k\,dx = 1$.

$$\int_2^7 dx = [x]_2^7 = 7 - 2 = 5$$

Thus $k = \frac{1}{5}$, and $f(x) = \frac{1}{5}$.

19. $f(x) = k(2 - x)$, $[0,2]$

Find k such that $\int_0^2 k(2 - x)\,dx = 1$.

$$\int_0^2 (2 - x)\,dx = \left[2x - \frac{x^2}{2}\right]_0^2$$
$$= \left(2\cdot 2 - \frac{2^2}{2}\right) - \left(2\cdot 0 - \frac{0^2}{2}\right)$$
$$= (4 - 2)$$
$$= 2$$

Thus $k = \frac{1}{2}$, and $f(x) = \frac{1}{2}(2 - x)$, or $\frac{2 - x}{2}$.

21. $f(x) = \frac{k}{x}$, $[1,3]$

Find k such that $\int_1^3 k \cdot \frac{1}{x}\,dx = 1$.

$$\int_1^3 \frac{1}{x}\,dx = [\ln x]_1^3$$
$$= \ln 3 - \ln 1$$
$$= \ln 3 \qquad (\ln 1 = 0)$$

Thus $k = \frac{1}{\ln 3}$, and $f(x) = \frac{1}{\ln 3} \cdot \frac{1}{x} = \frac{1}{x \ln 3}$.

23. $f(x) = k\,e^x$, $[0,3]$

Find k such that $\int_0^3 k\,e^x\,dx = 1$.

$$\int_0^3 e^x\,dx = [e^x]_0^3 = e^3 - e^0 = e^3 - 1$$

Thus $k = \frac{1}{e^3 - 1}$, and $f(x) = \frac{1}{e^3 - 1} \cdot e^x$, or $\frac{e^x}{e^3 - 1}$.

25. $f(x) = \frac{1}{50}x$, for $0 \leqslant x \leqslant 10$

$$P(2 \leqslant x \leqslant 6) = \int_2^6 \frac{1}{50}x\,dx$$
$$= \left[\frac{1}{50} \cdot \frac{x^2}{2}\right]_2^6$$
$$= \left[\frac{x^2}{100}\right]_2^6$$
$$= \frac{6^2}{100} - \frac{2^2}{100}$$
$$= \frac{36 - 4}{100}$$
$$= \frac{32}{100}$$
$$= \frac{8}{25} \text{ or } 0.32$$

The probability that the dart lands in [2,6] is 0.32.

27. $f(x) = \frac{1}{16}$, for $4 \leqslant x \leqslant 20$

$$P(9 \leqslant x \leqslant 17) = \int_9^{17} \frac{1}{16}\,dx$$
$$= \frac{1}{16}[x]_9^{17}$$
$$= \frac{1}{16}(17 - 9)$$
$$= \frac{8}{16}$$
$$= \frac{1}{2}, \text{ or } 0.5$$

The probability that a number selected is in the subinterval [9,17] is 0.5.

29. $f(x) = k\,e^{-kx}$, for $0 \leqslant x < \infty$, where $k = 1/a$ and a = the average distance between successive cars over some period of time.

We first determine k:

$$k = \frac{1}{100} = 0.01.$$

The probability density function for x is $f(x) = 0.01\,e^{-0.01x}$, for $0 \leqslant x < \infty$.

The probability that the distance between cars is 40 ft or less is

$$P(0 \leqslant x \leqslant 40) = \int_0^{40} 0.01\,e^{-0.01x}\,dx$$
$$= \left[\frac{0.01}{-0.01}\,e^{-0.01x}\right]_0^{40}$$
$$= [-e^{-0.01x}]_0^{40}$$
$$= \left(-e^{-0.01(40)}\right) - \left(-e^{-0.01(0)}\right)$$
$$= -e^{-0.4} + 1$$
$$= 1 - e^{-0.4}$$
$$\approx 1 - 0.670320$$
$$\approx 0.329680$$
$$\approx 0.3297$$

31. $f(t) = 2\,e^{-2t}$, $0 \leqslant t < \infty$

$$P([0,5]) = \int_0^5 2\,e^{-2t}\,dt$$
$$= \left[\frac{2}{-2}\,e^{-2t}\right]_0^5$$
$$= [-e^{-2t}]_0^5$$
$$= (-e^{-2\cdot 5}) - (-e^{-2\cdot 0})$$
$$= -e^{-10} + 1$$
$$= 1 - e^{-10}$$
$$\approx 1 - 0.0000454$$
$$\approx 0.9999546$$
$$\approx 0.99995$$

The probability that a phone call will last no more than 5 minutes is 0.99995.

33. $f(x) = k\,e^{-kt}$, $0 \leq t < \infty$, where $k = 1/a$ and a = the average time that will pass before a failure occurs.

We first find k:

$$k = \frac{1}{100} = 0.01.$$

The probability density function for x is $f(x) = 0.01\,e^{-0.01x}$, for $0 \leq x < \infty$.

The probability that a failure will occur in 50 hours or less is

$$P(0 \leq x \leq 50) = \int_0^{50} 0.01\,e^{-0.01x}\,dx$$
$$= \left[\frac{0.01}{-0.01}\,e^{-0.01x}\right]_0^{50}$$
$$= \left[-e^{-0.01x}\right]_0^{50}$$
$$= \left(-e^{-0.01(50)}\right) - \left(-e^{-0.01(0)}\right)$$
$$= -e^{-0.5} + 1$$
$$= 1 - e^{-0.5}$$
$$\approx 1 - 0.606531$$
$$\approx 0.393469$$
$$\approx 0.3935$$

The probability that a failure will occur in 50 hours or less is 0.3935.

35. $f(t) = 0.02\,e^{-0.02t}$, $0 \leq t < \infty$

$$P(0 \leq t \leq 150) = \int_0^{150} 0.02\,e^{-0.02t}\,dt$$
$$= \left[\frac{0.02}{-0.02}\,e^{-0.02t}\right]_0^{150}$$
$$= \left[-e^{-0.02t}\right]_0^{150}$$
$$= \left(-e^{-0.02(150)}\right) - \left(-e^{-0.02(0)}\right)$$
$$= -e^{-3} + 1$$
$$\approx -0.049787 + 1$$
$$\approx 0.950213$$
$$\approx 0.9502$$

The probability that a rat will learn its way through a maze in 150 seconds, or less, is 0.9502.

37. $f(x) = x^3$ is a probability function on $[0,b]$. Thus,

$$\int_0^b f(x)\,dx = 1$$
$$\int_0^b x^3\,dx = 1$$
$$\left[\frac{x^4}{4}\right]_0^b = 1$$
$$\frac{b^4}{4} - \frac{0^4}{4} = 1$$
$$\frac{b^4}{4} = 1$$
$$b^4 = 4$$
$$b = \sqrt[4]{4}$$
$$b = \sqrt[4]{2^2}$$
$$b = \sqrt{2}$$

Exercise Set 6.5

1. $f(x) = \frac{1}{3}$, $[2,5]$

$$E(x) = \int_a^b x\,f(x)\,dx \qquad \text{(Expected value of x)}$$
$$E(x) = \int_2^5 x \cdot \frac{1}{3}\,dx$$
$$= \frac{1}{3}\left[\frac{x^2}{2}\right]_2^5$$
$$= \frac{1}{3}\left(\frac{5^2}{2} - \frac{2^2}{2}\right)$$
$$= \frac{1}{3}\left(\frac{25}{2} - \frac{4}{2}\right)$$
$$= \frac{1}{3} \cdot \frac{21}{2}$$
$$= \frac{7}{2}$$

$$E(x^2) = \int_a^b x^2\,f(x)\,dx \qquad \text{(Expected value of } x^2\text{)}$$
$$E(x^2) = \int_2^5 x^2 \cdot \frac{1}{3}\,dx$$
$$= \frac{1}{3}\left[\frac{x^3}{3}\right]_2^5$$
$$= \frac{1}{3}\left(\frac{5^3}{3} - \frac{2^3}{3}\right)$$
$$= \frac{1}{3}\left(\frac{125}{3} - \frac{8}{3}\right)$$
$$= \frac{1}{3} \cdot \frac{117}{3}$$
$$= \frac{117}{9}$$
$$= 13$$

$$\mu = E(x) = \frac{7}{2} \qquad \text{(Mean)}$$

$$\sigma^2 = E(x^2) - [E(x)]^2$$
$$= 13 - \left(\frac{7}{2}\right)^2 \qquad \text{[Substituting 13 for } E(x^2) \text{ and } \tfrac{7}{2} \text{ for } E(x)\text{]}$$
$$= \frac{52}{4} - \frac{49}{4}$$
$$= \frac{3}{4} \qquad \text{(Variance)}$$

$$\sigma = \sqrt{\text{variance}}$$
$$= \sqrt{\frac{3}{4}} \qquad \text{(Substituting } \tfrac{3}{4} \text{ for the variance)}$$
$$= \frac{1}{2}\sqrt{3} \qquad \text{(Standard deviation)}$$

3. $f(x) = \frac{2}{9}x, \quad [0,3]$

$E(x) = \int_a^b x\, f(x)\, dx$ (Expected value of x)

$E(x) = \int_0^3 x \cdot \frac{2}{9}x\, dx$

$= \int_0^3 \frac{2}{9}x^2\, dx$

$= \frac{2}{9}\left[\frac{x^3}{3}\right]_0^3$

$= \frac{2}{9}\left(\frac{3^3}{3} - \frac{0^3}{3}\right)$

$= \frac{2}{9} \cdot 9$

$= 2$

$E(x^2) = \int_a^b x^2 f(x)\, dx$ (Expected value of x^2)

$E(x^2) = \int_0^3 x^2 \cdot \frac{2}{9}x\, dx$

$= \int_0^3 \frac{2}{9}x^3\, dx$

$= \frac{2}{9}\left[\frac{x^4}{4}\right]_0^3$

$= \frac{2}{9}\left(\frac{3^4}{4} - \frac{0^4}{4}\right)$

$= \frac{2}{9} \cdot \frac{81}{4}$

$= \frac{9}{2}$

$\mu = E(x) = 2$ (Mean)

$\sigma^2 = E(x^2) - [E(x)]^2$

$= \frac{9}{2} - 2^2$ [Substituting $\frac{9}{2}$ for $E(x^2)$ and 2 for $E(x)$]

$= \frac{9}{2} - \frac{8}{2}$

$= \frac{1}{2}$ (Variance)

5. $f(x) = \frac{2}{3}x, \quad [1,2]$

$E(x) = \int_a^b x\, f(x)\, dx$ (Expected value of x)

$E(x) = \int_1^2 x \cdot \frac{2}{3}x\, dx$

$= \int_1^2 \frac{2}{3}x^2\, dx$

$= \frac{2}{3}\left[\frac{x^3}{3}\right]_1^2$

$= \frac{2}{3}\left(\frac{2^3}{3} - \frac{1^3}{3}\right)$

$= \frac{2}{3}\left(\frac{8}{3} - \frac{1}{3}\right)$

$= \frac{2}{3} \cdot \frac{7}{3}$

$= \frac{14}{9}$

$E(x^2) = \int_a^b x^2 f(x)\, dx$ (Expected value of x^2)

$E(x^2) = \int_1^2 x^2 \cdot \frac{2}{3}x\, dx$

$= \int_1^2 \frac{2}{3}x^3\, dx$

$= \frac{2}{3}\left[\frac{x^4}{4}\right]_1^2$

$= \frac{2}{3}\left(\frac{2^4}{4} - \frac{1^4}{4}\right)$

$= \frac{2}{3}\left(\frac{16}{4} - \frac{1}{4}\right)$

$= \frac{2}{3} \cdot \frac{15}{4}$

$= \frac{5}{2}$

$\mu = E(x) = \frac{14}{9}$ (Mean)

$\sigma^2 = E(x^2) - [E(x)]^2$

$= \frac{5}{2} - \left(\frac{14}{9}\right)^2$ [Substituting $\frac{5}{2}$ for $E(x^2)$ and $\frac{14}{9}$ for $E(x)$]

$= \frac{5}{2} - \frac{196}{81}$

$= \frac{405}{162} - \frac{392}{162}$

$= \frac{13}{162}$ (Variance)

$\sigma = \sqrt{\text{variance}}$

$= \sqrt{\frac{13}{162}}$ (Substituting $\frac{13}{162}$ for the variance)

$= \frac{1}{9}\sqrt{\frac{13}{2}}$ (Standard deviation)

7. $f(x) = \frac{1}{3}x^2, \quad [-2,1]$

$E(x) = \int_a^b x\, f(x)\, dx$ (Expected value of x)

$E(x) = \int_{-2}^1 x \cdot \frac{1}{3}x^2\, dx$

$= \int_{-2}^1 \frac{1}{3}x^3\, dx$

$= \frac{1}{3}\left[\frac{x^4}{4}\right]_{-2}^1$

$= \frac{1}{3}\left(\frac{1^4}{4} - \frac{(-2)^4}{4}\right)$

$= \frac{1}{3}\left(\frac{1}{4} - \frac{16}{4}\right)$

$= \frac{1}{3}\left(-\frac{15}{4}\right)$

$= -\frac{5}{4}$

$E(x^2) = \int_a^b x^2 f(x)\,dx$ (Expected value of x^2)

$E(x^2) = \int_{-2}^{1} x^2 \cdot \frac{1}{3} x^2\,dx$

$= \int_{-2}^{1} \frac{1}{3} x^4\,dx$

$= \frac{1}{3}\left[\frac{x^5}{5}\right]_{-2}^{1}$

$= \frac{1}{3}\left[\frac{1^5}{5} - \frac{(-2)^5}{5}\right]$

$= \frac{1}{3}\left(\frac{1}{5} + \frac{32}{5}\right)$

$= \frac{1}{3}\left(\frac{33}{5}\right)$

$= \frac{11}{5}$

$\mu = E(x) = -\frac{5}{4}$ (Mean)

$\sigma^2 = E(x^2) - [E(x)]^2$

$= \frac{11}{5} - \left(-\frac{5}{4}\right)^2$ [Substituting $\frac{11}{5}$ for $E(x^2)$ and $-\frac{5}{4}$ for $E(x)$]

$= \frac{11}{5} - \frac{25}{16}$

$= \frac{176}{80} - \frac{125}{80}$

$= \frac{51}{80}$ (Variance)

$\sigma = \sqrt{\text{variance}}$

$= \sqrt{\frac{51}{80}}$ (Substituting $\frac{51}{80}$ for the variance)

$= \frac{1}{4}\sqrt{\frac{51}{5}}$ (Standard deviation)

9. $f(x) = \frac{1}{\ln 3} \cdot \frac{1}{x}$, [1,3]

$E(x) = \int_a^b x\,f(x)\,dx$ (Expected value of x)

$E(x) = \int_1^3 x\left[\frac{1}{\ln 3} \cdot \frac{1}{x}\right] dx$

$= \int_1^3 \frac{1}{\ln 3}\,dx$

$= \frac{1}{\ln 3}\int_1^3 dx$

$= \frac{1}{\ln 3}[x]_1^3$

$= \frac{1}{\ln 3}(3 - 1)$

$= \frac{2}{\ln 3}$

$E(x^2) = \int_a^b x^2 f(x)\,dx$ (Expected value of x^2)

$E(x^2) = \int_1^3 x^2\left[\frac{1}{\ln 3} \cdot \frac{1}{x}\right] dx$

$= \int_1^3 \frac{1}{\ln 3} \cdot x\,dx$

$= \frac{1}{\ln 3}\left[\frac{x^2}{2}\right]_1^3$

$= \frac{1}{\ln 3}\left(\frac{3^2}{2} - \frac{1^2}{2}\right)$

$= \frac{1}{\ln 3}\left(\frac{9}{2} - \frac{1}{2}\right)$

$= \frac{4}{\ln 3}$

$\mu = E(x) = \frac{2}{\ln 3}$ (Mean)

$\sigma^2 = E(x^2) - [E(x)]^2$

$= \frac{4}{\ln 3} - \left(\frac{2}{\ln 3}\right)^2$ [Substituting $\frac{4}{\ln 3}$ for $E(x^2)$ and $\frac{2}{\ln 3}$ for $E(x)$]

$= \frac{4 \ln 3}{(\ln 3)^2} - \frac{4}{(\ln 3)^2}$

$= \frac{4 \ln 3 - 4}{(\ln 3)^2}$

$= \frac{4(\ln 3 - 1)}{(\ln 3)^2}$ (Variance)

$\sigma = \sqrt{\text{variance}}$

$= \sqrt{\frac{4(\ln 3 - 1)}{(\ln 3)^2}}$ [Substituting $\frac{4(\ln 3 - 1)}{(\ln 3)^2}$ for the variance]

$= \frac{2}{\ln 3}\sqrt{\ln 3 - 1}$ (Standard deviation)

11.

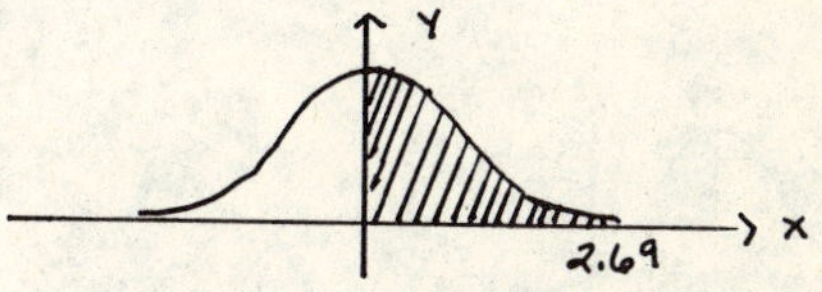

$P(0 \le x \le 2.69)$
$= 0.4964$ (Using Table 2)

13.

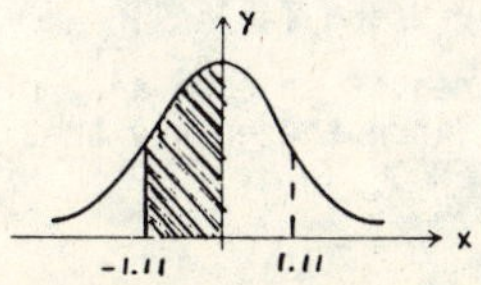

$P(-1.11 \le x \le 0)$
$= P(0 \le x \le 1.11)$ (Symmetry of the graph)
$= 0.3665$ (Using Table 2)

15.

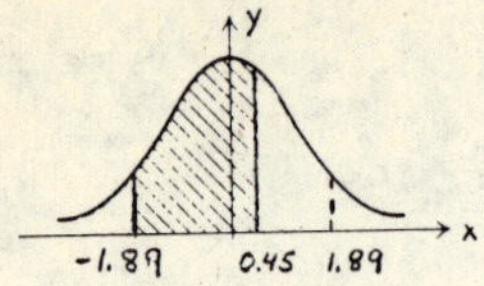

$P(-1.89 \leqslant x \leqslant 0.45)$
$= P(-1.89 \leqslant x \leqslant 0) + P(0 \leqslant x \leqslant 0.45)$
$= P(0 \leqslant x \leqslant 1.89) + P(0 \leqslant x \leqslant 0.45)$ (Symmetry of the graph)
$= 0.4706 + 0.1736$ (Using Table 2)
$= 0.6442$

17.

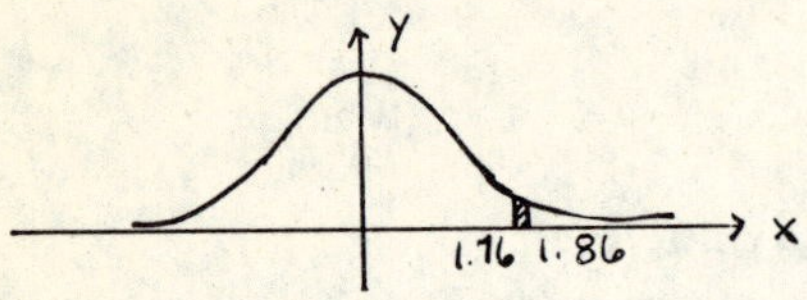

$P(1.76 \leqslant x \leqslant 1.86)$
$= P(0 \leqslant x \leqslant 1.86) - P(0 \leqslant x \leqslant 1.76)$
$= 0.4686 - 0.4608$ (Using Table 2)
$= 0.0078$

19.

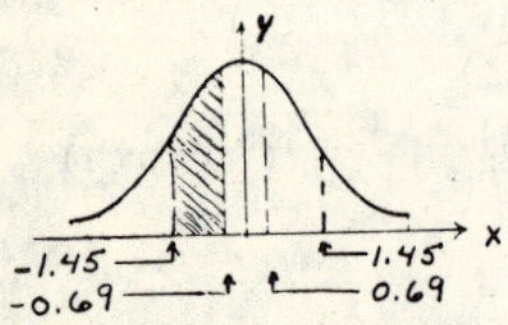

$P(-1.45 \leqslant x \leqslant -0.69)$
$= P(0.69 \leqslant x \leqslant 1.45)$ (Symmetry of the graph)
$= P(0 \leqslant x \leqslant 1.45) - P(0 \leqslant x \leqslant 0.69)$
$= 0.4265 - 0.2549$ (Using Table 2)
$= 0.1716$

21.

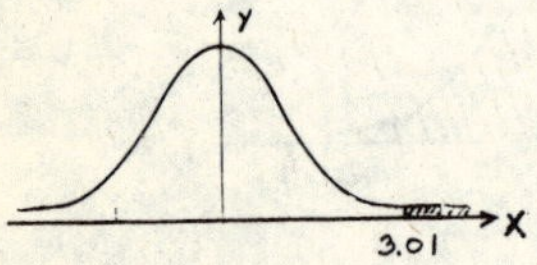

$P(x \geqslant 3.01)$
$= P(x \geqslant 0) - P(0 \leqslant x \leqslant 3.01)$
$= 0.5000 - 0.4987$ (Using Table 2)
[Half of the area is on each side of the line x = 0. Since the entire area is 1, $P(x \geqslant 0) = 0.5000$]
$= 0.0013$

23. a)

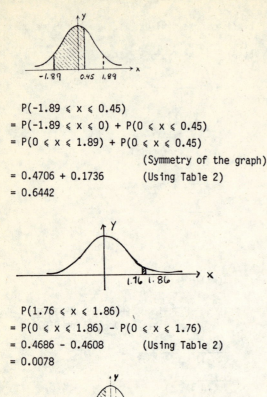

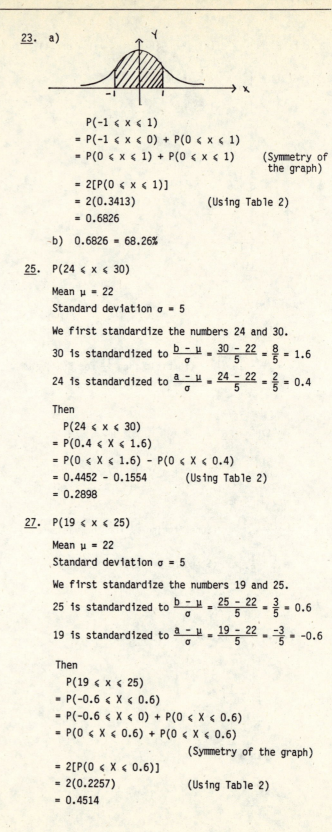

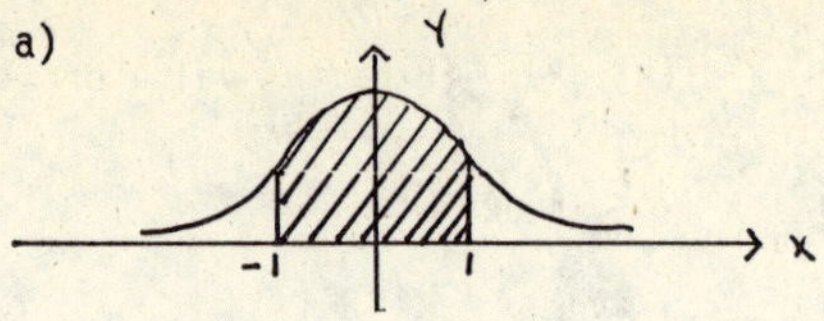

$P(-1 \leqslant x \leqslant 1)$
$= P(-1 \leqslant x \leqslant 0) + P(0 \leqslant x \leqslant 1)$
$= P(0 \leqslant x \leqslant 1) + P(0 \leqslant x \leqslant 1)$ (Symmetry of the graph)
$= 2[P(0 \leqslant x \leqslant 1)]$
$= 2(0.3413)$ (Using Table 2)
$= 0.6826$

b) 0.6826 = 68.26%

25. $P(24 \leqslant x \leqslant 30)$

Mean $\mu = 22$
Standard deviation $\sigma = 5$

We first standardize the numbers 24 and 30.

30 is standardized to $\frac{b - \mu}{\sigma} = \frac{30 - 22}{5} = \frac{8}{5} = 1.6$

24 is standardized to $\frac{a - \mu}{\sigma} = \frac{24 - 22}{5} = \frac{2}{5} = 0.4$

Then
$P(24 \leqslant x \leqslant 30)$
$= P(0.4 \leqslant X \leqslant 1.6)$
$= P(0 \leqslant X \leqslant 1.6) - P(0 \leqslant X \leqslant 0.4)$
$= 0.4452 - 0.1554$ (Using Table 2)
$= 0.2898$

27. $P(19 \leqslant x \leqslant 25)$

Mean $\mu = 22$
Standard deviation $\sigma = 5$

We first standardize the numbers 19 and 25.

25 is standardized to $\frac{b - \mu}{\sigma} = \frac{25 - 22}{5} = \frac{3}{5} = 0.6$

19 is standardized to $\frac{a - \mu}{\sigma} = \frac{19 - 22}{5} = \frac{-3}{5} = -0.6$

Then
$P(19 \leqslant x \leqslant 25)$
$= P(-0.6 \leqslant X \leqslant 0.6)$
$= P(-0.6 \leqslant X \leqslant 0) + P(0 \leqslant X \leqslant 0.6)$
$= P(0 \leqslant X \leqslant 0.6) + P(0 \leqslant X \leqslant 0.6)$ (Symmetry of the graph)
$= 2[P(0 \leqslant X \leqslant 0.6)]$
$= 2(0.2257)$ (Using Table 2)
$= 0.4514$

29. We first standardize the number 300.

300 is standardized to

$\frac{b - \mu}{\sigma} = \frac{300 - 250}{20}$ (Substituting 300 for b, 250 for μ, and 20 for σ)

$= \frac{50}{20}$

$= \frac{5}{2}$, or 2.5

$P(x \geqslant 300)$
$= P(X \geqslant 2.5)$
$= P(X \geqslant 0) - P(0 \leqslant X \leqslant 2.5)$
$= 0.5000 - 0.4938$ [Entire area is 1; $P(X \geqslant 0) = 0.5000$] (Using Table 2)
$= 0.0062$
$= 0.62\%$

Thus the company will have to hire extra help or pay overtime 0.62% of the days.

31. $\mu = 65$, $\sigma = 20$

We first standardize 80 and 89.

80 is standardized to $\frac{a - \mu}{\sigma} = \frac{80 - 65}{20} = \frac{15}{20} = 0.75$

89 is standardized to $\frac{b - \mu}{\sigma} = \frac{89 - 65}{20} = \frac{24}{20} = 1.2$

Then $P(80 \leqslant S \leqslant 89)$
$= P(0.75 \leqslant X \leqslant 1.2)$
$= P(0 \leqslant X \leqslant 1.2) - P(0 \leqslant X \leqslant 0.75)$
$= 0.3849 - 0.2734$
$= 0.1115$

33. $f(x) = \frac{1}{b - a}$, [a,b]

$E(x) = \int_a^b x\, f(x)\, dx$ (Expected value of x)

$E(x) = \int_a^b x \cdot \frac{1}{b - a}\, dx$

$= \int_a^b \frac{1}{b - a} \cdot x\, dx$

$= \frac{1}{b - a}\left[\frac{x^2}{2}\right]_a^b$

$= \frac{1}{b - a}\left(\frac{b^2}{2} - \frac{a^2}{2}\right)$

$= \frac{1}{b - a}\left(\frac{b^2 - a^2}{2}\right)$

$= \frac{1}{b - a} \cdot \frac{(b - a)(b + a)}{2}$

$= \frac{b + a}{2}$

$E(x^2) = \int_a^b x^2 f(x)\, dx$ (Expected value of x^2)

$E(x^2) = \int_a^b x^2 \cdot \frac{1}{b - a}\, dx$

$= \int_a^b \frac{1}{b - a} \cdot x^2\, dx$

$= \frac{1}{b - a}\left[\frac{x^3}{3}\right]_a^b$

$= \frac{1}{b - a}\left(\frac{b^3}{3} - \frac{a^3}{3}\right)$

$= \frac{1}{b - a}\left(\frac{b^3 - a^3}{3}\right)$

$= \frac{1}{b - a} \cdot \frac{(b - a)(b^2 + ba + a^2)}{3}$

$= \frac{b^2 + ba + a^2}{3}$

$\mu = E(x) = \frac{b + a}{2}$ (Mean)

$\sigma^2 = E(x^2) - [E(x)]^2$

$= \frac{b^2 + ba + a^2}{3} - \left(\frac{b + a}{2}\right)^2$ (Substituting)

$= \frac{b^2 + ba + a^2}{3} - \frac{b^2 + 2ba + a^2}{4}$

$= \frac{4b^2 + 4ba + 4a^2}{12} - \frac{3b^2 + 6ba + 3a^2}{12}$

$= \frac{4b^2 + 4ba + 4a^2 - 3b^2 - 6ba - 3a^2}{12}$

$= \frac{b^2 - 2ba + a^2}{12}$

$= \frac{(b - a)^2}{12}$ (Variance)

$\sigma = \sqrt{\text{variance}}$

$= \sqrt{\frac{(b - a)^2}{12}}$ (Substituting)

$= \frac{b - a}{2\sqrt{3}}$ (Standard deviation)

35. $f(x) = \frac{1}{2}x$, [0,2]

$\int_0^m f(x)\, dx = \frac{1}{2}$

$\int_0^m \frac{1}{2} x\, dx = \frac{1}{2}$

$\frac{1}{2}\int_0^m x\, dx = \frac{1}{2}$

$\int_0^m x\, dx = 1$

$\left[\frac{x^2}{2}\right]_0^m = 1$

$\frac{m^2}{2} - \frac{0^2}{2} = 1$

$\frac{m^2}{2} - 0 = 1$

$\frac{m^2}{2} = 1$

$m^2 = 2$

$m = \sqrt{2}$

37. $f(x) = k\,e^{-kx},\ [0,\infty)$

$$\int_0^m f(x)\,dx = \frac{1}{2}$$

$$\int_0^m k\,e^{-kx} = \frac{1}{2}$$

$$\left[\frac{k}{-k}\,e^{-kx}\right]_0^m = \frac{1}{2}$$

$$\left[-e^{-kx}\right]_0^m = \frac{1}{2}$$

$$(-e^{-km}) - (-e^{-k\cdot 0}) = \frac{1}{2}$$

$$-e^{-km} + e^0 = \frac{1}{2}$$

$$-e^{-km} + 1 = \frac{1}{2}$$

$$\frac{1}{2} = e^{-km}$$

$$\ln\frac{1}{2} = \ln e^{-km}$$

$$\ln 1 - \ln 2 = -km$$

$$-\ln 2 = -km \qquad (\ln 1 = 0)$$

$$\frac{-\ln 2}{-k} = m$$

$$\frac{\ln 2}{k} = m$$

39. Standardize 6.5.

$$X = \frac{6.5 - \mu}{\sigma} = \frac{6.5 - \mu}{0.3}$$

We are looking for a value of t for which $P(x > t) = \frac{1}{100} = 0.01$. Now if $t \geqslant 0$, then

$$P(X > t) = P(X \geqslant 0) - P(0 \leqslant X \leqslant t)$$
$$= 0.5000 - P(0 \leqslant X \leqslant t)$$

Then
$$P(0 \leqslant X \leqslant t) = 0.5000 - P(X > t)$$
$$= 0.5000 - 0.01$$
$$= 0.4900$$

Looking in the body of the table for the number closest to 0.4900, we find 0.4901. The value of t that corresponds to 0.4901 is 2.33. Set X equal to 2.33 and solve for μ.

$$2.33 = \frac{6.5 - \mu}{0.3}$$
$$0.699 = 6.5 - \mu$$
$$\mu = 5.801$$

The setting of μ should be 5.801 oz.

Exercise Set 6.6

1. Find the volume of the solid of revolution generated by rotating about the x-axis the region under the graph of

$$y = \sqrt{x}$$

from x = 0 to x = 3.

$$V = \int_a^b \pi\,[f(x)]^2\,dx \qquad \text{(Volume of a solid of revolution)}$$

$$V = \int_0^3 \pi\,[\sqrt{x}\,]^2\,dx \qquad \text{[Substituting 0 for a, 3 for b, and } \sqrt{x} \text{ for f(x)]}$$

$$= \int_0^3 \pi\,x\,dx$$

$$= \left[\pi\cdot\frac{x^2}{2}\right]_0^3$$

$$= \frac{\pi}{2}\,[x^2]_0^3$$

$$= \frac{\pi}{2}\,(3^2 - 0^2)$$

$$= \frac{\pi}{2}\cdot 9$$

$$= \frac{9}{2}\,\pi$$

3. Find the volume of the solid of revolution generated by rotating about the x-axis the region under the graph of

$$y = x$$

from x = 1 to x = 2.

$$V = \int_a^b \pi\,[f(x)]^2\,dx \qquad \text{(Volume of a solid of revolution)}$$

$$V = \int_1^2 \pi\,[x]^2\,dx \qquad \text{[Substituting 1 for a, 2 for b, and x for f(x)]}$$

$$= \int_1^2 \pi\,x^2\,dx$$

$$= \left[\pi\cdot\frac{x^3}{3}\right]_1^2$$

$$= \frac{\pi}{3}\,[x^3]_1^2$$

$$= \frac{\pi}{3}\,(2^3 - 1^3)$$

$$= \frac{\pi}{3}\,(8 - 1)$$

$$= \frac{7}{3}\,\pi$$

5. Find the volume of the solid of revolution generated by rotating about the x-axis the region under the graph of

$$y = e^x$$

from x = -2 to x = 5.

$$V = \int_a^b \pi\,[f(x)]^2\,dx \qquad \text{(Volume of a solid of revolution)}$$

$$V = \int_{-2}^5 \pi\,[e^x]^2\,dx \qquad \text{[Substituting -2 for a, 5 for b, and } e^x \text{ for f(x)]}$$

$$= \int_{-2}^5 \pi\,e^{2x}\,dx$$

$= \left[\pi \cdot \frac{1}{2} e^{2x}\right]_{-2}^{5}$

$= \frac{\pi}{2} [e^{2x}]_{-2}^{5}$

$= \frac{\pi}{2} \left(e^{2\cdot 5} - e^{2(-2)}\right)$

$= \frac{\pi}{2} (e^{10} - e^{-4})$

7. Find the volume of the solid of revolution generated by rotating about the x-axis the region under the graph of

$y = \frac{1}{x}$

from x = 1 to x = 3.

$V = \int_a^b \pi\, [f(x)]^2\, dx$ (Volume of a solid of revolution)

$= \int_1^3 \pi \left[\frac{1}{x}\right]^2 dx$ [Substituting 1 for a, 3 for b, and $\frac{1}{x}$ for f(x)]

$= \int_1^3 \pi \cdot \frac{1}{x^2}\, dx$

$= \int_1^3 \pi\, x^{-2}\, dx$

$= \left[\pi\, \frac{x^{-1}}{-1}\right]_1^3$

$= -\pi \left[\frac{1}{x}\right]_1^3$

$= -\pi \left(\frac{1}{3} - \frac{1}{1}\right)$

$= -\pi \cdot \left(-\frac{2}{3}\right)$

$= \frac{2}{3}\pi$

9. Find the volume of the solid of revolution generated by rotating about the x-axis the region under the graph of

$y = \frac{1}{\sqrt{x}}$

from x = 1 to x = 3.

$V = \int_a^b \pi\, [f(x)]^2\, dx$ (Volume of a solid of revolution)

$V = \int_1^3 \pi \left[\frac{1}{\sqrt{x}}\right]^2 dx$ [Substituting 1 for a, 3 for b, and $\frac{1}{\sqrt{x}}$ for f(x)]

$= \int_1^3 \pi \cdot \frac{1}{x}\, dx$

$= \pi\, [\ln x]_1^3$

$= \pi(\ln 3 - \ln 1)$

$= \pi \ln 3$ (ln 1 = 0)

11. Find the volume of the solid of revolution generated by rotating about the x-axis the region under the graph of

$y = 4$

from x = 1 to x = 3.

$V = \int_a^b \pi\, [f(x)]^2\, dx$ (Volume of a solid of revolution)

$V = \int_1^3 \pi\, [4]^2\, dx$ [Substituting 1 for a, 3 for b, and 4 for f(x)]

$= \int_1^3 16\pi\, dx$

$= 16\pi\, [x]_1^3$

$= 16\pi(3 - 1)$

$= 32\pi$

13. Find the volume of the solid of revolution generated by rotating about the x-axis the region under the graph of

$y = x^2$

from x = 0 to x = 2.

$V = \int_a^b \pi\, [f(x)]^2\, dx$ (Volume of a solid of revolution)

$V = \int_0^2 \pi\, [x^2]^2\, dx$ [Substituting 0 for a, 2 for b, and x^2 for f(x)]

$= \int_0^2 \pi\, x^4\, dx$

$= \left[\pi \cdot \frac{x^5}{5}\right]_0^2$

$= \frac{\pi}{5} (2^5 - 0^5)$

$= \frac{32}{5}\pi$

15. Find the volume of the solid of revolution generated by rotating about the x-axis the region under the graph of

$y = \sqrt{1 + x}$

from x = 2 to x = 10.

$V = \int_a^b \pi\, [f(x)]^2\, dx$ (Volume of a solid of revolution)

$V = \int_2^{10} \pi \left[\sqrt{1 + x}\right]^2 dx$ [Substituting 2 for a, 10 for b, and $\sqrt{1 + x}$ for f(x)]

$= \int_2^{10} \pi(1 + x)\, dx$

$= \left[\pi \cdot \frac{1}{2} (1 + x)^2\right]_2^{10}$

$= \frac{\pi}{2} [(1 + x)^2]_2^{10}$

$= \frac{\pi}{2} [(1 + 10)^2 - (1 + 2)^2]$

$= \frac{\pi}{2} (11^2 - 3^2)$

$= \frac{\pi}{2} (121 - 9)$

$= \frac{\pi}{2} \cdot 112$

$= 56\pi$

17. Find the volume of the solid of revolution generated by rotating about the x-axis the region under the graph of

$y = \sqrt{4 - x^2}$

from x = -2 to x = 2.

$V = \int_a^b \pi\, [f(x)]^2\, dx$ (Volume of a solid of revolution)

$V = \int_{-2}^{2} \pi \left[\sqrt{4 - x^2}\right]^2 dx$ [Substituting -2 for a, 2 for b, and $\sqrt{4 - x^2}$ for f(x)]

$= \int_{-2}^{2} \pi(4 - x^2)\, dx$

$= \pi \left[4x - \frac{x^3}{3}\right]_{-2}^{2}$

$= \pi \left[\left(4\cdot 2 - \frac{2^3}{3}\right) - \left(4\cdot(-2) - \frac{(-2)^3}{3}\right)\right]$

$= \pi \left(8 - \frac{8}{3} + 8 - \frac{8}{3}\right)$

$= \pi \left(16 - \frac{16}{3}\right)$

$= \pi \left(\frac{48}{3} - \frac{16}{3}\right)$

$= \frac{32}{3}\pi$

19. Find the volume of the solid of revolution generated by rotating about the x-axis the region under the graph of

$y = \sqrt{\ln x}$

from x = e to x = e^3.

$V = \int_a^b \pi\, [f(x)]^2\, dx$ (Volume of a solid of revolution)

$V = \int_e^{e^3} \pi \left[\sqrt{\ln x}\right]^2 dx$ (Substituting)

$= \int_e^{e^3} \pi \ln x\, dx$

$= \pi\, [x \ln x - x]_e^{e^3}$ (Using Formula 8)

$= \pi[(e^3 \ln e^3 - e^3) - (e \ln e - e)]$

$= \pi[(3e^3 - e^3) - (e - e)]$

$= 2\pi e^3$

21. $y = \frac{1}{x}$

a) $\int_1^\infty \frac{1}{x}\, dx$

$= \lim_{b\to\infty} \int_1^b \frac{1}{x}\, dx$

$= \lim_{b\to\infty} [\ln x]_1^b$

$= \lim_{b\to\infty} (\ln b - \ln 1)$

$= \lim_{b\to\infty} (\ln b - 0)$

The limit does not exist. As b increases, ln b increases indefinitely. The integral is divergent.

b) $V = \int_a^b \pi\, [f(x)]^2\, dx$

$V = \int_1^\infty \pi \left(\frac{1}{x}\right)^2 dx$ (Substituting)

$= \int_1^\infty \pi \cdot \frac{1}{x^2}\, dx$

$= \int_1^\infty \pi x^{-2}\, dx$

$= \lim_{b\to\infty} \int_1^b \pi x^{-2}\, dx$

$= \lim_{b\to\infty} \left[\pi \frac{x^{-1}}{-1}\right]_1^b$

$= \lim_{b\to\infty} \left[-\frac{\pi}{x}\right]_1^b$

$= \lim_{b\to\infty} \left[-\frac{\pi}{b} - \left(-\frac{\pi}{1}\right)\right]$

$= \lim_{b\to\infty} \left(-\frac{\pi}{b} + \pi\right)$

$= \pi$ $\left(\text{As } b \to \infty, \frac{\pi}{b} \to 0\right)$

Exercise Set 6.7

1. Solve: $y' = 4x^3$.

$y = \int 4x^3\, dx$

Solution: $y = x^4 + C$

This solution is called a general solution because taking all values of C gives all the solutions. Taking specific values of C gives particular solutions. The following are particular solutions of $y' = 4x^3$. Answers may vary.

$y = x^4 - 13$

$y = x^4$

$y = x^4 + \frac{7}{8}$

3. Solve: $y' = e^{2x} + x$

$y = \int (e^{2x} + x)\, dx$

Solution: $y = \frac{1}{2} e^{2x} + \frac{1}{2} x^2 + C$

This solution is called a general solution because taking all values of C gives all the solutions. Taking specific values of C gives particular solutions. The following are particular solutions of $y' = e^{2x} + x$. Answers may vary.

$y = \frac{1}{2} e^{2x} + \frac{1}{2} x^2 + \frac{5}{8}$

$y = \frac{1}{2} e^{2x} + \frac{1}{2} x^2 - 7$

$y = \frac{1}{2} e^{2x} + \frac{1}{2} x^2 + 83$

5. Solve: $y' = \frac{3}{x} - x^2 + x^5$

$$y = \int \left(\frac{3}{x} - x^2 + x^5\right) dx$$

Solution: $y = 3 \ln x - \frac{1}{3} x^3 + \frac{1}{6} x^6 + C$

This solution is called a general solution because taking all values of C gives all the solutions. Taking specific values of C gives particular solutions. The following are particular solutions of $y' = \frac{3}{x} - x^2 + x^5$. Answers may vary.

$$y = 3 \ln x - \frac{1}{3} x^3 + \frac{1}{6} x^6$$

$$y = 3 \ln x - \frac{1}{3} x^3 + \frac{1}{6} x^6 - 10$$

$$y = 3 \ln x - \frac{1}{3} x^3 + \frac{1}{6} x^6 + \frac{11}{16}$$

7. Solve $y' = x^2 + 2x - 3$, given that $y = 4$ when $x = 0$.

First find the general solution.

$$y = \int y' \, dx$$

$$y = \frac{1}{3} x^3 + x^2 - 3x + C$$

We substitute to find C.

$$4 = \frac{1}{3} \cdot 0^3 + 0^2 - 3 \cdot 0 + C$$ (Substituting 4 for y and 0 for x)

$$4 = C$$

Thus the solution is

$$y = \frac{1}{3} x^3 + x^2 - 3x + 4.$$

9. Solve $f'(x) = x^{2/3} - x$, given that $f(1) = -6$.

First find the general solution.

$$f(x) = \int f'(x) \, dx + C$$

$$f(x) = \frac{3}{5} x^{5/3} - \frac{1}{2} x^2 + C$$

We substitute to find C.

$$-6 = \frac{3}{5} \cdot 1^{5/3} - \frac{1}{2} \cdot 1^2 + C$$

[Substituting -6 for f(x) and 1 for x]

$$-6 = \frac{3}{5} - \frac{1}{2} + C$$

$$-6 - \frac{3}{5} + \frac{1}{2} = C$$

$$-\frac{60}{10} - \frac{6}{10} + \frac{5}{10} = C$$

$$-\frac{61}{10} = C$$

Thus the solution is

$$f(x) = \frac{3}{5} x^{5/3} - \frac{1}{2} x^2 - \frac{61}{10}.$$

11. Show that $y = x \ln x + 3x - 2$ is a solution to $y'' - \frac{1}{x} = 0$.

First find y' and y''.

$$y = x \ln x + 3x - 2$$

$$y' = x \cdot \frac{1}{x} + 1 \cdot \ln x + 3$$

$$= \ln x + 4$$

$$y'' = \frac{1}{x}$$

Substitute as follows in the differential equation.

$y'' - \frac{1}{x}$	$= 0$
$\frac{1}{x} - \frac{1}{x}$	0
0	

Thus, $y = x \ln x + 3x - 2$ is a solution of $y'' - \frac{1}{x} = 0$.

13. Show that $y = e^x + 3xe^x$ is a solution to $y'' - 2y' + y = 0$.

First find y' and y''.

$$y = e^x + 3xe^x$$

$$y' = e^x + 3(xe^x + e^x)$$

$$= e^x + 3xe^x + 3e^x$$

$$= 4e^x + 3xe^x$$

$$y'' = 4e^x + 3(xe^x + e^x)$$

$$= 4e^x + 3xe^x + 3e^x$$

$$= 7e^x + 3xe^x$$

Substitute as follows in the differential equation.

$y'' - 2y' + y$	$= 0$
$(7e^x + 3xe^x) - 2(4e^x + 3xe^x) + (e^x + 3xe^x)$	0
$7e^x + 3xe^x - 8e^x - 6xe^x + e^x + 3xe^x$	
$(7 - 8 + 1)e^x + (3 - 6 + 3)xe^x$	
$0 + 0$	
0	

Thus, $y = e^x + 3xe^x$ is a solution of $y'' - 2y' + y = 0$.

15. Solve $\frac{dy}{dx} = 4x^3y$.

We first separate variables as follows.

$dy = 4x^3y\ dx$ (Multiplying by dx)

$\frac{dy}{y} = 4x^3\ dx$ $\left[\text{Multiplying by } \frac{1}{y}\right]$

We then integrate both sides.

$\int \frac{dy}{y} = \int 4x^3\ dx$

$\ln y = x^4 + C$ $(y > 0)$

$y = e^{x^4+C}$ (Definition of logarithms)

$y = e^{x^4} \cdot e^C$

Thus

$y = C_1\ e^{x^4}$, where $C_1 = e^C$.

17. Solve $3y^2 \frac{dy}{dx} = 5x$

We first separate variables as follows.

$3y^2\ dy = 5x\ dx$ (Multiplying by dx)

We then integrate both sides.

$\int 3y^2\ dy = \int 5x\ dx$

$3 \cdot \frac{y^3}{3} = 5 \cdot \frac{x^2}{2} + C$

$y^3 = \frac{5}{2} x^2 + C$

$y = \sqrt[3]{\frac{5}{2} x^2 + C}$

19. Solve $\frac{dy}{dx} = \frac{2x}{y}$.

We first separate variables as follows.

$dy = \frac{2x}{y}\ dx$ (Multiplying by dx)

$y\ dy = 2x\ dx$ (Multiplying by y)

We then integrate both sides.

$\int y\ dy = \int 2x\ dx$

$\frac{y^2}{2} = x^2 + C$

$y^2 = 2x^2 + 2C$

$y^2 = 2x^2 + C_1$ $(C_1 = 2C)$

Thus,

$y = \sqrt{2x^2 + C_1}$ and $y = -\sqrt{2x^2 + C_1}$, where $C_1 = 2C$.

21. Solve $\frac{dy}{dx} = \frac{3}{y}$.

We first separate variables as follows.

$dy = \frac{3}{y}\ dx$

$y\ dy = 3\ dx$

We then integrate both sides.

$\int y\ dy = \int 3\ dx$

$\frac{y^2}{2} = 3x + C$

$y^2 = 6x + 2C$

$y^2 = 6x + C_1$ $(C_1 = 2C)$

Thus,

$y = \sqrt{6x + C_1}$ and $y = -\sqrt{6x + C_1}$, where $C_1 = 2C$.

23. Solve $y' = 3x + xy$.

We first separate variables as follows.

$\frac{dy}{dx} = 3x + xy$ $\left[\text{Replacing } y' \text{ with } \frac{dy}{dx}\right]$

$\frac{dy}{dx} = x(3 + y)$

$dy = x(3 + y)\ dx$ (Multiplying by dx)

$\frac{dy}{3 + y} = x\ dx$ $\left[\text{Multiplying by } \frac{1}{3 + y}\right]$

We then integrate both sides.

$\int \frac{dy}{3 + y} = \int x\ dx$

$\ln (3 + y) = \frac{x^2}{2} + C$ $(3 + y > 0)$

$3 + y = e^{x^2/2+C}$

$3 + y = e^{x^2/2} \cdot e^C$

$y = e^{x^2/2} \cdot e^C - 3$

$y = C_1\ e^{x^2/2} - 3$ $(C_1 = e^C)$

We substitute to find C_1.

$5 = C_1\ e^{0^2/2} - 3$ (Substituting 0 for x and 5 for y)

$8 = C_1\ e^0$

$8 = C_1 \cdot 1$

$8 = C_1$

Thus the solution is

$y = 8\ e^{x^2/2} - 3$.

25. Solve $y' = 5y^{-2}$.

We first separate variables as follows.

$\frac{dy}{dx} = \frac{5}{y^2}$ $\left[\text{Replacing } y' \text{ with } \frac{dy}{dx}\right]$

$y^2 \frac{dy}{dx} = 5$

$y^2\ dy = 5\ dx$

We then integrate both sides.

$\int y^2\ dy = \int 5\ dx$

$\frac{y^3}{3} = 5x + C$

$y^3 = 15x + 3C$

$y^3 = 15x + C_1$ $(C_1 = 3C)$

We substitute to find C.

$3^3 = 15\cdot 2 + C_1$ (Substituting 3 for y and 2 for x)

$27 = 30 + C_1$

$-3 = C_1$

Thus,

$y^3 = 15x - 3$

$y = \sqrt[3]{15x - 3}$

27. Solve $\frac{dy}{dx} = 3y$.

We first separate variables as follows.

$dy = 3y\ dx$

$\frac{dy}{y} = 3\ dx$

We then integrate both sides.

$\int \frac{dy}{y} = \int 3\ dx$

$\ln y = 3x + C$ $(y > 0)$

$y = e^{3x+C}$ (Definition of logarithms)

$y = e^{3x} \cdot e^C$

Thus $y = C_1 e^{3x}$, where $C_1 = e^C$.

29. Solve $\frac{dP}{dt} = 2P$.

We first separate variables as follows.

$dP = 2P\ dt$

$\frac{dP}{P} = 2\ dt$

We then integrate both sides.

$\int \frac{dP}{P} = \int 2\ dt$

$\ln P = 2t + C$ $(P > 0)$

$P = e^{2t+C}$

$P = e^{2t} \cdot e^C$

Thus $P = C_1 e^{2t}$, where $C_1 = e^C$.

31. Solve $f'(x) = \frac{1}{x} - 4x + \sqrt{x}$, given that $f(1) = \frac{23}{3}$.

$f(x) = \int f'(x)\ dx$

$f(x) = \ln x - 2x^2 + \frac{2}{3}x^{3/2} + C$

Substitute to find C.

$\frac{23}{3} = \ln 1 - 2\cdot 1^2 + \frac{2}{3}\cdot 1^{3/2} + C$

$\frac{23}{3} = 0 - 2 + \frac{2}{3} + C$

$\frac{23}{3} = -\frac{4}{3} + C$

$\frac{27}{3} = C$

$9 = C$

Then $f(x) = \ln x - 2x^2 + \frac{2}{3}x^{3/2} + 9$.

33. $C'(x) = 2.6 - 0.02x$

$C(x) = \int C'(x)\ dx = 2.6x - 0.01x^2 + K.$ (Using K for the constant)

We substitute to find K.

$C(0) = 2.6(0) - 0.01(0^2) + K = 120$

$K = 120$

Thus,

$C(x) = 2.6x - 0.01x^2 + 120$ and

$A(x)$ = the average cost of producing x units

$= \frac{C(x)}{x}$

$= \frac{2.6x - 0.01x^2 + 120}{x}$

$= 2.6 - 0.01x + \frac{120}{x}$

35. a) $\frac{dP}{dC} = \frac{-200}{(C+3)^{3/2}}$

$P(C) = \int \frac{-200}{(C+3)^{3/2}}\ dC$

$= -200 \int (C+3)^{-3/2}\ dC$

$= -200 \frac{(C+3)^{-1/2}}{-1/2} + K$

$= \frac{400}{\sqrt{C+3}} + K$ (Using K for the constant)

We substitute to find K.

$P(C) = \frac{400}{\sqrt{C+3}} + K$

$10 = \frac{400}{\sqrt{61+3}} + K$ (Substituting 10 for P and 61 for C)

$10 = \frac{400}{8} + K$

$10 = 50 + K$

$-40 = K$

Thus the solution is

$P(C) = \frac{400}{\sqrt{C+3}} - 40.$

b) The firm will break even when $P = 0$.

$P = \frac{400}{\sqrt{C+3}} - 40$

$0 = \frac{400}{\sqrt{C+3}} - 40$ (Substituting 0 for P)

$40 = \frac{400}{\sqrt{C+3}}$

$40\sqrt{C+3} = 400$

$\sqrt{C+3} = 10$

$C + 3 = 100$

$C = 97$

When the total cost is \$97, the firm breaks even.

37. a) $\frac{dR}{dS} = \frac{k}{S + 1}$

We first separate variables.

$dR = \frac{k}{S + 1}\, dS$ (Multiplying by dS)

We then integrate both sides.

$\int dR = \int \frac{k}{S + 1}\, dS$

$R = k \ln (S + 1) + C \qquad (S + 1) > 0$

b) Substitute to find C.

$R = k \ln (S + 1) + C$

$0 = k \ln (0 + 1) + C$

$0 = k \ln 1 + C$

$0 = k \cdot 0 + C$

$0 = C$

thus, $R = k \ln (S + 1)$.

c) $R(0) = 0$ is reasonable because there is no pleasure from 0 units.

39. $E(p) = \frac{p}{200 - p}$; $x = 190$ when $p = 10$

$E(p) = -\frac{p}{x} \cdot \frac{dx}{dp}$

$-\frac{p}{x} \cdot \frac{dx}{dp} = \frac{p}{200 - p}$

We separate variables.

$\frac{dx}{x} = \frac{p}{200 - p} \cdot \left(-\frac{dp}{p}\right)$

$\frac{dx}{x} = -\frac{dp}{200 - p}$

We then integrate both sides.

$\int \frac{dx}{x} = \int -\frac{dp}{200 - p}$

$\ln x = \ln (200 - p) + C$

We substitute to find C.

$\ln 190 = \ln (200 - 10) + C$

$\ln 190 = \ln 190 + C$

$0 = C$

Thus, $\ln x = \ln (200 - p)$

$x = 200 - p$

The demand function is $x = 200 - p$.

41. $E(p) = n$ for some constant n and all $p > 0$

$E(p) = -\frac{p}{x} \cdot \frac{dx}{dp}$

$-\frac{p}{x} \cdot \frac{dx}{dp} = n$

We separate variables.

$\frac{dx}{x} = n\left(-\frac{dp}{p}\right)$

$\frac{dx}{x} = -\frac{n}{p}\, dp$

We then integrate both sides.

$\int \frac{dx}{x} = \int -\frac{n}{p}\, dp$

$\ln x = -n \ln p + C$

$\ln x = -n \ln p + \ln C_1 \qquad (\ln C_1 = C)$

$\ln x = \ln C_1 - \ln p^n$

$\ln x = \ln \left(\frac{C_1}{p^n}\right)$

$x = \frac{C_1}{p^n}$

Exercise Set 7.1

1. $f(x,y) = x^2 - 2xy$

$f(0,-2) = 0^2 - 2\cdot 0\cdot(-2)$ (Substituting 0 for x and -2 for y)

$= 0 - 0$

$= 0$

$f(2,3) = 2^2 - 2\cdot 2\cdot 3$ (Substituting 2 for x and 3 for y)

$= 4 - 12$

$= -8$

$f(10,-5) = 10^2 - 2\cdot 10\cdot(-5)$ (Substituting 10 for x and -5 for y)

$= 100 + 100$

$= 200$

3. $f(x,y) = 3^x + 7xy$

$f(0,-2) = 3^0 + 7\cdot 0\cdot(-2)$ (Substituting 0 for x and -2 for y)

$= 1 + 0$

$= 1$

$f(-2,1) = 3^{-2} + 7\cdot(-2)\cdot 1$ (Substituting -2 for x and 1 for y)

$= \frac{1}{9} - 14$

$= -13\frac{8}{9}$, or $-\frac{125}{9}$

$f(2,1) = 3^2 + 7\cdot 2\cdot 1$ (Substituting 2 for x and 1 for y)

$= 9 + 14$

$= 23$

5. $f(x,y) = \ln x + y^3$

$f(e,2) = \ln e + 2^3$ (Substituting e for x and 2 for y)

$= 1 + 8$

$= 9$

$f(e^2,4) = \ln e^2 + 4^3$ (Substituting e^2 for x and 4 for y)

$= 2 + 64$

$= 66$

$f(e^3,5) = \ln e^3 + 5^3$ (Substituting e^3 for x and 5 for y)

$= 3 + 125$

$= 128$

7. $f(x,y,z) = x^2 - y^2 + z^2$

$f(-1,2,3) = (-1)^2 - 2^2 + 3^2$

(Substituting -1 for x, 2 for y and 3 for z)

$= 1 - 4 + 9$

$= 6$

$f(2,-1,3) = 2^2 - (-1)^2 + 3^2$

(Substituting 2 for x, -1 for y, and 3 for z)

$= 4 - 1 + 9$

$= 12$

9. $C_2 = \left[\frac{V_2}{V_1}\right]^{0.6} C_1$

$C_2 = \left[\frac{160{,}000}{80{,}000}\right]^{0.6} \cdot 100{,}000$

(Substituting 100,000 for C_1, 80,000 for V_1, and 160,000 for V_2)

$= 2^{0.6}(100{,}000)$

$\approx (1.5157166)(100{,}000)$

$\approx \$151{,}571.66$

11. $Y(D,P) = \frac{D}{P}$

$Y\left[\$1.30,\ \$16\frac{7}{8}\right] = \frac{\$1.30}{\$16\frac{7}{8}}$

$\left[\text{Substituting } \$1.30 \text{ for D and } \$16\frac{7}{8} \text{ for P}\right]$

$= \frac{1.30}{16.875}$

$= 0.07\overline{703}$

$\approx 7.7\%$

13. $S = \frac{aV}{0.51d^2}$

$S = \frac{0.78(1{,}600{,}000)}{0.51(100)^2}$ (Substituting 0.78 for a, 1,600,000 for V, and 100 for d)

$= \frac{1{,}248{,}000}{5100}$

≈ 244.7 mph

15. $W(v,T) = 91.4 - \frac{(10.45 + 6.68\sqrt{v} - 0.447v)(457-5T)}{110}$

$W(25,30)$

$= 91.4 - \frac{(10.45 + 6.68\sqrt{25} - 0.447\cdot 25)(457 - 5\cdot 30)}{110}$

$= 91.4 - \frac{(10.45 + 33.4 - 11.175)(307)}{110}$

$= 91.4 - \frac{(32.675)(307)}{110}$

$= 91.4 - \frac{10{,}031.225}{110}$

$= 91.4 - 91.2$

$\approx 0^0$ F (To the nearest degree)

17. $W(v,T) = 91.4 - \frac{(10.45 + 6.68\sqrt{v} - 0.447v)(457-5T)}{110}$

$W(40,20)$

$= 91.4 - \frac{(10.45 + 6.68\sqrt{40} - 0.447\cdot 40)(457 - 5\cdot 20)}{110}$

$= 91.4 - \frac{(10.45 + 42.248 - 17.88)(357)}{110}$

$= 91.4 - \frac{(34.818)(357)}{110}$

$= 91.4 - \frac{12,430.026}{110}$

$= 91.4 - 113.0$

$\approx -22° \text{ F}$ (To the nearest degree)

19. See the answer section in the text.

21. See the answer section in the text.

23. See the answer section in the text.

Exercise Set 7.2

1. $z = 2x - 3xy$

Find $\frac{\partial z}{\partial x}$.

$z = 2\underline{x} - 3\underline{x}y$ (The variable is underlined; y is treated as a constant.)

$\frac{\partial z}{\partial x} = 2 - 3y$

Find $\frac{\partial z}{\partial y}$.

$z = 2x - 3x\underline{y}$ (The variable is underlined; x is treated as a constant.)

$\frac{\partial z}{\partial y} = -3x$

Find: $\left.\frac{\partial z}{\partial x}\right|_{(-2,-3)}$

$\frac{\partial z}{\partial x} = 2 - 3y$

$\left.\frac{\partial z}{\partial x}\right|_{(-2,-3)} = 2 - 3(-3)$ (Substituting -3 for y)

$= 2 + 9$

$= 11$

Find: $\left.\frac{\partial z}{\partial y}\right|_{(0,-5)}$

$\frac{\partial z}{\partial y} = -3x$

$\left.\frac{\partial z}{\partial y}\right|_{(0,-5)} = -3(0)$ (Substituting 0 for x)

$= 0$

3. $z = 3x^2 - 2xy + y$

Find $\frac{\partial z}{\partial x}$.

$z = 3\underline{x}^2 - 2\underline{x}y + y$ (The variable is underlined; y is treated as a constant.)

$\frac{\partial z}{\partial x} = 6x - 2y$

Find $\frac{\partial z}{\partial y}$.

$z = 3x^2 - 2x\underline{y} + \underline{y}$ (The variable is underlined; x is treated as a constant.)

$\frac{\partial z}{\partial y} = -2x + 1$

Find: $\left.\frac{\partial z}{\partial x}\right|_{(-2,-3)}$

$\frac{\partial z}{\partial x} = 6x - 2y$

$\left.\frac{\partial z}{\partial x}\right|_{(-2,-3)} = 6(-2) - 2(-3)$ (Substituting -2 for x and -3 for y)

$= -12 + 6$

$= -6$

Find: $\left.\frac{\partial z}{\partial y}\right|_{(0,-5)}$

$\frac{\partial z}{\partial y} = -2x + 1$

$\left.\frac{\partial z}{\partial y}\right|_{(0,-5)} = -2\cdot 0 + 1$ (Substituting 0 for x)

$= 1$

5. $f(x,y) = 2x - 3y$

Find f_x.

$f(x,y) = 2\underline{x} - 3y$ (The variable is underlined; y is treated as a constant.)

$f_x = 2$

Find f_y.

$f(x,y) = 2x - 3\underline{y}$ (The variable is underlined; x is treated as a constant.)

$f_y = -3$

Find $f_x(-2,4)$.

$f_x = 2$

f_x has the value 2 for any x and y.

$f_x(-2,4) = 2$

Find $f_y(4,-3)$.

$f_y = -3$

f_y has the value -3 for any x and y.

$f_y(4,-3) = -3$.

7. $f(x,y) = \sqrt{x^2 + y^2}$
$= (x^2 + y^2)^{1/2}$

Find f_x.

$f(x,y) = (\underline{x}^2 + y^2)^{1/2}$ (The variable is underlined.)

$f_x = \frac{1}{2}(x^2 + y^2)^{-1/2} \cdot 2x$

$= x(x^2 + y^2)^{-1/2}$, or $\frac{x}{\sqrt{x^2 + y^2}}$

Find f_y.

$f(x,y) = (x^2 + \underline{y}^2)^{1/2}$ (The variable is underlined.)

$f_y = \frac{1}{2}(x^2 + y^2)^{-1/2} \cdot 2y$

$= y(x^2 + y^2)^{-1/2}$, or $\frac{y}{\sqrt{x^2 + y^2}}$

Find $f_x(-2,1)$.

$f_x = \frac{x}{\sqrt{x^2 + y^2}}$

$f_x(-2,1) = \frac{-2}{\sqrt{(-2)^2 + 1^2}}$ (Substituting -2 for x and 1 for y)

$= \frac{-2}{\sqrt{5}}$

Find $f_y(-3,-2)$.

$f_y = \frac{y}{\sqrt{x^2 + y^2}}$

$f_y(-3,-2) = \frac{-2}{\sqrt{(-3)^2 + (-2)^2}}$ (Substituting -3 for x and -2 for y)

$= \frac{-2}{\sqrt{13}}$

9. $f(x,y) = e^{2x+3y}$

Find f_x.

$f(x,y) = e^{2\underline{x}+3y}$ (The variable is underlined.)

$f_x = 2e^{2x+3y}$

Find f_y.

$f(x,y) = e^{2x+3\underline{y}}$ (The variable is underlined.)

$f_y = 3e^{2x+3y}$

11. $f(x,y) = e^{xy}$

Find f_x.

$f(x,y) = e^{\underline{x}y}$ (The variable is underlined.)

$f_x = y\,e^{xy}$

Find f_y.

$f(x,y) = e^{x\underline{y}}$ (The variable is underlined.)

$f_y = x\,e^{xy}$

13. $f(x,y) = y \ln(x + y)$

Find f_x.

$f(x,y) = y \ln(\underline{x} + y)$ (The variable is underlined.)

$f_x = y \cdot \frac{1}{x + y}$

$= \frac{y}{x + y}$

Find f_y.

$f(x,y) = \underline{y} \ln(x + \underline{y})$ (The variable is underlined.)

$f_y = y \cdot \frac{1}{x + y} + 1 \cdot \ln(x + y)$

$= \frac{y}{x + y} + \ln(x + y)$

15. $f(x,y) = x \ln(xy)$

Find f_x.

$f(x,y) = \underline{x} \ln(\underline{x}y)$ (The variable is underlined.)

$f_x = x \cdot \left[y \cdot \frac{1}{xy}\right] + 1 \cdot \ln(xy)$

$= 1 + \ln(xy)$

Find f_y.

$f(x,y) = x \ln(x\underline{y})$ (The variable is underlined.)

$f_y = x \cdot \left[x \cdot \frac{1}{xy}\right]$

$= \frac{x}{y}$

17. $f(x,y) = \frac{x}{y} - \frac{y}{x}$

Find f_x.

$f(x,y) = \frac{x}{y} - \frac{y}{x}$

$= \frac{1}{y} \cdot \underline{x} - y \cdot \underline{x}^{-1}$ (The variable is underlined.)

$f_x = \frac{1}{y} - y \cdot (-1) \cdot x^{-2}$

$= \frac{1}{y} + \frac{y}{x^2}$

Find f_y.

$f(x,y) = \frac{x}{y} - \frac{y}{x}$

$= x \cdot \underline{y}^{-1} - \frac{1}{x} \cdot \underline{y}$ (The variable is underlined.)

$f_y = x \cdot (-1) \cdot y^{-2} - \frac{1}{x}$

$= -\frac{x}{y^2} - \frac{1}{x}$

19. $f(x,y) = 3(2x + y - 5)^2$

Find f_x.

$f(x,y) = 3(2\underline{x} + y - 5)^2$ (The variable is underlined.)

$$f_x = 6(2x + y - 5)\cdot 2 = 12(2x + y - 5)$$

Find f_y.

$f(x,y) = 3(2x + \underline{y} - 5)^2$ (The variable is underlined.)

$$f_y = 6(2x + y - 5)\cdot 1 = 6(2x + y - 5)$$

21. $f(b,m) = (m + b - 4)^2 + (2m + b - 5)^2 + (3m + b - 6)^2$

Find $\frac{\partial f}{\partial b}$.

$f(b,m) = (m + \underline{b} - 4)^2 + (2m + \underline{b} - 5)^2 + (3m + \underline{b} - 6)^2$ (The variable is underlined.)

$$\frac{\partial f}{\partial b} = 2(m + b - 4)\cdot 1 + 2(2m + b - 5)\cdot 1 + 2(3m + b - 6)\cdot 1$$
$$= 2m + 2b - 8 + 4m + 2b - 10 + 6m + 2b - 12$$
$$= 12m + 6b - 30$$

Find $\frac{\partial f}{\partial m}$.

$f(b,m) = (\underline{m} + b - 4)^2 + (2\underline{m} + b - 5)^2 + (3\underline{m} + b - 6)^2$ (The variable is underlined.)

$$\frac{\partial f}{\partial m} = 2(m + b - 4)\cdot 1 + 2(2m + b - 5)\cdot 2 + 2(3m + b - 6)\cdot 3$$
$$= 2m + 2b - 8 + 8m + 4b - 20 + 18m + 6b - 36$$
$$= 28m + 12b - 64$$

23. $f(x,y,\lambda) = 3xy - \lambda(2x + y - 8)$

Find f_x.

$f(x,y,\lambda) = 3\underline{x}y - \lambda(2\underline{x} + y - 8)$ (The variable is underlined.)

$$f_x = 3y - \lambda\cdot 2 = 3y - 2\lambda$$

Find f_y.

$f(x,y,\lambda) = 3x\underline{y} - \lambda(2x + \underline{y} - 8)$ (The variable is underlined.)

$$f_y = 3x - \lambda\cdot 1 = 3x - \lambda$$

Find f_λ.

$f(x,y,\lambda) = 3xy - \underline{\lambda}(2x + y - 8)$ (The variable is underlined.)

$$f_\lambda = -(2x + y - 8) = -2x - y + 8$$

25. $f(x,y,\lambda) = x^2 + y^2 - \lambda(10x + 2y - 4)$

Find f_x.

$f(x,y,\lambda) = \underline{x}^2 + y^2 - \lambda(10\underline{x} + 2y - 4)$ (The variable is underlined.)

$$f_x = 2x - \lambda\cdot 10 = 2x - 10\lambda$$

Find f_y.

$f(x,y,\lambda) = x^2 + \underline{y}^2 - \lambda(10x + 2\underline{y} - 4)$ (The variable is underlined.)

$$f_y = 2y - \lambda\cdot 2 = 2y - 2\lambda$$

Find f_λ.

$f(x,y,\lambda) = x^2 + y^2 - \underline{\lambda}(10x + 2y - 4)$ (The variable is underlined.)

$$f_\lambda = -(10x + 2y - 4) = -10x - 2y + 4$$

27. a) $p(x,y) = 1800\, x^{0.621} y^{0.379}$

$p(2500,1700) = 1800(2500)^{0.621}(1700)^{0.379}$ (Substituting 2500 for x and 1700 for y)

$$= 1800(128.860777)(16.762552)$$
$$\approx 3{,}888{,}064 \text{ units}$$

b) Find $\frac{\partial p}{\partial x}$.

$p(x,y) = 1800\, \underline{x}^{0.621} y^{0.379}$ (The variable is underlined.)

$$\frac{\partial p}{\partial x} = 1800(0.621\, x^{-0.379})\, y^{0.379} = 1117.8\, x^{-0.379} y^{0.379} = 1117.8 \left(\frac{y}{x}\right)^{0.379}$$

Find $\frac{\partial p}{\partial y}$.

$p(x,y) = 1800\, x^{0.621} \underline{y}^{0.379}$ (The variable is underlined.)

$$\frac{\partial p}{\partial y} = 1800\, x^{0.621} (0.379\, y^{-0.621}) = 682.2\, x^{0.621} y^{-0.621} = 682.2 \left(\frac{x}{y}\right)^{0.621}$$

c) $\frac{\partial p}{\partial x} = 1117.8 \left(\frac{y}{x}\right)^{0.379}$

$$\left.\frac{\partial p}{\partial x}\right|_{(2500,1700)} = 1117.8\left(\frac{1700}{2500}\right)^{0.379}$$ (Substituting)

$$= 1117.8(0.68)^{0.379}$$
$$= 1117.8(0.864014)$$
$$\approx 965.8$$

$$\frac{\partial p}{\partial y} = 682.2\left(\frac{x}{y}\right)^{0.621}$$

$$\left.\frac{\partial p}{\partial y}\right|_{(2500,1700)} = 682.2\left[\frac{2500}{1700}\right]^{0.621}$$ (Substituting)

$$= 682.2(1.470588)^{0.621}$$
$$= 682.2(1.270609)$$
$$\approx 866.8$$

29. $T_h = 1.98T - 1.09(1 - H)(T - 58) - 56.9$

a) When $T = 85°$ and $H = 60\%$, or 0.6:

$$T_h = 1.98(85) - 1.09(1 - 0.6)(85 - 58) - 56.9$$
$$= 168.3 - 1.09(0.4)(27) - 56.9$$
$$= 168.3 - 11.772 - 56.9$$
$$= 99.628$$
$$\approx 99.6°$$ (To the nearest tenth)

b) When $T = 90°$ and $H = 90\%$, or 0.9:

$$T_h = 1.98(90) - 1.09(1 - 0.9)(90 - 58) - 56.9$$
$$= 178.2 - 1.09(0.1)(32) - 56.9$$
$$= 178.2 - 3.488 - 56.9$$
$$= 117.812$$
$$\approx 117.8°$$ (To the nearest tenth)

c) When $T = 90°$ and $H = 100\%$, or 1:

$$T_h = 1.98(90) - 1.09(1 - 1)(90 - 58) - 56.9$$
$$= 178.2 - 1.09(0)(32) - 56.9$$
$$= 178.2 - 0 - 56.9$$
$$= 121.3°$$

d) When $T = 78°$ and $H = 100\%$, or 1:

$$T_h = 1.98(78) - 1.09(1 - 1)(78 - 58) - 56.9$$
$$= 154.44 - 1.09(0)(20) - 56.9$$
$$= 154.44 - 0 - 56.9$$
$$= 97.54$$
$$\approx 97.5°$$ (To the nearest tenth)

e) $$\frac{\partial T_h}{\partial H} = -1.09(-1)(T - 58)$$
$$= 1.09(T - 58)$$

f) $$\frac{\partial T_h}{\partial T} = 1.98 - 1.09(1 - H)$$

31. $f(x,t) = \dfrac{x^2 + t^2}{x^2 - t^2}$

Find f_x.

$f(x,t) = \dfrac{\underline{x}^2 + t^2}{\underline{x}^2 - t^2}$ (The variable is underlined.)

$$f_x = \frac{(x^2 - t^2)(2x) - 2x(x^2 + t^2)}{(x^2 - t^2)^2}$$
$$= \frac{2x^3 - 2xt^2 - 2x^3 - 2xt^2}{(x^2 - t^2)^2}$$
$$= \frac{-4xt^2}{(x^2 - t^2)^2}$$

Find f_t.

$f(x,t) = \dfrac{x^2 + \underline{t}^2}{x^2 - \underline{t}^2}$ (The variable is underlined.)

$$f_t = \frac{(x^2 - t^2)2t - (-2t)(x^2 + t^2)}{(x^2 - t^2)^2}$$
$$= \frac{2x^2t - 2t^3 + 2x^2t + 2t^3}{(x^2 - t^2)^2}$$
$$= \frac{4x^2t}{(x^2 - t^2)^2}$$

33. $f(x,t) = \dfrac{2\sqrt{x} - 2\sqrt{t}}{1 + 2\sqrt{t}}$

Find f_x.

$f(x,t) = \dfrac{2\underline{x}^{1/2} - 2t^{1/2}}{1 + 2t^{1/2}}$ (The variable is underlined.)

$$f_x = \frac{(1 + 2t^{1/2})\left[2 \cdot \frac{1}{2} x^{-1/2}\right] - 0(2x^{1/2} - 2t^{1/2})}{(1 + 2t^{1/2})^2}$$
$$= \frac{1 + 2\sqrt{t}}{\sqrt{x}(1 + 2\sqrt{t})^2}$$
$$= \frac{1}{\sqrt{x}(1 + 2\sqrt{t})}$$

Find f_t.

$f(x,t) = \dfrac{2x^{1/2} - 2\underline{t}^{1/2}}{1 + 2\underline{t}^{1/2}}$ (The variable is underlined.)

$f_t =$

$$\frac{(1+2t^{1/2})\left[-2 \cdot \frac{1}{2} t^{-1/2}\right] - \left[2 \cdot \frac{1}{2} t^{-1/2}\right](2x^{1/2} - 2t^{1/2})}{(1 + 2t^{1/2})^2}$$
$$= \frac{(-t^{-1/2} - 2t^0) - (2x^{1/2}t^{-1/2} - 2t^0)}{(1 + 2t^{1/2})^2}$$
$$= \frac{-t^{-1/2} - 2 - 2x^{1/2}t^{-1/2} + 2}{(1 + 2t^{1/2})^2}$$
$$= \frac{t^{-1/2}(-1 - 2x^{1/2})}{(1 + 2t^{1/2})^2}$$
$$= \frac{-1 - 2\sqrt{x}}{\sqrt{t}(1 + 2\sqrt{t})^2}$$

35. $f(x,t) = 6x^{2/3} - 8x^{1/4} t^{1/2} - 12x^{-1/2} t^{3/2}$

Find f_x.

$f(x,t) = 6\underline{x}^{2/3} - 8\underline{x}^{1/4} t^{1/2} - 12\underline{x}^{-1/2} t^{3/2}$ (The variable is underlined.)

$$f_x = 6 \cdot \frac{2}{3} x^{-1/3} - 8 \cdot \frac{1}{4} x^{-3/4} \cdot t^{1/2} - 12 \cdot \left(-\frac{1}{2}\right) x^{-3/2} \cdot t^{3/2}$$
$$= 4x^{-1/3} - 2x^{-3/4} t^{1/2} + 6x^{-3/2} t^{3/2}$$

Find f_t.

$f(x,t) = 6x^{2/3} - 8x^{1/4} \underline{t}^{1/2} - 12x^{-1/2}\underline{t}^{3/2}$ (The variable is underlined.)

$$= -8x^{1/4} \cdot \frac{1}{2} t^{-1/2} - 12x^{-1/2} \cdot \frac{3}{2} t^{1/2}$$
$$= -4x^{1/4} t^{-1/2} - 18x^{-1/2} t^{1/2}$$

Exercise Set 7.3

1. $f(x,y) = 3x^2 - xy + y$

Find f_x.

$f(x,y) = 3\underline{x}^2 - \underline{x}y + y$ (The variable is underlined.)

$f_x = 6x - y$

Find f_{xx}.

$f_x = 6\underline{x} - y$ (The variable is underlined.)

$f_{xx} = 6$

Find f_{xy}.

$f_x = 6x - \underline{y}$ (The variable is underlined.)

$f_{xy} = -1$

Find f_y.

$f(x,y) = 3x^2 - x\underline{y} + \underline{y}$ (The variable is underlined.)

$f_y = -x + 1$

Find f_{yx}.

$f_y = -\underline{x} + 1$ (The variable is underlined.)

$f_{yx} = -1$

Find f_{yy}.

$f_y = -x + 1$

$f_{yy} = 0$ ($-x + 1$ is treated as a constant.)

3. $f(x,y) = 3xy$

Find f_x.

$f(x,y) = 3\underline{x}y$

$f_x = 3y$

Find f_{xx}.

$f_x = 3y$

$f_{xx} = 0$ (3y is treated as a constant.)

Find f_{xy}.

$f_x = 3\underline{y}$

$f_{xy} = 3$

Find f_y.

$f(x,y) = 3x\underline{y}$

$f_y = 3x$

Find f_{yx}.

$f_y = 3\underline{x}$

$f_{yx} = 3$

Find f_{yy}.

$f_y = 3x$

$f_{yy} = 0$ (3x is treated as a constant.)

5. $f(x,y) = x^5y^4 + x^3y^2$

Find f_x.

$f(x,y) = \underline{x}^5y^4 + \underline{x}^3y^2$

$f_x = 5x^4\cdot y^4 + 3x^2\cdot y^2$

$= 5x^4y^4 + 3x^2y^2$

Find f_{xx}.

$f_x = 5\underline{x}^4y^4 + 3\underline{x}^2y^2$

$f_{xx} = 5\cdot 4x^3\cdot y^4 + 3\cdot 2x\cdot y^2$

$= 20x^3y^4 + 6xy^2$

Find f_{xy}.

$f_x = 5x^4\underline{y}^4 + 3x^2\underline{y}^2$

$f_{xy} = 5x^4\cdot 4y^3 + 3x^2\cdot 2y$

$= 20x^4y^3 + 6x^2y$

Find f_y.

$f(x,y) = x^5\underline{y}^4 + x^3\underline{y}^2$

$f_y = x^5\cdot 4y^3 + x^3\cdot 2y$

$= 4x^5y^3 + 2x^3y$

Find f_{yx}.

$f_y = 4\underline{x}^5y^3 + 2\underline{x}^3y$

$f_{yx} = 4\cdot 5x^4\cdot y^3 + 2\cdot 3x^2\cdot y$

$= 20x^4y^3 + 6x^2y$

Find f_{yy}.

$f_y = 4x^5\underline{y}^3 + 2x^3\underline{y}$

$f_{yy} = 4x^5\cdot 3y^2 + 2x^3$

$= 12x^5y^2 + 2x^3$

7. $f(x,y) = 2x - 3y$

Find f_x.

$f(x,y) = 2\underline{x} - 3y$

$f_x = 2$

Find f_{xx}.

$f_x = 2$

$f_{xx} = 0$ (2 is a constant.)

Find f_{xy}.

$f_x = 2$

$f_{xy} = 0$

Find f_y.

$f(x,y) = 2x - 3\underline{y}$

$f_y = -3$

Find f_{yx}.

$f_y = -3$

$f_{yx} = 0$ (-3 is a constant.)

Find f_{yy}.

$f_y = -3$

$f_{yy} = 0$

<u>9</u>. $f(x,y) = e^{2xy}$

Find f_x.

$f(x,y) = e^{2\underline{x}y}$

$\quad f_x = 2y\ e^{2xy}$

Find f_{xx}.

$\quad f_x = 2y\ e^{2\underline{x}y}$

$\quad f_{xx} = 2y \cdot 2y\ e^{2xy}$

$\qquad = 4y^2\ e^{2xy}$

Find f_{xy}.

$\quad f_x = 2\underline{y}\ e^{2x\underline{y}}$

$\quad f_{xy} = 2(y \cdot 2x\ e^{2xy} + e^{2xy})$

$\qquad = 4xy\ e^{2xy} + 2\ e^{2xy}$

Find f_y.

$f(x,y) = e^{2x\underline{y}}$

$\quad f_y = 2x\ e^{2xy}$

Find f_{yx}.

$\quad f_y = 2\underline{x}\ e^{2\underline{x}y}$

$\quad f_{yx} = 2(x \cdot 2y\ e^{2xy} + e^{2xy})$

$\qquad = 4xy\ e^{2xy} + 2\ e^{2xy}$

Find f_{yy}.

$\quad f_y = 2x\ e^{2x\underline{y}}$

$\quad f_{yy} = 2x \cdot 2x\ e^{2xy}$

$\qquad = 4x^2\ e^{2xy}$

<u>11</u>. $f(x,y) = x + e^y$

Find f_x.

$f(x,y) = \underline{x} + e^y$

$\quad f_x = 1$

Find f_{xx}.

$\quad f_x = 1$

$\quad f_{xx} = 0$

Find f_{xy}.

$\quad f_x = 1$

$\quad f_{xy} = 0$

Find f_y.

$f(x,y) = x + e^{\underline{y}}$

$\quad f_y = e^y$

Find f_{yx}.

$\quad f_y = e^y$

$\quad f_{yx} = 0$ (e^y is treated as a constant.)

Find f_{yy}.

$\quad f_y = e^{\underline{y}}$

$\quad f_{yy} = e^y$

<u>13</u>. $f(x,y) = y\ \ln x$

Find f_x.

$f(x,y) = y\ \ln \underline{x}$

$\quad f_x = y \cdot \dfrac{1}{x}$

$\qquad = \dfrac{y}{x}$, or yx^{-1}

Find f_{xx}.

$\quad f_x = \dfrac{y}{\underline{x}}$, or $y\underline{x}^{-1}$

$\quad f_{xx} = y \cdot (-1)x^{-2}$

$\qquad = -\dfrac{y}{x^2}$

Find f_{xy}.

$\quad f_x = \dfrac{\underline{y}}{x}$, or $\dfrac{1}{x} \cdot \underline{y}$

$\quad f_{xy} = \dfrac{1}{x}$

Find f_y.

$f(x,y) = \underline{y}\ \ln x$

$\quad f_y = \ln x$

Find f_{yx}.

$\quad f_y = \ln \underline{x}$

$\quad f_{yx} = \dfrac{1}{x}$

Find f_{yy}.

$\quad f_y = \ln x$

$\quad f_{yy} = 0$ ($\ln x$ is treated as a constant.)

<u>15</u>. $f(x,y) = \dfrac{x}{y^2} - \dfrac{y}{x^2}$

Find f_x.

$f(x,y) = \underline{x}y^{-2} - \underline{x}^{-2}y$

$\quad f_x = y^{-2} - (-2)x^{-3}y$

$\qquad = y^{-2} + 2x^{-3}y$

Find f_{xx}.

$\quad f_x = y^{-2} + 2\underline{x}^{-3}y$

$\quad f_{xx} = 2(-3)x^{-4}y$

$\qquad = -6x^{-4}y$

$\qquad = \dfrac{-6y}{x^4}$

Find f_{xy}.

$\quad f_x = \underline{y}^{-2} + 2x^{-3}\underline{y}$

$\quad f_{xy} = -2y^{-3} + 2x^{-3}$

$\qquad = -\dfrac{2}{y^3} + \dfrac{2}{x^3}$

$\qquad = \dfrac{-2x^3 + 2y^3}{y^3x^3}$

$\qquad = \dfrac{2(y^3 - x^3)}{y^3x^3}$

Find f_y.

$f(x,y) = x\underline{y}^{-2} - x^{-2}\underline{y}$

$f_y = x(-2)y^{-3} - x^{-2}$

$= -2xy^{-3} - x^{-2}$

Find f_{yx}.

$f_y = -2\underline{x}y^{-3} - \underline{x}^{-2}$

$f_{yx} = -2y^{-3} - (-2)x^{-3}$

$= -2y^{-3} + 2x^{-3}$

$= -\frac{2}{y^3} + \frac{2}{x^3}$

$= \frac{2(y^3 - x^3)}{y^3x^3}$ (See the simplification of f_{xy}.)

Find f_{yy}.

$f_y = -2x\underline{y}^{-3} - x^{-2}$

$f_{yy} = -2x\cdot(-3)y^{-4}$

$= 6xy^{-4}$

$= \frac{6x}{y^4}$

17. $f(x,y) = \ln(x^2 + y^2)$

Find f_x.

$f(x,y) = \ln(\underline{x}^2 + y^2)$

$f_x = \frac{2x}{x^2 + y^2}$

Find f_{xx}.

$f_x = \frac{2\underline{x}}{\underline{x}^2 + y^2}$

$f_{xx} = \frac{(x^2 + y^2)2 - 2x\cdot 2x}{(x^2 + y^2)^2}$

$= \frac{2x^2 + 2y^2 - 4x^2}{(x^2 + y^2)^2}$

$= \frac{2y^2 - 2x^2}{(x^2 + y^2)^2}$

Find f_y.

$f(x,y) = \ln(x^2 + \underline{y}^2)$

$f_y = \frac{2y}{x^2 + y^2}$

Find f_{yy}.

$f_y = \frac{2\underline{y}}{x^2 + \underline{y}^2}$

$f_{yy} = \frac{(x^2 + y^2)2 - 2y\cdot 2y}{(x^2 + y^2)^2}$

$= \frac{2x^2 + 2y^2 - 4y^2}{(x^2 + y^2)^2}$

$= \frac{2x^2 - 2y^2}{(x^2 + y^2)^2}$

$\frac{\partial^2 f}{\partial x^2} + \frac{\partial^2 f}{\partial y^2}$	0
$\frac{2y^2 - 2x^2}{(x^2 + y^2)^2} + \frac{2x^2 - 2y^2}{(x^2 + y^2)^2}$	0
$\frac{0}{(x^2 + y^2)^2}$	
0	

Thus, f is a solution to $\frac{\partial^2 f}{\partial x^2} + \frac{\partial^2 f}{\partial y^2} = 0$.

19. $f(x,y) = \begin{cases} \frac{xy(x^2 - y^2)}{x^2 + y^2}, & \text{for } (x,y) \neq (0,0), \\ 0, & \text{for } (x,y) = (0,0). \end{cases}$

a) Find $f_x(0,y)$.

$\lim_{h\to 0} \frac{f(h,y) - f(0,y)}{h}$

$= \lim_{h\to 0} \frac{\frac{hy(h^2 - y^2)}{h^2 + y^2} - \frac{0\cdot y(0^2 - y^2)}{0^2 + y^2}}{h}$

(Substituting)

$= \lim_{h\to 0} \frac{hy(h^2 - y^2)}{h(h^2 + y^2)}$

$= \lim_{h\to 0} \frac{y(h^2 - y^2)}{(h^2 + y^2)}$

$= \frac{y(-y^2)}{y^2}$

$= -\frac{y^3}{y^2}$

$= -y$

Thus $f_x(0,y) = -y$.

b) Find $f_y(x,0)$.

$\lim_{h\to 0} \frac{f(x,h) - f(x,0)}{h}$

$= \lim_{h\to 0} \frac{\frac{xh(x^2 - h^2)}{x^2 + h^2} - \frac{x\cdot 0(x^2 - 0^2)}{x^2 + 0^2}}{h}$

(Substituting)

$= \lim_{h\to 0} \frac{xh(x^2 - h^2)}{h(x^2 + h^2)}$

$= \lim_{h\to 0} \frac{x(x^2 - h^2)}{x^2 + h^2}$

$= \frac{x(x^2)}{x^2}$

$= x$

Thus, $f_y(x,0) = x$.

c) Find $f_{yx}(0,0)$.

$\lim_{h\to 0} \frac{f_y(h,0) - f_y(0,0)}{h}$

$= \lim_{h\to 0} \frac{h - 0}{h}$ (Substituting) $[f_y(x,0) = x]$

$= \lim_{h\to 0} 1$

$= 1$

Find $f_{xy}(0,0)$.

$$\lim_{h\to 0} \frac{f_x(0,h) - f_x(0,0)}{h}$$

$$= \lim_{h\to 0} \frac{-h - 0}{h} \qquad [f_x(0,y) = -y]$$

$$= \lim_{h\to 0} -1$$

$$= -1$$

Thus, $f_{yx}(0,0) \neq f_{xy}(0,0)$.

Exercise Set 7.4

1. $f(x,y) = x^2 + xy + y^2 - y$

Find f_x.

$f(x,y) = \underline{x}^2 + \underline{x}y + y^2 - y$ (The variable is underlined.)

$f_x = 2x + y$

Find f_y.

$f(x,y) = x^2 + x\underline{y} + \underline{y}^2 - \underline{y}$ (The variable is underlined.)

$f_y = x + 2y - 1$

Find f_{xx} and f_{xy}.

$f_x = 2\underline{x} + y \qquad f_x = 2x + \underline{y}$

$f_{xx} = 2 \qquad f_{xy} = 1$

Find f_{yy}.

$f_y = x + 2\underline{y} - 1$

$f_{yy} = 2$

Solve the system of equations $f_x = 0$ and $f_y = 0$:

$2x + y = 0 \qquad (1)$

$x + 2y - 1 = 0 \qquad (2)$

Solving Eq. (1) for y we get $y = -2x$. Substituting $-2x$ for y in Eq. (2) and solving we get

$$x + 2(-2x) - 1 = 0$$
$$x - 4x - 1 = 0$$
$$-3x = 1$$
$$x = -\frac{1}{3}$$

To find y when $x = -\frac{1}{3}$, we substitute $-\frac{1}{3}$ for x in in either Eq. (1) or Eq. (2). We use Eq. (1).

$$2\left(-\frac{1}{3}\right) + y = 0$$
$$-\frac{2}{3} + y = 0$$
$$y = \frac{2}{3}$$

Thus, $\left(-\frac{1}{3}, \frac{2}{3}\right)$ is our candidate for a maximum or minimum.

We have to check to see if $f\left(-\frac{1}{3}, \frac{2}{3}\right)$ is a maximum or minimum.

$$D = f_{xx}(a,b)\cdot f_{yy}(a,b) - [f_{xy}(a,b)]^2$$

$$D = f_{xx}\left(-\frac{1}{3}, \frac{2}{3}\right)\cdot f_{yy}\left(-\frac{1}{3}, \frac{2}{3}\right) - \left[f_{xy}\left(-\frac{1}{3}, \frac{2}{3}\right)\right]^2$$

$D = 2\cdot 2 - 1^2$ (For all values of x and y, $f_{xx} = 2$, $f_{yy} = 2$, and $f_{xy} = 1$)

$= 4 - 1$

$= 3$

Thus $D = 3$ and $f_{xx}\left(-\frac{1}{3}, \frac{2}{3}\right) = 2$. Since $D > 0$ and $f_{xx}\left(-\frac{1}{3}, \frac{2}{3}\right) > 0$, it follows that f has a relative minimum at $\left(-\frac{1}{3}, \frac{2}{3}\right)$ and that the minimum is found as follows:

$$f(x,y) = x^2 + xy + y^2 - y$$

$$f\left(-\frac{1}{3}, \frac{2}{3}\right) = \left(-\frac{1}{3}\right)^2 + \left(-\frac{1}{3}\right)\cdot\frac{2}{3} + \left(\frac{2}{3}\right)^2 - \frac{2}{3}$$

$$= \frac{1}{9} - \frac{2}{9} + \frac{4}{9} - \frac{2}{3}$$

$$= \frac{1}{9} - \frac{2}{9} + \frac{4}{9} - \frac{6}{9}$$

$$= -\frac{3}{9}$$

$$= -\frac{1}{3}$$

The relative minimum value of f is $-\frac{1}{3}$ at $\left(-\frac{1}{3},\frac{2}{3}\right)$.

3. $f(x,y) = 2xy - x^3 - y^2$

Find f_x.

$f(x,y) = 2\underline{x}y - \underline{x}^3 - y^2$

$f_x = 2y - 3x^2$

Find f_y.

$f(x,y) = 2x\underline{y} - x^3 - \underline{y}^2$

$f_y = 2x - 2y$

Find f_{xx} and f_{xy}.

$f_x = 2y - 3\underline{x}^2 \qquad f_x = 2\underline{y} - 3x^2$

$f_{xx} = -6x \qquad f_{xy} = 2$

Find f_{yy}.

$f_y = 2x - 2\underline{y}$

$f_{yy} = -2$

Solve the system of equations $f_x = 0$ and $f_y = 0$:

$2y - 3x^2 = 0 \qquad (1)$

$2x - 2y = 0 \qquad (2)$

Solving Eq. (1) for 2y we get $2y = 3x^2$. Substituting $3x^2$ for 2y in Eq. (2) and solving we get

$$2x - 3x^2 = 0$$
$$x(2 - 3x) = 0$$

$x = 0$ or $2 - 3x = 0$ (Principle of zero products)

$x = 0$ or $2 = 3x$

$x = 0$ or $\frac{2}{3} = x$

To find y when x = 0 we substitute 0 for x in either Eq. (1) or Eq. (2). We use Eq. (2).

$2\cdot 0 - 2y = 0$

$-2y = 0$

$y = 0$

Thus (0,0) is one critical value (candidate for a maximum or minimum). To find the other critical value we substitute $\frac{2}{3}$ for x in either Eq. (1) or Eq. (2). We use Eq. (2).

$2 \cdot \frac{2}{3} - 2y = 0$

$\frac{4}{3} - 2y = 0$

$-2y = -\frac{4}{3}$

$y = \frac{2}{3}$

Thus $\left(\frac{2}{3}, \frac{2}{3}\right)$ is the other critical point.

We have to check both (0,0) and $\left(\frac{2}{3}, \frac{2}{3}\right)$ as to whether they yield maximum or minimum values.

For (0,0):

$D = f_{xx}(0,0)\cdot f_{yy}(0,0) - [f_{xy}(0,0)]^2$

$= 0\cdot(-2) - 2^2$ $\quad \left[\begin{array}{l} f_{xx}(0,0) = -6\cdot 0 = 0 \\ f_{yy}(0,0) = -2 \\ f_{xy}(0,0) = 2 \end{array}\right]$

$= -4$

Since D < 0, it follows that f(0,0) is neither a maximum nor a minimum, but a saddle point.

For $\left(\frac{2}{3}, \frac{2}{3}\right)$:

$D = f_{xx}\left(\frac{2}{3}, \frac{2}{3}\right)\cdot f_{yy}\left(\frac{2}{3}, \frac{2}{3}\right) - \left[f_{xy}\left(\frac{2}{3}, \frac{2}{3}\right)\right]^2$

$= -4\cdot(-2) - 2^2$ $\quad \left[\begin{array}{l} f_{xx}\left(\frac{2}{3}, \frac{2}{3}\right) = -6 \cdot \frac{2}{3} = -4 \\ f_{yy}\left(\frac{2}{3}, \frac{2}{3}\right) = -2 \\ f_{xy}\left(\frac{2}{3}, \frac{2}{3}\right) = 2 \end{array}\right]$

$= 8 - 4$

$= 4$

Thus D = 4 and $f_{xx}\left(\frac{2}{3}, \frac{2}{3}\right) = -4$. Since D > 0 and $f_{xx}\left(\frac{2}{3}, \frac{2}{3}\right) < 0$, it follows that f has a relative maximum at $\left(\frac{2}{3}, \frac{2}{3}\right)$ and that maximum is found as follows:

$f(x,y) = 2xy - x^3 - y^2$

$f\left(\frac{2}{3}, \frac{2}{3}\right) = 2 \cdot \frac{2}{3} \cdot \frac{2}{3} - \left(\frac{2}{3}\right)^3 - \left(\frac{2}{3}\right)^2$

$= \frac{8}{9} - \frac{8}{27} - \frac{4}{9}$

$= \frac{4}{9} - \frac{8}{27}$

$= \frac{12}{27} - \frac{8}{27}$

$= \frac{4}{27}$

The relative maximum value of f is $\frac{4}{27}$ at $\left(\frac{2}{3}, \frac{2}{3}\right)$.

5. $f(x,y) = x^3 + y^3 - 3xy$

Find f_x.

$f(x,y) = \underline{x}^3 + y^3 - 3\underline{x}y$

$f_x = 3x^2 - 3y$

Find f_y.

$f(x,y) = x^3 + \underline{y}^3 - 3x\underline{y}$

$f_y = 3y^2 - 3x$

Find f_{xx} and f_{xy}.

$f_x = 3\underline{x}^2 - 3y$ $\qquad f_x = 3x^2 - 3\underline{y}$

$f_{xx} = 6x$ $\qquad f_{xy} = -3$

Find f_{yy}.

$f_y = 3\underline{y}^2 - 3x$

$f_{yy} = 6y$

Solve the system of equations $f_x = 0$ and $f_y = 0$:

$3x^2 - 3y = 0$ (1)

$3y^2 - 3x = 0$ (2)

$x^2 - y = 0$ (1) $\left[\text{Multiplying by } \frac{1}{3}\right]$

$y^2 - x = 0$ (2) $\left[\text{Multiplying by } \frac{1}{3}\right]$

Solving Eq. (1) for y we get $y = x^2$. Substituting x^2 for y in Eq. (2) and solving we get

$(x^2)^2 - x = 0$

$x^4 - x = 0$

$x(x^3 - 1) = 0$

$x = 0$ or $x^3 - 1 = 0$ (Principle of zero products)

$x = 0$ or $x^3 = 1$

$x = 0$ or $x = 1$

To find y when x = 0, substitute 0 for x in either Eq. (1) or Eq. (2). We use Eq. (2).

$y^2 - 0 = 0$

$y^2 = 0$

$y = 0$

Thus (0,0) is one critical value. To find the other critical values, we substitute 1 for x in either Eq. (1) or Eq. (2). We use Eq. (2).

$y^2 - 1 = 0$

$y^2 = 1$

$y = \pm 1$

Thus (1,-1) and (1,1) are also critical points.

We have to check (0,0), (1,-1) and (1,1) as to whether they yield maximum or minimum values.

For (0,0):

$D = f_{xx}(0,0) \cdot f_{yy}(0,0) - [f_{xy}(0,0)]^2$

$= 0 \cdot 0 - (-3)^2 \quad \begin{bmatrix} f_{xx}(0,0) = -6 \cdot 0 = 0 \\ f_{yy}(0,0) = 6 \cdot 0 = 0 \\ f_{xy}(0,0) = -3 \end{bmatrix}$

$= 0 - 9$

$= -9$

Since $D < 0$, it follows that f(0,0) is neither a maximum nor a minimum, but a saddle point.

For (1,-1):

$D = f_{xx}(1,-1) \cdot f_{yy}(1,-1) - [f_{xy}(1,-1)]^2$

$= 6 \cdot (-6) - (-3)^2 \quad \begin{bmatrix} f_{xx}(1,-1) = 6 \cdot 1 = 6 \\ f_{yy}(1,-1) = 6 \cdot (-1) = -6 \\ f_{xy}(1,-1) = -3 \end{bmatrix}$

$= -36 - 9$

$= -45$

Since $D < 0$, it follows that f(1,-1) is neither a maximum nor a minimum, but a saddle point.

For (1,1):

$D = f_{xx}(1,1) \cdot f_{yy}(1,1) - [f_{xy}(1,1)]^2$

$= 6 \cdot 6 - (-3)^2 \quad \begin{bmatrix} f_{xx}(1,1) = 6 \cdot 1 = 6 \\ f_{yy}(1,1) = 6 \cdot 1 = 6 \\ f_{xy}(1,1) = -3 \end{bmatrix}$

$= 36 - 9$

$= 27$

Thus $D = 27$ and $f_{xx}(1,1) = 6$. Since $D > 0$ and $f_{xx}(1,1) > 0$, it follows that f has a relative minimum at (1,1) and that minimum is found as follows:

$f(x,y) = x^3 + y^3 - 3xy$

$f(1,1) = 1^3 + 1^3 - 3 \cdot 1 \cdot 1$

$= 1 + 1 - 3$

$= -1$

The relative minimum value of f is -1 at (1,1).

7. $f(x,y) = x^2 + y^2 - 2x + 4y - 2$

Find f_x.

$f(x,y) = \underline{x}^2 + y^2 - 2\underline{x} + 4y - 2$

$f_x = 2x - 2$

Find f_y.

$f(x,y) = x^2 + \underline{y}^2 - 2x + 4\underline{y} - 2$

$f_y = 2y + 4$

Find f_{xx} and f_{xy}.

$f_x = 2\underline{x} - 2 \qquad f_x = 2x - 2$

$f_{xx} = 2 \qquad f_{xy} = 0$

Find f_{yy}.

$f_y = 2\underline{y} + 4$

$f_{yy} = 2$

Solve the system of equations $f_x = 0$ and $f_y = 0$.

$2x - 2 = 0 \qquad 2y + 4 = 0$

$2x = 2 \qquad 2y = -4$

$x = 1 \qquad y = -2$

The only critical value is (1,-2). We have to check to see if f(1,-2) is a maximum or a minimum.

$D = f_{xx}(1,-2) \cdot f_{yy}(1,-2) - [f_{xy}(1,-2)]^2$

$= 2 \cdot 2 - 0^2 \quad \begin{bmatrix} f_{xx}(1,-2) = 2 \\ f_{yy}(1,-2) = 2 \\ f_{xy}(1,-2) = 0 \end{bmatrix}$

$= 4$

Thus $D = 4$ and $f_{xx}(1,-2) = 2$. Since $D > 0$ and $f_{xx}(1,-2) > 0$, it follows that f has a relative minimum at (1,-2) and that the minimum is found as follows:

$f(x,y) = x^2 + y^2 - 2x + 4y - 2$

$f(1,-2) = 1^2 + (-2)^2 - 2 \cdot 1 + 4 \cdot (-2) - 2$

$= 1 + 4 - 2 - 8 - 2$

$= -7$

The relative minimum value of f is -7 at (1,-2).

9. $f(x,y) = x^2 + y^2 + 2x - 4y$

Find f_x.

$f(x,y) = \underline{x}^2 + y^2 + 2\underline{x} - 4y$

$f_x = 2x + 2$

Find f_y.

$f(x,y) = x^2 + \underline{y}^2 + 2x - 4\underline{y}$

$f_y = 2y - 4$

Find f_{xx} and f_{xy}.

$f_x = 2\underline{x} + 2 \qquad f_x = 2x + 2$

$f_{xx} = 2 \qquad f_{xy} = 0$

Find f_{yy}.

$f_y = 2\underline{y} - 4$

$f_{yy} = 2$

Solve the system of equations $f_x = 0$ and $f_y = 0$.

$2x + 2 = 0 \qquad 2y - 4 = 0$

$2x = -2 \qquad 2y = 4$

$x = -1 \qquad y = 2$

The only critical value is $(-1,2)$. We have to check to see if $f(-1,2)$ is a maximum or minimum.

$D = f_{xx}(-1,2)\cdot f_{yy}(-1,2) - [f_{xy}(-1,2)]^2$

$= 2\cdot 2 - 0^2$ $\quad \begin{bmatrix} f_{xx}(-1,2) = 2 \\ f_{yy}(-1,2) = 2 \\ f_{xy}(-1,2) = 0 \end{bmatrix}$

$= 4$

Thus $D = 4$ and $f_{xx}(-1,2) = 2$. Since $D > 0$ and $f_{xx}(-1,2) > 0$, it follows that f has a relative minimum at $(-1,2)$ and that the minimum is found as follows:

$f(x,y) = x^2 + y^2 + 2x - 4y$

$f(-1,2) = (-1)^2 + 2^2 + 2(-1) - 4\cdot 2$

$= 1 + 4 - 2 - 8$

$= -5$

The relative minimum value of f is -5 at $(-1,2)$.

11. $f(x,y) = 4x^2 - y^2$

Find f_x.

$f(x,y) = 4\underline{x}^2 - y^2$

$f_x = 8x$

Find f_y.

$f(x,y) = 4x^2 - \underline{y}^2$

$f_y = -2y$

Find f_{xx} and f_{xy}.

$f_x = 8\underline{x}$ $\qquad f_x = 8x$

$f_{xx} = 8$ $\qquad f_{xy} = 0$

Find f_{yy}.

$f_y = -2\underline{y}$

$f_{yy} = -2$

Solve the system of equations $f_x = 0$ and $f_y = 0$.

$8x = 0$ $\qquad -2y = 0$

$x = 0$ $\qquad y = 0$

The only critical value is $(0,0)$. We have to check to see if $f(0,0)$ is a maximum or a minimum.

$D = f_{xx}(0,0)\cdot f_{yy}(0,0) - [f_{xy}(0,0)]^2$

$= 8\cdot(-2) - 0^2$ $\quad \begin{bmatrix} f_{xx}(0,0) = 8 \\ f_{yy}(0,0) = -2 \\ f_{xy}(0,0) = 0 \end{bmatrix}$

$= -16$

Since $D < 0$, it follows that $(0,0)$ is neither a maximum nor a minimum, but a saddle point.

Thus f has no relative maximum value or relative minimum value.

13. $R(x,y) = 17x + 21y$

$C(x,y) = 4x^2 - 4xy + 2y^2 - 11x + 25y - 3$

Total profit, $P(x,y)$, is given by

$P(x,y)$

$= R(x,y) - C(x,y)$

$= (17x + 21y) - (4x^2 - 4xy + 2y^2 - 11x + 25y - 3)$

$= 17x + 21y - 4x^2 + 4xy - 2y^2 + 11x - 25y + 3$

$= -4x^2 - 2y^2 + 4xy + 28x - 4y + 3$

Find P_x.

$P(x,y) = -4\underline{x}^2 - 2y^2 + 4\underline{x}y + 28\underline{x} - 4y + 3$

$P_x = -8x + 4y + 28$

Find P_y.

$P(x,y) = -4x^2 - 2\underline{y}^2 + 4x\underline{y} + 28x - 4\underline{y} + 3$

$P_y = -4y + 4x - 4$

Find P_{xx} and P_{xy}.

$P_x = -8\underline{x} + 4y + 28$ $\qquad P_x = -8x + 4\underline{y} + 28$

$P_{xx} = -8$ $\qquad P_{xy} = 4$

Find P_{yy}.

$P_y = -4\underline{y} + 4x - 4$

$P_{yy} = -4$

Solve the system of equations $P_x = 0$ and $P_y = 0$.

$-8x + 4y + 28 = 0$ $\qquad (1)$

$-4y + 4x - 4 = 0$ $\qquad (2)$

Adding these equations, we get

$-4x + 24 = 0.$

Then

$-4x = -24$

$x = 6.$

To find y when $x = 6$, we substitute 6 for x in either Eq. (1) or Eq. (2). We use Eq. (1).

$-8x + 4y + 28 = 0$

$-8\cdot 6 + 4y + 28 = 0$ $\qquad$ (Substituting)

$-48 + 4y + 28 = 0$

$4y - 20 = 0$

$4y = 20$

$y = 5$

Thus, $(6,5)$ is our candidate for a maximum or minimum.

We have to check to see if $P(6,5)$ is a maximum.

$D = P_{xx}(6,5)\cdot P_{yy}(6,5) - [P_{xy}(6,5)]^2$

$= (-8)(-4) - 4^2$ $\quad \begin{bmatrix} P_{xx}(6,5) = -8 \\ P_{yy}(6,5) = -4 \\ P_{xy}(6,5) = 4 \end{bmatrix}$

$= 32 - 16$

$= 16$

Thus D = 16 and $P_{xx} = -8$. Since D > 0 and $P_{xx}(6,5) < 0$, it follows that P has a relative maximum at (6,5). Thus to maximize profit, the company must produce and sell 6 thousand of the \$17 radio and 5 thousand of the \$21 radio.

15. $P(a,p) = 2ap + 80p - 15p^2 - \frac{1}{10}a^2p - 100$

Find P_a.

$P(a,p) = 2\underline{a}p + 80p - 15p^2 - \frac{1}{10}\underline{a}^2p - 100$

$P_a = 2p - \frac{1}{5}ap$

Find P_p.

$P(a,p) = 2a\underline{p} + 80\underline{p} - 15\underline{p}^2 - \frac{1}{10}a^2\underline{p} - 100$

$P_p = 2a + 80 - 30p - \frac{1}{10}a^2$

Find P_{aa} and P_{ap}.

$P_a = 2p - \frac{1}{5}\underline{a}p$ $\qquad$ $P_a = 2\underline{p} - \frac{1}{5}a\underline{p}$

$P_{aa} = -\frac{1}{5}p$ $\qquad$ $P_{ap} = 2 - \frac{1}{5}a$

Find P_{pp}.

$P_p = 2a + 80 - 30\underline{p} - \frac{1}{10}a^2$

$P_{pp} = -30$

Solve the system of equations $P_a = 0$ and $P_p = 0$.

$2p - \frac{1}{5}ap = 0 \qquad (1)$

$2a + 80 - 30p - \frac{1}{10}a^2 = 0 \qquad (2)$

Solving Eq. (1) for a we get a = 10. Substituting 10 for a in Eq. (2) and solving we get

$2\cdot 10 + 80 - 30p - \frac{1}{10}(10)^2 = 0$ (Substituting 10 for a)

$20 + 80 - 10 = 30p$

$90 = 30p$

$3 = p$

Thus, (10,3) is a candidate for a maximum or a minimum.

We have to check to see if P(10,3) is a maximum.

$D = P_{aa}(10,3)\cdot P_{pp}(10,3) - [P_{ap}(10,3)]^2$

$= \left(-\frac{3}{5}\right)(-30) - 0^2$ $\quad \left[\begin{array}{l} P_{aa}(10,3) = -\frac{1}{5}\cdot 3 = -\frac{3}{5} \\ P_{pp}(10,3) = -30 \\ P_{ap}(10,3) = 2 - \frac{1}{5}\cdot 10 = 0 \end{array}\right]$

$= 18$

Thus D = 18 and $P_{aa} = -\frac{3}{5}$. Since D > 0 and $P_{aa}(10,3) < 0$, it follows that P has a relative maximum at (10,3). Thus to maximize profit, the company must spend 10 million dollars on advertising and charge \$3 per item. The maximum profit is found as follows:

$P(a,p) = 2ap + 80p - 15p^2 - \frac{1}{10}a^2p - 100$

$P(10,3) = 2\cdot 10\cdot 3 + 80\cdot 3 - 15\cdot 3^2 - \frac{1}{10}\cdot 10^2\cdot 3 - 100$ (Substituting)

$= 60 + 240 - 135 - 30 - 100$

$= 35$

The maximum profit is \$35 million.

17.

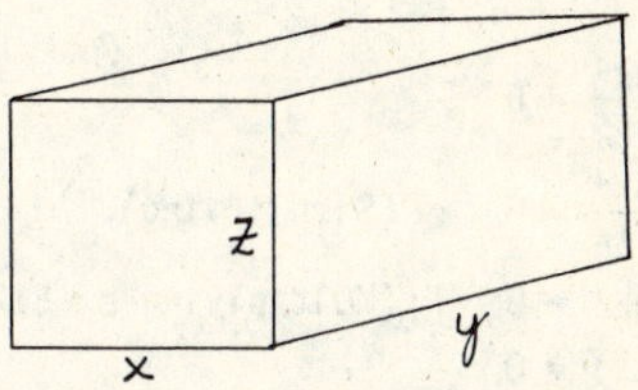

Let x, y, and z represent the dimensions of the container as shown in the drawing.

$V = xyz$

$320 = xyz \qquad (V = 320\ \text{ft}^3)$

$\frac{320}{xy} = z$

Now we can express the cost as a function of two variables, x and y. The area of the bottom is xy ft², so the cost of the bottom is 5xy. The area of each of two of the sides is xz, or $x\left(\frac{320}{xy}\right) = \frac{320}{y}$, so the area of both such sides is $2\cdot\frac{320}{y}$, or $\frac{640}{y}$. The area of each of the remaining sides is yz, or $y\left(\frac{320}{xy}\right) = \frac{320}{x}$, so the area of both such sides is $2\cdot\frac{320}{x}$, or $\frac{640}{x}$. Then the total area of the four sides is $\frac{640}{y} + \frac{640}{x}$, and the cost of the four sides is $4\left(\frac{640}{y} + \frac{640}{x}\right)$, or $\frac{2560}{y} + \frac{2560}{x}$. Now we can write the cost function.

Total cost	=	Cost of bottom	+	Cost of sides
$C(x,y)$	=	$5xy$	+	$\left(\frac{2560}{y} + \frac{2560}{x}\right)$

We try to find a minimum value for C(x,y).

$C_x = 5y - \frac{2560}{x^2} \qquad C_y = 5x - \frac{2560}{y^2}$

$C_{xx} = \frac{5120}{x^3} \qquad C_{yy} = \frac{5120}{y^3}$

$C_{xy} = 5$

Solve the system of equations $C_x = 0$, $C_y = 0$.

$5y - \frac{2560}{x^2} = 0$ (1)

$5x - \frac{2560}{y^2} = 0$ (2)

Solve Eq. (1) for y:

$$5y - \frac{2560}{x^2} = 0$$

$$5y = \frac{2560}{x^2}$$

$$y = \frac{512}{x^2}$$

Substitute $\frac{512}{x^2}$ for y in Eq. (2) and solve for x:

$$5x - \frac{2560}{\left[\frac{512}{x^2}\right]^2} = 0$$

$$5x - \frac{2560}{\frac{262,144}{x^4}} = 0$$

$$5x - \frac{2560x^4}{262,144} = 0$$

$$5x - \frac{5x^4}{512} = 0 \quad \text{(Simplifying)}$$

$$2560x - 5x^4 = 0 \quad \text{(Multiplying by 512)}$$

$$5x(512 - x^3) = 0$$

$5x = 0$ or $512 - x^3 = 0$

$x = 0$ or $512 = x^3$

$x = 0$ or $8 = x$

Since none of the dimensions can be 0, only x = 8 has meaning in this application. Substitute 8 for x in Eq. (1) to find y:

$$5y - \frac{2560}{x^2} = 0$$

$$5y - \frac{2560}{8^2} = 0$$

$$5y - \frac{2560}{64} = 8$$

$$5y - 40 = 0$$

$$5y = 40$$

$$y = 8$$

Check to determine if C(8,8) is a minimum.

$D = C_{xx}(8,8) \cdot C_{yy}(8,8) - [C_{xy}(8,8)]^2$

$= \frac{5120}{8^3} \cdot \frac{5120}{8^3} - 5^2$

$= \frac{5120}{512} \cdot \frac{5120}{512} - 25$

$= 10 \cdot 10 - 25$

$= 100 - 25$

$= 75$

Since $D > 0$ and $C_{xx}(8,8) = 10 > 0$, C has a minimum at (8,8). Thus, the dimensions of the bottom are 8 ft by 8 ft. The height is $\frac{320}{8 \cdot 8}$, or 5 ft.

19. a) $q_1 = 64 - 4p_1 - 2p_2$ (1)

$q_2 = 56 - 2p_1 - 4p_2$ (2)

$R = p_1q_1 + p_2q_2$ (Total revenue)

$R = p_1(64 - 4p_1 - 2p_2) + p_2(56 - 2p_1 - 4p_2)$ (Substituting)

$= 64p_1 - 4p_1^2 - 2p_1p_2 + 56p_2 - 2p_1p_2 - 4p_2^2$

$= 64p_1 - 4p_1^2 - 4p_1p_2 + 56p_2 - 4p_2^2$

b) We now find the values of p_1 and p_2 to maximize total revenue.

$R_{p_1} = 64 - 8p_1 - 4p_2$ $\quad R_{p_2} = -4p_1 + 56 - 8p_2$

$R_{p_1p_1} = -8$ $\quad R_{p_2p_2} = -8$

$R_{p_1p_2} = -4$

Solve the system of equations $R_{p_1} = 0$, $R_{p_2} = 0$.

$$64 - 8p_1 - 4p_2 = 0$$

$$-4p_1 + 56 - 8p_2 = 0$$

The solution of the system is $p_1 = 6$ and $p_2 = 4$.

We check to determine if R(6,4) is a maximum.

$D = R_{p_1p_1}(6,4) \cdot R_{p_2p_2}(6,4) - [R_{p_1p_2}(6,4)]^2$

$= -8 \cdot (-8) - (-4)^2$

$= 64 - 16$

$= 48$

Since $D > 0$ and $R_{p_1p_2}(6, 4) = -8 < 0$, R has a maximum at (6,4). Then to maximize revenue p_1 must be 6×10, or \$60 and p_2 must be 4×10, or \$40.

c) We substitute 6 for p_1 and 4 for p_2 in Eqs. (1) and (2) to find q_1 and q_2.

$q_1 = 64 - 4p_1 - 2p_2$

$= 64 - 4 \cdot 6 - 2 \cdot 4$

$= 64 - 24 - 8$

$= 32$ (hundreds)

$q_2 = 56 - 2p_1 - 4p_2$

$= 56 - 2 \cdot 6 - 4 \cdot 4$

$= 56 - 12 - 16$

$= 28$ (hundreds)

d) To maximize revenue 3200 units of the \$60 calculator and 2800 units of the \$40 calculator must be produced and sold. The maximum revenue is found as follows.

R = \$60·3200 + \$40·2800

= \$192,000 + \$112,000

= \$304,000

21. $f(x,y) = e^x + e^y - e^{x+y}$

We first find f_x, f_y, f_{xx}, f_{yy}, and f_{xy}.

$f_x = e^x - e^{x+y}$

$f_y = e^y - e^{x+y}$

$f_{xx} = e^x - e^{x+y}$

$f_{yy} = e^y - e^{x+y}$

$f_{xy} = -e^{x+y}$

Next we solve the system of equations $f_x = 0$ and $f_y = 0$.

$e^x - e^{x+y} = 0$ (1)

$e^y - e^{x+y} = 0$ (2)

We can solve the first equation for y.

$e^x - e^{x+y} = 0$

$e^x = e^{x+y}$

$x = x + y$

$0 = y$

We can solve the second equation for x.

$e^y - e^{x+y} = 0$

$e^y = e^{x+y}$

$y = x + y$

$0 = x$

Thus, (0,0) is a candidate for a maximum or a minimum.

We use the D-test to check f(0,0).

$D = f_{xx}(0,0)\cdot f_{yy}(0,0) - [f_{xy}(0,0)]^2$

$= 0\cdot 0 - (-1)^2$

$= -1$

Since D < 0, it follows that f(0,0) is neither a maximum nor a minimum, but a saddle point.

23. $f(x,y) = 2y^2 + x^2 - x^2y$

Find f_x.

$f(x,y) = 2y^2 + \underline{x}^2 - \underline{x}^2y$

$f_x = 2x - 2xy$

Find f_y.

$f(x,y) = 2\underline{y}^2 + x^2 - x^2\underline{y}$

$f_y = 4y - x^2$

Find f_{xx} and f_{xy}.

$f_x = 2\underline{x} - 2\underline{x}y$ $\qquad f_x = 2x - 2x\underline{y}$

$f_{xx} = 2 - 2y$ $\qquad f_{xy} = -2x$

Find f_{yy}.

$f_y = 4\underline{y} - x^2$

$f_{yy} = 4$

Solve the system of equations $f_x = 0$ and $f_y = 0$.

$2x - 2xy = 0$ (1)

$4y - x^2 = 0$ (2)

Solve Eq. (2) for y.

$4y - x^2 = 0$

$4y = x^2$

$y = \frac{x^2}{4}$

Substitute $\frac{x^2}{4}$ for y in Eq. (1) and solve for x.

$2x - 2x\cdot\frac{x^2}{4} = 0$

$2x - \frac{x^3}{2} = 0$

$4x - x^3 = 0$

$x(4 - x^2) = 0$

$x(2 - x)(2 + x) = 0$

$x = 0$ or $2 - x = 0$ or $2 + x = 0$

$x = 0$ or $x = 2$ or $x = -2$

When $x = 0$, $y = \frac{0^2}{4} = 0$.

When $x = 2$, $y = \frac{2^2}{4} = 1$.

When $x = -2$, $y = \frac{(-2)^2}{4} = 1$.

The critical values are (0,0), (2,1), and (-2,1).

We check all critical points to see if they yield maximum or minimum values.

For (0,0):

$D = f_{xx}(0,0)\cdot f_{yy}(0,0) - [f_{xy}(0,0)]^2$

$= 2\cdot 4 - 0^2$ $\qquad \begin{bmatrix} f_{xx}(0,0) = 2 \\ f_{yy}(0,0) = 4 \\ f_{xy}(0,0) = 0 \end{bmatrix}$

$= 8$

Since D > 0 and $f_{xx} > 0$, f has a relative minimum at (0,0) and that minimum value is found as follows:

$f(x,y) = 2y^2 + x^2 - x^2y$

$f(0,0) = 2\cdot 0^2 + 0^2 - 0^2\cdot 0 = 0$

The relative minimum of f is 0 at (0,0).

For (2,1):

$D = f_{xx}(2,1)\cdot f_{yy}(2,1) - [f_{xy}(2,1)]^2$

$= 0\cdot 4 - (-4)^2$ $\qquad \begin{bmatrix} f_{xx}(2,1) = 0 \\ f_{yy}(2,1) = 4 \\ f_{xy}(2,1) = -4 \end{bmatrix}$

$= -16$

For (-2,1):

$D = f_{xx}(-2,1)\cdot f_{yy}(-2,1) - [f_{xy}(-2,1)]^2$

$= 0\cdot 4 - 4^2$ $\quad \begin{bmatrix} f_{xx}(-2,1) = 0 \\ f_{yy}(-2,1) = 4 \\ f_{xy}(-2,1) = 4 \end{bmatrix}$

$= -16$

Since $D < 0$ for both (2,1) and (-2,1), f has neither a maximum nor a minimum at these points. Both (2,1) and (-2,1) are saddle points.

Exercise Set 7.5

1. a) The data points are

(1,30), (2,45), (3,40), (4,45), (5,45), and (6,45).

The points on the regression line are

$(1,y_1)$, $(2,y_2)$, $(3,y_3)$, $(4,y_4)$, $(5,y_5)$, and $(6,y_6)$.

The y-deviations are

$y_1 - 30$, $y_2 - 45$, $y_3 - 40$, $y_4 - 45$, $y_5 - 45$, and $y_6 - 45$.

Using the least squares assumption, we need to minimize

$$S = (y_1 - 30)^2 + (y_2 - 45)^2 + (y_3 - 40)^2 + (y_4 - 45)^2 + (y_5 - 45)^2 + (y_6 - 45)^2$$

and since the points $(1,y_1)$, $(2,y_2)$, $(3,y_3)$, $(4,y_4)$, $(5,y_5)$ and $(6,y_6)$ must be solutions of $y = mx + b$, it follows that

$y_1 = m\cdot 1 + b = m + b$

$y_2 = m\cdot 2 + b = 2m + b$

$y_3 = m\cdot 3 + b = 3m + b$

$y_4 = m\cdot 4 + b = 4m + b$

$y_5 = m\cdot 5 + b = 5m + b$

$y_6 = m\cdot 6 + b = 6m + b$

Substituting $m + b$ for y_1, $2m + b$ for y_2, $3m + b$ for y_3, $4m + b$ for y_4, $5m + b$ for y_5, and $6m + b$ for y_6, we have

$$S = (m + b - 30)^2 + (2m + b - 45)^2 + (3m + b - 40)^2 + (4m + b - 45)^2 + (5m + b - 45)^2 + (6m + b - 45)^2$$

Thus to find the regression line for the given data, we must find the values m and b that minimize the function S given by the sum above.

To apply the D-test, we first find the partial derivatives $\partial S/\partial b$ and $\partial S/\partial m$:

$$\frac{\partial S}{\partial b} = 2(m + b - 30) + 2(2m + b - 45) + 2(3m + b - 40) + 2(4m + b - 45) + 2(5m + b - 45) + 2(6m + b - 45)$$

$= 2m + 2b - 60 + 4m + 2b - 90 + 6m + 2b - 80 + 8m + 2b - 90 + 10m + 2b - 90 + 12m + 2b - 90$

$= 42m + 12b - 500$

$$\frac{\partial S}{\partial m} = 2(m + b - 30) + 2(2m + b - 45)2 + 2(3m + b - 40)3 + 2(4m + b - 45)4 + 2(5m + b - 45)5 + 2(6m + b - 45)6$$

$= 2m + 2b - 60 + 8m + 4b - 180 + 18m + 6b - 240 + 32m + 8b - 360 + 50m + 10b - 450 + 72m + 12b - 540$

$= 182m + 42b - 1830$

We set these derivatives equal to 0 and solve the resulting system.

$42m + 12b - 500 = 0$ (1)

$182m + 42b - 1830 = 0$ (2)

$21m + 6b - 250 = 0$ (1) $\left[\text{Multiplying by } \frac{1}{2}\right]$

$91m + 21b - 915 = 0$ (2) $\left[\text{Multiplying by } \frac{1}{2}\right]$

$441m + 126b - 5250 = 0$ (1) (Multiplying by 21)

$-546m - 126b + 5490 = 0$ (2) (Multiplying by -6)

$-105m + 240 = 0$ (Adding)

$-105m = -240$

$m = \frac{-240}{-105}$

$m = \frac{16}{7}$

Substitute $\frac{16}{7}$ for m in either Eq. (1) or Eq. (2) and solve for b. Here we use Eq. (1).

$21m + 6b - 250 = 0$ (1)

$21 \cdot \frac{16}{7} + 6b - 250 = 0$ (Substituting)

$48 + 6b - 250 = 0$

$6b - 202 = 0$

$6b = 202$

$b = \frac{202}{6}$

$b = \frac{101}{3}$

The solution of the system is $b = \frac{101}{3}$, $m = \frac{16}{7}$.

We use the D-test to verify that $\left(\frac{101}{3}, \frac{16}{7}\right)$ does yield the minimum of S.

We first find S_{bb}, S_{mm}, and S_{bm}.

$S_b = 42m + 12\underline{b} - 500$

$S_{bb} = 12$

$S_m = 182\underline{m} + 42b - 1830$

$S_{mm} = 182$

$S_b = 42\underline{m} + 12b - 500$

$S_{bm} = 42$

$$D = S_{bb}\left(\frac{101}{3}, \frac{16}{7}\right)\cdot S_{mm}\left(\frac{101}{3}, \frac{16}{7}\right) - \left[S_{bm}\left(\frac{101}{3}, \frac{16}{7}\right)\right]^2$$

$= 12\cdot 182 - 42^2$

$= 2184 - 1764$

$= 420$

Thus $D = 420$ and $S_{bb}\left(\frac{101}{3}, \frac{16}{7}\right) = 12$. Since $D > 0$ and $S_{bb}\left(\frac{101}{3}, \frac{16}{7}\right) > 0$, it follows that S has a relative minimum at $\left(\frac{101}{3}, \frac{16}{7}\right)$.

The regression line is

$y = \frac{16}{7}x + \frac{101}{3}$ $\qquad \left(b = \frac{101}{3}, m = \frac{16}{7}\right)$

or

$y = 2.29x + 33.67$

b) In 1994, $x = 15$.

$y = 2.29x + 33.67$

$y = 2.29(15) + 33.67$

$= 34.35 + 33.67$

$= \$68.02$

In 2000, $x = 21$,

$y = 2.29x + 33.67$

$y = 2.29(21) + 33.67$

$= 48.09 + 33.67$

$= \$81.76$

3. a) The data points are (1,457), (2,722), and (3,862).

The points on the regression line are $(1,y_1)$, $(2,y_2)$, and $(3,y_3)$.

The y-deviations are $y_1 - 457$, $y_2 - 722$, and $y_3 - 862$.

We want to minimize

$S = (y_1 - 457)^2 + (y_2 - 722)^2 + (y_3 - 862)^2$,

where $y_1 = m\cdot 1 + b = m + b$,

$y_2 = m\cdot 2 + b = 2m + b$,

$y_3 = m\cdot 3 + b = 3m + b$.

Substituting, we get

$S = (m + b - 457)^2 + (2m + b - 722)^2 + (3m + b - 862)^2$.

In order to minimize S we find the first partial derivatives.

$\frac{\partial S}{\partial b} = 2(m + b - 457) + 2(2m + b - 722) + 2(3m + b - 862)$

$= 2m + 2b - 914 + 4m + 2b - 1444 + 6m + 2b - 1724$

$= 12m + 6b - 4082$

$\frac{\partial S}{\partial m} = 2(m + b - 457) + 2(2m + b - 722)(2) + 2(3m + b - 862)(3)$

$= 2m + 2b - 914 + 8m + 4b - 2888 + 18m + 6b - 5172$

$= 28m + 12b - 8974$

We set these derivatives equal to 0 and solve the resulting system of equations.

$12m + 6b - 4082 = 0$

$28m + 12b - 8974 = 0$

The solution of the system is $b = \frac{1652}{6}$, $m = \frac{405}{2}$, or $b \approx 275.33$, $m = 202.50$.

We use the D-test to verify that S(275.3, 202.50) is a relative minimum. We first find the second-order partial derivatives.

$S_{bb} = 6$, $S_{bm} = 12$, $S_{mm} = 28$

$D = S_{bb}(275.33, 202.50)\cdot S_{mm}(275.33, 202.50) - [S_{bm}(275.33, 202.50)]^2$

$= 6\cdot 28 - 12^2$

$= 168 - 144$

$= 24$

Since $D > 0$ and $S_{bb}(275.33, 202.50) = 6 > 0$, S has a relative minimum at (275.33, 202.50).

The regression line is $y = 202.50x + 275.33$.

b) In 2000, $x = 13$.

$y = 202.50(13) + 275.33$

$= 2632.5 + 275.33$

$\approx \$2908$ million

In 2020, $x = 33$.

$y = 202.50(33) + 275.33$

$= 6682.5 + 275.33$

$\approx \$6958$ million

5. a) The data points are (70,75), (60,62) and (85,89).

The points on the regression line are $(70,y_1)$, $(60,y_2)$ and $(85,y_3)$.

The y-deviations are

$y_1 - 75$, $y_2 - 62$, and $y_3 - 89$.

Using the least squares assumption, we need to minimize

$S = (y_1 - 75)^2 + (y_2 - 62)^2 + (y_3 - 89)^3$

and since the points $(70,y_1)$, $(60,y_2)$, and $(85,y_3)$ must be solutions of $y = mx + b$, it follows that

$y_1 = m\cdot 70 + b = 70m + b$

$y_2 = m\cdot 60 + b = 60m + b$

$y_3 = m\cdot 85 + b = 85m + b$

Substituting $70m + b$ for y_1, $60m + b$ for y_2, and $85m + b$ for y_3, we have

$$S = (70m + b - 75)^2 + (60m + b - 62)^2 + (85m + b - 89)^2.$$

Thus to find the regression line for the given data, we must find the values m and b that minimize the function S given by the sum above.

To apply the D-test, we first find the partial derivatives $\partial S/\partial b$ and $\partial S/\partial m$:

$$\frac{\partial S}{\partial b} = 2(70m + b - 75) + 2(60m + b - 62) + 2(85m + b - 89)$$
$$= 140m + 2b - 150 + 120m + 2b - 124 + 170m + 2b - 178$$
$$= 430m + 6b - 452$$

$$\frac{\partial S}{\partial m} = 2(70m + b - 75)\cdot 70 + 2(60m + b - 62)\cdot 60 + 2(85m + b - 89)\cdot 85$$
$$= 9800m + 140b - 10{,}500 + 7200m + 120b - 7440 + 14{,}450m + 170b - 15{,}130$$
$$= 31{,}450m + 430b - 33{,}070$$

We set these derivatives equal to 0 and solve the resulting system

$430m + 6b - 452 = 0$ (1)

$31{,}450m + 430b - 33{,}070 = 0$ (2)

$215m + 3b - 226 = 0$ (1) $\left[\text{Multiplying by } \frac{1}{2}\right]$

$3145m + 43b - 3307 = 0$ (2) $\left[\text{Multiplying by } \frac{1}{10}\right]$

$-9245m - 129b + 9718 = 0$ (1) (Multiplying by -43)

$9435m + 129b - 9921 = 0$ (2) (Multiplying by 3)

$190m \quad - 203 = 0$ (Adding)

$190m = 203$

$m = \frac{203}{190}$

Substitute $\frac{203}{190}$ for m in either Eq. (1) or Eq. (2) and solve for b. Here we use (1).

$$215m + 3b - 226 = 0$$
$$215 \cdot \frac{203}{190} + 3b - 226 = 0$$
$$\frac{43{,}645}{190} + 3b - \frac{42{,}940}{190} = 0$$
$$3b = \frac{42{,}940 - 43{,}645}{190}$$
$$3b = -\frac{705}{190}$$
$$b = \frac{1}{3} \cdot \left(-\frac{705}{190}\right)$$
$$b = -\frac{235}{190}$$
$$b = -\frac{47}{38}$$

The solution of the system is $b = -\frac{47}{38}$, $m = \frac{203}{190}$.

We use the D-test to verify that $\left(-\frac{47}{38}, \frac{203}{190}\right)$ does yield the minimum of S.

We first find S_{bb}, S_{mm}, and S_{bm}.

$S_b = 430m + 6\underline{b} - 452$

$S_{bb} = 6$

$S_m = 31{,}450\underline{m} + 430b - 33{,}070$

$S_{mm} = 31{,}450$

$S_b = 430\underline{m} + 6b - 452$

$S_{bm} = 430$

$$D = S_{bb}\left(-\frac{47}{38},\frac{203}{190}\right)\cdot S_{mm}\left(-\frac{47}{38},\frac{203}{190}\right) - \left[S_{bm}\left(-\frac{47}{38},\frac{203}{190}\right)\right]^2$$
$$= 6\cdot(31{,}450) - 430^2$$
$$= 188{,}700 - 184{,}900$$
$$= 3800$$

Thus $D = 3800$ and $S_{bb}\left(-\frac{47}{38},\frac{203}{190}\right) = 6$. Since $D > 0$ and $S_{bb}\left(-\frac{47}{38},\frac{203}{190}\right) > 0$, it follows that S has a relative minimum at $\left(-\frac{47}{38},\frac{203}{190}\right)$.

The regression line is

$y = \frac{203}{190}x - \frac{47}{38}$ $\left(m = \frac{203}{190},\ b = -\frac{47}{38}\right)$

or

$y = 1.068421x - 1.236842$

b) $y = 1.068421x - 1.236842$

$y = 1.068421(81) - 1.236842$ (Substituting 81 for x)

$y = 86.542101 - 1.236842$

$y = 85.305259$

$y \approx 85.3$

7. a)

c_i	d_i	$c_i - \bar{x}$	$(c_i - \bar{x})^2$
1875	4.4083	-69.5833	4841.8356
1894	4.3033	-50.5833	2558.6702
1923	4.1733	-21.5833	465.8388
1937	4.1067	-7.5833	57.5064
1942	4.1033	-2.5833	6.6734
1945	4.0233	0.4167	0.1736
1954	3.9900	9.4167	88.6742
1964	3.9067	19.4167	377.0082
1967	3.8517	22.4167	502.5084
1975	3.8233	30.4167	925.1756
1979	3.8167	34.4167	1184.5092
1980	3.8133	35.4167	1254.3426

$$\sum_{i=1}^{12} c_i = 23{,}335 \qquad \sum_{i=1}^{12} d_i = 48.3199$$

$$\bar{x} = \frac{23{,}335}{12} \approx 1944.5833 \qquad \bar{y} = \frac{48.3199}{12} \approx 4.0267$$

$(d_i - \bar{y})$	$(c_i - \bar{x})(d_i - \bar{y})$
0.3816	-26.5530
0.2766	-13.9913
0.1466	-3.1641
0.0800	-0.6067
0.0766	-0.1979
-0.0034	-0.0014
-0.0367	-0.3456
-0.1200	-2.3300
-0.1750	-3.9229
-0.2034	-6.1868
-0.2100	-7.2275
-0.2134	-7.5579

$$\sum_{i=1}^{12} (c_i - \bar{x})^2 = 1021.9097$$

$$\sum_{i=1}^{12} (c_i - \bar{x})(d_i - \bar{y}) = -6.0071$$

$$m = \frac{-6.0071}{1021.9097}$$

$m \approx -0.0059$

The regression line is

$y - 4.0267 = -0.0059(x - 1944.5833)$

$y - 4.0267 = -0.0059x + 11.4730$

$y = -0.0059x + 15.4997$

(Answers may vary slightly due to rounding differences.)

b) Substitute 1996 for x.

$y = -0.0059(1996) + 15.4997$
$= -11.7764 + 15.4997$
$= 3.7233$
$= 3{:}43.4 \qquad (0.7233 \times 60 \approx 43.4)$

Substitute 2000 for x.

$y = -0.0059(2000) + 15.4997$
$= -11.8000 + 15.4997$
$= 3.6997$
$\approx 3{:}42.0 \qquad (0.6997 \times 60 \approx 42.0)$

c) Substitute 1985 for x.

$y = -0.0059(1985) + 15.4997$
$= -11.7115 + 15.4997$
$= 3.7882$
$= 3{:}47.3 \qquad (0.7882 \times 60 \approx 47.3)$

According to the regression line, the record should have been 3:47.3, so Steve Cram beat the regression line prediction.

9. $y = B e^{kx}$

$\ln y = \ln B e^{kx}$

$\ln y = \ln B + \ln e^{kx}$

$\ln y = \ln B + kx$

$Y = mx + b$ (Linear function)

($Y = \ln y$, $m = k$, $b = \ln B$)

a) Consider the data points (1,30), (2,45), (3,40), (4,45), (5,45), and (6,45).

First we find the logarithms of the y-values.

x	1	2	3	4	5	6
Y=ℓn y	3.4012	3.8067	3.6889	3.8067	3.8067	3.8067

To find the regression line we use the abbreviated procedure.

c_i	d_i	$c_i - \bar{x}$	$(c_i - \bar{x})^2$
1	3.4012	-2.5	6.25
2	3.8067	-1.5	2.25
3	3.6889	-0.5	0.25
4	3.8067	0.5	0.25
5	3.8067	1.5	2.25
6	3.8067	2.5	6.25

$$\sum_{i=1}^{6} c_i = 21 \qquad \sum_{i=1}^{6} d_i = 22.3169$$

$$\bar{x} = \frac{21}{6} = 3.5 \qquad \bar{Y} = \frac{22.3169}{6} \approx 3.7195$$

$d_i - \overline{Y}$	$(c_i - \overline{x})(d_i - \overline{Y})$
-0.3183	0.7958
0.0872	-0.1308
-0.0306	0.0153
0.0872	0.0436
0.0872	0.1308
0.0872	0.2180

$$\sum_{i=1}^{6} (c_i - \overline{x})^2 = 17.5$$

$$\sum_{i=1}^{6} (c_i - \overline{x})(d_i - \overline{Y}) = 1.0727$$

$$m = \frac{1.0727}{17.5} \approx 0.0613$$

Thus the regression line is

$Y - 3.7195 = 0.0613(x - 3.5)$

$Y - 3.7195 = 0.0613x - 0.2146$

$Y = 0.0613x + 3.5049$

From this equation we know that

$m = k = 0.0613$ and $b = \ln B = 3.5049$.

Thus, $B = e^{3.5049} \approx 33.2781$.

Then the desired equation is

$y = 33.2781\, e^{0.0613x}$.

(Answers may vary slightly due to rounding differences.)

b) In 1994, $x = 15$.

$y = 33.2781\, e^{0.0613(15)}$

$y = 33.2781\, e^{0.9195}$

$= 33.2781(2.5080)$

$\approx \$83.46$

In 2000, $x = 21$.

$y = 33.2781\, e^{0.0613(21)}$

$= 33.2781\, e^{1.2873}$

$= 33.2781(3.6230)$

$\approx \$120.57$

c) This model predicts higher prices than the model in Exercise 1.

Exercise Set 7.6

1. Find the maximum value of

$f(x,y) = xy$

subject to the constraint

$2x + y = 8$.

We form the new function F given by

$F(x,y,\lambda) = xy - \lambda(2x + y - 8)$.

(Expressing $2x + y = 8$ as $2x + y - 8 = 0$)

We find the first partial derivatives.

$F(x,y,\lambda) = \underline{x}y - \lambda(2\underline{x} + y - 8)$

$F_x = y - 2\lambda$

$F(x,y,\lambda) = x\underline{y} - \lambda(2x + \underline{y} - 8)$

$F_y = x - \lambda$

$F(x,y,\lambda) = xy - \underline{\lambda}(2x + y - 8)$

$F_\lambda = -(2x + y - 8)$

We set these derivatives equal to 0 and solve the resulting system.

$y - 2\lambda = 0$ (1)

$x - \lambda = 0$ (2)

$2x + y - 8 = 0$ (3) [$-(2x + y - 8) = 0$ or $2x + y - 8 = 0$]

Solving Eq. (1) for y, we get

$y - 2\lambda = 0$

$y = 2\lambda$.

Solving Eq. (2) for x, we get

$x - \lambda = 0$

$x = \lambda$.

Substituting 2λ for y and λ for x in Eq. (3), we get

$2x + y - 8 = 0$

$2\lambda + 2\lambda - 8 = 0$

$4\lambda = 8$

$\lambda = 2$.

Then

$x = \lambda = 2$

and

$y = 2\lambda = 2 \cdot 2 = 4$.

The maximum of f subject to the constraint occurs at (2,4) and is

$f(2,4) = 2 \cdot 4 = 8$. [$f(x,y) = xy$]

3. Find the maximum value of

$f(x,y) = 4 - x^2 - y^2$

subject to the constraint

$x + 2y = 10$.

We form the new function F given by

$F(x,y,\lambda) = 4 - x^2 - y^2 - \lambda(x + 2y - 10)$.

(Expressing $x + 2y = 10$ as $x + 2y - 10 = 0$)

We find the first partial derivatives.

$F(x,y,\lambda) = 4 - \underline{x}^2 - y^2 - \lambda(\underline{x} + 2y - 10)$

$F_x = -2x - \lambda$

$F(x,y,\lambda) = 4 - x^2 - \underline{y}^2 - \lambda(x + 2\underline{y} - 10)$

$F_y = -2y - 2\lambda$

$F(x,y,\lambda) = 4 - x^2 - y^2 - \underline{\lambda}(x + 2y - 10)$

$F_\lambda = -(x + 2y - 10)$

We set these derivatives equal to 0 and solve the resulting system.

$2x + \lambda = 0$ (1) $(-2x - \lambda = 0$ or $2x + \lambda = 0)$

$y + \lambda = 0$ (2) $(-2y - 2\lambda = 0$ or $y + \lambda = 0)$

$x + 2y - 10 = 0$ (3) $[-(x + 2y - 10) = 0$ or $x + 2y - 10 = 0]$

Solving Eq. (1) for x, we get

$$2x + \lambda = 0$$
$$2x = -\lambda$$
$$x = -\frac{1}{2}\lambda$$

Solving Eq. (2) for y, we get

$$y + \lambda = 0$$
$$y = -\lambda.$$

Substituting $-\frac{1}{2}\lambda$ for x and $-\lambda$ for y in Eq. (3), we get,

$$x + 2y - 10 = 0$$
$$-\frac{1}{2}\lambda + 2(-\lambda) - 10 = 0$$
$$-\frac{5}{2}\lambda - 10 = 0$$
$$-\frac{5}{2}\lambda = 10$$
$$\lambda = -4.$$

Then

$$x = -\frac{1}{2}\lambda = -\frac{1}{2}(-4) = 2$$

and

$$y = -\lambda = -(-4) = 4.$$

The maximum of f subject to the constraint occurs at (2,4) and is

$f(2,4) = 4 - 2^2 - 4^2$ $[f(x,y) = 4 - x^2 - y^2]$

$= 4 - 4 - 16$

$= -16.$

5. Find the minimum value of

$$f(x,y) = x^2 + y^2$$

subject to the constraint

$$2x + y = 10.$$

We form the new function F given by

$$F(x,y,\lambda) = x^2 + y^2 - \lambda(2x + y - 10).$$

(Expressing $2x + y = 10$ as $2x + y - 10 = 0$)

We find the first partial derivatives.

$F(x,y,\lambda) = \underline{x}^2 + y^2 - \lambda(2\underline{x} + y - 10)$

$F_x = 2x - 2\lambda$

$F(x,y,\lambda) = x^2 + \underline{y}^2 - \lambda(2x + \underline{y} - 10)$

$F_y = 2y - \lambda$

$F(x,y,\lambda) = x^2 + y^2 - \underline{\lambda}(2x + y - 10)$

$F_\lambda = -(2x + y - 10)$

We set these derivatives equal to 0 and solve the resulting system.

$x - \lambda = 0$ (1) $(2x - 2\lambda = 0$ or $x - \lambda = 0)$

$2y - \lambda = 0$ (2)

$2x + y - 10 = 0$ (3) $[-(2x + y - 10) = 0$ or $2x + y - 10 = 0]$

Solving Eq. (1) for x, we get

$$x - \lambda = 0$$
$$x = \lambda.$$

Solving Eq. (2) for y, we get

$$2y - \lambda = 0$$
$$2y = \lambda$$
$$y = \frac{1}{2}\lambda.$$

Substituting λ for x and $\frac{1}{2}\lambda$ for y in Eq. (3), we get

$$2x + y - 10 = 0$$

$2\lambda + \frac{1}{2}\lambda - 10 = 0$ (Substituting)

$$\frac{5}{2}\lambda = 10$$
$$\lambda = 4.$$

Then

$$x = \lambda = 4$$

and

$$y = \frac{1}{2}\lambda = \frac{1}{2}\cdot 4 = 2.$$

The minimum of f subject to the constraint occurs at (4,2) and is

$f(4,2) = 4^2 + 2^2$ $[f(x,y) = x^2 + y^2]$

$= 16 + 4$

$= 20.$

7. Find the minimum value of

$$f(x,y) = 2y^2 - 6x^2$$

subject to the constraint

$$2x + y = 4.$$

We form the new function F given by

$$F(x,y,\lambda) = 2y^2 - 6x^2 - \lambda(2x + y - 4).$$

(Expressing $2x + y = 4$ as $2x + y - 4 = 0$)

We find the first partial derivatives.

$F(x,y,\lambda) = 2y^2 - 6\underline{x}^2 - \lambda(2\underline{x} + y - 4)$

$F_x = -12x - 2\lambda$

$F(x,y,\lambda) = 2\underline{y}^2 - 6x^2 - \lambda(2x + \underline{y} - 4)$

$F_y = 4y - \lambda$

$F(x,y,\lambda) = 2y^2 - 6x^2 - \underline{\lambda}(2x + y - 4)$

$F_\lambda = -(2x + y - 4)$

We set these derivatives equal to 0 and solve the resulting system.

$12x + 2\lambda = 0$ (1) $(-12x - 2\lambda = 0$ or $12x + 2\lambda = 0)$

$4y - \lambda = 0$ (2)

$2x + y - 4 = 0$ (3) $[-(2x + y - 4) = 0$ or $2x + y - 4 = 0]$

Solving Eq. (1) for x, we get

$$12x + 2\lambda = 0$$
$$12x = -2\lambda$$
$$x = -\frac{\lambda}{6}.$$

Solving Eq. (2) for y, we get

$$4y - \lambda = 0$$
$$4y = \lambda$$
$$y = \frac{\lambda}{4}.$$

Substituting $-\frac{\lambda}{6}$ for x and $\frac{\lambda}{4}$ for y in Eq. (3), we get

$$2x + y - 4 = 0$$
$$2\left(-\frac{\lambda}{6}\right) + \frac{\lambda}{4} - 4 = 0$$
$$-\frac{\lambda}{3} + \frac{\lambda}{4} = 4$$
$$-4\lambda + 3\lambda = 48$$
$$-\lambda = 48$$
$$\lambda = -48.$$

Then

$$x = -\frac{\lambda}{6} = -\frac{(-48)}{6} = 8$$

and

$$y = \frac{\lambda}{4} = \frac{-48}{4} = -12.$$

The minimum of f subject to the constraint occurs at (8,-12) and is

$$f(8,-12) = 2(-12)^2 - 6\cdot 8^2 \qquad [f(x,y) = 2y^2 - 6x^2]$$
$$= 2\cdot 144 - 6\cdot 64$$
$$= 288 - 384$$
$$= -96.$$

9. Find the minimum value of

$$f(x,y,z) = x^2 + y^2 + z^2$$

subject to the constraint

$$y + 2x - z = 3.$$

We form the new function F given by

$$F(x,y,z,\lambda) = x^2 + y^2 + z^2 - \lambda(y + 2x - z - 3).$$

(Expressing $y + 2x - z = 3$ as $y + 2x - z - 3 = 0$)

We find the first partial derivatives.

$$F(x,y,z,\lambda) = \underline{x}^2 + y^2 + z^2 - \lambda(y + 2\underline{x} - z - 3)$$
$$F_x = 2x - 2\lambda$$

$$F(x,y,z,\lambda) = x^2 + \underline{y}^2 + z^2 - \lambda(\underline{y} + 2x - z - 3)$$
$$F_y = 2y - \lambda$$

$$F(x,y,z,\lambda) = x^2 + y^2 + \underline{z}^2 - \lambda(y + 2x - \underline{z} - 3)$$
$$F_z = 2z + \lambda$$

$$F(x,y,z,\lambda) = x^2 + y^2 + z^2 - \lambda(y + 2x - z - 3)$$
$$F_\lambda = -(y + 2x - z - 3)$$

We set these derivatives equal to 0 and solve the resulting system.

$2x - 2\lambda = 0$ (1)

$2y - \lambda = 0$ (2)

$2z + \lambda = 0$ (3)

$y + 2x - z - 3 = 0$ (4) $[-(y + 2x - z - 3) = 0$ or $y + 2x - z - 3 = 0]$

Solving Eq. (1) for x, we get

$$2x - 2\lambda = 0$$
$$2x = 2\lambda$$
$$x = \lambda$$

Solving Eq. (2) for y, we get

$$2y - \lambda = 0$$
$$2y = \lambda$$
$$y = \frac{1}{2}\lambda.$$

Solving Eq. (3) for z, we get

$$2z + \lambda = 0$$
$$2z = -\lambda$$
$$z = -\frac{1}{2}\lambda.$$

Substituting λ for x, $\frac{1}{2}\lambda$ for y, and $-\frac{1}{2}\lambda$ for z in Eq. (4), we get

$$y + 2x - z - 3 = 0$$
$$\frac{1}{2}\lambda + 2\lambda - \left(-\frac{1}{2}\lambda\right) - 3 = 0$$
$$3\lambda - 3 = 0$$
$$3\lambda = 3$$
$$\lambda = 1.$$

Then

$$x = \lambda = 1$$
$$y = \frac{1}{2}\lambda = \frac{1}{2}\cdot 1 = \frac{1}{2}$$

and

$$z = -\frac{1}{2}\lambda = -\frac{1}{2}\cdot 1 = -\frac{1}{2}.$$

The minimum of f subject to the constraint occurs at $\left(1, \frac{1}{2}, -\frac{1}{2}\right)$ and is

$$f\left(1, \frac{1}{2}, -\frac{1}{2}\right) = 1^2 + \left(\frac{1}{2}\right)^2 + \left(-\frac{1}{2}\right)^2 \qquad [f(x,y,z) = x^2 + y^2 + z^2]$$
$$= 1 + \frac{1}{4} + \frac{1}{4}$$
$$= 1\frac{1}{2}, \text{ or } \frac{3}{2}.$$

11. Find the maximum value of

$f(x,y) = xy$ (Product is xy)

subject to the constraint

$x + y = 70.$ (Sum is 70)

We form the new function F given by

$F(x,y,\lambda) = xy - \lambda(x + y - 70).$

(Expressing $x + y = 70$ as $x + y - 70 = 0$)

We find the first partial derivatives.

$F_x = y - \lambda$

$F_y = x - \lambda$

$F_\lambda = -(x + y - 70)$

We set these derivatives equal to 0 and solve the resulting system.

$y - \lambda = 0$ (1)

$x - \lambda = 0$ (2)

$x + y - 70 = 0$ (3) $[-(x + y - 70) = 0$ or $x + y - 70 = 0]$

Solving Eq. (1) for y, we get

$y - \lambda = 0$

$y = \lambda.$

Solving Eq. (2) for x, we get

$x - \lambda = 0$

$x = \lambda.$

Substituting λ for x and for y in Eq. (3), we get

$x + y - 70 = 0$

$\lambda + \lambda - 70 = 0$

$2\lambda = 70$

$\lambda = 35$

Then $x = y = \lambda = 35.$

The maximum of f subject to the constraint occurs at (35,35). Thus, the two numbers whose sum is 70 that have the maximum product are 35 and 35.

13. Find the minimum value of

$f(x,y) = xy$ (Product is xy)

subject to the constraint

$x - y = 6.$ (Difference is 6)

We form the new function F given by

$F(x,y,\lambda) = xy - \lambda(x - y - 6).$

(Expressing $x - y = 6$ as $x - y - 6 = 0$)

We find the first partial derivatives.

$F_x = y - \lambda$

$F_y = x + \lambda$

$F_\lambda = -(x - y - 6)$

We set these derivatives equal to 0 and solve the resulting system.

$y - \lambda = 0$ (1)

$x + \lambda = 0$ (2)

$x - y - 6 = 0$ (3) $[-(x - y - 6) = 0$ or $x - y - 6 = 0]$

Solving Eq. (1) for y, we get

$y - \lambda = 0$

$y = \lambda.$

Solving Eq. (2) for x, we get

$x + \lambda = 0$

$x = -\lambda.$

Substituting λ for y and -λ for x in Eq. (3), we get

$x - y - 6 = 0$

$-\lambda - \lambda - 6 = 0$

$-2\lambda = 6$

$\lambda = -3.$

Then

$x = -\lambda = -(-3) = 3$

and

$y = \lambda = -3.$

The minimum of f subject to the constraint occurs at (3,-3). Thus, the two numbers whose difference is 6 that have the minimum product are 3 and -3.

15.

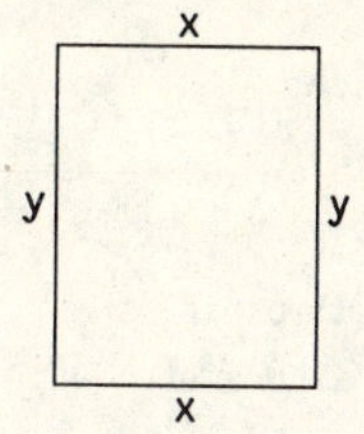

$A = xy$ (Area)

$P = 2x + 2y = 39$ (Perimeter)

We need to maximize the area A

$A = xy$

subject to the constraint

$2x + 2y = 39.$

We form the new function F given by

$F(x,y,\lambda) = xy - \lambda(2x + 2y - 39).$

(Expressing $2x + 2y = 39$ as $2x + 2y - 39 = 0$)

We find the first partial derivatives.

$F_x = y - 2\lambda$

$F_y = x - 2\lambda$

$F_\lambda = -(2x + 2y - 39)$

We set these derivatives equal to 0 and solve the resulting system.

$$y - 2\lambda = 0 \quad (1)$$
$$x - 2\lambda = 0 \quad (2)$$
$$2x + 2y - 39 = 0 \quad (3) \quad [-(2x + 2y - 39) = 0 \text{ or } 2x + 2y - 39 = 0]$$

Solving Eq. (1) for y, we get

$y - 2\lambda = 0$

$y = 2\lambda.$

Solving Eq. (2) for x, we get

$x - 2\lambda = 0$

$x = 2\lambda.$

Substituting 2λ for x and for y in Eq. (3), we get

$$2x + 2y - 39 = 0$$
$$2(2\lambda) + 2(2\lambda) - 39 = 0$$
$$8\lambda = 39$$
$$\lambda = \frac{39}{8}$$

Then

$$x = y = 2\lambda = 2 \cdot \frac{39}{8} = \frac{39}{4} \text{ or } 9\tfrac{3}{4}.$$

The maximum value of A subject to the constraint occurs at $\left(\frac{39}{4}, \frac{39}{4}\right)$. Thus, the dimensions of the paper that will give the most typing area, subject to the perimeter constraint of 39 in. are $9\frac{3}{4}$ in. $\times$ $9\frac{3}{4}$ in.

The maximum area is

$$A = \frac{39}{4} \cdot \frac{39}{4} = \frac{1521}{16} = 95\tfrac{1}{16} \text{ in}^2.$$

The area of the standard $8\frac{1}{2} \times 11$ paper is $\frac{17}{2} \cdot \frac{11}{1} = \frac{187}{2}$, or $93\frac{1}{2}$ in^2. The perimeter of an $8\frac{1}{2} \times 11$ sheet of paper is 39 in., but its area is less than the area of a sheet whose perimeter is also 39 in. but whose dimensions are $9\frac{3}{4} \times 9\frac{3}{4}$.

17. We want to minimize the function s given by

$s(h,r) = 2\pi rh + 2\pi r^2$

subject to the volume constraint

$\pi r^2 h = 27$, or $\pi r^2 h - 27 = 0$.

We form the new function S given by

$S(h,r,\lambda) = 2\pi rh + 2\pi r^2 - \lambda(\pi r^2 h - 27)$.

We find the first partial derivatives.

$$\frac{\partial S}{\partial h} = 2\pi r - \lambda\pi r^2,$$

$$\frac{\partial S}{\partial r} = 2\pi h + 4\pi r - 2\lambda\pi rh,$$

$$\frac{\partial S}{\partial \lambda} = -(\pi r^2 h - 27).$$

We set these derivatives equal to 0 and solve the resulting system.

$$2\pi r - \lambda\pi r^2 = 0 \quad (1)$$
$$2\pi h + 4\pi r - 2\lambda\pi rh = 0 \quad (2)$$
$$\pi r^2 h - 27 = 0 \quad (3) \quad [-(\pi r^2 h - 27) = 0 \text{ or } \pi r^2 h - 27 = 0]$$

Note that we can solve Eq. (1) for r:

$\pi r(2 - \lambda r) = 0$

$\pi r = 0$ or $2 - \lambda r = 0$

$r = 0$ or $\quad r = \frac{2}{\lambda}$

Note r = 0 cannot be a solution to the original problem, so we continue by substituting $2/\lambda$ for r in Eq. (2).

$$2\pi h + 4\pi \cdot \frac{2}{\lambda} - 2\lambda\pi \cdot \frac{2}{\lambda} \cdot h = 0$$
$$2\pi h + \frac{8\pi}{\lambda} - 4\pi h = 0$$
$$\frac{8\pi}{\lambda} - 2\pi h = 0$$
$$-2\pi h = -\frac{8\pi}{\lambda}$$
$$h = \frac{4}{\lambda}$$

Since $h = 4/\lambda$ and $r = 2/\lambda$, it follows that $h = 2r$. Substituting 2r for h in Eq. (3) yields

$$\pi r^2 \cdot 2r - 27 = 0$$
$$2\pi r^3 - 27 = 0$$
$$2\pi r^3 = 27$$
$$\pi r^3 = \frac{27}{2}$$
$$r^3 = \frac{27}{2\pi}$$
$$r = \sqrt[3]{\frac{27}{2\pi}} \approx 1.6 \text{ ft.}$$

So when r = 1.6 ft, h = 3.2 ft, the surface area is a minimum and is about $2\pi(1.6)(3.2) + 2\pi(1.6)^2$, or 48.3 ft^2.

19. We need to find the maximum value of

$S(L,M) = ML - L^2$

subject to the constraint

$M + L = 80$.

We form the new function F given by

$F(L,M,\lambda) = ML - L^2 - \lambda(M + L - 80)$.

(Expressing $M + L = 80$ as $M + L - 80 = 0$)

We find the first partial derivatives.

$F_L = M - 2L - \lambda$

$F_M = L - \lambda$

$F_\lambda = -(M + L - 80)$

We set these derivatives equal to 0 and solve the resulting system.

$M - 2L - \lambda = 0$ (1)

$L - \lambda = 0$ (2)

$M + L - 80 = 0$ (3) $[-(M + L - 80) = 0$ or $M + L - 80 = 0]$

Solving Eq. (2) for L we get

$L - \lambda = 0$

$L = \lambda.$

Substituting λ for L in Eq. (1) and solving for M, we get

$M - 2L - \lambda = 0$

$M - 2\lambda - \lambda = 0$

$M = 3\lambda$

Substituting λ for L and 3λ for M in Eq. (3), we get

$M + L - 80 = 0$

$3\lambda + \lambda - 80 = 0$

$4\lambda = 80$

$\lambda = 20$

Then

$L = \lambda = 20$

and

$M = 3\lambda = 3\cdot 20 = 60.$

The maximum value of S subject to the constraint occurs at (20,60) and is

$S(20,60) = 60\cdot 20 - 20^2$ $[S(L,M) = ML - L^2]$

$= 1200 - 400$

$= 800.$

21. a) The area of the floor is xy.

The cost of the floor is 4xy.

The area of the walls is 2xz + 2yz.

The cost of the walls is 3(2xz + 2yz).

The area of the ceiling is xy.

The cost of the ceiling is 3xy.

$C(x,y,z) = 4xy + 3(2xz + 2yz) + 3xy$

$= 7xy + 6xz + 6yz$

b) We need to find the minimum value of

$C(x,y,z) = 7xy + 6xz + 6yz$

subject to the constraint

$xyz = 252{,}000.$ (Volume = $\ell\cdot w\cdot h$)

We form the new function N given by

$N(x,y,z,\lambda) = 7xy + 6xz + 6yz - \lambda(xyz-252{,}000)$

(Expressing $xyz = 252{,}000$ as $xyz - 252{,}000 = 0$)

We find the first partial derivatives.

$N_x = 7y + 6z - \lambda yz$

$N_y = 7x + 6z - \lambda xz$

$N_z = 6x + 6y - \lambda xy$

$N_\lambda = -(xyz - 252{,}000)$

We set these derivatives equal to 0 and solve the resulting system.

$7y + 6z - \lambda yz = 0$ (1)

$7x + 6z - \lambda xz = 0$ (2)

$6x + 6y - \lambda xy = 0$ (3)

$xyz - 252{,}000 = 0$ (4)

Solving Eq. (2) for x and Eq. (1) for y, we get

$x = \frac{6z}{\lambda z - 7}$ and $y = \frac{6z}{\lambda z - 7}.$

Thus, x = y.

Substituting x for y we get the following system.

$7x + 6z - \lambda xz = 0$ or $7x + 6z - \lambda xz = 0$ (2)

$6x + 6x - \lambda xx = 0$ $12x - \lambda x^2 = 0$ (3)

$xxz - 252{,}000 = 0$ $x^2z - 252{,}000 = 0$ (4)

Solving Eq. (3) for x, we get

$12x - \lambda x^2 = 0$

$x(12 - \lambda x) = 0$

$x = 0$ or $12 - \lambda x = 0$

$x = 0$ or $x = \frac{12}{\lambda}$

We only consider $x = \frac{12}{\lambda}$ since x cannot be 0 in the original problem. We continue by substituting $\frac{12}{\lambda}$ for x in Eq. (4) and solve for z.

$\left(\frac{12}{\lambda}\right)^2 z - 252{,}000 = 0$

$\frac{144}{\lambda^2} z = 252{,}000$

$z = \frac{252{,}000}{144}\lambda^2$

$z = 1750\lambda^2$

Next we substitute $\frac{12}{\lambda}$ for x and $1750\lambda^2$ for z in Eq. (2) and solve for λ.

$7\cdot\frac{12}{\lambda} + 6\cdot 1750\lambda^2 - \lambda\cdot\frac{12}{\lambda}\cdot 1750\lambda^2 = 0$

$\frac{84}{\lambda} + 10{,}500\lambda^2 - 21{,}000\lambda^2 = 0$

$\frac{84}{\lambda} = 10{,}500\lambda^2$

$84 = 10{,}500\lambda^3$

$\frac{84}{10{,}500} = \lambda^3$

$\frac{1}{125} = \lambda^3$

$\frac{1}{5} = \lambda$

Thus,

$$x = \frac{12}{\lambda} = \frac{12}{\frac{1}{5}} = 12 \cdot \frac{5}{1} = 60$$

$$y = \frac{12}{\lambda} = \frac{12}{\frac{1}{5}} = 60$$

$$z = 1750\lambda^2 = 1750\left(\frac{1}{5}\right)^2 = \frac{1750}{25} = 70.$$

The minimum total cost is obtained when the dimensions are 60 ft by 60 ft by 70 ft. The minimum cost is found as follows:

$$C(x,y,z) = 7xy + 6xz + 6yz$$

$$C(60,60,70) = 7\cdot 60\cdot 60 + 6\cdot 60\cdot 70 + 6\cdot 60\cdot 70$$

$$= 25{,}200 + 25{,}200 + 25{,}200$$

$$= \$75{,}600$$

23. We need to find the minimum value of

$$C(x,y) = 10 + \frac{x^2}{6} + 200 + \frac{y^3}{9} \quad [C(x,y) = C(x)+C(y)]$$

$$= \frac{x^2}{6} + \frac{y^3}{9} + 210$$

subject to the constraint

$x + y = 10{,}100.$

We form the new function N given by

$$N(x,y,\lambda) = \frac{x^2}{6} + \frac{y^3}{9} + 210 - \lambda(x + y - 10{,}100)$$

(Expressing $x + y = 10{,}100$ as $x + y - 10{,}100 = 0$)

We find the first partial derivatives.

$$N_x = \frac{1}{3}x - \lambda$$

$$N_y = \frac{1}{3}y^2 - \lambda$$

$$N_\lambda = -(x + y - 10{,}100)$$

We set these derivatives equal to 0 and solve the resulting system.

$$\frac{1}{3}x - \lambda = 0 \quad (1)$$

$$\frac{1}{3}y^2 - \lambda = 0 \quad (2)$$

$$x + y - 10{,}100 = 0 \quad (3) \quad [-(x + y - 10{,}100) = 0 \text{ or } x + y - 10{,}100 = 0]$$

Solve Eq. (1) for x.

$$\frac{1}{3}x - \lambda = 0$$

$$\frac{1}{3}x = \lambda$$

$$x = 3\lambda$$

Solve Eq. (2) for y^2.

$$\frac{1}{3}y^2 - \lambda = 0$$

$$\frac{1}{3}y^2 = \lambda$$

$$y^2 = 3\lambda$$

Thus, $x = y^2$.

Substitute y^2 for x in Eq. (3) and solve for y.

$$y^2 + y - 10{,}100 = 0$$

$$(y + 101)(y - 100) = 0$$

$$y + 101 = 0 \quad \text{or} \quad y - 100 = 0$$

$$y = -101 \quad \text{or} \quad y = 100$$

Since y cannot be -101 in the original problem, we only consider y = 100. If y = 100, x = 10,100 - 100, or 10,000. To minimize total costs, 10,000 units on A and 100 units on B should be made.

25. Find the minimum value of

$$f(x,y) = 2x^2 + y^2 + 2xy + 3x + 2y$$

subject to the constraint

$y^2 = x + 1.$

We form the new function F given by

$$F(x,y,\lambda) = 2x^2 + y^2 + 2xy + 3x + 2y - \lambda(y^2 - x - 1)$$

(Expressing $y^2 = x + 1$ as $y^2 - x - 1 = 0$)

We find the first partial derivatives.

$$F_x = 4x + 2y + 3 + \lambda$$

$$F_y = 2y + 2x + 2 - 2\lambda y$$

$$F_\lambda = -(y^2 - x - 1)$$

We set these derivatives equal to 0 and solve the resulting system.

$$4x + 2y + 3 + \lambda = 0 \text{ or } 4x + 2y + 3 + \lambda = 0 \quad (1)$$

$$2y + 2x + 2 - 2\lambda y = 0 \qquad y + x + 1 - \lambda y = 0 \quad (2)$$

$$-(y^2 - x - 1) = 0 \qquad y^2 - x - 1 = 0 \quad (3)$$

Multiply Eq. (1) by y and add the result to Eq. (2).

$$4xy + 2y^2 + 3y + \lambda y = 0 \quad (1)$$

$$\underline{y + x + 1 - \lambda y = 0} \quad (2)$$

$$4y + x + 4xy + 2y^2 + 1 = 0 \quad \text{(Adding)} \quad (4)$$

Solve Eq. (3) for x.

$$y^2 - x - 1 = 0$$

$$y^2 - 1 = x$$

Substitute $y^2 - 1$ for x in Eq. (4) and solve for y.

$$4y + (y^2 - 1) + 4(y^2 - 1)y + 2y^2 + 1 = 0$$

$$4y + y^2 - 1 + 4y^3 - 4y + 2y^2 + 1 = 0$$

$$4y^3 + 3y^2 = 0$$

$$y^2(4y + 3) = 0$$

$$y^2 = 0 \text{ or } 4y + 3 = 0$$

$$y = 0 \text{ or } \quad y = -\frac{3}{4}$$

For y = 0, $x = 0^2 - 1$, or -1.

For $y = -\frac{3}{4}$, $x = \left(-\frac{3}{4}\right)^2 - 1 = \frac{9}{16} - \frac{16}{16} = -\frac{7}{16}$.

When $x = -1$ and $y = 0$,

$$f(-1,0) = 2(-1)^2 + (0)^2 + 2(-1)(0) + 3(-1) + 2(0)$$
$$= 2 + 0 + 0 - 3 + 0$$
$$= -1$$

When $x = -\frac{7}{16}$ and $y = -\frac{3}{4}$,

$$f\left(-\frac{7}{16}, -\frac{3}{4}\right) = 2\left(-\frac{7}{16}\right)^2 + \left(-\frac{3}{4}\right)^2 + 2\left(-\frac{7}{16}\right)\left(-\frac{3}{4}\right) + 3\left(-\frac{7}{16}\right) + 2\left(-\frac{3}{4}\right)$$
$$= \frac{98}{256} + \frac{9}{16} + \frac{42}{64} - \frac{21}{16} - \frac{6}{4}$$
$$= \frac{49}{128} - \frac{12}{16} + \frac{21}{32} - \frac{3}{2}$$
$$= \frac{49}{128} - \frac{24}{32} + \frac{21}{32} - \frac{48}{32}$$
$$= \frac{49}{128} - \frac{51}{32}$$
$$= \frac{49}{128} - \frac{204}{128}$$
$$= -\frac{155}{128} \qquad \text{(Minimum)}$$

Thus, the minimum value of f is $-\frac{155}{128}$ at $\left(-\frac{7}{16}, -\frac{3}{4}\right)$.

27. Find the maximum value of
$f(x,y,z) = x^2y^2z^2$
subject to the constraint
$x^2 + y^2 + z^2 = 1$.

We form the new function F given by

$$F(x,y,z,\lambda) = x^2y^2z^2 - \lambda(x^2 + y^2 + z^2 - 1)$$

(Expressing $x^2 + y^2 + z^2 = 1$ as $x^2 + y^2 + z^2 - 1 = 0$)

We find the first partial derivatives.

$$F_x = 2xy^2z^2 - 2\lambda x$$
$$F_y = 2x^2yz^2 - 2\lambda y$$
$$F_z = 2x^2y^2z - 2\lambda z$$
$$F_\lambda = -(x^2 + y^2 + z^2 - 1)$$

We set these derivatives equal to 0 and solve the resulting system.

$$2xy^2z^2 - 2\lambda x = 0 \text{ or } x(y^2z^2 - \lambda)=0 \quad (1)$$
$$2x^2yz^2 - 2\lambda y = 0 \text{ or } y(x^2z^2 - \lambda)=0 \quad (2)$$
$$2x^2y^2z - 2\lambda z = 0 \text{ or } z(x^2y^2 - \lambda)=0 \quad (3)$$
$$-(x^2 + y^2 + z^2 - 1) = 0 \text{ or } x^2 + y^2 + z^2 - 1=0 \quad (4)$$

Note that for $x = 0$, $y = 0$, or $z = 0$, $f(x,y,z) = 0$. For all values of x, y, and $z \neq 0$, $f(x,y,z) > 0$. Thus the maximum value of f cannot occur when any or all of the variables is 0. Thus, we will only consider nonzero values of x, y, and z.

Using the Principle of Zero Products, we get:

From Eq. (1)	From Eq. (2)	From Eq. (3)
$y^2z^2 - \lambda = 0$	$x^2y^2 - \lambda = 0$	$x^2y^2 - \lambda = 0$
$y^2z^2 = \lambda$	$x^2z^2 = \lambda$	$x^2y^2 = \lambda$

Thus, $y^2z^2 = x^2z^2 = x^2y^2$ and $x^2 = y^2 = z^2$.

Substitute x^2 for y^2 and z^2 in Eq. (4) and solve for x.

$$x^2 + y^2 + z^2 - 1 = 0$$
$$x^2 + x^2 + x^2 - 1 = 0$$
$$3x^2 = 1$$
$$x^2 = \frac{1}{3}$$
$$x = \pm\sqrt{\frac{1}{3}}, \text{ or } \pm\frac{1}{\sqrt{3}}$$

Now $x^2 = y^2 = z^2$ so we can find y and z.

$$y^2 = \frac{1}{3} \qquad z^2 = \frac{1}{3}$$
$$y = \pm\frac{1}{\sqrt{3}} \qquad z = \pm\frac{1}{\sqrt{3}}$$

For $\left(\pm\frac{1}{\sqrt{3}}, \pm\frac{1}{\sqrt{3}}, \pm\frac{1}{\sqrt{3}}\right)$:

$$f(x,y,z) = \left(\pm\frac{1}{\sqrt{3}}\right)^2 \cdot \left(\pm\frac{1}{\sqrt{3}}\right)^2 \cdot \left(\pm\frac{1}{\sqrt{3}}\right)^2$$
$$= \frac{1}{3} \cdot \frac{1}{3} \cdot \frac{1}{3} = \frac{1}{27}.$$

Thus f has a maximum value of $\frac{1}{27}$ at $\left(\pm\frac{1}{\sqrt{3}}, \pm\frac{1}{\sqrt{3}}, \pm\frac{1}{\sqrt{3}}\right)$.

29. Find the maximum value of
$f(x,y,z,t) = x + y + z + t$
subject to the constraint
$x^2 + y^2 + z^2 + t^2 = 1$.

We form the new function F given by

$$F(x,y,z,t,\lambda) = x + y + z + t - \lambda(x^2+y^2+z^2+t^2-1)$$

(Expressing $x^2 + y^2 + z^2 + t^2 = 1$ as $x^2 + y^2 + z^2 + t^2 - 1 = 0$)

We find the first partial derivatives.

$$F_x = 1 - 2\lambda x$$
$$F_y = 1 - 2\lambda y$$
$$F_z = 1 - 2\lambda z$$
$$F_t = 1 - 2\lambda t$$
$$F_\lambda = -(x^2 + y^2 + z^2 + t^2 - 1)$$

We set these derivatives equal to 0 and solve the resulting system.

$$1 - 2\lambda x = 0 \quad (1)$$
$$1 - 2\lambda y = 0 \quad (2)$$
$$1 - 2\lambda z = 0 \quad (3)$$
$$1 - 2\lambda t = 0 \quad (4)$$
$$x^2 + y^2 + z^2 + t^2 - 1 = 0 \quad (5)$$

$[-(x^2 + y^2 + z^2 + t^2 - 1) = 0$ or $x^2 + y^2 + z^2 + t^2 - 1 = 0]$

Solving the first four equations for x, y, z, and t respectively, we get

$x = \frac{1}{2\lambda}$

$y = \frac{1}{2\lambda}$

$z = \frac{1}{2\lambda}$

$t = \frac{1}{2\lambda}$

Thus, $x = y = z = t$.

Next we substitute $\frac{1}{2\lambda}$ for x, y, z, and t in Eq. (5) and solve for λ.

$$x^2 + y^2 + z^2 + t^2 - 1 = 0 \qquad (5)$$

$$\left(\frac{1}{2\lambda}\right)^2 + \left(\frac{1}{2\lambda}\right)^2 + \left(\frac{1}{2\lambda}\right)^2 + \left(\frac{1}{2\lambda}\right)^2 - 1 = 0$$

(Substituting)

$$4\left(\frac{1}{4\lambda^2}\right) - 1 = 0$$

$$\frac{1}{\lambda^2} = 1$$

$$1 = \lambda^2$$

$$\pm 1 = \lambda$$

When $\lambda = 1$, $x = y = z = t = \frac{1}{2}$ and

$f\left(\frac{1}{2}, \frac{1}{2}, \frac{1}{2}, \frac{1}{2}\right) = \frac{1}{2} + \frac{1}{2} + \frac{1}{2} + \frac{1}{2} = 2.$

When $\lambda = -1$, $x = y = z = t = -\frac{1}{2}$ and

$f\left(-\frac{1}{2}, -\frac{1}{2}, -\frac{1}{2}, -\frac{1}{2}\right) = -\frac{1}{2} + \left(-\frac{1}{2}\right) + \left(-\frac{1}{2}\right) + \left(-\frac{1}{2}\right) = -2.$

Thus f has a maximum value of 2 at $\left(\frac{1}{2}, \frac{1}{2}, \frac{1}{2}, \frac{1}{2}\right)$.

31. $P(x,y,\lambda) = p(x,y) - \lambda(c_1x + c_2y - B)$

(Expressing $c_1x + c_2y = B$ as $c_1x + c_2y - B = 0$)

$P_x = p_x - \lambda c_1$

$P_y = p_y - \lambda c_2$

We set these partial derivatives equal to 0 and solve for λ.

$p_x - \lambda c_1 = 0$ $\qquad$ $p_y - \lambda c_2 = 0$

$p_x = \lambda c_1$ $\qquad$ $p_y = \lambda c_2$

$\frac{p_x}{c_1} = \lambda$ $\qquad$ $\frac{p_y}{c_2} = \lambda$

Thus, $\lambda = \frac{p_x}{c_1} = \frac{p_y}{c_2}$.

Exercise Set 7.7

1. $\int_0^1 \int_0^1 2y\, dx\, dy$

$= \int_0^1 \left(\int_0^1 2y\, dx\right) dy$

We first evaluate the inside integral.

$\int_0^1 2y\, dx$

$= 2y \int_0^1 dx$ $\qquad$ (2y is a constant)

$= 2y\,[x]_0^1$ $\qquad$ (Integrating with respect to x)

$= 2y(1 - 0)$

$= 2y$

Then we evaluate the outside integral.

$\int_0^1 \left[\int_0^1 2y\, dx\right] dy$

$= \int_0^1 2y\, dy$ $\qquad$ $\left[\int_0^1 2y\, dx = 2y\right]$

$= \left[y^2\right]_0^1$ $\qquad$ (Integrating with respect to y)

$= 1^2 - 0^2$

$= 1$

3. $\int_{-1}^1 \int_x^1 xy\, dy\, dx$

$= \int_{-1}^1 \left(\int_x^1 xy\, dy\right) dx$

We first evaluate the inside integral.

$\int_x^1 xy\, dy$

$= x \int_x^1 y\, dy$ $\qquad$ (x is a constant)

$= x\left[\frac{y^2}{2}\right]_x^1$ $\qquad$ (Integrating with respect to y)

$= x\left(\frac{1^2}{2} - \frac{x^2}{2}\right)$

$= \frac{x}{2} - \frac{x^3}{2}$

$= \frac{1}{2}(x - x^3)$

Then we evaluate the outside integral.

$\int_{-1}^1 \left(\int_x^1 xy\, dy\right) dx$

$= \int_{-1}^1 \frac{1}{2}(x - x^3)\, dx$ $\qquad$ $\left[\int_x^1 xy\, dy = \frac{1}{2}(x - x^3)\right]$

$= \frac{1}{2}\int_{-1}^1 (x - x^3)\, dx$

$= \frac{1}{2}\left[\frac{x^2}{2} - \frac{x^4}{4}\right]_{-1}^1$ $\qquad$ (Integrating with respect to x)

$= \frac{1}{2}\left[\left(\frac{1^2}{2} - \frac{1^4}{4}\right) - \left(\frac{(-1)^2}{2} - \frac{(-1)^4}{4}\right)\right]$

$= \frac{1}{2}\left[\left(\frac{1}{2} - \frac{1}{4}\right) - \left(\frac{1}{2} - \frac{1}{4}\right)\right]$

$= \frac{1}{2}\left[\frac{1}{2} - \frac{1}{4} - \frac{1}{2} + \frac{1}{4}\right]$

$= \frac{1}{2} \cdot 0$

$= 0$

5. $\int_0^1 \int_{-1}^3 (x + y)\,dy\,dx$

$= \int_0^1 \left[\int_{-1}^3 (x + y)\,dy\right] dx$

We first evaluate the inside integral.

$\int_{-1}^3 (x + y)\,dy$

$= \left[xy + \frac{y^2}{2}\right]_{-1}^3$ (Integrating with respect to y; x is a constant)

$= \left(3x + \frac{3^2}{2}\right) - \left(-x + \frac{(-1)^2}{2}\right)$

$= 3x + \frac{9}{2} + x - \frac{1}{2}$

$= 4x + 4$

Then we evaluate the outside integral.

$\int_0^1 \left[\int_{-1}^3 (x + y)\,dy\right] dx$

$= \int_0^1 (4x + 4)\,dx$ $\left[\int_{-1}^3 (x + y)\,dy = 4x + 4\right]$

$= 4\int_0^1 (x + 1)\,dx$

$= 4\left[\frac{x^2}{2} + x\right]_0^1$ (Integrating with respect to x)

$= 4\left[\left(\frac{1^2}{2} + 1\right) - \left(\frac{0^2}{2} + 0\right)\right]$

$= 4\left(\frac{1}{2} + 1\right)$

$= 6$

7. $\int_0^1 \int_{x^2}^x (x + y)\,dy\,dx$

$= \int_0^1 \left[\int_{x^2}^x (x + y)\,dy\right] dx$

We first evaluate the inside integral.

$\int_{x^2}^x (x + y)\,dy$

$= \left[xy + \frac{y^2}{2}\right]_{x^2}^x$ (Integrating with respect to y; x is a constant)

$= \left[\left(x \cdot x + \frac{x^2}{2}\right) - \left(x \cdot x^2 + \frac{(x^2)^2}{2}\right)\right]$

$= \left[x^2 + \frac{x^2}{2} - x^3 - \frac{x^4}{2}\right]$

$= \frac{3}{2}x^2 - x^3 - \frac{1}{2}x^4$

Then we evaluate the outside integral.

$\int_0^1 \left[\int_{x^2}^x (x + y)\,dy\right] dx$

$= \int_0^1 \left(\frac{3}{2}x^2 - x^3 - \frac{1}{2}x^4\right) dx$

$\left[\int_{x^2}^x (x + y)\,dy = \frac{3}{2}x^2 - x^3 - \frac{1}{2}x^4\right]$

$= \left[\frac{x^3}{2} - \frac{x^4}{4} - \frac{x^5}{10}\right]_0^1$ (Integrating with respect to x)

$= \left(\frac{1^3}{2} - \frac{1^4}{4} - \frac{1^5}{10}\right) - \left(\frac{0^3}{2} - \frac{0^4}{4} - \frac{0^5}{10}\right)$

$= \left(\frac{1}{2} - \frac{1}{4} - \frac{1}{10}\right) - 0$

$= \frac{10}{20} - \frac{5}{20} - \frac{2}{20}$

$= \frac{3}{20}$

9. $\int_0^2 \int_0^x (x + y^2)\,dy\,dx$

$= \int_0^2 \left[\int_0^x (x + y^2)\,dy\right] dx$

We first evaluate the inside integral.

$\int_0^x (x + y^2)\,dy$

$= \left[xy + \frac{y^3}{3}\right]_0^x$ (Integrating with respect to y; x is a constant)

$= \left(x \cdot x + \frac{x^3}{3}\right) - \left(x \cdot 0 + \frac{0^3}{3}\right)$

$= x^2 + \frac{x^3}{3}$

Then we evaluate the outside integral.

$\int_0^2 \left[\int_0^x (x + y^2)\,dy\right] dx$

$= \int_0^2 \left(x^2 + \frac{x^3}{3}\right) dx$ $\left[\int_0^x (x + y^2)\,dy = x^2 + \frac{x^3}{3}\right]$

$= \left[\frac{x^3}{3} + \frac{x^4}{12}\right]_0^2$ (Integrating with respect to x)

$= \left(\frac{2^3}{3} + \frac{2^4}{12}\right) - \left(\frac{0^3}{3} + \frac{0^4}{12}\right)$

$= \frac{8}{3} + \frac{16}{12}$

$= \frac{8}{3} + \frac{4}{3}$

$= \frac{12}{3}$, or 4

11. $\int_0^1 \int_0^{1-x^2} (1 - y - x^2)\,dy\,dx$

$= \int_0^1 \left[\int_0^{1-x^2} (1 - y - x^2)\,dy\right] dx$

We first evaluate the inside integral.

$$\int_0^{1-x^2} (1 - y - x^2)\,dy$$

$= \left[y - \frac{y^2}{2} - x^2y\right]_0^{1-x^2}$ (Integrating with respect to y; x is a constant)

$= \left[(1-x^2) - \frac{(1-x^2)^2}{2} - x^2(1-x^2)\right] - \left[0 - \frac{0^2}{2} - x^2\cdot 0\right]$

$= 1 - x^2 - \frac{1 - 2x^2 + x^4}{2} - x^2 + x^4$

$= 1 - x^2 - \frac{1}{2} + x^2 - \frac{1}{2}x^4 - x^2 + x^4$

$= \frac{1}{2}x^4 - x^2 + \frac{1}{2}$

Then we evaluate the outside integral.

$$\int_0^1 \left[\int_0^{1-x^2} (1 - y - x^2)\,dy\right] dx$$

$= \int_0^1 \left(\frac{1}{2}x^4 - x^2 + \frac{1}{2}\right) dx$

$= \left[\frac{x^5}{10} - \frac{x^3}{3} + \frac{1}{2}x\right]_0^1$ (Integrating with respect to x)

$= \left(\frac{1^5}{10} - \frac{1^3}{3} + \frac{1}{2}\cdot 1\right) - \left(\frac{0^5}{10} - \frac{0^3}{3} + \frac{1}{2}\cdot 0\right)$

$= \frac{1}{10} - \frac{1}{3} + \frac{1}{2}$

$= \frac{3}{30} - \frac{10}{30} + \frac{15}{30}$

$= \frac{8}{30}$

$= \frac{4}{15}$

13. $f(x,y) = x^2 + \frac{1}{3}xy$

$0 \leqslant x \leqslant 1$

$0 \leqslant y \leqslant 2$

Find

$$\int_0^2 \int_0^1 f(x,y)\,dx\,dy$$

$= \int_0^2 \left[\int_0^1 \left(x^2 + \frac{1}{3}xy\right) dx\right] dy$

We first evaluate the inside integral.

$$\int_0^1 \left(x^2 + \frac{1}{3}xy\right) dx$$

$= \left[\frac{x^3}{3} + \frac{1}{3}y\cdot\frac{x^2}{2}\right]_0^1$ (Integrating with respect to x; y is a constant)

$= \left(\frac{1^3}{3} + \frac{1}{3}y\cdot\frac{1^2}{2}\right) - \left(\frac{0^3}{3} + \frac{1}{3}y\cdot\frac{0^2}{2}\right)$

$= \frac{1}{3} + \frac{1}{6}y$

Then we evaluate the outside integral.

$$\int_0^2 \left[\int_0^1 \left(x^2 + \frac{1}{3}xy\right) dx\right] dy$$

$= \int_0^2 \left(\frac{1}{3} + \frac{1}{6}y\right) dy$

$= \left[\frac{1}{3}y + \frac{1}{6}\cdot\frac{y^2}{2}\right]_0^2$ (Integrating with respect to y)

$= \left[\frac{1}{3}y + \frac{1}{12}y^2\right]_0^2$

$= \left(\frac{1}{3}\cdot 2 + \frac{1}{12}\cdot 2^2\right) - \left(\frac{1}{3}\cdot 0 + \frac{1}{12}\cdot 0^2\right)$

$= \frac{2}{3} + \frac{1}{3}$

$= 1$

15. $\int_0^1 \int_1^3 \int_{-1}^2 (2x + 3y - z)\,dx\,dy\,dz$

We first evaluate the x-integral.

$$\int_{-1}^2 (2x + 3y - z)\,dx$$

$= \left[x^2 + 3xy - xz\right]_{-1}^2$ (Integrating with respect to x; y and z are constants)

$= [2^2 + 3\cdot 2\cdot y - 2\cdot z] - [(-1)^2 + 3\cdot(-1)\cdot y - (-1)z]$

$= 4 + 6y - 2z - 1 + 3y - z$

$= 9y - 3z + 3$

Then we evaluate the y-integral.

$$\int_1^3 \left[\int_{-1}^2 (2x + 3y - z)\,dx\right] dy$$

$= \int_1^3 (9y - 3z + 3)\,dy$

$= \left[\frac{9}{2}y^2 - 3zy + 3y\right]_1^3$ (Integrating with respect to y; z is a constant)

$= \left(\frac{9}{2}\cdot 3^2 - 3z\cdot 3 + 3\cdot 3\right) - \left(\frac{9}{2}\cdot 1^2 - 3z\cdot 1 + 3\cdot 1\right)$

$= \frac{81}{2} - 9z + 9 - \frac{9}{2} + 3z - 3$

$= 42 - 6z$

Next we evaluate the z-integral.

$$\int_0^1 \left[\int_1^3 \int_{-1}^2 (2x + 3y - z)\,dx\,dy\right] dz$$

$= \int_0^1 (42 - 6z)\,dz$

$= 6\int_0^1 (7 - z)\,dz$

$= 6\left[7z - \frac{z^2}{2}\right]_0^1$ (Integrating with respect to z)

$= 6\left[\left(7\cdot 1 - \frac{1^2}{2}\right) - \left(7\cdot 0 - \frac{0^2}{2}\right)\right]$

$= 6\left(7 - \frac{1}{2}\right)$

$= 6\cdot\frac{13}{2}$

$= 39$

17. $\int_0^1 \int_0^{1-x} \int_0^{2-x} xyz\, dz\, dy\, dx$

We first evaluate the z-integral.

$$\int_0^{2-x} xyz\, dz$$

$$= xy \int_0^{2-x} z\, dz \qquad \text{(x and y are constants)}$$

$$= xy \left[\frac{z^2}{2}\right]_0^{2-x} \qquad \text{(Integrating with respect to z)}$$

$$= xy \left[\frac{(2 - x)^2}{2} - \frac{0^2}{2}\right]$$

$$= xy \cdot \frac{4 - 4x + x^2}{2}$$

$$= 2xy - 2x^2y + \frac{1}{2} x^3y$$

Then we evaluate the y-integral.

$$\int_0^{1-x} \left[\int_0^{2-x} xyz\, dz\right] dy$$

$$= \int_0^{1-x} \left[2xy - 2x^2y + \frac{1}{2} x^3y\right] dy$$

$$= \left[xy^2 - x^2y^2 + \frac{x^3y^2}{4}\right]_0^{1-x} \qquad \text{(Integrating with respect to y; x is a constant)}$$

$$= \left[x(1 - x)^2 - x^2(1 - x)^2 + \frac{x^3(1 - x)^2}{4}\right] - \left[x \cdot 0^2 - x^2 \cdot 0^2 + \frac{x^3 \cdot 0^2}{4}\right]$$

$$= x(1 - 2x + x^2) - x^2(1 - 2x + x^2) + \frac{x^3(1-2x+x^2)}{4}$$

$$= x - 2x^2 + x^3 - x^2 + 2x^3 - x^4 + \frac{1}{4} x^3 - \frac{1}{2} x^4 + \frac{1}{4} x^5$$

$$= \frac{1}{4} x^5 - \frac{3}{2} x^4 + \frac{13}{4} x^3 - 3x^2 + x$$

Next we evaluate the x-integral.

$$\int_0^1 \left[\int_0^{1-x} \int_0^{2-x} xyz\, dz\, dy\right] dx$$

$$= \int_0^1 \left[\frac{1}{4} x^5 - \frac{3}{2} x^4 + \frac{13}{4} x^3 - 3x^2 + x\right] dx$$

$$= \left[\frac{x^6}{24} - \frac{3x^5}{10} + \frac{13x^4}{16} - x^3 + \frac{x^2}{2}\right]_0^1 \qquad \text{(Integrating with respect to x)}$$

$$= \left[\frac{1}{24} - \frac{3}{10} + \frac{13}{16} - 1 + \frac{1}{2}\right] - 0$$

$$= \frac{10}{240} - \frac{72}{240} + \frac{195}{240} - \frac{240}{240} + \frac{120}{240}$$

$$= \frac{13}{240}$$

Exercise Set 8.1

1.

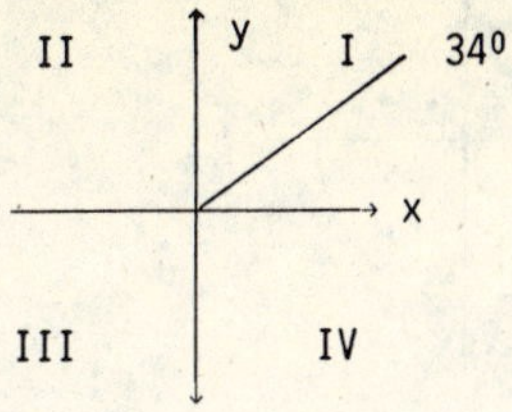

The terminal side of an angle whose measure is 34^0 lies in the first quadrant.

3.

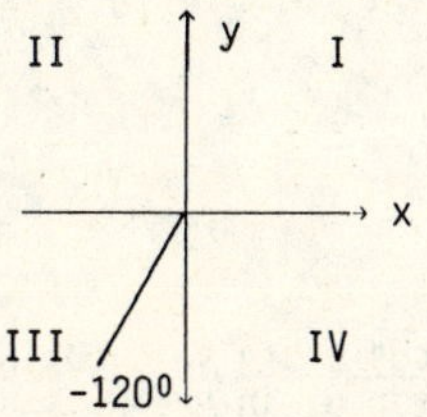

The terminal side of an angle whose measure is -120° lies in the third quadrant.

5. $\frac{\text{Radian measure}}{\pi} = \frac{\text{Degree measure}}{180}$ (Theorem 1)

$\frac{\text{Radian measure}}{\pi} = \frac{30}{180}$ (Substituting 30 for degree measure)

$\text{Radian measure} = \frac{30}{180} \cdot \pi$

$= \frac{\pi}{6}$

7. $\frac{\text{Radian measure}}{\pi} = \frac{\text{Degree measure}}{180}$ (Theorem 1)

$\frac{\text{Radian measure}}{\pi} = \frac{60}{180}$ (Substituting 60 for degree measure)

$\text{Radian measure} = \frac{60}{180} \cdot \pi$

$= \frac{\pi}{3}$

9. $\frac{\text{Radian measure}}{\pi} = \frac{\text{Degree measure}}{180}$ (Theorem 1)

$\frac{\text{Radian measure}}{\pi} = \frac{75}{180}$ (Substituting 75 for degree measure)

$\text{Radian measure} = \frac{75}{180} \cdot \pi$

$= \frac{5\pi}{12}$

11. $\frac{\text{Radian measure}}{\pi} = \frac{\text{Degree measure}}{180}$ (Theorem 1)

$\frac{\frac{3\pi}{2}}{\pi} = \frac{\text{Degree measure}}{180}$ $\left[\text{Substituting } \frac{3\pi}{2} \text{ for radian measure}\right]$

$\frac{3\pi}{2} \cdot \frac{1}{\pi} = \frac{\text{Degree measure}}{180}$

$180 \cdot \frac{3}{2} = \text{Degree measure}$

$270° = \text{Degree measure}$

13. $\frac{\text{Radian measure}}{\pi} = \frac{\text{Degree measure}}{180}$ (Theorem 1)

$\frac{-\frac{\pi}{4}}{\pi} = \frac{\text{Degree measure}}{180}$ $\left[\text{Substituting } -\frac{\pi}{4} \text{ for radian measure}\right]$

$-\frac{\pi}{4} \cdot \frac{1}{\pi} = \frac{\text{Degree measure}}{180}$

$180 \cdot \left(-\frac{1}{4}\right) = \text{Degree measure}$

$-45° = \text{Degree measure}$

15. $\frac{\text{Radian measure}}{\pi} = \frac{\text{Degree measure}}{180}$ (Theorem 1)

$\frac{8\pi}{\pi} = \frac{\text{Degree measure}}{180}$ (Substituting 8π for radian measure)

$8 = \frac{\text{Degree measure}}{180}$

$8 \cdot 180 = \text{Degree measure}$

$1440° = \text{Degree measure}$

17. $\frac{\text{Radian measure}}{\pi} = \frac{\text{Degree measure}}{180}$ (Theorem 1)

$\frac{1}{\pi} = \frac{\text{Degree measure}}{180}$

$180 \cdot \frac{1}{\pi} = \text{Degree measure}$

$\left(\frac{180}{\pi}\right)^\circ = \text{Degree measure}$

$57.3° \approx \text{Degree measure}$

19.

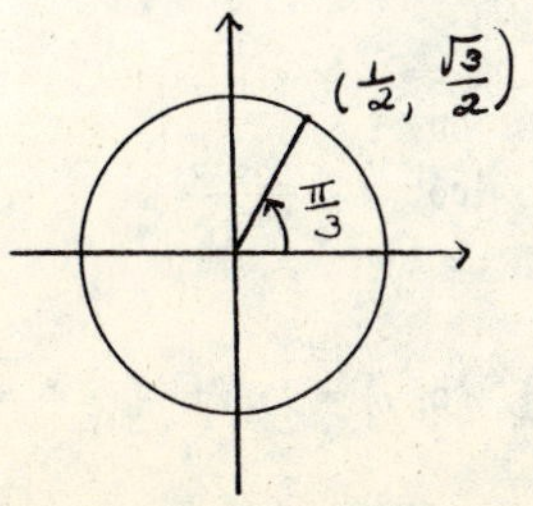

$\sin \frac{\pi}{3} = \frac{\sqrt{3}}{2}$

21.

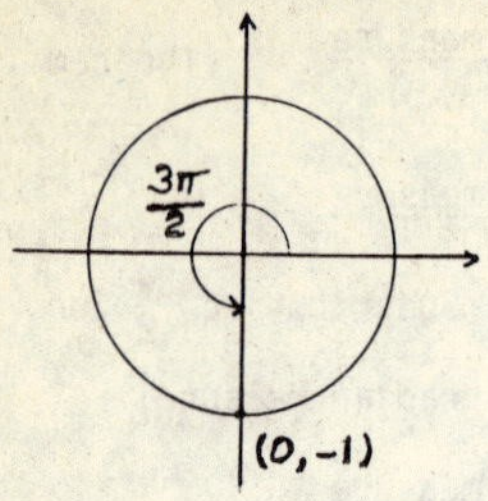

$\cos \frac{3\pi}{2} = 0$

23.

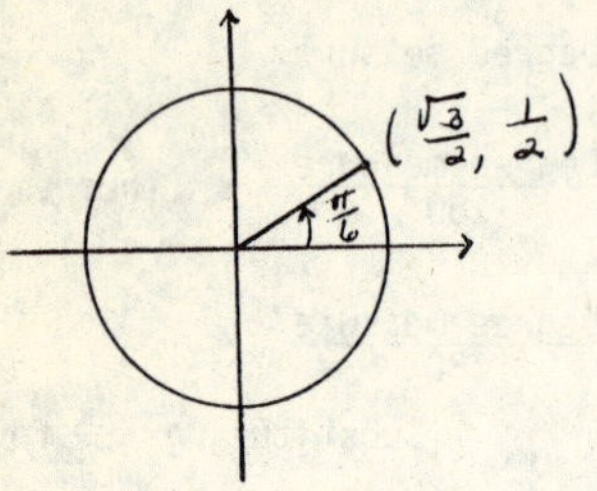

$\sin \frac{\pi}{6} = \frac{1}{2}$

25.

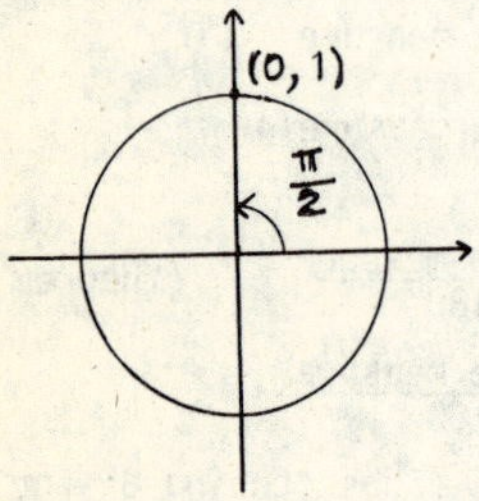

$\tan \frac{\pi}{2} = \dfrac{\sin \frac{\pi}{2}}{\cos \frac{\pi}{2}}$ $\left[\tan x = \frac{\sin x}{\cos x}\right]$

$= \frac{1}{0}$ $\left[\sin \frac{\pi}{2} = 1;\ \cos \frac{\pi}{2} = 0\right]$

Since division by 0 is undefined, $\tan \frac{\pi}{2}$ is <u>undefined</u>.

27.

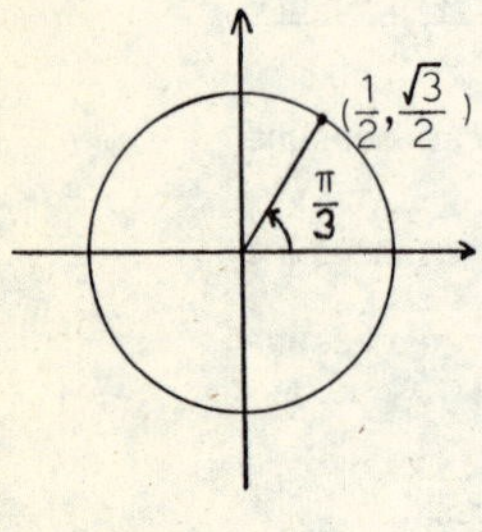

$\cot \frac{\pi}{3} = \dfrac{\cos \frac{\pi}{3}}{\sin \frac{\pi}{3}}$ $\left[\cot x = \frac{\cos x}{\sin x}\right]$

$= \dfrac{\frac{1}{2}}{\frac{\sqrt{3}}{2}}$ $\left[\cos \frac{\pi}{3} = \frac{1}{2};\ \sin \frac{\pi}{3} = \frac{\sqrt{3}}{2}\right]$

$= \frac{1}{2} \cdot \frac{2}{\sqrt{3}} = \frac{1}{\sqrt{3}}$ or $\frac{\sqrt{3}}{3}$

29.

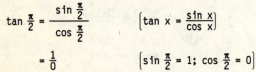

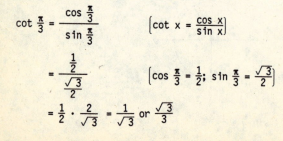

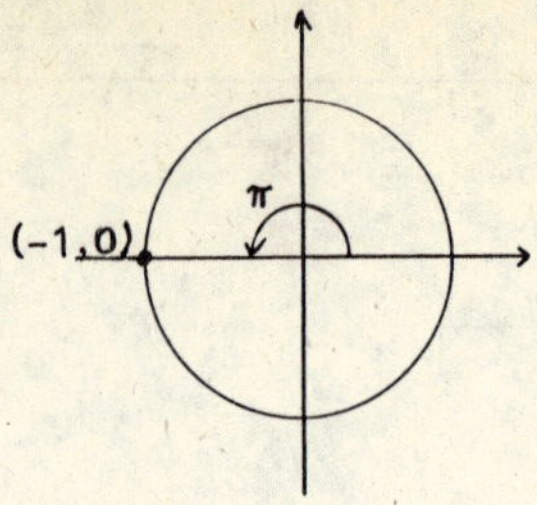

$\sec \pi = \frac{1}{\cos \pi}$ $\left[\sec x = \frac{1}{\cos x}\right]$

$= \frac{1}{-1}$ $(\cos \pi = -1)$

$= -1$

31. $\tan (u + v)$

$= \dfrac{\sin (u + v)}{\cos (u + v)}$

$= \dfrac{\sin u \cos v + \cos u \sin v}{\cos u \cos v - \sin u \sin v}$ (Using Identity 6) (Using Identity 4)

$= \dfrac{\sin u \cos v + \cos u \sin v}{\cos u \cos v - \sin u \sin v} \cdot \dfrac{\frac{1}{\cos u \cos v}}{\frac{1}{\cos u \cos v}}$

(Multiplying by a form of 1)

$= \dfrac{\frac{\sin u}{\cos u} + \frac{\sin v}{\cos v}}{1 - \frac{\sin u}{\cos u} \cdot \frac{\sin v}{\cos v}}$

$= \dfrac{\tan u + \tan v}{1 - \tan u \tan v}$

33. $\sin 31.4° \approx 0.5210$ (Using a calculator)

35. $\tan 139.2° \approx -0.8632$ (Using a calculator)

37. $\cos (-1.91) = \cos (-109.4349°)$ (-1.91 radians $= 109.4349°$)

≈ -0.3327 (Using a calculator)

39. $\cot 49\pi = \cot \pi$

$= \frac{\cos \pi}{\sin \pi}$

$= \frac{-1}{0}$

Since division by 0 is undefined, $\cot \pi$ is undefined.

41. $S(t) = 7\left[1 - \cos \frac{\pi}{6} t\right]$

$S(0) = 7\left[1 - \cos \frac{\pi}{6} \cdot 0\right]$ (Substituting 0 for t)

$= 7(1 - \cos 0)$

$= 7(1 - 1)$

$= 7 \cdot 0$

$= \$0$

$S(1) = 4\left[1 - \cos \frac{\pi}{6} \cdot 1\right]$ (Substituting 1 for t)

$= 7\left[1 - \cos \frac{\pi}{6}\right]$

$= 7\left[1 - \frac{\sqrt{3}}{2}\right]$

$\approx 7(1 - 0.866)$

$\approx 7(0.134)$

$\approx$ \$0.9 thousand

$S(2) = 7\left[1 - \cos \frac{\pi}{6} \cdot 2\right]$ (Substituting 2 for t)

$= 7\left[1 - \cos \frac{\pi}{3}\right]$

$\approx 7(1 - 0.5)$

$\approx 7(0.5)$

$\approx$ \$3.5 thousand

$S(3) = 7\left[1 - \cos \frac{\pi}{6} \cdot 3\right]$ (Substituting 3 for t)

$= 7\left[1 - \cos \frac{\pi}{2}\right]$

$= 7(1 - 0)$

$= 7 \cdot 1$

= \$7 thousand

$S(6) = 7\left[1 - \cos \frac{\pi}{6} \cdot 6\right]$ (Substituting 6 for t)

$= 7(1 - \cos \pi)$

$= 7[1 - (-1)]$

$= 7(1 + 1)$

$= 7 \cdot 2$

= \$14 thousand

$S(12) = 7\left[1 - \cos \frac{\pi}{6} \cdot 12\right]$ (Substituting 12 for t)

$= 7(1 - \cos 2\pi)$

$= 7(1 - 1)$

$= 7 \cdot 0$

= \$0

$S(15) = 7\left[1 - \cos \frac{\pi}{6} \cdot 15\right]$ (Substituting 15 for t)

$= 7\left[1 - \cos \frac{5\pi}{2}\right]$

$= 7(1 - 0)$

$= 7 \cdot 1$

= \$7 thousand

Exercise Set 8.2

1. Prove $\frac{d}{dx} \sec x = \tan x \sec x$.

$\frac{d}{dx} \sec x$

$= \frac{d}{dx} \frac{1}{\cos x}$ $\left[\sec x = \frac{1}{\cos x}\right]$

$= \frac{(\cos x) \cdot 0 - (-\sin x) \cdot 1}{\cos^2 x}$ (Quotient Rule)

$= \frac{\sin x}{\cos^2 x}$

$= \frac{\sin x}{\cos x} \cdot \frac{1}{\cos x}$

$= \tan x \cdot \sec x$ $\left[\frac{\sin x}{\cos x} = \tan x;\ \frac{1}{\cos x} = \sec x\right]$

3. $y = x \sin x$

$\frac{dy}{dx} = x \cdot \cos x + 1 \cdot \sin x$ (Product Rule)

$= x \cos x + \sin x$

5. $f(x) = e^x \sin x$

$f'(x) = e^x \cdot \cos x + e^x \cdot \sin x$ (Product Rule)

$= e^x(\cos x + \sin x)$

7. $y = \frac{\sin x}{x}$

$\frac{dy}{dx} = \frac{x \cdot \cos x - 1 \cdot \sin x}{x^2}$ (Quotient Rule)

$= \frac{x \cos x - \sin x}{x^2}$

9. $f(x) = \sin^2 x$

$f'(x) = 2 \sin x \cdot \left[\frac{d}{dx} \sin x\right]$ (Extended Power Rule)

$= 2 \sin x \cos x$

11. $y = \sin x \cos x$

$\frac{dy}{dx} = \sin x \cdot (-\sin x) + \cos x \cdot \cos x$ (Product Rule)

$= -\sin^2 x + \cos^2 x$

13. $f(x) = \frac{\sin x}{1 + \cos x}$

$f'(x) = \frac{(1 + \cos x) \cdot \cos x - (-\sin x) \cdot \sin x}{(1 + \cos x)^2}$

(Quotient Rule)

$= \frac{\cos x + \cos^2 x + \sin^2 x}{(1 + \cos x)^2}$

$= \frac{\cos x + 1}{(1 + \cos x)^2}$ $(\cos^2 x + \sin^2 x = 1)$

$= \frac{(1 + \cos x)}{(1 + \cos x)(1 + \cos x)}$

$= \frac{1}{1 + \cos x}$

15. $y = \tan^2 x$

$$\frac{dy}{dx} = 2 \tan x \cdot \left[\frac{d}{dx} \tan x\right] \quad \text{(Extended Power Rule)}$$

$$= 2 \tan x \sec^2 x$$

$$= 2 \cdot \frac{\sin x}{\cos x} \cdot \frac{1}{\cos^2 x} \quad \left[\tan x = \frac{\sin x}{\cos x};\ \sec^2 x = \frac{1}{\cos^2 x}\right]$$

$$= \frac{2 \sin x}{\cos^3 x}$$

17. $f(x) = \sqrt{1 + \cos x}$

$$= (1 + \cos x)^{1/2}$$

$$f'(x) = \frac{1}{2}(1 + \cos x)^{-1/2} \cdot \left[\frac{d}{dx}(1 + \cos x)\right]$$

$$= \frac{1}{2}(1 + \cos x)^{-1/2}(-\sin x)$$

$$= -\frac{\sin x}{2\sqrt{1 + \cos x}}$$

19. $y = x^2 \cos x - 2x \sin x - 2 \cos x$

$$\frac{dy}{dx} = [x^2 \cdot (-\sin x) + 2x \cdot \cos x] - 2(x \cdot \cos x + 1 \cdot \sin x) - 2(-\sin x)$$

$$= -x^2 \sin x + 2x \cos x - 2x \cos x - 2 \sin x + 2 \sin x$$

$$= -x^2 \sin x$$

21. $y = e^{\sin x}$

$$\frac{dy}{dx} = \cos x \cdot e^{\sin x} \quad \left[\frac{d}{dx} e^{f(x)} = f'(x)\, e^{f(x)}\right]$$

23. $y = \sin x$

$$\frac{dy}{dx} = \cos x \quad \text{(1st derivative)}$$

$$\frac{d^2y}{dx^2} = -\sin x \quad \text{(2nd derivative)}$$

25. $y = \sin(x^2 + x^3)$

$$\frac{dy}{dx} = [\cos(x^2 + x^3)]\left[\frac{d}{dx}(x^2 + x^3)\right] \quad \text{(Using the Chain Rule)}$$

$$= [\cos(x^2 + x^3)] \cdot [2x + 3x^2]$$

$$= (2x + 3x^2)\cos(x^2 + x^3)$$

27. $f(x) = \cos(x^5 - x^4)$

$$f'(x) = [-\sin(x^5 - x^4)]\left[\frac{d}{dx}(x^5 - x^4)\right]$$

(Using the Chain Rule)

$$= [-\sin(x^5 - x^4)] \cdot [5x^4 - 4x^3]$$

$$= -(5x^4 - 4x^3)\sin(x^5 - x^4)$$

$$= (4x^3 - 5x^4)\sin(x^5 - x^4)$$

29. $f(x) = \cos\sqrt{x}$

$$= \cos x^{1/2}$$

$$f'(x) = (-\sin x^{1/2})\left[\frac{1}{2}x^{-1/2}\right]$$

$$= -\frac{1}{2}x^{-1/2}\sin x^{1/2}$$

$$= -\frac{\sin\sqrt{x}}{2\sqrt{x}}$$

31. $y = \sin(\cos x)$

$$\frac{dy}{dx} = [\cos(\cos x)]\left[\frac{d}{dx}\cos x\right]$$

$$= [\cos(\cos x)][-\sin x]$$

$$= -\sin x \cdot \cos(\cos x)$$

33. $y = \sqrt{\cos 4x}$

$$= (\cos 4x)^{1/2}$$

$$\frac{dy}{dx} = \frac{1}{2}(\cos 4x)^{-1/2} \cdot \frac{d}{dx}\cos 4x$$

$$= \frac{1}{2}(\cos 4x)^{-1/2} \cdot (-\sin 4x)(4)$$

$$= \frac{-2 \sin 4x}{\sqrt{\cos 4x}}$$

35. $f(x) = \cot\sqrt[3]{5 - 2x}$

$$= \cot(5 - 2x)^{1/3}$$

$$f'(x) = [-\csc^2(5 - 2x)^{1/3}]\left[\frac{d}{dx}(5 - 2x)^{1/3}\right]$$

$$= [-\csc^2(5 - 2x)^{1/3}] \cdot \left[\frac{1}{3}(5 - 2x)^{-2/3}(-2)\right]$$

$$= \frac{2}{3}(5 - 2x)^{-2/3}\csc^2(5 - 2x)^{1/3}$$

$$= \frac{2 \csc^2\sqrt[3]{5 - 2x}}{3\sqrt[3]{(5 - 2x)^2}}$$

37. $y = \tan^4 3x - \sec^4 3x$

$$\frac{dy}{dx} = 4 \tan^3 3x\left[\frac{d}{dx}\tan 3x\right] - 4 \sec^3 3x\left[\frac{d}{dx}\sec 3x\right]$$

$$= 4 \tan^3 3x[(\sec^2 3x) \cdot 3] - 4 \sec^3 3x[(\tan 3x \sec 3x) \cdot 3]$$

$$= 12 \sec^2 3x \tan^3 3x - 12 \tan 3x \sec^4 3x$$

$$= 12 \sec^2 3x \tan 3x (\tan^2 3x - \sec^2 3x)$$

$$= (12 \sec^2 3x \tan 3x) \cdot (-1)$$

$(\tan^2 t + 1 = \sec^2 t$ or $\tan^2 t - \sec^2 t = -1)$

$$= -12 \sec^2 3x \tan 3x$$

39. $y = \ell n\,|\sin x|$

$\frac{dy}{dx} = \frac{1}{\sin x}\left[\frac{d}{dx}\sin x\right]$

$\left[\frac{d}{dx}\ell n\,|f(x)| = \frac{1}{f(x)}\cdot f'(x)\right]$

$= \frac{1}{\sin x}\cdot\cos x$

$= \frac{\cos x}{\sin x}$

$= \cot x$

41. $S(t) = 40{,}000(\sin t + \cos t)$

$S'(t) = 40{,}000\left[\cos t + (-\sin t)\right]$

$= 40{,}000(\cos t - \sin t)$

43. $T(t) = 101.6° + 3\sin\frac{\pi}{8}t$

$T'(t) = 0 + 3\left[\cos\frac{\pi}{8}t\right]\cdot\frac{\pi}{8}$

$= \frac{3\pi}{8}\cos\frac{\pi}{8}t$

45. $y = 5000\left[\cos\frac{\pi}{45}(t - 10)\right]$

$\frac{dy}{dt} = 5000\cdot\left[-\sin\frac{\pi}{45}(t - 10)\right]\cdot\left[\frac{d}{dt}\frac{\pi}{45}(t - 10)\right]$

$= 5000\cdot\left[-\sin\frac{\pi}{45}(t - 10)\right]\cdot\frac{\pi}{45}$

$= -\frac{5000\pi}{45}\left[\sin\frac{\pi}{45}(t - 10)\right]$

$= -\frac{1000\pi}{9}\left[\sin\frac{\pi}{45}(t - 10)\right]$

47. $y = 5\sin(4t + \pi)$

$y = 5\sin[4t - (-\pi)]$

$\uparrow \quad \uparrow \quad \uparrow$

$y = A\sin[Bx - C] + D$

a) Amplitude $= |A|$

$= |5|$ (Substituting 5 for A)

$= 5$

Period $= \frac{2\pi}{B}$

$= \frac{2\pi}{4}$ (Substituting 4 for B)

$= \frac{\pi}{2}$

Phase shift $= \frac{C}{B}$

$= \frac{-\pi}{4}$ (Substituting $-\pi$ for C and 4 for B)

$= -\frac{\pi}{4}$

b) $y = 5\sin(4t + \pi)$

$\frac{dy}{dt} = 5\cos(4t + \pi)\cdot\frac{d}{dt}(4t + \pi)$

$= 5\cos(4t + \pi)\cdot 4$

$= 20\cos(4t + \pi)$

49. $L = (a^{2/3} + b^{2/3})^{3/2}$ (Result of Margin Exercise 8)

$= (8^{2/3} + 8^{2/3})^{3/2}$ (Substituting)

$= (4 + 4)^{3/2}$

$= 8^{3/2}$

$= \sqrt[2]{8^3}$

$= 8\sqrt{8} = 8\sqrt{4\cdot 2}$

$= 8\cdot 2\sqrt{2} = 16\sqrt{2}$

51. We first make a drawing.

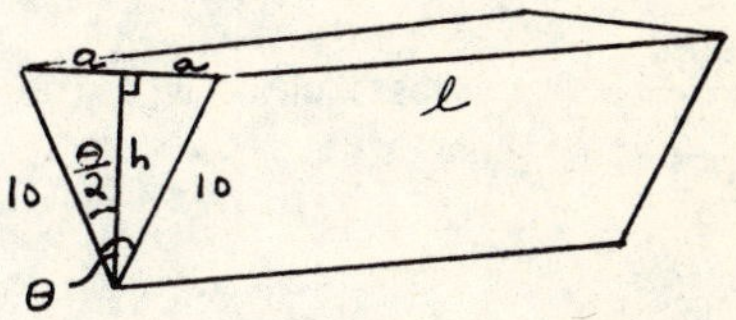

We let 2a represent the width of the trough, l the length, and h the height of the end.

The volume of the trough can be found by multiplying the area of the end and the length.

V = Area of end · Length

$= \left[\frac{1}{2}\cdot 2a\cdot h\right]\cdot l$

$= a\cdot h\cdot l$

$= a\cdot\sqrt{100 - a^2}\cdot l$

(Substituting $\sqrt{100 - a^2}$ for h; $a^2 + h^2 = 10^2$)

We now maximize V.

$V(a) = l\cdot a(100 - a^2)^{1/2}$ (l is a constant)

We first find V'(a).

$V'(a) = l\left[a\cdot\frac{1}{2}(100 - a^2)^{-1/2}\cdot(-2a) + 1\cdot(100 - a^2)^{1/2}\right]$

$= l\left[-\frac{a^2}{\sqrt{100 - a^2}} + \sqrt{100 - a^2}\right]$

Then we set V'(a) = 0 and solve for a.

$$0 = 1\left[-\frac{a^2}{\sqrt{100 - a^2}} + \sqrt{100 - a^2}\right]$$

$$0 = -\frac{a^2}{\sqrt{100 - a^2}} + \sqrt{100 - a^2}$$

(1 is a constant)

$$\frac{a^2}{\sqrt{100 - a^2}} = \sqrt{100 - a^2}$$

$$a^2 = 100 - a^2$$

$$2a^2 = 100$$

$$a^2 = 50$$

$$a = \sqrt{50}$$

$$a = 5\sqrt{2}$$

V will be a maximum when $a = 5\sqrt{2}$.

We now find θ. Note that

$\sin\frac{\theta}{2} = \frac{a}{10}$ (See figure above.)

$\sin\frac{\theta}{2} = \frac{5\sqrt{2}}{10}$ (Substituting $5\sqrt{2}$ for a)

$$\sin\frac{\theta}{2} = \frac{\sqrt{2}}{2}$$

$$\frac{\theta}{2} = 45^0$$

$$\theta = 90^0$$

Thus, θ must be 90^0 to maximize the volume of water that the trough can carry.

53. $f(x,y) = \sin 2y - ye^x$

Find f_x.

$f(x,y) = \sin 2y - ye^{\underline{x}}$ (The variable is underlined.)

$$f_x = -ye^x$$

Find f_y.

$f(x,y) = \sin 2\underline{y} - \underline{y}e^x$

$$f_y = (\cos 2y)\cdot 2 - e^x$$
$$= 2\cos 2y - e^x$$

Find f_{xx}.

$$f_x = -ye^{\underline{x}}$$
$$f_{xx} = -ye^x$$

Find f_{xy}.

$$f_x = -\underline{y}e^x$$
$$f_{xy} = -e^x$$

Find f_{yx}.

$$f_y = 2\cos 2y - e^{\underline{x}}$$
$$f_{yx} = -e^x$$

Find f_{yy}.

$$f_y = 2\cos 2\underline{y} - e^x$$
$$f_{yy} = 2(-\sin 2y)\cdot 2$$
$$= -4\sin 2y$$

55. $f(x,y) = \cos(2x + 3y)$

Find f_x.

$f(x,y) = \cos(2\underline{x} + 3y)$ (The variable is underlined.)

$$f_x = [-\sin(2x + 3y)]\cdot 2$$
$$= -2\sin(2x + 3y)$$

Find f_y.

$f(x,y) = \cos(2x + 3\underline{y})$

$$f_y = [-\sin(2x + 3y)]\cdot 3$$
$$= -3\sin(2x + 3y)$$

Find f_{xx}.

$$f_x = -2\sin(2\underline{x} + 3y)$$
$$f_{xx} = -2[\cos(2x + 3y)]\cdot 2$$
$$= -4\cos(2x + 3y)$$

Find f_{xy}.

$$f_x = -2\sin(2x + 3\underline{y})$$
$$f_{xy} = -2[\cos(2x + 3y)]\cdot 3$$
$$= -6\cos(2x + 3y)$$

Find f_{yx}.

$$f_y = -3\sin(2\underline{x} + 3y)$$
$$f_{yx} = -3[\cos(2x + 3y)]\cdot 2$$
$$= -6\cos(2x + 3y)$$

Find f_{yy}.

$$f_y = -3\sin(2x + 3\underline{y})$$
$$f_{yy} = -3[\cos(2x + 3y)]\cdot 3$$
$$= -9\cos(2x + 3y)$$

57. $f(x,y) = x^3 \tan(5xy)$

Find f_x.

$f(x,y) = \underline{x}^3 \tan(5\underline{x}y)$ (The variable is underlined.)

$$f_x = x^3\left[\sec^2(5xy)\right]5y + 3x^2\tan(5xy)$$
$$= 5x^3y\sec^2(5xy) + 3x^2\tan(5xy)$$

Find f_y.

$f(x,y) = x^3 \tan(5x\underline{y})$

$$f_y = x^3\left[\sec^2(5xy)\right]5x$$
$$= 5x^4\sec^2(5xy)$$

Find f_{xx}.

$$f_x = 5\underline{x}^3y\sec^2(5\underline{x}y) + 3\underline{x}^2\tan(5\underline{x}y)$$

$$f_x = 5y[x^3\sec^2(5xy)] + 3[x^2\tan(5xy)]$$

(Factoring out the constants)

$$f_{xx} = 5y[x^3\cdot 2\sec(5xy)\cdot\tan(5xy)\sec(5xy)\cdot 5y + 3x^2\cdot\sec^2(5xy)] + 3[x^2\cdot\left[\sec^2(5xy)\right]\cdot 5y + 2x\cdot\tan(5xy)]$$

$= 5y[10x^3y\ sec^2(5xy)\ tan\ (5xy) +$

$3x^2\ sec^2(5xy)] + 3[5x^2y\ sec^2(5xy) +$

$2x\ tan\ (5xy)]$

$= 50x^3y^2\ sec^2(5xy)\ tan\ (5xy) +$

$15x^2y\ sec^2(5xy) + 15x^2y\ sec^2(5xy) +$

$6x\ tan\ (5xy)$

$= 50x^3y^2\ sec^2(5xy)\ tan\ (5xy) +$

$30x^2y\ sec^2(5xy) + 6x\ tan\ (5xy)$

Find f_{xy}.

$f_x = 5x^3\underline{y}\ sec^2(5x\underline{y}) + 3x^2\ tan\ (5x\underline{y})$

$f_{xy} = 5x^3[y\cdot 2\ sec\ (5xy)\cdot tan\ (5xy)\ sec\ (5xy)\cdot 5x +$

$sec^2(5xy)] + 3x^2[sec^2(5xy)\cdot 5x]$

$= 50x^4y\ sec^2(5xy)\ tan\ (5xy) + 5x^3\ sec^2(5xy) +$

$15x^3\ sec^2(5xy)$

$= 50x^4y\ sec^2(5xy)\ tan\ (5xy) + 20x^3\ sec^2(5xy)$

Find f_{yx}.

$f_y = 5\underline{x}^4\ sec^2\ (5\underline{x}y)$

$f_{yx} = 5[x^4\cdot 2\ sec\ (5xy)\cdot tan\ (5xy)\ sec\ (5xy)\cdot 5y +$

$4x^3\cdot sec^2(5xy)]$

$= 50x^4y\ sec^2(5xy)\ tan\ (5xy) + 20x^3\ sec^2(5xy)$

Find f_{yy}.

$f_y = 5x^4\ sec^2(5x\underline{y})$

$f_{yy} = 5x^4\cdot 2\ sec\ (5xy)\cdot tan\ (5xy)\ sec(5xy)\cdot 5x$

$= 50x^5\ sec^2(5xy)\ tan\ (5xy)$

59.

$y = x\ cos\ y$

$\frac{d}{dx}\ y = \frac{d}{dx}\ x\ cos\ y$

$\frac{dy}{dx} = x\cdot(-\ sin\ y)\ \frac{dy}{dx} + 1\cdot cos\ y$

$\frac{dy}{dx} + x\ sin\ y\ \frac{dy}{dx} = cos\ y$

$(1 + x\ sin\ y)\ \frac{dy}{dx} = cos\ y$

$\frac{dy}{dx} = \frac{cos\ y}{1 + x\ sin\ y}$

61. See the answer section in the text.

63. See the answer section in the text.

65. See the answer section in the text.

67. See the answer section in the text.

Exercise Set 8.3

1. Find the area under
$y = sin\ x$
on the interval $\left[0, \frac{\pi}{3}\right]$.

$\int_0^{\pi/3} sin\ x\ dx$

$= [-\ cos\ x]_0^{\pi/3}$ $\left[\int sin\ x\ dx = -\ cos\ x + C\right]$

$= -\left[cos\ \frac{\pi}{3} - cos\ 0\right]$

$= -\left[\frac{1}{2} - 1\right]$

$= -\left[-\ \frac{1}{2}\right]$

$= \frac{1}{2}$

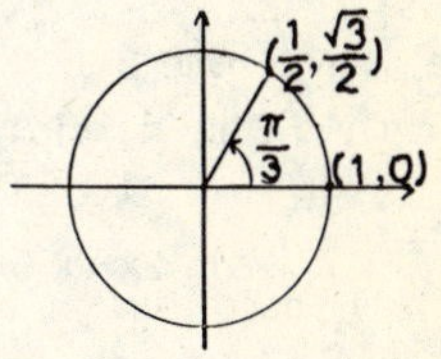

3. $\int_{-\pi}^{\pi} cos\ x\ dx$

$= [sin\ x]_{-\pi}^{\pi}$ $\left[\int cos\ x\ dx = sin\ x + C\right]$

$= sin\ \pi - sin(-\pi)$

$= 0 - 0$

$= 0$

5. $\int_{-\pi/4}^{2\pi} sec^2x\ dx$

$= [tan\ x]_{-\pi/4}^{2\pi}$ $\left[\int sec^2x\ dx = tan\ x + c\right]$

$= tan\ 2\pi - tan\ (-\pi/4)$

$= 0 - (-1)$

$= 1$

7. $\int sin^4x\ cos\ x\ dx$

Let $u = sin\ x$, then $du = cos\ x\ dx$.

$= \int u^4\ du$ (Substituting u for sin x and du for cos x dx)

$= \frac{u^5}{5} + C$

$= \frac{sin^5x}{5} + C$

9. $\int -cos^2\ x\ sin\ x\ dx$

Let $u = cos\ x$, then $du = -\ sin\ x\ dx$.

$= \int u^2\ du$ (Substituting u for cos x and du for - sin x dx)

$= \frac{u^3}{3} + C$

$= \frac{cos^3x}{3} + C$

11. $\int \cos (x + 3)\, dx$

Let $u = x + 3$, then $du = dx$.

$= \int \cos u\, du$ (Substituting u for x + 3 and du for dx)

$= \sin u + C$

$= \sin (x + 3) + C$

13. $\int \sin 2x\, dx$

Let $u = 2x$, then $du = 2\, dx$.

We do not have 2 dx. We only have dx and need to supply a 2 by multiplying by $\frac{1}{2} \cdot 2$ as follows:

$\frac{1}{2} \cdot 2 \int \sin 2x\, dx$ (Multiplying by 1)

$= \frac{1}{2} \int 2 \sin 2x\, dx$

$= \frac{1}{2} \int \sin 2x\, (2\, dx)$

$= \frac{1}{2} \int \sin u\, du$ (Substituting u for 2x and du for 2 dx)

$= \frac{1}{2} (- \cos u) + C$

$= - \frac{1}{2} \cos 2x + C$

15. $\int x \cos x^2\, dx$

Let $u = x^2$, then $du = 2x\, dx$.

We do not have 2x dx. We only have x dx and need to supply a 2 by multiplying by $\frac{1}{2} \cdot 2$ as follows:

$\frac{1}{2} \cdot 2 \int x \cos x^2\, dx$ (Multiplying by 1)

$= \frac{1}{2} \int 2x \cos x^2\, dx$

$= \frac{1}{2} \int (\cos x^2)(2x\, dx)$

$= \frac{1}{2} \int \cos u\, du$ (Substituting u for x^2 and du for 2x dx)

$= \frac{1}{2} \sin u + C$

$= \frac{1}{2} \sin x^2 + C$

17. $\int e^x \sin (e^x)\, dx$

Let $u = e^x$, then $du = e^x\, dx$.

$= \int \left[\sin (e^x)\right]\left[e^x\, dx\right]$

$= \int \sin u\, du$ (Substituting u for e^x and du for e^x dx)

$= - \cos u + C$

$= - \cos (e^x) + C$

19. $\int \tan x\, dx$

$= \int \frac{\sin x}{\cos x}\, dx$

$= \int \frac{1}{\cos x} \sin x\, dx$

Let $u = \cos x$, then $du = - \sin x\, dx$.

We do not have $- \sin x\, dx$. We only have $\sin x\, dx$ and need to supply a -1 by multiplying by (-1)(-1) as follows:

$(-1)(-1) \int \frac{1}{\cos x} \sin x\, dx$

$= - \int \frac{1}{\cos x} (- \sin x)\, dx$

$= - \int \frac{1}{u}\, du$ [Substituting u for cos x and du for (- sin x) dx]

$= - \ln |u| + C$

$= - \ln |\cos x| + C$

This answer can also be expressed as

$- \ln |\cos x| + C$

$= - \ln \left|\frac{1}{\sec x}\right| + C$

$= - \ln |\sec x|^{-1} + C$

$= - (-1) \ln |\sec x| + C$

$= \ln |\sec x| + C.$

21. $\int x \cos 4x\, dx$

Let

$u = x$ and $dv = \cos 4x\, dx$.

Then

$du = dx$ and $v = \frac{1}{4} \sin 4x$.

Using the integration-by-parts formula, we get:

$$\int \underset{u}{(x)}\underset{dv}{(\cos 4x\, dx)} = \underset{u}{(x)}\underset{v}{\left[\frac{1}{4} \sin 4x\right]} - \int \underset{v}{\left[\frac{1}{4} \sin 4x\right]} \underset{du}{dx}$$

$$= \frac{1}{4} x \sin 4x - \frac{1}{4} \int \sin 4x\, dx$$

$$= \frac{1}{4} x \sin 4x - \frac{1}{4} \left[- \frac{1}{4} \cos 4x\right] + C$$

$$= \frac{1}{4} x \sin 4x + \frac{1}{16} \cos 4x + C$$

23. $\int 3x \cos x\, dx$

Let

$u = 3x$ and $dv = \cos x\, dx$.

Then

$du = 3\, dx$ and $v = \sin x$.

Using the integration-by-parts formula, we get:

$$\int \overset{u}{(3x)}\overset{dv}{(\cos x\, dx)} = \overset{u}{(3x)}\overset{v}{(\sin x)} - \int \overset{v}{(\sin x)}\overset{du}{(3\, dx)}$$

$$= 3x \sin x - 3 \int \sin x\, dx$$

$$= 3x \sin x - 3\,(-\cos x) + C$$

$$= 3x \sin x + 3 \cos x + C$$

25. From Margin Exercise 5 we have

$$\int x \sin x\, dx = \sin x - x \cos x + C.$$

$$\int x^2 \sin x\, dx$$

Let

$u = x$ and $dv = x \sin x\, dx$.

Then

$du = dx$ and $v = \sin x - x \cos x$.

Using the integration-by-parts formula, we get:

$$\int \overset{u}{(x)}\overset{dv}{(x \sin x\, dx)}$$

$$= \overset{u}{x}\, \overset{v}{(\sin x - x \cos x)} - \int \overset{v}{(\sin x - x \cos x)}\, \overset{du}{dx}$$

$$= x \sin x - x^2 \cos x - \int \sin x\, dx + \int x \cos x\, dx$$

$$= x \sin x - x^2 \cos x + \cos x + \underbrace{x \sin x + \cos x}_{} + C$$

(Using the result of Example 5)

$$= 2x \sin x + 2 \cos x - x^2 \cos x + C$$

$$= 2x \sin x - (x^2 - 2) \cos x + C$$

27. $\int \tan^2 x\, dx$

$$= \int (\sec^2 x - 1)\, dx \qquad (\tan^2 t + 1 = \sec^2 t;\ \tan^2 t = \sec^2 t - 1)$$

$$= \int \sec^2 x\, dx - \int dx$$

$$= \tan x - x + C$$

29. $S(t) = 7\left[1 - \cos \frac{\pi}{6} t\right]$

$$\int_0^{12} S(t)\, dt$$

$$= \int_0^{12} 7\left[1 - \cos \frac{\pi}{6} t\right] dt$$

$$= 7\int_0^{12} \left[1 - \cos \frac{\pi}{6} t\right] dt$$

$$= 7\left[t - \frac{6}{\pi} \sin \frac{\pi}{6} t\right]_0^{12}$$

$$= 7\left[\left[12 - \frac{6}{\pi} \sin \left[\frac{\pi}{6} \cdot 12\right]\right] - \left[0 - \frac{6}{\pi} \sin \left[\frac{\pi}{6} \cdot 0\right]\right]\right]$$

$$= 7\left[12 - \frac{6}{\pi} \sin 2\pi + \frac{6}{\pi} \sin 0\right]$$

$$= 7\left[12 - \frac{6}{\pi} \cdot 0 + \frac{6}{\pi} \cdot 0\right]$$

$$= 7 \cdot 12$$

$$= 84$$

Thus the total sales for the year are $84 thousand.

31. $\int \sin(\ell n\, x)\, dx$

Let

$u = \sin(\ell n\, x)$ and $dv = dx$.

Then

$du = \frac{\cos(\ell n\, x)}{x}\, dx$ and $v = x$.

Using the integration-by-parts formula, we get:

$$\int \sin(\ell n\, x)\, dx$$

$$= [\sin(\ell n\, x)] \cdot x - \int x \cdot \frac{\cos(\ell n\, x)}{x}\, dx$$

$$= x \sin(\ell n\, x) - \int \cos(\ell n\, x)\, dx$$

Let

$u = \cos(\ell n\, x)$ and $dv = dx$.

Then

$du = \frac{-\sin(\ell n\, x)}{x}\, dx$ and $v = x$.

Again using the integration-by-parts formula, we get:

$$= x \sin(\ell n\, x) - \left[x \cos(\ell n\, x) - \int x \cdot \frac{-\sin(\ell n\, x)}{x}\, dx\right]$$

$$= x \sin(\ell n\, x) - x \cos(\ell n\, x) - \int \sin(\ell n\, x)\, dx$$

Now adding $\int \sin(\ell n\, x)\, dx$ on both sides, we get:

$$2 \int \sin(\ell n\, x)\, dx = x \sin(\ell n\, x) - x \cos(\ell n\, x)$$

Thus,

$$\int \sin(\ell n\, x)\, dx = \frac{1}{2}\, x[\sin(\ell n\, x) - \cos(\ell n\, x)] + C$$

33. $\int e^x \cos x\, dx$

Let

$u = e^x$ and $dv = \cos x\, dx$.

Then

$du = e^x\, dx$ and $v = \sin x$.

Using the integration-by-parts formula, we get:

$$\int e^x(\cos x\, dx)$$

$$= e^x \sin x - \int \sin x\, (e^x\, dx)$$

$$= e^x \sin x - \int e^x \sin x\, dx$$

Now let,

$u = e^x$ and $dv = \sin x\,dx$.

Then

$du = e^x\,dx$ and $v = -\cos x$.

Again using the integration-by-parts formula, we get:

$= e^x \sin x - \left[(e^x)(-\cos x) - \int (-\cos x)(e^x\,dx)\right]$

$= e^x \sin x + e^x \cos x - \int e^x \cos x\,dx$

Now adding $\int e^x \cos x\,dx$ on both sides, we get:

$2\int e^x \cos x\,dx = e^x \sin x + e^x \cos x$

Thus,

$\int e^x \cos x\,dx = \frac{e^x}{2}(\sin x + \cos x) + C$

35. $\int x^3 \sin x\,dx$

We use tabular integration.

f(x) and repeated derivatives		g(x) and repeated integrals
x^3	+	$\sin x$
$3x^2$	-	$-\cos x$
$6x$	+	$-\sin x$
6	-	$\cos x$
0		$\sin x$

$\int x^3 \sin x\,dx$

$= x^3(-\cos x) - 3x^2(-\sin x) + 6x\cos x - 6\sin x + C$

$= -x^3\cos x + 3x^2\sin x + 6x\cos x - 6\sin x + C$

37. $\int \sec^2 7x\,dx$

Let $u = 7x$, then $du = 7\,dx$.

We do not have $7\,dx$. We only have dx and need to supply a 7 by multiplying by $\frac{1}{7}\cdot 7$ as follows:

$\frac{1}{7}\cdot 7\int \sec^2 7x\,dx$ (Multiplying by 1)

$= \frac{1}{7}\int (\sec^2 7x)\,7\,dx$

$= \frac{1}{7}\int \sec^2 u\,du$ (Substituting u for $7x$ and du for $7\,dx$)

$= \frac{1}{7}\tan u + C$

$= \frac{1}{7}\tan 7x + C$

39. $\int \sec x(\sec x + \tan x)\,dx$

$= \int (\sec^2 x + \sec x \tan x)\,dx$

$= \int \sec^2 x\,dx + \int \sec x \tan x\,dx$

$= \tan x + \sec x + C$

41. $\int \csc u \cot u\,du$

$= \int \frac{1}{\sin u}\cdot\frac{\cos u}{\sin u}\,du$

$= \int \frac{\cos u}{\sin^2 u}\,du$

Let $x = \sin u$, then $dx = \cos u\,du$.

$= \int \frac{1}{x^2}\,dx$

$= \int x^{-2}\,dx$

$= \frac{x^{-1}}{-1} + C$

$= -\frac{1}{x} + C$

$= -\frac{1}{\sin u} + C$

$= -\csc u + C$

43. $\int \frac{\cos^2 x}{\sin x}\,dx$

$= \int \frac{1 - \sin^2 x}{\sin x}\,dx$

$= \int \left(\frac{1}{\sin x} - \frac{\sin^2 x}{\sin x}\right)dx$

$= \int (\csc x - \sin x)\,dx$

$= \int \csc x\,dx - \int \sin x\,dx$

$= \underbrace{-\ell n\,|\csc x + \cot x|}\, - (-\cos x) + C$

(Using the result of Margin Exercise 6)

$= \cos x - \ell n\,|\csc x + \cot x| + C$

45. $\int (1 + \sec x)^2\,dx$

$= \int (1 + 2\sec x + \sec^2 x)\,dx$

$= \int dx + 2\int \sec x\,dx + \int \sec^2 x\,dx$

$= x + 2\underbrace{\ln|\sec x + \tan x|} + \tan x + C$

(Using the result of Example 6)

47. $\int_{\pi/4}^{\pi/2}\int_{0}^{\pi/2} \sin x \cos y \, dy \, dx$

Evaluate the inside integral.

$$\int_0^{\pi/2} \sin x \cos y \, dy$$

$= \sin x[\sin y]_0^{\pi/2}$

$= \sin x(\sin \pi/2 - \sin 0)$

$= \sin x(1 - 0)$

$= \sin x$

Then evaluate the outside integral.

$$\int_{\pi/4}^{\pi/2} \sin x \, dx$$

$= [-\cos x]_{\pi/4}^{\pi/2}$

$= -\cos \pi/2 - (-\cos \pi/4)$

$= -0 - \left[-\frac{\sqrt{2}}{2}\right]$

$= \frac{\sqrt{2}}{2}$

49. $\int_0^{\pi}\int_0^{x} x \cos y \, dy \, dx$

Evaluate the inside integral.

$\int_0^x x \cos y \, dy = x[\sin y]_0^x$

$= x(\sin x - \sin 0)$

$= x(\sin x - 0)$

$= x \sin x$

Then evaluate the outside integral.

$$\int_0^{\pi} x \sin x \, dx$$

Use integration by parts.

Let

$u = x$ and $dv = \sin x \, dx$.

Then

$du = dx$ and $v = -\cos x$.

$$\int_0^{\pi} x \sin x \, dx$$

$= [x(-\cos x)]_0^{\pi} - \int_0^{\pi} -\cos x \, dx$

$= [-x \cos x]_0^{\pi} + \int_0^{\pi} \cos x \, dx$

$= [-x \cos x]_0^{\pi} + [\sin x]_0^{\pi}$

$= [-\pi \cos \pi - (-0 \cos 0)] + (\sin \pi - \sin 0)$

$= [-\pi(-1) + 0\cdot 1] + (0 - 0)$

$= \pi$

Exercise Set 8.4

1. Find $\sin^{-1}\left[\frac{\sqrt{2}}{2}\right]$.

Think:

The number from $-\frac{\pi}{2}$ to $\frac{\pi}{2}$ whose sine is $\frac{\sqrt{2}}{2}$ is $\frac{\pi}{4}$.

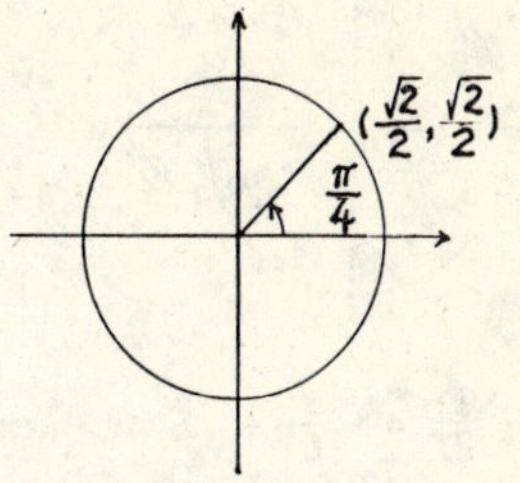

Thus $\sin^{-1} \frac{\sqrt{2}}{2} = \frac{\pi}{4}$.

3. Find $\cos^{-1}(0)$.

Think:

The number from 0 to π whose cosine is 0 is $\frac{\pi}{2}$.

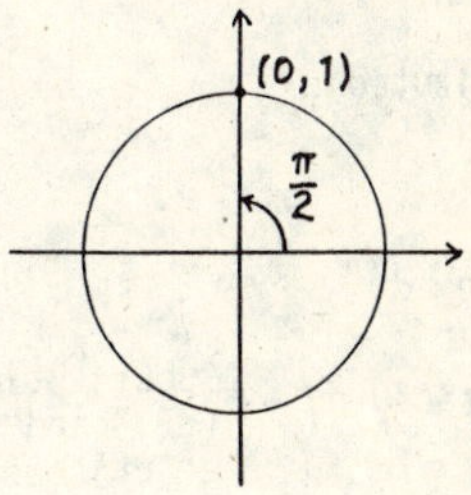

Thus $\cos^{-1}(0) = \frac{\pi}{2}$.

5. Find $\tan^{-1}(1)$.

Think:

The number from $-\frac{\pi}{2}$ to $\frac{\pi}{2}$ whose tangent is 1 is $\frac{\pi}{4}$.

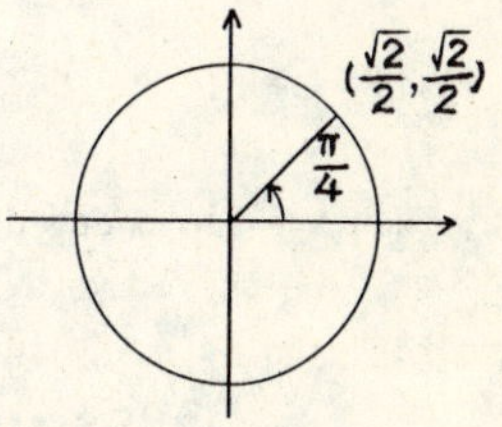

$\tan \frac{\pi}{4} = \frac{\sin \pi/4}{\cos \pi/4} = \frac{\sqrt{2}/2}{\sqrt{2}/2} = 1.$

Thus $\tan^{-1}(1) = \frac{\pi}{4}$.

7. Find $\sin^{-1}\left(-\frac{1}{2}\right)$.

Think:

The number from $-\frac{\pi}{2}$ to $\frac{\pi}{2}$ whose sine is $-\frac{1}{2}$ is $-\frac{\pi}{6}$.

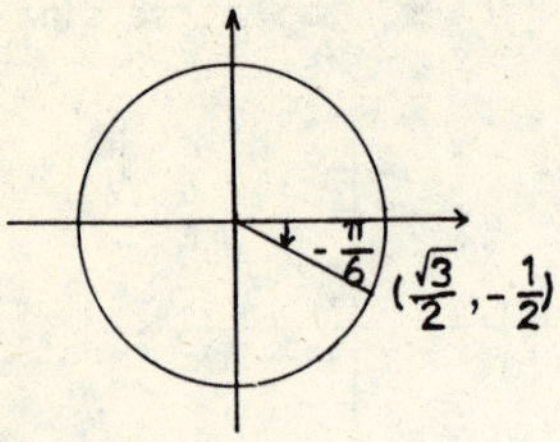

Thus $\sin^{-1}\left(-\frac{1}{2}\right) = -\frac{\pi}{6}$.

9. $\int \frac{e^t}{1 + e^{2t}}\,dt$

Let

$u = \tan^{-1}(e^t)$

$e^t = \tan u$

$e^t\,dt = \sec^2 u\,du$.

Then use substitution.

$\int \frac{e^t}{1 + e^{2t}}\,dt$

$= \int \frac{1}{1 + (e^t)^2}\,e^t\,dt$

$= \int \frac{1}{1 + \tan^2 u}\sec^2 u\,du$ [Substituting $\tan^2 u$ for $(e^t)^2$ and $\sec^2 u\,du$ for $e^t\,dt$]

$= \int \frac{1}{\sec^2 u}\sec^2 u\,du$ ($\tan^2 u + 1 = \sec^2 u$)

$= \int du$

$= u + C$

$= \tan^{-1}(e^t) + C$

11. Show that

$\int \frac{1}{\sqrt{1 - x^2}}\,dx = \sin^{-1}x + C$.

Let $u = \sin^{-1}x$, then $x = \sin u$ and $dx = \cos u\,du$.

$\int \frac{1}{\sqrt{1 - x^2}}\,dx$

$= \int \frac{1}{\sqrt{1 - \sin^2 u}}\cos u\,du$ (Substituting $\sin^2 u$ for x^2 and $\cos u\,du$ for dx)

$= \int \frac{1}{\sqrt{\cos^2 u}}\cos u\,du$ ($\sin^2 u + \cos^2 u = 1$)

$= \int \frac{1}{\cos u}\cos u\,du$

$= \int du$

$= u + C$

$= \sin^{-1}x + C$

13. $\sin^{-1}(0.9874) = 1.412$ radians

15. $\cos^{-1}(0.9988) = 0.049$ radians

17. $\int \tan^{-1} x\,dx$

Let

$u = \tan^{-1} x$ and $dv = dx$.

Then

$du = \frac{1}{1 + x^2}\,dx$ and $v = x$.

Using Integration by Parts, we get:

$\int \tan^{-1} x\,dx = x\tan^{-1} x - \int x \cdot \frac{1}{1 + x^2}\,dx$

$= x\tan^{-1} x - \int \frac{1}{2} \cdot 2 \cdot x \cdot \frac{1}{1 + x^2}\,dx$

(Multiplying by 1)

$= x\tan^{-1} x - \frac{1}{2}\int \frac{2x}{1 + x^2}\,dx$

Let $u = 1 + x^2$, then $du = 2x\,dx$. Using substitution we get:

$= x\tan^{-1} x - \frac{1}{2}\int \frac{1}{u}\,du$

$= x\tan^{-1} x - \frac{1}{2}\ln u + C$

$= x\tan^{-1} x - \frac{1}{2}\ln(1 + x^2) + C$